Science

R00067 31756

CHICAGO PUBLIC LIBRARY
HAROLD WASHINGTON LIBRARY CENTER

R0006731756

Form 178 rev. 11-00

BUSINESS/SCIENCE/TECHNOLOGY DIVISION
CHICAGO PUBLIC LIBRARY
400 SOUTH STATE STREET
CHICAGO, IL 60605

PREFACE
TO
THE FIFTH EDITION

The first edition of this work appeared in 1943, to meet the needs mainly of the undergraduate and graduate students of the region. Subsequent editions were published in 1949, 1956 and 1960. The text has been thoroughly revised for the present (fifth) edition, to include the results of work in the Indo-Pakistan-Burma region up to 1966.

The help received from various colleagues — Dr. A. P. Subramaniam, Mr. N. A. Vemban, Dr. K. Jacob, Prof. G. W. Chiplonkar, Prof. F. Ahmad, and others — during the revision of the work is gratefully acknowledged.

I wish to express my appreciation of the help rendered by Messrs. Higginbothams Ltd. and especially by Mr. V. Balaraman, Director and Manager of the firm, during publication. My thanks are also due to Mr. S. NATARAJAN, Proprietor of Sankar Printers (Madras) and his STAFF for the keen interest they evinced in expeditiously handling the printing of the book.

MADRAS,
February 15, 1968.

M. S. KRISHNAN.

CONTENTS

Chapter		Pages
I.	INTRODUCTION AND PHYSICAL GEOLOGY	1—47

    Physiographic Divisions. Climate. Mountains. Glaciers. Peninsular Rivers. Waterfalls. Extra-Peninsular Rivers. Lakes. Earthquakes. Volcanoes. Mud Volcanoes.

II.	STRUCTURE AND TECTONICS OF INDIA	48—75

    Structural trends. Assam and Kashmir Wedges. The Indo-Gangetic Basin. The Himalayas. Burmese Arc. Baluchistan Arc. Eastern Coast. Western Coast. Orogenic belts and Ultramafic rocks. Seismic Phenomena. Geodetic observations.

III.	REVIEW OF INDIAN STRATIGRAPHY	76—86

    Principles of Stratigraphy. Geologic Time-Scale. Standard geological divisions. Review of Indian Stratigraphy.

IV.	THE ARCHAEAN GROUP : PENINSULA	87—136

    Introduction. Mysore-Dharwar System. Peninsular Gneiss. Charnockites. Andhra. Madras. Ceylon. Eastern Ghats. Jeypore-Bastar. Raipur-Drug. Bilaspur-Balaghat-Nagpur-Bhandara-Chhindwara. Jabalpur. Gujarat. Rajasthan. Bundelkhand. Singhbhum. Gangpur. Son Valley. Bengal. Assam.

V.	THE ARCHAEAN GROUP : EXTRA-PENINSULA	137—146

    Northwest Himalaya. Spiti Valley. Simla-Garhwal. Nepal. Sikkim. Bhutan. Burma.

VI.	MINERAL RICHES OF THE ARCHAEANS	147—162

    Antimony. Arsenic. Chromite. Columbite. Copper. Gold. Iron. Lead-Zinc. Manganese. Molybdenite. Monazite. Nickel-Cobalt. Tin. Titanium. Tungsten. Uranium. Vanadium. Apatite. Asbestos. Beryl. Building Stones. Clays. Corundum. Feldspar. Gemstones. Graphite. Kyanite-Sillimanite. Magnesite. Mica. Steatite.

VII.	THE CUDDAPAH SYSTEM	163—181

    Cuddapah Basin. Pakhal Series. Penganga Beds. Kolhan Series. Gwalior System. Bijawar Series. Himalayas and Burma.

VIII.	THE VINDHYAN SYSTEM	182—196

    Semri Series. Upper Vindhyans. Kurnool System. Bhima and Sullavai Series.

Chapter		Pages
IX.	THE PALAEOZOIC GROUP : CAMBRIAN TO CARBONIFEROUS ...	197—238
	Cambrian, Ordovician, Silurian, Devonian and Carboniferous Systems. Unfossiliferous Palaeozoic Strata.	
X.	THE GONDWANA GROUP ...	239—291
	Nomenclature. Talchir, Barakar, Barren Measures, Raniganj, Panchet, Mahadeva, Rajmahal, Jabalpur, Umia Series. Igneous Rocks. Gondwanas in other continents.	
XI.	THE UPPER CARBONIFEROUS AND PERMIAN SYSTEMS	292—315
XII.	THE TRIASSIC SYSTEM	316—346
XIII.	THE JURASSIC SYSTEM	347—369
XIV.	THE CRETACEOUS SYSTEM	370—404
XV.	THE DECCAN TRAPS	405—421
	Structural features. Dykes and Sills. Petrology and Petrography. Chemical Characters, Alteration and Weathering. Lametas. Infra- and Inter-Trappean Beds. Economic Geology.	
XVI.	THE TERTIARY GROUP (SUMMARY) ...	422—438
XVII.	THE EOCENE SYSTEM	439—462
XVIII.	THE OLIGOCENE AND LOWER MIOCENE SYSTEMS...	463—481
XIX.	MIDDLE MIOCENE TO LOWER PLEISTOCENE	482—496
XX.	THE PLEISTOCENE AND RECENT	497—525
	Human evolution. Glaciation. Karewa Formations. Potwar Silts. Loess. Alluvium. Aeolian and Desert Deposits. Laterite. Soils. Changes of Level.	
	INDEX ...	526—536

LIST OF ILLUSTRATIONS

SKETCH MAPS

Map		Page
I	Mountain Arcs of Southern Asia	10
II	Himalayan and Central Asian Ranges	11
III	Structural Trends in the Archaeans	50
IV	Himalayan Tectonic Units	60
V	Main Tectonic Features of India	65
VI	The Dharwars of South India (face)	88
VII	Geological Map of Ceylon	106
VIII	The Cuddapah Basin	166
IX	Lower Gondwana Coalfields of India (face)	284
X	The Raniganj Coalfield	287
XI	The Jharia Coalfield	288
XII	Geological map of Malla Johar	298
XIII	The Jurassic Rocks of Kutch	366
XIV	The Cretaceous Rocks of Trichinopoly	387
XV	Geological Sketch-map of Kashmir Himalayas (face)	484

TEXT FIGURES

Figure		Page
1.	Schematic Section of the Salt Range	200
2.	Section across the Nilawan Ravine	205
3.	Section across the Dandot Scarp	209
4.	Section on the Parahio River, Spiti	233
5.	Section across the Lidar Valley anticline	234
6.	Section through Naubug Valley and Margan Pass	303
7.	Section across the Makrach Valley, Salt Range	305
8.	Section near Lilang, Spiti	320
9.	Section northwest of Kalapani, Byans	332
10.	Triassic Section near Pastannah, Kashmir	338
11.	Section in the Chichali Pass, Trans-Indus Region	357
12.	Section through the Bakh Ravine, Salt Range	445
13.	Section across Pir Panjal from Tatakuti to Nilnag	503
14.	Section through the Narmada Pleistocene	507

PLATES

(Drawings of Fossils)

		Page
I	Cambrian	217
II	Ordovician and Silurian	224
III	Devonian	229
IV	Devonian	(Face) 232
V—VIII	Gondwana	Between 248–249
IX	Permo-Carboniferous	297
X, XI	Permian	306, 309
XII	Permian and Triassic	312
XIII—XV	Triassic	323, 326, 330
XVI	Triassic and Jurassic	333
XVII, XVIII	Jurassic	353, 359
XIX, XX	Cretaceous	389, 391
XXI, XXII, XXIII	Cretaceous	396, 398, 400
XXIV	Cretaceous and Tertiary	442
XXV, XXVI, XXVII	Tertiary	451, 454, 457

ABBREVIATIONS FOR BIBLIOGRAPHY

Bull.	...	Bulletin
G. S. I.	...	Geological Survey of India
Mem.	...	Memoirs of the G. S. I.
M. G. D.	...	Mysore Geological Department
Q. J. G. S.	...	Quarterly Journal of the Geological Society of London
Pal. Ind.	...	Palaeonotologia Indica (G. S. I.)
Rec.	...	Records of the G. S. I.
Ser.	...	Series. *N. S.* New Series
Q. J. G. M. M. S.	...	Quarterly Journal of the Geological, Mining and Metallurgical Society of India
T. M. G. I. I.	...	Transactions of the Mining, Geological and Metallurgical Institute of India.

ERRATA

Page, Line	In Place of	Read
52, 20	Dauki Fauki	Dauki Fault
60	Map VI	Map IV
74, 8	Peninsular	Peninsula
94, 20	3.73	3.33
96, 7 (Table)	Fe O	$Fe_2 O_3$
139, 5	Chal Series	Chail Series
142, 21	Acquamarine	Aquamarine
164, 10 (from bottom)	Cumbum Shale (*Below* Tadpatri Shale)	Pulivendla (Nagari) Quartzite
191, Table	Berinag Guartzite	Berinag Quartzite
192, 13	Smeri Series	Semri Series
199, 5, 6	Products	Productus
293, 5	Rotheliegende	Rothliegende
483, 15	uetta Qexhibits	Quetta exhibits
493, 11	As out	About
494, 15	resorcded	recorded

CHAPTER I
INTRODUCTION AND PHYSICAL GEOLOGY
The Physical Divisions of India

A physical map of India shows strikingly that the country can be divided into three well-marked regions each having distinguishing characters of its own. The first is the *Peninsula* or Peninsular Shield ('shield' being a term used for geologically very old and stable parts of the crust) lying to the south of the plains of the Indus and Ganges river systems. The second division comprises these *Indo-Gangetic alluvial plains* stretching across northern India from Assam and Bengal on the east, through Bihar and Uttar Pradesh, to the Punjab and Sind on the west. The third is the *Extra-peninsula*, the mountainous region of the mighty Himalayan ranges and their extensions into Baluchistan on the one hand and Burma and Arakan on the other. These three divisions exhibit marked contrast in physical features, stratigraphy and structure.

Physiographically, the Peninsula is an ancient plateau exposed for long ages to denudation and approaching peneplanation. Its mountains are of the relict type, *i.e.*, they represent the survival of the harder masses of rock, which have escaped weathering and removal; their topographical expression may not therefore be directly attributable to their structure. Its rivers traverse, for the most part, a comparatively flat country with low gradients and have built up shallow and broad valleys. The Extra-Peninsula, on the other hand, is a region of folded and overthrust mountain chains, of geologically recent origin. Its rivers are youthful and are actively eroding their beds in their precipitous courses and carving out deep and steep-sided gorges. The Indo-Gangetic plains are broad, monotonous, level expanses built up of recent alluvium through which the rivers flow sluggishly towards the seas.

Stratigraphically, the Peninsula is a 'shield' area composed of geologically ancient rocks of diverse origin, most of which have undergone much crushing and metamorphism. Over these ancient rocks lie a few areas of Pre-Cambrian and later sediments and extensive sheets of horizontally bedded lavas of the Deccan Trap formation. Some Mesozoic and Tertiary sediments are found mainly along the coastal regions. The Extra-Peninsula, though containing some very old rocks, is predominantly a region in which the sediments, laid down in a vast geosyncline continuously from the Cambrian to early Tertiary, have been compressed, overthrust and elevated into dry land only since the end of the Mesozoic times. The core of the mountains is composed of granitic intrusions of presumably Tertiary age. The southern fringe, bordering on the plains, consists of fresh-water and estuarine deposits of Mio-Pliocene age derived largely from the erosion of the rising Himalayas. The Gangetic Plains are built up of layers of sands, clays and occasional organic debris (peat-beds, etc.) of geologically very recent date (Pleistocene and Recent), filling up a deep depression between the two other units.

Structurally, the Peninsula represents a stable block of the earth's crust which has remained unaffected by mountain building movements since practically the close of the Pre-Cambrian era. The later changes which it suffered have been mainly of the nature of normal and block faulting because of which some parts have sunk down relative to others. Along its coasts, there have been marine transgressions which have laid down sedimentary beds of Upper Gondwana, Cretaceous or Tertiary ages, but not of great thickness or extent. In contrast with this, the Extra-Peninsula has recently undergone earth movements of stupendous magnitude. Its strata are marked by complex folds, reverse faults, overthrusts and nappes of great dimensions. There is reason to believe that these movements have not yet completely died down, for this region is still unstable and is frequently visited by earthquakes of varying intensities. The Gangetic plains owe their origin to a sag in the crust, probably formed contemporaneously with the uplift of the Himalayas. This sag or depression has since been filled up by sediments derived from both sides, and especially from the lofty chains of the Himalayas which are actively being eroded by the many rivers traversing them. The little geological interest which these plains hold is confined to the rich soils and to the history of the river systems; indeed, the alluvium effectively conceals the solid geology of its floor, a knowledge of which would be highly interesting and probably even profitable. These alluvial plains are, however, of absorbing interest in human history, being thickly populated, and the scene of many important developments and events in the cultural and social history of Hindustan.

CLIMATE

India, Pakistan and Burma together have an area of over 1,899,000 sq. miles (India 1,271,300 ; Pakistan 365,900; Burma 261,600 sq. miles). India and Pakistan stretch between N. latitudes 8° and 37° and E. longitudes 61° and 97°, Burma extending further east to a little beyond 100°. From Cape Comorin to the north of Kashmir the distance is about 2,000 miles, (3,200 km.) this being exceeded by the distance between the western border of Baluchistan and eastern border of Burma by some 400 miles (650 km).

Within this extensive domain are present a variety of climatic conditions, but the dominant feature is the tropical monsoon. The northern part of the country, that beyond the latitude of Calcutta and Ahmedabad, lies to the north of the Tropic of Cancer. The interior, owing to its inland or continental nature, is subject to extremes of temperature. The mountain barrier of the Himalayas plays an important part not only in influencing the distribution of rain in Northern India, but also in preventing this region from experiencing the very cold winters characterising the territories to their north.

The south-west monsoon reigns from the end of May to December, the earlier half being the general rainy season. The latter half marks the 'retreating monsoon' during which some parts of the eastern coast, particularly the Madras coast, receive some rain. The north-east monsoon is active during the cold weather but the winds are dry before they blow over the Bay of Bengal.

During the cold weather (December to February) the temperature reaches a minimum, especially in the Punjab and the north-west which show mean temperatures below 55°F. In Upper India there is a region of high pressure from which winds radiate to the south and south-east. North-east winds are experienced in Bengal but they are dry until they blow over the sea when they pick up moisture and precipitate it on the Madras coast and Ceylon. Some cyclonic storms are also experienced in N.W. India during this period, but these are due to winds travelling eastwards from the Mediterranean.

During the succeeding months of March to May, the temperature rises steadily to a maximum, the interior of the country registering 110° to 120°F. (42° to 49°C) in early May. Strong winds blow from the north-west down the Ganges valley, familiarly known as 'norwesters'. Though during this period, there is a low pressure region in Northern India, there is no flow of moisture-bearing winds from the Indian Ocean as there is obstruction to such a flow in the intervening low pressure equatorial belt. It is only towards the end of May that this latter is wiped out and the south-west winds establish themselves.

The south-west monsoon strikes the Malabar and Arakan coasts and is deflected northwards by the hills present along these coasts. The Deccan plateau falls in the rain-shadow of the Western Ghats and hence receives only a small amount of rain. The Western Ghats receive over 250 cm of rain during the monsoon whereas the 'shadow' region gets only 60 cm or less. The winds sweeping up through the Bay of Bengal strike the Arakan and Assam hills, the latter forcing the winds up to an altitude of some 5,000 feet when all their moisture content is precipitated as rain. The neighbourhood of Cherrapunji is known to receive the highest rainfall in the world, averaging about 450 inches (1150 cm) per year, the maximum recorded being 905 inches (2300 cm) in 1861. Part of the monsoon winds is deflected up the Ganges valley to as far as western Punjab, bringing rain to these regions between the middle of June and the end of August. There is of course more rain along and near the Himalayan foothills than away from them and hence Southern Punjab and Rajasthan are regions of low rainfall.

A broad and rather irregular belt of low rainfall (20-30 inches) stretches from the interior of Madras in a northerly direction through Bombay and Central India to the Punjab. But South-western Punjab, Western Rajputana, Sind and Baluchistan constitute a region of very low rainfall (below 10 inches per annum) and enclose the tract known as the Thar which is a semi-desert.

The latter part of the south-west monsoon season is marked by a gradual rise of pressure in Northern India which has the effect of obstructing and relatively pushing back the monsoon current. The winds therefore appear to 'retreat' and precipitate the moisture content along the east coast of Madras, during October and November. This is in fact the chief rainy season of this part of India. The real north-east monsoon begins to be effective only at a later period but actually contributes less rain than the retreating south-west monsoon.

PHYSIOGRAPHY

MOUNTAINS

Peninsular India

The chief mountains of Peninsular India are the Western and Eastern Ghats, Vindhyas, Satpuras, the Aravallis and those forming the Assam plateau.

The Western Ghats : These form a well-marked feature along the western coast of India from the Tapi Valley down to Cape Comorin. The northern part, down to Dharwar and Ratnagiri in Bombay, is composed of the Deccan Traps, while the southern part consists of Archaean gneisses, schists and charnockites. The Western Ghats are nearly 1,600 km long. Their average elevation is from 1000 to 1,300 m but many peaks rise to over 2,400 m (e.g. Doddabetta (2,636 m) and Makurti (2,554 m) in the Nilgiris, Anaimudi (2,693 m) in the Anaimalais and Vembadi Shola (2,505 m) in the Palni hills.

The Western Ghats in Bombay are composed of flat-topped ridges which are due to the more resistant flows of the Deccan Traps forming a series of step-like terraces. In this portion rise the Godavari, Bhima and Krishna rivers. The Satmala ridge branches off to the east between the Tapi and the Godavari, while Mahadeo ridge branches off between the Bhima and Krishna rivers. Mahabaleswar is a well-known hill station (1,438 m) near the source of the Krishna river. A few passes in the ghats provide lines of communication between the coastal plains and the interior and have had great strategic significance in the past history of the country. The Western Ghats are generally known as "Sahyadris" which is the name by which they have been described in the epic *Ramayana*. As we proceed to the south into Mysore, the Ghats tend to recede 50 to 65 km from the coast. In the Nilgiris the Eastern Ghats join them to form a mountain knot whose highest point is Doddabetta. Further south the Western Ghats recede still further, to a distance of 60-80 km from the coast.

To the south of the Nilgiri hills is the remarkable pass or gap in the Western Ghats, known as the Palghat Gap. This has always provided a major line of communication between the coastal plains of Malabar and Kanara on the one hand and the plains of South Madras on the other. The Palghat Gap is only 300 m in elevation, but has a maximum width of 24 km. It might represent the valley of a westerly flowing river of the Tertiary times as suggested by Jacob and Narayanaswami (1954).

South of the Palghat Gap the Western Ghats rise again to form the Anaimalai hills whose highest peak is Anaimudi. The top of these hills is an undulating plateau which is well-forested, with several useful timber species, e.g., teak, ebony, rose wood, etc., and a large variety of bamboos. The Mysore and Travancore parts of the Western Ghats also support large plantations of tea, coffee, cinchona, cardamom, while the coastal plains grow much black pepper, cocoanut and cashew. One of the eastern spurs from these hills forms the Palni hills. Further south in Travancore they decrease in height and disappear finally a few miles to the north of Cape Comorin.

Though situated close to the Arabian Sea, the Western Ghats form the real watershed of the Peninsula. Their easterly slopes are gentle and grade into the Mysore plateau and the plains of the interior of Madras. The western slopes are much steeper and often precipitous. As they are exposed to the full vehemence of the south-west monsoon, they receive around 150 inches (380 cm) of rain per annum. All the important peninsular rivers, *viz.*, the Godavari, Krishna, Cauvery and Tambraparni and their important tributaries rise on the Western Ghats and flow eastwards into the Bay of Bengal.

The Eastern Ghats: These are a series of rather detached hill ranges of heterogeneous composition which stretch intermittently from the northern border of Orissa through the coastal regions of the Andhra Pradesh to join the Nilgiris in the western part of Madras. They are uniform in their character in Orissa and in the northern part of Andhra down to the valley of the Krishna river, being composed of garnetiferous sillimanite gneisses (khondalite) and large masses of charnockite. Their average elevation here is about 750 m, but a few points rise to over 1500 m, for example Korlapat (1,213 m) and Banksamo (1,274 m) in Kalahandi ; Nimaigiri (1,515 m) in Koraput ; Malayagiri (1,186 m) in Pal Lahara ; Meghasani (1,164 m) in Mayurbhanj ; Mankarnacha (1,109 m) in Bonai and Mahendragiri (1,500 m) in Ganjam. In a few places the garnetiferous gneisses are capped by laterite which is generally too poor in alumina to form commercial bauxite.

South of the Krishna Valley, they continue into the Kondavidu hills which are composed of charnockites. A part of the Ghats in the Krishna district strikes into the Bay of Bengal to emerge near Madras City. The Nallamalai and Palkonda hills are composed of Cuddapah and Kurnool formations. Their continuation is to be seen in the Javadi, Shevaroy and the Biligiri Rangan hills of Salem and Coimbatore which finally join the Nilgiris. All these are made up of charnockite and granitoid gneisses. Of these, the Shevaroys, which rise to a height of 1,650 m, contain a few flat-topped peaks on which bauxite deposits occur.

The Eastern Ghats to the north of the Krishna river have been considered by Fermor to have been uplifted in late Pre-Cambrian times. This is supported by the fact that the majority of the rocks have the impress of high grade metamorphism. They contain intrusive masses of granitic rocks which, when invading manganiferous sediments, have given rise to hybrid rocks originally described as *Kodurites* by Fermor. In a few places such as the Baula hills in Keonjhar, in the Kondapalle hills in the Krishna district and in the Chalk hills of Salem there are intrusive masses of ultrabasic rock associated with chromite.

The Vindhya Mountains: The Vindhyas, which separate Southern from Northern India are a fairly continuous group of hill ranges, or rather a series of plateaux, lying to the north of the Narmada river and extending from

Jobat (22°27′ : 74°35′) in Gujarat to Sasaram (24°57′ : 84°02′) in Bihar, through Indore, Bhopal, Baghelkhand and Bundelkhand. Their general elevation is 450 to 600 m, but a few points rise above 900 m. The majority of the ranges are composed of sandstones and quartzites of the Vindhyan System, these being relict mountains. The western part, to the west of Jabalpur, forms the northern boundary of the Narmada Valley and consists mainly of Deccan Trap. The eastern part, including the Kaimur Range, is composed of the Vindhyan sandstones. The Maikal Range, forming so to say a connecting link between the Vindhyas and Satpuras, is a large plateau which was once well populated but is now highly forested. Together with the Satpuras, the Vindhya mountains form the watershed of Central India from which rise the Narmada, Chambal, Betwa, Tons, Ken, Sone and other streams, some of which flow into the Ganges and the others into the Godavari and Mahanadi.

The Satpura Mountains: This name was applied originally to the hills in the Nimar district of Madhya Pradesh which separated the Narmada and the Tapi Rivers. Their western termination is in the Rajpipla Hills in Gujarat while in the east they comprise the Pachmarhi Hills, the Maikal Range and the hills of Surguja, Ranchi and Hazaribagh. They have a general E.N.E.-W.S.W. trend. The peaks in the Mahadeva Hills are over 1,200 m in height (Pachmarhi : 1,335 m) while the Amarkantak peak is 1,064 m high. The highest point is Dhupgarh (1,348 m). In the eastern part, the Satpuras are composed of Gondwanas and Archaean gneisses. In Berar, the Satpuras occupy a broad zone, 112 to 160 km wide, composed of several more or less parallel ridges of Deccan Trap lava flows. Their northern slopes are drained by the Narmada River and the southern slopes by the Wainganga, Wardha and Tapi.

The Rajmahal Hills at the head of the Ganges Delta were once regarded as part of the Vindhyas or the Satpuras. They are, however, not connected with either, being composed of lava flows occupying the area between latitudes 24°30′ and 25°15′ roughly along longitude 87°30′.

The Aravalli Mountains: These are now the remnants of once great mountain ranges of tectonic origin. They cross Rajasthan from south-west to north-east separating the arid semi-desert of the Bikaner, Jodhpur, and Jaisalmer area on the west from the more fertile region of Udaipur and Jaipur on the east. They are composed of rocks of the Aravalli, Delhi and Vindhyan Systems.

The small detached quartzite ridges near Delhi are their northernmost stumps. They continue to Khetri north of Jaipur where the first well-marked ridge appears. They gradually rise higher towards the south-west forming the peaks of Kho (979 m), Raghunathpur (1,052 m) and Harisnath (905 m). They pass to the west of the Sambhar Lake and open out to form several parallel ranges, the highest point here being Taragarh fort hill (870 m). From Beawar onwards they form conspicuous ridges while beyond Merwara they spread out into a zone of hills 40 to 48 km. wide. The highest point attained by the Aravalli Mountains is Gurusikhar (1,722 m) in Mt. Abu.

Further south-west they gradually become straggling hills forming the rugged country extending from south-west Mewar into Dungarpur and Banswara. They may be said to terminate in the district of Sirohi. The south-eastern flanks extending into Udaipur are less steep than the north-western flanks, the latter being better wooded, because of the slightly higher rainfall on that side.

Though the Aravallis terminate in Gujarat in the south and near Delhi in the north, there are indications that they extend in both directions. Pre-Cambrian rocks with the Aravalli trend are noticed in Garhwal in the U.P. Himalayas and are considered to represent their former extension into this region. In the south of Rajasthan they tend to splay out, the different parts being continued probably into the Laccadives in the Arabian Sea on the one hand and into Mysore on the other.

The Aravallis are thought to constitute a true tectonic range. They were formed in Pre-Cambrian (post-Aravalli and post-Delhi) times and were probably uplifted again in post-Vindhyan times. The last movement may have been merely a block uplift as suggested by Fermor (*Records*, *G.S.I.*, Vol. LXII, pp. 391-409, 1930). They form the major watershed of Northern India, separating the drainage of the Ganges River system from that of the Indus, which are destined respectively for the Bay of Bengal and the Arabian Sea.

Extra-Peninsular Ranges

Arcuate Disposition: The mountains surrounding Indo-Pakistan on the north, north-west and north-east are, as mentioned already, tectonic ranges formed during the Tertiary. Their curvilinear disposition is very striking. They consist mainly of circular arcs which are convex towards the Peninsula, *i.e.*, towards the rigid crust against which they appear to have been thrust.

Of these, the Himalayan and the Burmese arcs are of immense radius. The Himalayas extend with a smooth sweep from Assam to Kashmir, for a length of about 2,400 km. The Baluchistan arm however consists of arcs of smaller radii which succeed one another at short intervals. The three main segments here are the Hazara mountains with the Samana Range and Safed Koh; the Sulaiman ranges which terminate near Quetta; and the system composed of the Bugti hills, and the Kirthar and Mekran ranges.

The Himalayan arc is followed to the north by a succession of ranges across the great Tibetan table-land, their trend being more or less parallel to the Himalayas but their curvature gradually decreasing in the more northern ranges. Thus the Aling Kangri, the Karakoram and the Kun Lun become progressively straighter, the last being practically a straight mountain system (See Map II). In the case of the Baluchistan arc, the transition from the strongly curved outer ranges to the slightly curved inner ones is more rapid than in the Himalaya as will be clearly seen, for instance, in the case of the Sulaiman and associated ranges. The convexity of the arcs is in all cases towards the rigid mass of the Peninsular shield and indicates the apparent direction of thrust movements.

Tibet: The Tibetan plateau has an average altitude of 4,200 m. To its north-west is the Pamir plateau (3,600 m) which connects up with the Tien Shan plateau farther north. The Tibetan plateau is now generally covered to a large extent by alluvium and loess. It is studded with a large number of lakes which were formerly much more extensive and probably connected with some system of drainage. Now, however, they are mostly brackish and are, together with the whole region, becoming desiccated, consequent upon the rise of the Himalayas which have effectively shut off the moisture-bearing winds from the Indian Ocean. In the mountains on its southern and eastern border, rise all the great rivers of Southern and South-eastern Asia.

Karakoram: The Karakoram range forms, so to say, the back-bone of the Tibetan region and is continuous with the Hindukush range to its west. The Karakoram carries the peaks K^2 (8,640 m), Gasherbrum (8,068 m). Masherbrum (7,821 m), etc. It forms the chief water parting between Central Asia and South Asia. To the south of the watershed it is some 50 to 100 km wide, carrying peaks over 6,100 m high. It contains also several important glaciers — Baltoro, Biafo, Hispar, etc. The valley of the Hunza river, at an altitude of 4,700 m, constitutes a pass to Central Asia while the pass between Leh and Yarkand farther east is 5,600 m high. South of the Karakoram in Tibet is a range of snow-clad mountains named Aling Kangri. How far to the east this range extends is not known. Between Aling Kangri and the Kailas Range lies another range called the Trans-Himalaya by Sven Hedin, the great Scandinavian explorer to whose explorations in Tibet we owe a great deal of our knowledge. This Trans-Himalaya Range is the real watershed between the northerly drainage flowing into Tibet and the southerly drainage destined for the Indian Ocean.

Kailas and Ladakh Ranges: Some distance south of the Trans-Himalaya is the Kailas Range, the latter being parallel to, and some 80 km north of, the Ladakh Range. About 30 km north of the sacred Manasarowar lake, the Kailas Range contains a cluster of peaks of which the chief is Mount Kailas (6,715 m). South of the Kailas Range comes the Ladakh Range which can be followed from Baltistan to Eastern Tibet, and forms the watershed between the latter and Nepal. To the west, it probably merges into the Haramosh Range on which the peak Rakaposhi (7,788 m) is situated. The highest peak of the Ladakh Range is Gurla Mandhata (7,728 m). There are several gaps in the Ladakh Range; one of them is traversed by the Sutlej; a second, some 24 km wide, is seen south-west of Manasarowar; the third and largest is a gap 100 km wide, north of Chomo Lhari, which is drained by the Nyang, a tributary of the Tsang-po. The Indus river is inextricably connected with this range; it first flows on the northern side of the range for 200 km from its source, then crosses it to the south near Thangra, flowing W.N.W. for nearly 480 km along the southern flank; it again cuts across the range northward just before it is joined by the Shyok river.

The Zanskar Range : This is really a northerly branch of the Himalaya lying between the Ladakh Range on the north and the Great Himalaya on the south. A good part of this range is unexplored territory. Its best known peak is Kamet (7,766 m). It is traversed by the Dras and the Zanskar rivers.

There are several passes over this range some of the well-known ones being Dharma (5,490 m), Kungri Bingri (5,490 m) and Shalshal (4,940 m).

The Pir Panjal: This forms the southern boundary of the Kashmir valley and extends from Muzaffarabad on the Jhelum to Kishtwar on the Chenab and farther east. The high central part is 130 km long, with peaks rising to over 4,300 m. It is the youngest chain of the Himalayas raised up during the early Pleistocene.

THE HIMALAYAS PROPER

The Himalayas can be divided longitudinally into four zones, parallel to each other :—

1. **The Siwalik foot-hills,** 10 to 50 km wide, whose altitude rarely exceeds 900 m. This region is generally covered with a damp and unhealthy forest. The rainfall varies between 120 cm in the west to 250 cm in the east.

2. **The Lesser Himalayan Zone,** 60 to 80 km wide and of an average altitude of about 3,000 m. This consists of parallel ranges in Nepal and Punjab but of scattered mountains in Kumaon. In this are found remnants of the fringes of the old Gondwanaland. The zone between 1,500 and 2,400 m is covered by evergreen and oak forests and that between 2,400 to 3,000 m by coniferous forests. In the lower slopes are found magnificent forests of chir (*Pinus longifolia*), deodar (*Cedrus deodara*), the blue pine (*Pinus excelsa*), oaks and magnolias, whereas above 2,500 m are found birch, spruce, silver fir and other species.

3. **The Great Himalaya** or Central Himalaya, comprising the zone of high snow-capped peaks which are about 140 km from the edge of the plains. This zone shows both sedimentary and old metamorphosed rocks which have been intruded by large masses of granite, probably of different ages. This consists of a lower, alpine zone up to 4,800 m and an upper, snow-bound zone usually above 5,000 m. The alpine zone contains rhododendrons, trees with crooked and twisted stems, thick shrubs with a variety of beautiful flowers, and grass.

4. **The Trans-Himalayan Zone,** about 40 km in width, containing the valleys of the rivers rising behind the Great Himalayas. These river basins are at an altitude of 3,600 to 4,200 m and consist of rocks of the geosynclinal or Tibetan facies.

In the Darjeeling-Nepal region, the Himalayas have an E.-W.-trend. Farther to the east, they have an E.N.E. or N.E. course, while to the west of Nepal they first have a west-north-westerly course and then a north-westerly course. The main range throws off minor ranges (all on the convex side, except in one case) which first proceed in the original direction of the main range at the point of branching but gradually swing parallel to the main range. The best known of these are the Nag Tibba, the Mahabharat and Dhauladhar ranges in Nepal and U.P. and the Pir Panjal in Punjab and Kashmir.

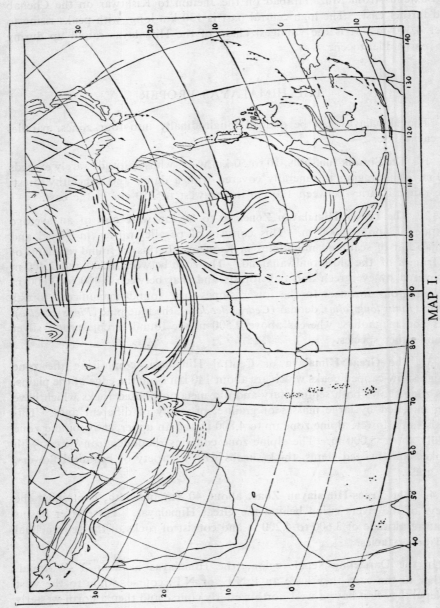

MAP I.
THE MOUNTAIN ARCS OF SOUTHERN ASIA.

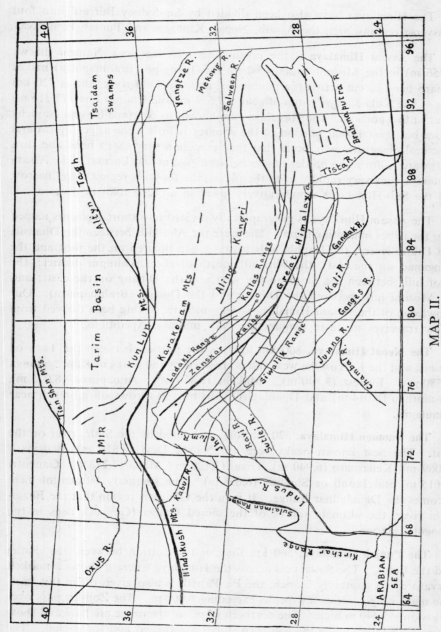

MAP II. HIMALAYAN AND CENTRAL ASIAN RANGES (AFTER BURRARD).

REGIONAL DESCRIPTION OF THE HIMALAYAS

The Himalayas have also been divided by Sir Sidney Burrard into four transverse regions, *viz.*, the Assam, Nepal, Kumaon and Punjab Himalayas.

The Assam Himalaya is the portion between the peak Namcha Barwa (7,756 m) in the Mishmi country where the Tsang-po (Brahmaputra) makes a sharp bend to cut across the mountains, and the Tista river, and is 720 km long. The Kula Kangri group of peaks (7,554 m) and Chomo Lhari (7,316 m) occur in this portion. The Himalayas are known to continue beyond Namcha Barwa but though that portion of the country is little known, geological and structural observations indicate that they execute a very sharp bend and turn southward to form the border ranges between Assam and Burma. The Assam Himalaya rises very rapidly from the plains, the foothills region being narrow and the Sub-Himalayas comparatively lower in altitude than in other areas.

The Assam Himalaya is geographically divided into short segments named after the tribes inhabiting them. These are the Aka hills between the Dhansiri and Dikrai rivers; the Daphla hills between the Bhaireli on the west and the Ranganad on the east: the Miri hills north of the Lakhimpur district; the Abor hills between the Siom on the west and the Dibang on the east ; and the Mishmi hills between the Dibang and the Dihang (Brahmaputra). Our knowledge of the Assam Himalaya is very meagre, having been derived from a few traverses made in conjunction with military expeditions.

The Nepal Himalaya, 800 km long, is the portion between the Tista on the east and the Kali on the west and is crowned by the peaks of Kanchenjunga (8,580 m), Everest (8,840 m), Makalu (8,470 m), Annapurna (8,075 m), Gosainthan (8,014 m) and Dhaulagiri (8,168 m). It throws off a branch near Dhaulagiri.

The Kumaon Himalaya, 320 km long, is limited by the Sutlej river on the west. The best known peaks here are Nanda Devi (7,816 m), Badrinath (7,069 m), Kedarnath (6,940 m), Trisul (7,120 m), Mana (7,273 m), Gangotri (6,615 m) and Jaonli or Shivling (6,638 m). The southerly bifurcation here becomes the Dhauladhar Range. It is in the Gangotri region that the Bhagirathi river, the ultimate source of the sacred Ganga (Ganges), rises in the Gangotri glacier.

The Punjab Himalaya, 560 km long, is the portion between the Sutlej and the Indus. The Sutlej cuts across the Himalaya where it shows a marked curvature. A southerly branch, the Pir Panjal, is also given off in this part. The main range carries few peaks exceeding 6,000 m. The Zoji-La pass over this is only 3,440 m high. The northern slopes of the range are bare and show plains with lakes, while the southern slopes are rugged and forest-clad. The Punjab Himalaya is not cut across by any river. Topographically it culminates in the Indus valley just beyond Nanga Parbat (8,114 m), but geologically and structurally it makes a sharp southerly bend and merges into the Safed Koh range on the one hand and the mountains of the Baluchistan arc on the other.

THE BALUCHISTAN ARC

The ranges corresponding to the Himalayas and their foothills are to be found in the Black Mountains of Hazara, the Kala Chitta and Margala hills, the Salt Range, the Sulaiman Range, Bugti and Mari hills, Kirthar Range and the Mekran Ranges, the last sweeping westwards into the mountains of South Iran.

The northernmost ranges of this arc start from the Hindu Kush mountains and proceed through Chitral and Swat. Farther south are the Muhmand hills while south of the Khyber pass are the hills of Tirah inhabited by the Afridi tribes. The Safed Koh Range branches off from the northern end of this area. Further south are the hills of Waziristan which merge into the Sulaiman Range part of which traverses the Kohat district. The Kala Chitta — Margala hills and the Salt Range join the Sulaiman Range from the east. The Kohat hills are associated with the curving hills of Maidan, Khasor and Marwat Ranges to the west of the Indus opposite the Salt Range.

The Sulaiman Range runs from the Gumal river to Quetta in a large loop. The inner ranges of this loop are the Bugti and Mari hills. South of Quetta are the Kirthar and Laki hills. The former, composed of several parallel ridges, form the boundary between Sind and Baluchistan. One of these stretches straight on to Cape Monze. The highest point in the Kirthar Ranges is Mt. Zardac (2,265 m). The Laki hills may be considered as merely an offshoot of the Kirthar hills, being the easternmost ridge bordering on the Indus plains, and terminating at Sehwan on the Indus. To the west of the Kirthar hills are the ranges of Brahui and Mekran, the latter turning westwards in a broad curve and proceeding into Southern Iran.

The Kirthars themselves apparently continue into the Arabian Sea and turn parallel to the Mekran Ranges. They are thought to connect with the coastal ranges of Oman in Arabia.

THE BURMESE ARC

At the north-eastern extremity of the Himalayas, a little beyond the peak Namcha Barwa, the geological formations turn sharply southwards and form the conspicuous arc forming the border of India and Burma, continuing into the Andaman and Nicobar islands and the Indonesian Archipelago. The Burmese arc on the Indo-Burma border is an area about which very little is known, except for the meagre information obtained from a few traverses. It is composed of the Patkoi, Naga, Manipur, Lushai, Chin, Arakan and other ranges. The Median Tertiary belt of Burma lying to the east of these ranges corresponds to the Tertiary zone of Baluchistan, while the zone of the Shan plateau, with its Pre-Cambrian, Palaeozoic and Mesozoic rocks, is a foreign element beyond the original Tethyan zone, and belongs to South-east Asia, belonging to the Indo-China province.

GLACIERS

Snow-line and limit of Glaciers : The 'snow-line' or the lowest limit of perpetual snow and ice is at different altitudes in different parts of the Himalayas and associated ranges. In the Assam Himalaya the snow-line is at about 4,400 m or higher, whereas in the Kashmir Himalaya it varies from 5,100 to 5,800 m. This is probably due to the scantiness of moisture in the region of the north-western Himalaya and Tibet. It is an interesting fact that in all the ranges north of the Great Himalayas, the snow-line is at a higher elevation on the southern than on the northern slopes, because the sun's rays affect the southern slopes more than the northern. But the reverse is the case in the Great Himalayas since the southern slope receives much greater precipitation than the northern and is also steeper, the slope helping the gliding down of the ice quickly to low levels. The glaciers also descend to lower levels in the Punjab Himalaya than in the Assam Himalaya ; this is due partly to the lower latitude and greater condensation of atmospheric moisture as rain (rather than snow) in the Assam Himalaya.

The glaciers are now confined to the higher ranges. The more important ones are valley-glaciers flowing through longitudinal valleys, and having large dimensions. The hanging glaciers along short transverse valleys are less important and more affected by variations of temperature, seasonal snowfall, etc.

The following glaciers may be mentioned as important :—

Table I.—Length of Important Glaciers

Name	Location	Length Km
Siachen	Karakoram	72
Hispar	do.	61
Biafo	do.	60
Baltoro	do.	58
Batura	do.	58
Rimo	Punjab Himalaya	40
Punmah	do.	27
Rupal	do.	16
Diamir	do.	11
Sonapani	do.	11
Gangotri	Kumaon Himalaya	26
Milam	do.	16
Kosa	do.	11
Kedarnath	do.	14
Zemu	Nepal Himalaya	26
Kanchenjunga	do.	16

The glaciers of the Himalayas and Central Asia have been studied by the Geological Survey of India as well as by many explorers, among whom may be mentioned Montgomerie, Conway, Longstaff, De Filippi, Visser, Dainelli and others.

Many of the Himalayan glaciers are much smaller than those listed above, being generally 3 to 8 km long. In the largest glaciers the thickness of ice amounts to hundreds of metres *e.g.*, Fedchenko 550 m; Zemu 180 m; Baltoro 120 m. The hanging and transverse valley glaciers are small and have a more rapid movement than the longitudinal valley glaciers. The daily movement varies from 2 to 108 cm and rarely higher, depending on the topography.

The Himalayan glaciers are definitely receding gradually. In many of them large amounts of moraine cover the ice near the snout. During the summer they melt and the water escapes through crevasses forming englacial streams issuing out of tunnel-like caves.

The Himalayan rivers and their tributaries are to a large extent fed by glaciers. They also receive affluents and important streams traversing the Sub-Himalayan region. The larger, snow-fed rivers are often full in the latter part of the summer because of the water contributed by melting snow and ice at their head-waters. Rains in the Sub-Himalayas also contribute to the waters of these rivers.

The present glaciers are mere remnants of the extensive glaciation of the Pleistocene period when very large areas of the mountainous tract must have been covered by snow and ice. Terminal moraines are found at as low altitudes as 1,800 m in the region of the Lesser Himalayas. Erratics or large boulders, brought down from the mountains over long distances, are found in the plains of the Punjab far from the source of the rocks. These and other glacial features such as fluvioglacial deposits and moraine-filled glacial lakes suggest that the Pleistocene glaciation covered very large areas in the Himalayas and extended to very low altitudes.

RIVERS

Rivers of the Peninsula

There are numerous rivers traversing the Indian Peninsula, the more important ones being the Damodar, Subarnarekha, Brahmani, Mahanadi, Godavari, Krishna, Penner, Cauvery and Tambraparni, which flow into the Bay of Bengal, while the Narmada and the Tapi flow into the Arabian Sea. The Banas, Luni, Chambal, Sindh, Betwa, Southern Tons, Ken and the Sone are Peninsular rivers of Northern India belonging to the Ganges system while there are also a few others rising in Central India and the Aravallis and flowing into the Rann of Kutch or the Gulf of Cambay.

Most of the Peninsular Rivers can be said to have reached a mature state of development, particularly in the lower portions of their valleys. The larger rivers have built up deltas at their mouths, but delta formation is prevented in the case of the smaller rivers by the strong ocean currents which flow along the coasts. In the Western Ghats the rivers still show an early stage of development probably because there may have been an upward tilt as well as an uplift of the western part of the Peninsula in the Tertiary era, as indicated by the presence of Upper Tertiary rocks along the western coast which were laid down when the western coast was faulted down in early Miocene.

The Damodar rises in the Chota Nagpur plateau near Tori in the Palamau district of Bihar. Its tributaries are the Garhi, Konar, Jamunia and the Barakar. It becomes a large river after its confluence with the Barakar. It joins the Hooghly after a course of nearly 600 km, a few kilometres below Calcutta.

The Subarnarekha rises a little to the south-west of Ranchi, and flows in a general easterly direction through Singhbhum, Mayurbhanj and Midnapore districts. It is about 480 km long and drains an area of about 28,000 sq. km.

The Brahmani, which is formed by the confluence of the Koel and Sankh which join together near Panposh and Rourkela in Gangpur, flows through Bonai, Talchar and Balasore districts, and is finally joined by the Baitarani River before it enters the Bay of Bengal. Its total length is about 420 km.

The Mahanadi rises near Sihawa in the Raipur district in Madhya Pradesh. It flows towards the north-east, and after being joined by the Seonath, turns to the east and then south-east. It is a large river at Sambalpur below which it flows through the Eastern Ghats, entering the sea through several channels in its own delta. Its total length is 885 km, and the area of its drainage basin 114,000 sq. km.

The Godavari is the largest river of the Peninsula. It has a total length of 1,500 km and its drainage basin covers 290,000 sq. km. It rises in the Western Ghats in the Nasik district and on its way receives several important tributaries including the Purna, Maner, Pranhita (Wardha and Wainganga), Indravati, Tal, and Sabari. A large part of the area through which it flows is densely forested. Its passage through the Eastern Ghats is through a picturesque gorge above Polavaram, where it is proposed to build a high dam. It is 2,750 m wide at Rajahmundry below which it splits up into several branches which have formed a large delta.

The Krishna rises near Mahabaleswar in the Western Ghats. Its length is 1,300 km and its drainage basin about 260,000 sq. km. Its chief tributaries are the Koyna, Ghatprabha, Malprabha, Bhima and Tungabhadra. It flows through Southern Maharashtra and Andhra. Its delta commences a little below Bezwada (Vijayawada). Its largest tributary, the Tungabhadra, is composed of the Tunga and Bhadra Rivers which rise in western Mysore and join just below Shimoga in Mysore. On its banks near Hospet are the ruins of Hampi, the capital of the once great Vijayanagar Kingdom. The Tungabhadra joins the Krishna River near Kurnool town after a course of nearly 640 km from the sources in Mysore.

The Penner rises in the Kolar district of Mysore, its chief tributaries being the Chitravati and the Papaghni. It flows through a gorge of Cuddapah quartzites near Gandikota in the Cuddapah district. It enters the sea near Nellore Town and has no delta worth the name.

The Cauvery (Kaveri) rises in Coorg and flows across the Mysore plateau before reaching the plains. Its total length is 765 km, and its drainage basin is 72,000 sq. km in area. Its chief tributaries are the Bhavani, Noyil and the Amaravati. It is a comparatively small river in Mysore and its descent from the plateau is marked by a few cascades and falls. At the head of its considerable and fertile delta is the city of Trichinopoly; while at its mouth once stood, several centuries ago, the flourishing port of Kaveri-pumpattinam the ruins of which now lie buried in alluvium.

The Tambraparni, the river of the Tirunelveli district, rises on the slopes of Agastyamalai in the Western Ghats, and flows into the Gulf of Mannar. Eight kilometres inland from its present mouth are the remnants of Korkai which was at one time the capital of the Pandya Kingdom and a great seaport. At a later date its place as a sea-port was taken by Kayalpatnam, which was visited by Marco Polo during his voyages.

The Narmada rises on the western flanks of the Amarkantak plateau at about $22°40'$ $81°45'$ in Rewa, the actual source being a small pond surrounded by a group of temples. It flows alongside Ramnagar and Mandla as a deep placid stream and then turns towards Jabalpur where it cascades down, through Marble Rocks, to a depth of 9 m forming the well-known Dhuandhara Falls. From Jabalpur westwards it flows for 320 km between the Vindhya and Satpura mountains. After passing Handia and Punasa it enters the alluvial plains of Indore. Below Broach, it widens into a 27 km broad estuary and enters the Gulf of Cambay. The total length of the river is 1,300 km and the drainage basin about 93,000 sq. km.

The Tapi (Tapti) rises on a plateau in the Satpuras at $21°48'$: $78°15'$. It flows through the Betul district and Berar, part of its valley being very narrow and overhung by steep cliffs. It is joined by the Purna tributary just before entering Khandesh. The last part of the course is through the plains of Surat. The total length of the river is only 700 km, the last 48 km of which are tidal.

The Luni (Lavanavari) rises in the Aravallis to the south-west of Ajmer, and flows through a semi-arid tract more or less parallel to and west of the Aravallis. Its length is 320 km, ending on the Sahni marshes north of the Rann of Kutch. There are several tributaries, including the Sarsuti which rises from the Pushkar Lake. The river contains water only during the rains. The water is sweet only down to Balotra a few miles from Pachbhadra lake, but becomes increasingly saline thereafter. Its waters are conserved by means of several dams of which the largest is near Bilara.

The Banas rises north-west of Mt. Abu and flows through Palanpur into the Little Rann after a course of 270 km.

The Sabarmati rises in the Mewar hills and enters the sea at the head of the Gulf of Cambay after a course of 320 km. Its major tributaries are the Sabar and the Hathmati coming from Idar and Mahikantha respectively.

The Mahi rises in Gwalior and flows through Dhar, Jhabua, Ratlam and Gujarat into the Gulf of Cambay. It is 560 km long, the last 65 km being tidal.

Notes on the other rivers of the Peninsula, which flow into the Ganges in the plains of Uttar Pradesh and Bihar, will be found under the Ganges river system.

WATERFALLS

There are numerous waterfalls in the Western Ghats, many of them small, only 6-9 m high and generally found in the courses of the westerly flowing streams. The *Jog falls* (Gersoppa) on the Sharavati river in Mysore comprise four magnificent falls called Raja, Rocket, Roarer and Dame Blanche arranged on a curve and having a sheer drop of 255 m. The *Sivasamudram falls* on the Cauvery, a series of cascades about 90 m high, are well-known since they were the first falls in India to be harnessed for power. The *Pykara falls* in the Nilgiris are a series of cascades which have also been utilised for hydro-electric power. The *Gokak falls* (54 m) on the Gokak River are near Belgaum. The *Yenna falls* near Mahabaleshwar have a drop of 180 m.

The southern Tons (Tamasa) leaves the Vindhyan plateau in a series of beautiful waterfalls the best known of which is the *Behar falls* which when in flood presents a solid sheet of water 180 m wide and 110 m high. There are several cascades in the course of the Chambal above Kotah, the largest being 18 m high. Various cascades and waterfalls mark the course of the Sone and the Betwa, particularly where they pass the Vindhyan sandstones. The *Dhuandhara falls*, in the well-known Marble Rocks near Jabalpur, though only 9 m high, are very picturesque and well-known. Two other falls in this river, each 12 m high are found near Mandhar and Punasa. The Krishna river during the floods is a rushing torrent descending some 125 m within a horizontal distance of 5 km near Echampet in Raichur district, Andhra Pradesh. The waters form a series of cascades over granite ledges and are churned into huge clouds of spray, providing a wonderful sight.

The Western Ghats form the major watershed of the Peninsula but the Vindhyan plateau in the north acts as another watershed which separates the Ganges basin from the Peninsula. The rivers in the Western Ghats show a comparatively early stage of development marked by cascades and waterfalls, but in the plains they show maturity of development. This is thought to be due to the rejuvenation of the Ghats by an upward tilt connected in some way with the faulting of the Arabian Sea Coast, and probably also by block uplift of the region during the Middle Tertiary.

The larger rivers flowing into the Bay of Bengal have built up deltas but the smaller ones do not bring in enough sediments to resist the removal by the prevalent coastal currents. Those flowing into the Arabian sea are short and do not contribute much sediment to be able to form deltas. The Narmada and Tapi are the only large rivers to flow westwards though rising in the central part of India. Their courses are determined by a rift or fault zone and contain a large thickness of Pleistocene and recent alluvium. There is evidence that the Narmada formerly flowed more towards the south-west in Kandesh and was joined by the Tapi. The courses of these two rivers are now separate. The absence of an extensive delta at their mouths is due to the fact that the sediments brought by them are removed by the strong S.W. monsoon currents partly to silt up the Gulf of Cambay and partly to spread over the wide continental shelf in front of the Bombay coast.

Extra-Peninsular Rivers

Between Hazara in the north-west and the Chinese Frontier in the north-east, the Himalayas give rise to some 20 important rivers. They rise from the Great Himalaya, Karakoram, Ladakh, Zanskar, Kailas and Trans-Himalaya ranges and ultimately join together to form the three great river systems of the Indus, Ganges and Brahmaputra, after cutting through the mountains. The head-streams are generally snow-fed from the great glaciers of these mountains and have precipitous and picturesque courses. It is remarkable that, though the Indus, Sutlej and Tsang-po rise in the Mount Kailas neighbourhood, they flow in very different directions to reach the plains. They are described here briefly and the reader may obtain much interesting information on them from the work of Burrard and Hayden.

THE INDUS SYSTEM

The Indus (Sanskrit : *Sindhu*) is the westernmost of the Himalayan rivers and its Sanskrit name, which also means the sea, is perhaps in allusion to its vast size when in flood. It is also possible that a large part of its present delta in Sind and south-western Punjab was, during the Vedic times, an arm of the Sea or a broad estuary, which was gradually filled up and became dry land only in recent centuries. It is one of the mightiest rivers of the world, draining the glaciers and mountain slopes of many famous peaks like Aling Kangri (7,315 m), Tirich Mir (7,690 m), Gasherbrum (8,068 m), Masherbrum (7,821 m), K^2 (8,610 m), Rakaposhi (7,788 m), Nanga Parbat (8,114 m), etc., and receiving a galaxy of great rivers as its tributaries. It has a total length of over 2,900 km and the drainage basin is estimated to have an area of 966,000 sq. km. Its name in Tibet is *Singi Khambab* (Lion's mouth) which is also the name of the original northern tributary at the source, at an altitude of 4,164 m in a glacier near BokharChu (31°15′ : 81°40′) in the Trans-Himalaya Range northeast of Mt. Kailas (6,714 m). The southern tributary which also rises in this region is called *Gartang Chu*. It flows for about 290 km over a flat country along the inner (northern) flank of the Ladakh Range. Then it cuts across that Range at Thangra and flows along the outer flank for another 480 km. Near Skardu (altitude 2,610 m) it cuts the Ladakh Range again, resuming the general trend of its course on the other side. After circling round the Nanga Parbat (8,114 m) it flows south-west through Hazara towards the plains of the Punjab. The exit of the Indus from the main Himalayan range is through a comparatively broad valley in contrast with the great gorge of the Brahmaputra at the other end in Assam.

The chief tributaries of the Indus are the *Zanskar* which rises on the Zanskar Range, cuts through it and joins below Leh; the *Dras River* which comes from the northern side of Zojila Pass and the plains of Deosai ; the *Shyok* which rises on the northern side of the Karakoram Range and cuts across to the south and joins the Indus near Kiris; the *Shigar* which drains the southern slopes of K^2 and the Biafo and Baltoro glaciers; the *Gilgit* (with its Hunza tributary) which rises behind and cuts through the Karakoram range.

The Indus is a large river over 150 m wide and 3 m deep at Skardu, even in winter. It passes through Gut in Chitral, and then traverses 200 km of the wild territory of Kohistan, entering the N.W. Frontier Province near Darband. At Attock, 1,385 km from the source and 1,515 km from the mouth, it receives the Kabul tributary flowing in from Afghanistan and then flows due south. Here the bed of the Indus is 610 m above the sea level and its width varies from 90 to 230 m according to the season. Below Attock, the Haroh and Sohan (Soan) rivers join the Indus. At the confluence of the Punjab rivers near Mithankot it is 550 m wide and its normal discharge is about 25,000 cusecs which increases in flood time to about a million cusecs. This confluence is 85 m above sea level and 790 km from the sea. The river sometimes rises as much as 6 to 7 m in flood. The delta, which is 7,800 sq. km in extent, contains several old channels and is entirely flat with practically no large trees.

The main channel of the river formerly flowed down the middle of the Thal. In 1800, the Indus divided into two streams at the head of the delta. In 1819, one of the channels (the Khedewari) was closed as a result of an earthquake. Its channels have been shifting every few years and many towns which once flourished in its banks *e.g.*, Ghorabari, Keti, Bhimanjopura, have been abandoned. In fact, practically the whole of its course below Attock, except small stretches near Sukkur and Kotri, changes from time to time. The frequent and recurring floods, such as those of 1833 and 1858, cause much havoc. The barrage at Sukkur which was opened in 1932 has contributed much to the conservation of the waters and their distribution by a canal system which commands more than two million acres of cultivated land.

The Jhelum (Sanskrit : *Vitasta*) rises in a spring of deep blue water at Seshanag at the head of its Lidar tributary. It flows for 110 km in a north-west direction when it enters the Wular Lake. Its basin here lies between the Great Himalayan Range and the Pir Panjal. Below Srinagar it is joined by the Sind River. From Baramula it flows through a narrow defile known as Basmangal, 2,130 m deep with very steep sides. At Uri, below this defile, it runs parallel to the Pir Panjal to Muzaffarabad where it is joined by the Kishenganga which drains Hazara. After emerging from Jammu it flows past Pind Dadankhan and Bhera and is joined by the Chenab at Trimmu, the total length of the river being 725 km. The Jhelum is an important river in Kashmir for it is the main water-way and fosters much trade.

The Chenab (Sanskrit : *Asikni* or *Chandrabhaga*): The two tributaries of this river at the source are the Chandra and Bhaga which rise on the opposite sides of Baralacha Pass (4,880 m) in Lahaul. The Chandra is a stream of good size though it flows through a snow-clad barren uninhabited country. The Bhaga is a precipitous stream. They join at Tandi and then flow through Chamba State in a north-westerly direction for 160 km in the trough between the Great Himalaya and the Pir Panjal, on the same alignment as the Jhelum in the Kashmir Valley. De Terra states that there is evidence that the Jhelum originally flowed in a south-east direction (reverse to the present direction) into the Chenab valley. This is supported by the fact that several of the present tributaries of the Jhelum join it in a direction opposite to the present course. Moreover, the Chenab valley shows greater maturity of topography than the Jhelum valley, and is evidently older in age than the latest uplift of the Himalayas in Pleistocene. The Chenab makes a very sharp knee-bend at Kishtwar (similar to that of the Kishenganga at Shardi, the Indus at Bunji, the Jhelum at Wular and the Ravi at Chamba) and flows across the Pir Panjal through a fine gorge. It leaves the Himalayas at Akhnur, 290 km below Kishtwar and 640 km from the sources, the average gradient being 5 m per km. There is evidence that the Chenab flowed to the east of Multan before 1245 A.D. The Beas then occupied its old bed passing by Dipalpur. The Jhelum, Chenab, and Ravi met to the north-east of Multan and after flowing east of that place joined the Beas ultimately near Uch, 45 km south of Multan. By 1397 A.D. the Chenab had changed its course to the present one which is to the west of Multan.

The Ravi (Sanskrit : *Parushni* or *Iravati*), though the smallest river of the Punjab, is well-known as the river of Lahore. It rises in the mountains of the Bangahal basin draining the northern slopes of the Dhauladhar Range and the southern slopes of Pir Panjal. It leaves the basin through an inaccessible gorge with perpendicular sides and flows through Chamba State in a north-westerly direction parallel to the Dhauladhar Range cutting through that range a few kilometres to the north-west of Dalhousie. It leaves the mountains at Basaoli after a course of 210 km during which it drops by 4,570 m in altitude (22 m per km). It finally joins the Chenab at 30°31′: 71°51′, the total length up to this junction being 720 km.

The Beas (Sanskrit : *Vipasa* or *Argikiya*) rises on the southern face of the Rohtang Pass in Kulu not far from the source of the Ravi. Barely 10 km from its source it passes through a gorge at Koti, which is a chasm barely 6 m wide and 275 m long. It cuts through the Dhauladhar Range by another gorge near Larji, and then flows through Kulu, Mandi and Kangra. It finally passes through Kapurthala and Amritsar and joins the Sutlej in the south-west corner of Kapurthala after a total length of 465 km. The old course of the Beas can be traced from its present junction with the Sutlej through the Lahore and Montgomery districts to where it originally used to join the Chenab near Shujabad before the Chenab turned westwards.

The Sutlej (Sanskrit : *Satadru* or *Satudri*) rises near the Manasarowar Lake at a height of 4,630 m. According to Swami Pranavananda, its name in Tibet is Langchen Khambab and it rises in the glacial springs of Dulchu Khambab, 35 km west of Parkha which is an important trading centre between Kailas and Manasarowar. In Tibet it has a very narrow basin between the Giri River on the east and the Beas on the west whose beds are, however, at an elevation of 180 to 215 m above the level of the Sutlej. It is very likely that the stream which periodically flows connecting the Manasarowar and Rakshas Lakes is connected with the Sutlej underground. The river also receives waters from the southern glaciers of Mt. Kailas and the northern glaciers of Mt. Kamet. The peak Riwo Phargyul is in its basin.

From Rakshas Lake to Shipki, the Sutlej flows north-westwards through the Province of Ngari Khorsum which is at an elevation of 4,270 to 4,570 m above sea level. The valley here is filled with thick recent alluvium and gravels. In this region the Sutlej has cut a deep and extraordinary canyon which is 900 m deep in places and bears comparison with the Grand Canyon of Colorado. As this region is dry, it has not been subjected to erosion and smoothening by water. It is joined by a few tributaries which have also cut steep-sided canyons through which they flow. It cuts through the Zanskar Range near Shipki, barely 7.2 km from the Peak of Riwo Phargyul (6,770 m), the bed being at an elevation of 3,000 m above the sea level. Sixteen kilometres below Shipki, the right bank of the Sutlej rises as a perpendicular wall of rock to a height of 1,800 to 2,100 m from the river bed.

The main tributary of the Sutlej is the Spiti River which drains a large area beyond the Central Himalayan Range. It has also cut deep into the rocks of the country of Kulu and Himachal Pradesh through which it flows. From the junction with the Spiti River in Kanaur (Bashahr), the Sutlej is a rushing torrent right down to the plains, for it descends through an elevation of something like 2,440 m in this stretch, with an average fall of 6 m per km. There are several river terraces in this region, which show that originally there were some lakes along the course of the river. The Sutlej crosses the Dhauladhar Range near Rampur through a narrow gorge. It is deflected several times in its course by the Siwalik Ranges and leaves them at Rupar. The river joins the Beas in the south-west corner of Kapurthala and the combined river joins the Indus near Mithankot. The total length of the Sutlej is about 1,440 km. It is known that formerly, before the 11th Century, it did not join the Indus system, but flowed into the Sarasvati (Hakra or Ghaggar) in Bikaner through one or more of its several old channels. The oldest was the old Sirhind channel between Sirsa and Bhatnair and the later ones the three Naiwals. From Sirsa the channel can be traced back to Tohana and thence very indistinctly to Rupar.

THE SARASVATI

The Sarasvati rises in the Siwalik hills of Sirmur on the borders of the Ambala district in the region of the junction of the talus fans of the Yamuna on the east and the Sutlej on the west, and enters the plains at Adhbadri. It disappears in the sands after passing by Bhawanipur and Balchhapar but reappears after a short distance, flowing through Karnal. The Ghaggar which also rises in the same region (near 30° 4′ : 77°12′) joins it at Rasula in Patiala after a course of 175 km. Further on, the river is called the Hakra or Sotar whose dry bed is of considerable size and must have once contained a large river. It is lost in the sands near Hanumangarh (Bhatnair) in Bikaner. Near Bhatnair, it is joined by the Chitrang (also mostly dry) from the east ; this stream, which can be traced north-eastward almost up to the Yamuna, was probably an old channel of the Yamuna flowing towards Bikaner to join the Sarasvati. Between Bhatnair and Sirsa (a corrupt form of 'Sarasvati') it is joined by the Sirhind or Wah which can be traced back nearly to Rupar near which place the Sutlej emerges from the Himalayan foothills. This must originally have been the main channel of the Sutlej until it was abandoned by that river. There are at least three other old channels from the north, all called the Naiwals, joining it near Kurrulwala (29°33' : 73°52'), which can also be traced back to the Sutlej.

The old bed of the Hakra (or Sotar) for over 160 km in Bikaner territory is 5 to 8 km wide and consists of dark, rich, loamy soil in marked contrast with the sandy soil on either side and on both the banks. The local people still call this channel the Sarasvati. The vegetation on the loamy soil on the banks has prevented the old bed from being overwhelmed by the drifting sands from the desert. On both banks of the river there are numerous mounds containing abundant evidences of prehistoric and early historic settlements. These mounds are small near Hanumangarh but larger and fewer near Sutargarh and further west. The mounds, which have been investigated by Sir Aurel Stein and by the officers of the Archaeologic 1 Survey of India, have revealed temples, dwellings, potsherds and other objects of the Mohenjo-Daro type (third millenium B.C.) and of later periods. This channel can be very clearly followed through Mirgarh, Dilawar, etc., in Bahawalpur (where it is known as Hakra or Wahind) into the channel of the Eastern Nara in Sind which leads into the Rann of Kutch. This is distinct from, and east of, the old channels of the Indus river. When the Sarasvati was a live river, it must have irrigated an area of perhaps over 18,000 sq. km of what is now practically a desert.

The Sarasvati has been described in Vedic literature (probably 5,000 B.C. or earlier) as a great river — greater even than the Indus and the Ganges. Between that and the time of Manu and the Mahabharata its upper course had dried up, probably because of the easterly diversion of the waters of the Yamuna. The lower course in Bikaner, Bahawalpur and Sind continued to be well watered, for during the invasion of Alexander of Macedon in the fourth century B.C. and of the Arabs in the ninth century A.D., the Rann of Kutch was a fairly deep gulf, and ships moved up the river into Sind. In Todd's 'Annals of Rajasthan' it is stated that the Hakra in Bikaner became dry for the first time about the year 1044 A.D.

There seems to be little doubt that the major part of the waters of the Sarasvati (Hakra, Sotar or Wahind) was derived from the Sutlej which as Sirhind originally joined it between Bhatnair and Sirsa, and later as one of the Naiwals near Kurrulwala and Wullur. There is historical evidence in the writings of the Greeks and the Arabs that the Sutlej was not a Punjab river until about 1200 A.D., when it abandoned its southerly course and joined the Beas. The Beas below this junction is still known to the local people as the Beas, though Sutlej is undoubtedly the greater river. Before that, during the earlier centuries of the Christian era, the Beas flowed in a channel now dry and lying to the north of the present joint course of the Beas-Sutlej.

Little is known about the details of the geological history of the Sirmur region where the Sarasvati and the Ghaggar rise. In view of the fact that the Himalayan region has been experiencing uplift in the Pleistocene and perhaps even after man had appeared, it would be interesting to know whether the Sarasvati originally derived its waters from the Himalayas beyond the Siwalik Zone and whether the old drainage was cut off by the subsequent rise of the Siwalik hills and by other changes.

THE GANGES SYSTEM

The term 'Ganges' is a corrupt form, used by the Greek historians, of the Sanskrit name 'Ganga' by which the river is known throughout India and in the lands where Indian civilization had spread. Its drainage basin covers one of the most thickly populated regions of the world where the Indo-Aryan civilisation has flourished for many centuries. It comprises the Ganges and many important affluents, such as the Yamuna (Jamuna), Kali, Karnali, Ramganga, Gandak and the Kosi, all of which rise in the Himalayas and are mainly snow-fed. On the side of the Peninsula the tributaries are the Chambal, Betwa, Tons, Ken, Sone, etc., which rise from the highlands of the central part of India.

The Ganges proper is formed of two tributaries called the Bhagirathi and the Alaknanda. The latter is the larger river and is itself formed from the confluence of the Dhauli, which rises in the Zanskar Range near Niti Pass and receives numerous streams draining the northern and the western slopes of Nanda Devi, and the Vishnu-Ganga which rises on Mount Kamet (7,756 m) behind Badrinath near the Mana Pass. The Dhauli and the Vishnu-Ganga join together at Joshimath, also called Vishnu Prayag (the word 'Prayag' denoting in Sanskrit the junction of two rivers). The Alaknanda then flows through a mighty gorge across the central Himalayan range between the peaks Nanda Devi (7,816 m) and Badrinath (7,068 m). The Pindar River, which gathers its waters from Nanda Devi and East Trisul (6,803 m) joins the Alaknanda at Karna Prayag. The Mandakini or Kali Ganga joins it at Rudra Prayag, south of Badrinath and Kedarnath (6,940 m). The Nandakna joins the Pindar River at Nanda Prayag to the west of the Trisul mountain. The junction of the Alaknanda and the Bhagirathi is called Deva Prayag, to the north of the Mussoorie Hills, which the river crosses before passing into the Siwalik Ranges of Dehra Dun and Hardwar.

The Bhagirathi is accepted traditionally as the original Ganga, though, as mentioned above, the Alaknanda is the larger riever. The actual source of the former is in the Gangotri glacier (which is 25 km long) some distance north of Kedarnath at a point called Gaumukh (30°56′ : 79°4½′) at an altitude of 3,900 m. The Gangotri shrine is a few kilometres down the stream from Gaumukh. The Bhagirathi joins its western tributary called the Jahnavi some distance to the north of main Himalayan range and about 11 km below the Gangotri temple. The combined river then cuts through the main Himalayan range between Bandarpunch (6,315 m) and Srikanta (6,133 m) through a magnificent gorge in which the river bed is 3,960 m below the peaks on either side (Griesbach, *Memoirs, G.S.I.*, Vol. XXIII, p. 197, 1891). This gorge is said to be a slitlike opening in the rocks with practically vertical sides reaching down 180 m to the bed of the river.

The Yamuna (Jamuna or Jumna), the western-most river of the Ganges system, rises on the western slopes of Bandarpunch in the Jamnotri glacier and passes by the Jamnotri shrine at the foot of that mountain. It is later joined by the Tons River behind the Mussoorie Hills and then breaks through these hills to be joined by the Giri and Asan Rivers, which drain the area between the Bandarpunch and Chor peaks. The Yamuna emerges from the Mussoorie Hills into the plains where it flows in a broad curve by Delhi, Mathura and Agra to join the Ganges at Allahabad (Prayag). Its tributaries in the plains are the Chambal and the Sind which join it below Etawah, the Betwa at Hamirpur and the Ken above Allahabad. Its total length to Allahabad (Prayag) where it joins the Ganges is 1,380 km. There is some reason to believe that the Yamuna might possibly have flowed to the south and south-west into Rajasthan or, at any rate, shared its waters with the Sarasvati, which was undoubtedly a large river during the Vedic times.

The Ramganga is a comparatively small river rising on the southern side of the main range between the Ganges and Kali basins. Its principal tributary is the Kosila which joins it only in the plains. The river is deflected to the south-east by the Siwalik Hills which it cuts through before emerging into the plains.

The Kali (Kaliganga or Sarda) rises in the Milam glacier where it is called the Goriganga. A few streams contributing to its waters rise in Kungri Bingri, Lipulekh and other peaks nearby. These are to the north of the main Himalayan range and drain the area between Nanda Devi and Apinampa (7,132 m). The two major tributaries of the Kali are the Dharma and Lissar which flow in a S.S.E. direction parallel to it before joining it. The combined stream is later joined by Sarju and the Eastern Ramganga at Pacheswar. The Sarju tributary flows from the N.W. to S.E. on the same line as the Pindar which is further to the north-west and is a tributary of the Ganges. From the junction at Pacheswar, the river is called the Sarda or Sarju. After emerging into the plains at Barmdeo, it splits up into two or more channels. The Sarda joins the Gogra at Bahramghat.

The Karnali which is called the Kauriala in the mountains, becomes the Gogra in the plains. It rises in the glaciers of Mapchachungo, north-west of Taklakot, draining the western and the southern flanks of Mount Gurla Mandhata (7,728 m), where its basin adjoins that of the Sutlej. It then flows in a south-easterly direction and cuts across the main Himalayan Range in a south-westerly direction. Before traversing the main range, this stream and its tributary drain a large tract 160 km long to the south of the Brahmaputra basin. Its gorge through the great Himalayan Range is deep and picturesque. After flowing 80 km in a south-westerly direction, it is joined by the Tila and executes a remarkable hair-pin bend turning towards the west at about Latitude 28°30′ and Longitude 81°30′. It is then joined by the Seti River just east of Api and cuts through the Mahabharat Range and receives the Beri tributary, which rises near the Diji Pass and drains the area to the west of Dhaulagiri (8,168 m). The passage through the Siwalik Hills is through a picturesque gorge known as Shishapani with precipitous sides 600 m high and through a series of fine rapids. In the plains it is joined by the Sarda (Sarju) at Bahramghat and acquires the name of Gogra, which is a corruption of the word 'Gharghara' meaning 'rattling' or 'laughing.' It passes through Oudh (Ayodhya) and finally joins the Ganges at Chapra a little above Dinapore. This is the Sarju (Sarayu) of the epic Ramayana. Its maximum flood discharge is said to be a little short of one million cusecs.

The Gandak (Sanskrit : *Sadanira*) is also called the Saligrami in Nepal, and the Narayani in the plains. The name in Nepal is apprently due to its bringing down a large number of *saligrams* (ammonite fossils) from the Spiti Shales of Jurassic age. In the mountains it drains the area between Dhaulagiri and Gosainthan. Its two main tributaries are the Kaligandak rising close to Photu Pass near Muktinath, and the Trisulganga. The Kaligandak cuts across the great Himalayas by a gorge and is later deflected east to west by the Mahabharat Range. The other tributary, Trisulganga, rises to the north of Gosainthan (8,013 m) and then flows south-west through the main Himalayan Range. It is then joined by the Buri Gandak and Marsyandi. The combined river (the Gandak) then cuts through the Mahabharat Range and finally emerges through the Siwalik Ranges at Tribeni. It joins the Ganges near Patna. In the plains there are evidences of the river having frequently changed its course.

The Kosi (Sanskrit : *Kausika*) is the largest of the tributaries of the Ganges, said to be next only to the Indus and the Brahmaputra in size and in the volume of its discharge. It drains the area between Gosainthan and Kanchenjunga in the Himalayas. The main stream, the Arun (called Phungchu in Tibet), rises to the north of Gosainthan and flows south-west for nearly 320 km in a fairly flat stretch just to the south of the Brahmaputra basin and of the Ladakh Range. This region is called the Dingri Maidan, composed of Spiti Shales, through which the river meanders. It is joined by the Yaru River from the east, the combined stream then flowing south between Mount Everest (8,840 m) on the west and the Kanchenjunga (8,579 m) on the east, receiving numerous small tributaries from the glaciers of these mountains. After cutting through the main Himalayan Range it is joined by the Sun Kosi from the west and the Tamur Kosi from the east. The former is composed of several tributaries, *viz.*, the Indravati, Bhote Kosi, Tamba Kosi, Likhu, Dudh Kosi and others. The Bhote Kosi rises at Thanglang, a few miles south of the upper reaches of the Arun. It cuts through the main Himalayan Range by a vertical-sided chasm 460 m deep and only 25 to 30 m wide. The Dudh Kosi rises between Mounts Gaurishankar and Everest. The Tamur Kosi rises on the western slopes of the Kanchenjunga and traverses the main Himalayan Range by a comparatively broad valley, in contrast with several other rivers which have cut narrow gorges.

The Sun Kosi and the Tamur Kosi run for a fairly long distance parallel to and north of the Mahabharat Range and join the Arun at Dangkera. The

Kosi cuts across the Mahabharat Range and the Siwalik Hills and emerges into the plains near Chatra. In the plains it is building up a large delta of its own through which its channels have wandered for centuries. It is believed that the Kosi originally joined the Mahananda, a river coming from the Darjeeling Himalayas. It is known that the Kosi flowed by Purnea 200 years ago, but its present course is about 160 km to the west of that place, having swept over an area of 10,500 sq. km. on which it has deposited huge quantities of sand and silt. It now joins the Ganges 32 km west of Manihari but formerly it used to join that river near Manihari itself. The Kosi is notorious for its frequent and disastrous floods and the vagaries of its channels. In high flood it is said to have a flow of nearly one million cusecs loaded with much gravel, sand and silt.

The Chambal rises in Central India near Mhow. It flows through Bundi, Kotah and Dholpur, finally joining the Yamuna 40 km east of Etawah. The total length of the river upto its junction with the Yamuna is 960 km. Burrard remarks that, if the criterion of the source of a river is the distance between the mouth and the farthest point amongst the sources of its tributaries, then the source of the Ganges will have to be regarded as identical with that of the Chambal; but as we have seen, tradition has decreed the sources of the Ganges to be Gangotri.

The Sind, which is one of the larger rivers of Northern India, takes its source near Nainwas in Tonk. It is probably the river Sindhu mentioned in the epic Vishnupurana. It is a perennial stream but subject to sudden floods. It joins the Yamuna a little to the north of Jagmanpur.

The Betwa (Sanskrit : *Vetravati*) rises a few miles south-west of Bhopal City and flows through Sanchi, Gwalior, Jhansi and Orchha. It flows through the Vindhyans in its upper reaches and through a granitic country further down. It joins the Yamuna at Hamirpur, some 50 km south of Kanpur (Cawnpore).

The Ken (Sanskrit : *Karnavati*) rises on the northern slopes of the Kaimur Hills and flows through Bundelkhand before it joins the Yamuna near Banda.

The Southern Tons (Sanskrit : *Tamasa*) rises in the Kaimur Range in Maihar. It leaves the plateau in a series of waterfalls, the largest of which is called Bihar which, when full, is a sheet of water 180 m broad and 113 m high. One of its tributaries the Belan, has also a waterfall 30 m high. The Tons joins the Ganges near Sirsa, a little below Allahabad.

The Sone (Sanskrit : *Swarna nadi*) is a large river rising on the Amarkantak plateau not far from the source of the Narmada. It flows in a northwesterly direction past Sohagpur through Rewa and Baghelkhand, along the foot of the Vindhyan scarp, in a narrow channel through a wild country. It has several tributaries — the Mahanadi, Banas, Gopat, Rihand, Kunhar, etc. In the plains of Bihar its bed is as wide as 4.8 km. During the rains the river rises to huge proportions and the maximum discharge has been estimated at three-quarter million cusecs.

About 1,000 years ago the Sone joined the Ganges a little below Patna. Since then it has been gradually receding westwards. About 1750 A.D. its junction was at Maner, but now it is 16 km further up the Ganges. The length of the river from its source to its junction with the Ganges is 775 km and its drainage basin has an area of 54,000 sq. km.

The Mahananda rises in the Darjeeling Himalayas near 26° 56′ : 88° 20′. It emerges into the plains near Siliguri and after passing through Purnea and Malda, joins the Ganges near Godagiri.

The delta of the Ganges begins near Gaur, a famous old historic city. The present main branch of the river flows in a south-easterly direction and is called the Padma. It flows past Pabna and Goalundo and is then joined by the Jamuna which is the major channel of the Brahmaputra. The total length of the Ganges from its source in Gangotri to the mouths in the Bay of Bengal is 2,490 km. A few centuries ago the main channel of the Ganges in Bengal was the Hooghly together with its feeders, the Bhagirathi, Jalangi and Mathabhanga, which are called the Nadia Rivers. The Bhagirathi is known to have been the main channel originally, for on its banks stand the various former capitals of Bengal such as Gaur, Pandua, Nabadwip and Satgaon (Saptagram). The former main course of the river, known as the Sarasvati, left the Hoogly at Satgaon and had a more westerly course and rejoined the Hooghly at Sankrail, a few miles below Calcutta. Ocean going ships used to sail up the Sarasvati to Satgaon and Tribeni until some 450 years ago. Formerly, the Damodar joined the Hooghly at Nayasarai, 63 km above Calcutta and contributed to the flushing of the Hooghly and keeping its channel deep. But by 1770 the Damodar changed its course and joined the Hooghly at Falta 56 km below Calcutta. It was after the Sarasvati channel was abandoned by the Hooghly that the trading settlements of the Portuguese, Dutch, French and Danes were established at Hooghly, Chinsurah, Chandranagore and Serampore (Srirampore). After the Damodar changed its course, the Hooghly at and above Calcutta has been shoaling up.

Formerly the Hooghly flowed south-east from Calcutta near the exit of the present Tolly's nullah and joined the sea near Sagar island. This channel, called the Adiganga, can be picked up now as a series of ponds and pools across the 24-Parganas up to near Sagar Island. On its banks stood the sacred Kali temple at Kalighat and also several other places which are still considered sacred. Now the Hooghly flows S.S.W. from Calcutta and then turns southeast towards Sagar island.

The Hooghly estuary is notorious for its sand banks and dangerous shoals of which the James and Mary Sands, 56 km below Calcutta and between the mouths of the Damodar and Rupnarain, are well-known. New areas are being reclaimed by the sediments brought down by the Ganges. These are known as the Sundarbans, which are extensive swampy flats forming the lowest portions of the delta.

THE BRAHMAPUTRA SYSTEM

The Brahmaputra : The source of the Brahmaputra, which is called Tsang-po in Tibet, is at Tamchok Khambab Chorten in the Chemayung-dung glacier, approximately at 31° 30′ : 82° 0′, some 145 km from Parkha (an important trading centre between the Manasarowar Lake and Mount Kailas) and near the source of the Karnali and the Sutlej. It has a long course through the comparatively dry and flat region of southern Tibet before it breaks through the Himalayas near the peak Namcha Barwa (7,755 m). Its chief tributaries in India are Amochu, Raidak, Sankosh, Manas, Bhareli, Subansiri, Dibang and Luhit (Zayul). The several tributaries in Tibet are derived partly from a low range between the main Himalaya and the Tsang-po and partly from the Nyen-Chen-Tanghla Range to the north of the Tsang-po. The total length of the river from the source in south-western Tibet to the mouth in the Bay of Bengal is 2,900 km.

The Brahmaputra is known as the Dihang in the Assam Himalaya before it comes into the plains. The Dibang and Luhit meet it from the east near Sadiya. The Dibang drains the Himalayas east of the Dihang while the Luhit drains an area between Assam and Burma.

The course of the Tsang-po in Tibet is through a plain but it has not been deeply cut into as is the case with the course of the Indus and Sutlej in western Tibet. It is quite a sluggish river south of Lhasa. The elevation of its bed is 4,523 m at Tradom ; 3,608 m at Shigatse ; 2,440 m at Gyela Sindong near Namcha Barwa ; but only 135 m at Sadiya in N.E. Assam.

Its course in Tibet from the source to where it enters the central Himalayan range near Namcha Barwa is 1,600 km long. It has three tributaries above Shamsang, *viz.,* Kubi Tsang-po, Chema Yungdung and Maryum Chu ; the first of these is the Tsang-po proper, which is also much larger than the other two. All the three rise in the watershed separating the Tsang-po basin from Lake Manasarowar. Several other tributaries join it further east. It is noticed that many of these flow in a direction opposite to the Tsang-po and this has led to the speculation that the Tsang-po might originally have flown westwards. Which route it chose is uncertain, for it may have followed what is now the course of the Kali Gandak or the Karnali or the Sutlej. According to Burrard, there is some evidence that the first may be the most likely, for the Phuto Pass (in Ladakh Range) separating the Kali Gandak and Tsang-po is an extraordinary depression barely 75 m higher than the Tsang-po valley in Tibet. The gorge of the Kali Gandak through the main range is also too deep to have been cut by the Kali Gandak which has a very small catchment above the gorge.

The chief westerly flowing tributaries of the Tsang-po are the Kyi (Lhasa river), Nyang (joining at Shigatse), Rang and Shang. Of those rising on the south Tibetan watershed, only the Nyang rises in the Central Himalayan Range north of Chomo Lhari (7,314 m) and cuts through the watershed before joining the Tsang-po. The Kyi-chu rises on the Nyen-Chen-Tanghla Range to the north of the Tsang-po valley. It is the river of Lhasa and the most important tributary of the Tsang-po. Other tributaries rise on the same range and some even on the Trans-Himalaya Range further to the north.

The Tsang-po flows for nearly 1,600 km in Tibet before it makes a great knee-bend encircling Namcha Barwa and then breaking through the Himalayas. The bend is just east of longitude 94°. The river has a waterfall 9 m high near Pemako Chung. A tributary which joins it near Gyela has a waterfall 45 m high called Shingchu-chogye. During its course through the Himalayas, the river (called here the Dihang) descends through 2,285 m, the altitude at Sadiya being only 135 m above sea level.

The Raidak rises in the Chomo Lhari mountains and joins the Brahmaputra at Kurigram, while the *Amochu* drains the Chumbi valley and joins the Brahmaputra at Alipur.

The Sankosh is larger and longer than the Amochu and Raidak. It drains the area between Kula Kangri South peak (7,540 m) and Chomo Lhari (7,314 m). It flows by Punaka where it is 365 m wide, but narrows down further below to flow through a gorge. It joins the Brahmaputra at Patamari below Dhubri.

The Manas is formed by the combination of several streams which join in the outer Himalayas. The main tributary is called Lhobrak or Kuru-chu by the Tibetans, and it rises on the north-western slopes of Kula Kangri. It breaks through the main Himalayan Range at Thunkar, south of Lhakhang Dzong. The bed of this gorge is at an altitude of 3,000 m and is impassable. After collecting all its tributaries in the lesser Himalayas, the Manas breaks through the foothill ranges and emerges into the plains to join the Brahmaputra.

The Subansiri : Very little is known about this river except that it has tributaries both from the north and south of the main Himalayan Range. It runs for a distance of 160 km in the plains before joining the Brahmaputra at the western end of Sibsagar district. It is said to have a long course in the Himalayas, flowing through a series of gorges and rapids. It separates the Abor from the Miri Hills of the outer Himalayas.

The Dhansiri rises in the Naga Hills and after a course of 290 km through Nowgong and Sibsagar, falls into the Brahmaputra below Golaghat.

The Torsa rises at roughly 27° 49' : 89° 11' below the Tang Pass. It flows through the Chumbi valley and Bhutan. It enters the plains in the Jalpaiguri district and splits into two branches the western one (Charla) joining the Brahmaputra at 25° 40' : 89° 44', after a course of 350 km, and the eastern one flowing into the Raidak.

The Tista (Sanskrit : *Trishna* or *Trisrota*) rises in Chitamu Lake in Tibet, (about 28° 2' : 88° 44'), and is said to have another source below Kanchenjunga. It joins the Brahmaputra in the Rangpur district of Bengal. Its tributaries are the Rangpo, Rangit, Rangjo, Ryeng and the Sivok. It flows through a magnificent gorge known as the Sivok Gola Pass in the Darjeeling district. It is a wild river in the Darjeeling Hills where its valley is clothed with dense forest, but its drainage area in the mountains is only 12,500 sq. km. In the

1950 floods its flow was estimated as nearly 0.66 million cusecs. Up to the close of the 18th century it flowed into the Ganges, but after the destructive floods of 1787 in which a large part of the Rangpur district was laid waste, it suddenly turned eastwards and joined the Brahmaputra. Its tributaries in the plains are the Lish, Gish, Saldanga, etc. There are many old channels especially in the western part of the Rangpur district which were occupied by this river formerly, and the Karatoya through which it joined the Ganges is still known as the Burhi Tista (Old Tista).

The Jamuna or Janai : This is the name of the present lower section of the Brahmaputra, from its entry into Bengal to its junction with the Ganges at 23° 15′ : 89° 45′. On its banks are the towns Tangail, Sirajganj and Bogra.

The Meghna : This is the name given to the course of the original Brahmaputra after its confluence with the Surma River, from Bhairab Bazar onwards. It enters the sea by four streams, enclosing several islands. During floods it expands, into a vast sheet of water. The various streams constantly shift their courses and navigation in them is always attended with danger. In the Noakhali district on its eastern side the land is steadily advancing towards the sea. The Meghna is subject to tides which rise 4 to 6 m, and also to bores which are often disastrous. The valley of the Meghna is frequently visited by severe cyclones, some of which work great havoc. It is said that about 100,000 people perished in the cyclone of 31st October, 1876.

The Feni is a short river, 115 km long, rising in the Chittagong hill tracts and Tripura. It falls into the Sandwip channel in Noakhali.

The Surma rises on the southern slopes of the mountain range to the north of Manipur. Its upper part, known as Barak, is marked by steep banks and several falls. It turns west in the Cachar district through which it flows down to Badarpur where it separates into two branches which rejoin at Habiganj. It has numerous small tributaries and its total length is 900 km before it joins the Brahmaputra. This river constitutes an important trade route and on it are situated Silchar, Sylhet, Badarpur, Sunamganj, etc.

The Brahmaputra in the plains is a mighty river and spreads into a vast expanse of water. When in flood it brings down much sediment and a large amount of vegetation including large trees. There are numerous islands in its bed and the streams change their course very often. After traversing the Assam Valley for 720 km, it sweeps round the Garo Hills, enters the Rangpur district and flows southwards for nearly 240 km before joining the Padma and the Sea. In Bengal, it is locally known as the Jamuna as far as its junction with the Padma near Goalundo. Originally the Brahmaputra flowed southeast across the Mymensingh district where it received the Surma River and directly united with the Meghna, as shown in Rennell's map of 1785. By the beginning of 19th century its bed had risen and it found an outlet farther west along its present course. The entire lower portion of the Brahmaputra consists of a vast network of channels, which are dry in the cold season but are inundated in the summer and in the rains. The river is more or less navigable as far up as Dibrugarh, 1,300 km from the mouth, but normally the larger boats can reach only up to Gauhati.

Most of the Himalayan rivers rise in the Great Himalaya or in the Trans-Himalaya. The fact that the chief watershed is beyond the line of great peaks is generally cited as evidence in favour of the drainage being *antecedent*, or having been in existence before the main phase of upheaval of the Himalayas occurred. The courses of the rivers, where they cross the high range, are at right angles to the latter, *i.e.*, they have a radial disposition with reference to the Himalayan arc.

There are several cases of the recession of the heads of streams in the mountains by the action both of the streams and the glaciers which feed them. This *head erosion* has, in some cases, led to the source going to the northern slopes of the Great Himalaya and capturing the drainage of other streams. An excellent example of *river capture* or *piracy* is furnished by the Kosi whose tributary, the Arun, drains the northern slopes of the Great Himalaya of Nepal region. Other examples are found in the Indus and Ganges systems. Rivers have also inherited the valleys of glaciers in many cases ; the tributary streams in these cases are often found in hanging valleys a few metres above the main streams, the junction being marked by cascades or waterfalls.

RIVERS OF BURMA

The chief rivers of Burma are the Irrawaddy, Chindwin, Sittang and Salween.

The Irrawddy rises in Upper Burma near about 29° N. latitude and has a drainage basin of 414,000 sq. km. Yet it is an immense river when in flood. There are two tributaries, the Nmai Hka and Mali Hka whose confluence is about 100 km above Myitkyina. There are three narrow portions or defiles along the course, the first below Sinbo, the second below Bhamo, and the third near Thabeitkyin near the Ruby Mines district. The river is however navigable up to Bhamo. It flows through Mandalay and then Pakokku where it is joined by the Chindwin. Below this junction is the dry belt of Burma and the course runs through sandstone country amidst its own older terraces. It is thought that the river entered the sea near Prome about a third of a million years ago, *i.e,*. about the end of the Tertiary era, and that the delta below Prome has been built up since.

The Irrawaddy river system is of Tertiary age, the lower portion being of post-Pegu age. Its chief tributaries are the Nam-tu and Chindwin. The Nam-tu is the chief river of the Northern Shan States, rising within a short distance of the course of the Salween at latitude 23° 20′ N. Its course south of Meng-tat is through a deep narrow valley. It is joined first by the Namma and further down by the Namhsin. The course, after the junction with the latter river, is through a deep gorge (Gogteik gorge) where a succession of rapids and pools is seen, and in which the stream may be 600 m below the top of the hills at the sides. After receiving the waters of the Nam-Hka and Nampanshe, it leaves the hills about 22 km to the south-east of Mandalay. The course of the river is entirely in the Plateau Limestone. There is abundant evidence that this river has captured its tributaries by head erosion.

The Chindwin river rises at latitude 25° 40′ N. and longitude 97° E. first flowing northward and then north-westward through the Hukawng valley and turning southward. In the Hukawng valley its tributaries are the Taron and Tawan Hka. It then flows through a rocky gorge and joins the Irrawaddy near Pakokku. The Uru river which flows through the jade mines tract is also a tributary of the Chindwin. It is interesting to note that the Lower Irrawaddy is the direct continuation of the valley of the Chindwin and that the former inherited this portion from the latter.

The Sittang rises in the Yamethin district and flows north for some distance before taking on its southerly course. There appears to be some support for the view that this river course was formerly that of the Lower Irrawaddy. As most of the lower part of the river is in a low plain, it meanders a good deal.

The Salween rises in the Tibetan mountains to the west of the head-waters of the Mekong and Yang-tse-Kiang all of which have parallel courses there in proximity to each other. The Salween is a wild river, and its course through the Shan States is marked by gorges. In the early part of its course it runs parallel to the Irrawaddy and the Mekong. It receives several tributaries in Yunnan and in the Shan States. The river seems to traverse Archaean and Palaeozoic rocks throughout its course. It joins the sea through two branches one to the west of Moulmein and the other several miles South of Moulmein.

The narrowness of the drainage basin of the Salween is noteworthy. It is probable that some of its tributaries have been captured by the other parallel rivers. Throughout its course, this river shows extraordinary vitality which has helped to carve its deep gorges.

GEOLOGICAL ACTION OF THE RIVERS

The Peninsular rivers are entirely rainfed so that they are full only during and immediately after the south-west monsoon, *i.e.*, from June to October. They are fairly active in their upper reaches where they generally flow over a rocky bed and actively erode the bed. They have built deltas near their mouths where they deposit the load of silt carried from above. The delta regions are frequently liable to floods where the distributaries are silted up and are not kept open for carrying away the surplus waters. The deltas of the Godavari, Krishna, Mahanadi and Cauvery are rich agricultural tracts and they are gradually reclaiming land from the sea as is proved by the fact that places which were used as ports for sailing ships centuries ago are now some distance inland.

Compared to the Peninsular rivers, the three main Himalayan river systems are mighty giants. The Indus carries to the sea an average of about a million tons of silt per day, the Ganges a little less and the Brahmaputra a little more. The Irrawaddy has been estimated to transport about two-third million tons of silt per day. The Himalayan rivers are fed both by rain and snow, by rain during June to September and by snow during the warmer half of the year. In their courses through the mountains they have good gradients and carry much coarse materials including pebbles and boulders, brought in by glaciers and also torn off from the beds and banks. They carry enormous quantities of fine sand and silt derived from the Himalayas as well as from the higher peninsular up-lands. Much of the coarse material is deposited near their debouchure into the plains. The aggrading action is confined mostly to the lower reaches of the rivers and the deltas.

CHANGES IN THE COURSE OF RIVERS

The Indus, Ganges and Brahmaputra have changed their courses in the plains frequently in historic and pre-historic times. Compared to these the changes in the courses of Peninsular rivers are not so marked.

In North-western India and the adjoining parts of Pakistan the desert is gradually encroaching on fertile lands in Rajasthan, Punjab and Sind. Part of the lower course of the Indus is really a desert. During historic times many cities and villages on the banks of the Indus have been flooded out or abandoned because the river had changed its course. There are numerous pre-historic settlements like Mohenjo-daro which have been buried in sands and alluvium.

The history of the Sarasvati is exceedingly interesting. The withdrawal of the waters from the Sutlej, and possibly from the Jamuna, dried up the Sarasvati which was once a great river which flowed through Bikaner, Bahawalpur and Sind. In Vedic literature the Sarasvati is given greater prominence than the Indus and the Ganges. Along its banks, especially in Bikaner, there are numerous mounds containing remnants of pre-historic and early historic settlements. It flowed into the Rann of Kutch which was at the time a fairly deep gulf of the Arabian Sea which permitted the entry of ocean-going vessels. The Sarasvati and the Ghaggar seem to have dried up finally about the middle of the 13th century, when there was a great migration of people as a result of the drought which followed. This was the third time it did so, for Todd states in his 'Annals of Rajasthan' that the first time the Ghaggar or Hakra became dry was in 1044 A.D.

The Sutlej was originally an independent river not belonging to the Indus system for it joined the Ghaggar in Bikaner. It abandoned its course finally in the 13th century and joined the Beas near the south-west corner of Kapurthala. The Beas also had an old channel which can still be seen. The combined Sutlej-Beas received the Chenab near Alipur and the combined stream joined the Indus at Mithankot.

The main channel of the Indus up to 1800 A.D. flowed through the middle of the Thal. In that year it split up into two channels and one of these (the Khedewari) which was the main waterway up to 1819 was blocked by an earthquake. By 1837 the Kakaiwari was the main stream but this also become blocked in 1867. In 1900 the chief channel was the Hajamro. Owing to these frequent changes, many flourishing cities on its banks have been abandoned or flooded out *e.g.*, Ghorabari, Keti and Bhimanjopura, etc.

Before the year 1245 A.D., the Jhelum, Chenab and Ravi joined near Multan, flowing just east of that place, and then joined the Beas east of Uch, 45 km south of Multan. But by the end of the fourteenth century the Chenab had changed its course to the one which it now occupies to the west of Multan.

All the Himalayan rivers that debouch into the plains of Northern India show ample evidence of building deltas in which their courses change very frequently. The Kosi, for instance, once flowed by the side of Purnea but is now many km further west, having swept over an area of 10,400 sq. km It now joins the Ganges about 32 km to the west of its original confluence near Manihari. The courses of the Kosi during the last 200 years are shown in the report of the Advisory Committee on Kosi Project (1952).

The original course of the Ganges in Bengal, about a couple of centuries ago, was through the Bhagirathi and the Hooghly. But the Hooghly is now a minor branch, while the Padma flowing through East Bengal has become the major one. Similarly the Damodar, which used to join the Hooghly some 56 km above Calcutta, now meets it several kms below Calcutta. The Bhagirathi originally flowed through the channel called Sarasvati whose course is seen to the west of the modern Hooghly. It left the Hooghly at Tribeni, 58 km above Calcutta and rejoined it at Sankrail, 10 km below Calcutta. It was an important waterway until the fifteenth century and on its banks stood Satgaon (Saptagram) which was a former capital of Bengal and a great centre of trade, as ocean-going vessels regularly called there. Satgaon lost its importance when the Sarasvati abandoned its course.

The Tista was, barely a century and half ago, a tributary of the Ganges, but after the disastrous floods of 1787 it left its old course and became a tributary of the Brahmaputra. The Brahmaputra itself originally flowed to the east of the Madhupur jungle but now joins the Padma much further west through the channel known as the Jamuna.

Pataliputra which was the capital of the great Maurya and Gupta empires for nearly 1,000 years until the fifth century A.D. now lies buried beneath the modern Patna. It is said to have been located at the junction of five rivers —the Ganges, Gogra, Gandak, Sone and Punpun. It was destroyed by a series of floods. The junctions of these rivers with the Ganges are widely separated from each other at the present day.

Gaur at the head of the Ganges delta was a very important and flourishing city between the fifth and sixteenth centuries A.D. A swamp is said to have formed around the city and caused a great epidemic in the year 1575 which led to its abandonment.

Floods : All the major rivers of India are liable to floods, especially in their lower course and in the deltas. Floods in the plains are due to the high precipitation of rain during a short period when the river channels are unable to carry the run-off and therefore spread out over the neighbouring area. In the case of the Himalayan rivers the floods are often due to the formation of barriers in their courses. This may be due to landslips, accumulation of vegetation, or glacial moraines. The following may be given as examples of great floods.

In December 1840 a part of the hillside near the base of the Nanga Parbat slipped down into the Indus and formed a dam nearly 300 m in height. A huge lake, 270 m deep, was said to have been created behind the dam. This led to the rise of the waters by nearly 90 m at Bunji and their extension up to Gilgit town. After six months the water overflowed the dam which suddenly gave way, emptying the reservoir within a couple of days. The resulting huge flood affected many places in the plains, including Attock. This catastrophe of 1841 is still remembered along the Indus Valley. Similar floods also occurred in 1833 and 1858 in the Indus. A flood in 1863 completely swept away more than 1,000 acres of the Dhareja forest in Sind.

The Shyok was in high floods in 1926, 1929 and 1932. It is stated that in the last mentioned year a dam, 120 m high and 400 m thick, was formed by landslip. When the dam burst, the river below it rose nearly 9 m in less than an hour.

In 1893 the Alaknanda tributary of the Ganges was dammed up by the slip of a hillside near Gohna. The dam burst later and caused a great flood. Disastrous floods are known to have occurred in the Sutlej in 1819, in the Chenab in 1790 and in the Ganges in 1893 and so on.

Floods are caused as a result of obstruction of river courses by landslips during earthquakes. The terrible floods which followed the 1934 earthquake in North Bihar and the 1950 earthquake in Assam are well-known.

Floods are brought about by exceptionally heavy rainfall during a short period when the river channel is unable to cope with the run-off and the water spills over the banks and spreads over the countryside. With the steady deforestation of the country the soil cover is stripped off by erosion, percolation is diminished and the run-off is greatly increased, establishing conditions favourable for floods. Large floods are of such frequent occurrence in most parts of India that it is scarcely necessary to enlarge on them.

LAKES

Peninsular area

For a country of the size of the Indian sub-continent, lakes are comparatively of infrequent occurrence and of little importance. There are only a a few lakes in the Peninsula, these being generally due to depressions on the surface or to obstruction of drainage. A few lakes have also been formed along the coast by bars and spits.

Along the western coast of Malabar, Cochin and Travancore, there are several large lakes or back-waters which are separated from the sea by narrow bars. They are called *Kayals* and are used as waterways for trade.

Pulicat Lake : Just north of the Madras City is a shallow salt water lagoon called the Pulicat Lake. It is 60 km long and 5-15 km wide, but the average depth is only 2 m. It is separated from the sea by a bar, on the sea-ward side of which is a shoal which may represent the sediments formerly deposited here by the Old Palar River. The area in which there are islands in this lake is inundated during the monsoons. The marine silt in the islands contains thin layers of gypsum deposited from sea water during recent times.

Colair (Kolleru) : This is a large freshwater lake in the Krishna district between latitudes 16°32′ and 16°37′ and between longitudes 18°04′ and 18°23′. It is due to the growth and coalescence of the Godavari and Krishna deltas on either side, leaving a body of sea water in the middle. It is very shallow and elliptical in shape with an area nearly 250 sq. km. during the monsoon. It is gradually being silted up by the small streams draining into it.

The Chilka Lake : This is a shallow gulf between latitudes 19°28′ and 19°56, now practically cut off from the sea by a sandy bar. It is pear-shaped, being 70 km long and 52 km broad at the maximum. The average depth is less than 3 m and the area occupied by water varies between 800 and 1,200 sq. km in different seasons. It has a narrow outlet to the sea through which tides flow in.

There are several lakes in Rajasthan which are apparently depressions in the topography now fed by inland drainage. As they are in an arid region, they are gradually being silted up and the waters are generally rich in saline content.

The Sambhar Lake lies on the border of Jodhpur and Jaipur between latitudes 26°01′ and 27°0′ and longitudes 74°54′ and 75°14′ about 64 km to the west of Jaipur City. It is at an elevation of 360 m above sea level and is 32 km long and 240 sq. km in area. It is generally dry in summer leaving a small marshy patch. Four or five small streams flow into it. The saline mud in the lake is 22 m deep and is underlain by Aravalli schists. It has been worked for common salt since at least the days of Emperor Akbar and possibly much earlier. It is estimated that in the upper 4 m of the mud in the lake, whose average salt content is about 6 per cent, there would be roughly one million tons of salt per sq. km of area.

The greater part of this lake in summer is covered by a thin white crust of glistening common salt. The sediments below are saturated with brine. Holland and Christie (*Rec.*, *G.S.I.*, Vol. 38, pp. 154-186, 1909) thought that the salt was derived from wind-borne saline matter coming from the Rann of Kutch. Dr. Pramanik of the Meteorological Department found, on investigation of the saline content of air samples in Rajasthan, that there is little or no support to that hypothesis (*Proc. Nat. Inst. Sci. Ind.*, XX, 265-273, 1954).

The other salt lakes of Rajasthan include Didwana, 65 km to the north-west of Sambhar ; Pachbadra, 100 km south-west of Jodhpur ; Chhapar near Sujangarh ; Lunkaransar in Bikaner, etc.

The following table gives the analyses of the saline material in the water of some of the Rajasthan lakes and wells.

TABLE 2—Composition of saline matter in Rajasthan Lakes

	Sea water	Sambhar	Kuchman	Khara Saghoda	Didwana	Pach-badra
NaCl ...	77.76	87.3	84.2	70.8	77.2	85.2
KCl ...	2.50	0.13	...	2.0	...	...
$MgBr_2$ ...	0.22	0.05	...	0.35	...	...
$MgCl_2$ ...	10.80	...	...	22.40	...	1.9
Na_2SO_4 ...	...	8.60	10.5	...	20.60	...
$CaSO_4$ ...	3.60	...	0.50	2.10	...	2.9
$MgSO_4$ ...	4.70	...	2.60	2.30	...	9.4
Na_2CO_3 ...	...	3.80	...	...	0.60	...
$NaHCO_3$ ...	...	...	...	...	1.60	...
$CaCO_3$ ...	0.35	...	0.50	0.06	...	...

It will be seen that the Sambhar Lake contains practically no magnesium salts while they are prominent in Pachbadra and Khara Saghoda. Potassium salts are absent in Pachbadra and Didwana. Sodium sulphate is present in Sambhar, Didwana and Kuchman but not in Pachbadra. These peculiarities are probably to be explained by the peculiarities in the salt content of the local soils and drainage.

There are also some fresh water lakes in Rajasthan like Jaisamand and Fatehsagar in Udaipur and Rajsamand in Kankroli. These occupy depressions in the surface.

The Lonar Lake : This is a circular depression near Lonar (19°57′ : 76°34′) in the Buldana district of Berar, now in Maharashtra, having a diameter of 1800 m at the rim and 1200 m. at the water level, the area of water-spread being a little over 1 million sq.m. The sides dip inward at 30°. It is about 100 m deep, the bottom 50 m being of highly weathered trap, overlain by silt and 5 to 8 m of brine. The saline water contains Cl, 30.87 ; CO_2, 7.52 ; HCO_3, 11.65; SO_4, 0.67 ; Na, 15.15 ; K, 2.05 ; Ca, 0.18 ; SiO_2, 1.46 ; and B, 0.13 parts per thousand. No appreciable amounts of Ni, Co, Cr or V were found in the water or in the rocks. The rim around the lake stands 5 or 6 m above the surrounding country, the trap appearing to dip away from the depression. No appreciable fracturing or shattering is noticeable in the rocks nor have any high-pressure minerals been detected in and around the rim, though the ratio of the diameter to depth is suggestive of meteorite impact. Gravity and magnetic surveys of the lake and surroundings by the Geological Survey of India did not reveal any appreciable anomalies. The lake may therefore be merely a collapse structure.

The Nal : The low area between Saurashtra and Gujarat in the Ahmedabad district was till recently an arm of the sea, the remnant of which is Nal. It is a large lake, 60 km south-west of Ahmedabad covering an area of 125 sq. km. The water is brackish when the lake is full but becomes markedly saline in the hot seasons when the level goes down.

The Manchhar Lake is a large shallow lake lying between latitudes 26°22′ and 26°28′ and longitudes 67°37′ and 67°47′ in the Sehwan Taluq of Larkana, Sind. It is a depression in the course of the Western Nara. When full, the lake is over 24 km long and 16 km wide. The margins are shallow but the centre is 4-6 m deep. It supports rich fisheries, and during the dry weather the borders of the lake are cultivated. During the season it is full of lotus flowers which lend it a beautiful appearance.

Lakes on Potwar Plateau : There are 4 or 5 important lakes as well as a few small depressions in the Salt Range and Potwar plateau in Pakistan. The largest of these is *Son-Sakesar* at an altitude of 770 m. It is 5 km long and 1.6 km wide and receives the drainage of an area of 125 sq. km around it. The greatest depth in the lake is about 15 m near the northern side which is marked by a fault scarp. After heavy rains the lake occupies about 10 sq. km. The *Kalar Kahar* is circular and has a diameter of about a km but is shallow with only 1 m of water. The *Khabaki Kahar* is found occupying a long and narrow synclinal fold at an elevation of 756 m. The *Khotaka Kahar* occupies the eastern end of the same syncline, being 1.6 km long and 0.8 km wide. The *Jalar Kahar* is a long valley in the eroded crest of an anticline. It is about 900 m long and 450 m wide and has been formed by the damming up of the outlet of the valley by loose deposits.

The above-mentioned lakes are saline in various degrees as they are in an area which receives low rainfall (about 40 cm a year) and have no outlets. They contain water throughout the year, but the water level fluctuates considerably.

The Rann of Kutch (Sanskrit : *Irina*) is the remnant of an arm of the sea which formerly connected the Narmada rift with Sind and separated Kutch from the mainland. It comprises the Great Rann and the Little Rann, with a total area of 24,000 sq. km and with a maximum width up to 55 km. It is now a saline desert for the greater part of the year and a marsh during the monsoon when a thin sheet of water inundates it. It was formerly a navigable lake and though it was silted up gradually, it was still deep enough for sailing ships in the fourth century B.C. when Alexander the Great invaded India. Local tradition still speaks of several ports on its shores in former days.

When the sea water which covers it during the monsoon dries up, the surface is covered by a hard layer of salt and shingle. In the summer heat its surface is blinding white due to the reflection of light by the crust of salt. It does not support vegetation except in a few small raised areas where some fresh water is available, and the only animals that occasionally traverse it are small herds of wild asses. If it receives any rain or if strong south-west winds carry some water over its surface, it at once becomes marshy and impassable. The great earthquake of 1819 is reported to have raised up the central area of the Rann by several feet. The eastern part, called the Little Rann, is also gradually silting up.

Dhands : There are a number of alkaline lakes called dhands in Sind which occupy small depressions amidst the sand hills. The saline water in them evaporates during the dry weather and yields sodium salts. There are also several small lakes in Baluchistan and Mekran which contain saline water.

EXTRA-PENINSULA AND TIBET

There are numerous lakes of different sizes in Tibet which are considered to have been formed by the obstruction caused by glacial moraines in river valleys. In the earlier part of the Tertiary era Tibet must have been a region of normal rainfall. But, after the rise of the Himalayas, the moisture-bearing winds from the Indian Ocean have been cut off, making it progressively arid. Lake terraces are found in many parts of Tibet several hundred feet above present water level, indicating that the water level was formerly very much higher. As a result of the concentration of saline matter in the waters, many of these lakes now deposit common salt and in some cases borax.

The largest lake in the Tibetan region is *Issik-Kul* (5,200 sq. km) in the Tien Shan area, while the *Koko-nor* is over 4,100 sq. km in area. The Tsaidam depression is a lake basin having an area of 31,000 sq. km at an elevation of 2,700 m ; it is for the most part a dry saline desert.

Several important lakes also occur in Southern Tibet, like the *Manasarowar* (500 sq. km), *Rakshas* (360 sq. km), *Yamdrok* (870 sq. km), *Gunchu* (100 sq. km), *Trigu* (130 sq. km) and others. Within the Himalayan ranges there are numerous lakes of comparatively small dimensions such as the *Tso-Morari, Naini Tal, Khewan Tal, Wular*, etc. The Wular lake, in the northern part of the Kashmir valley, lies at an elevation of 1,578 m in the course of the Jhelum river. Its ancient name is *Maha Padma Saras*. Its normal area of 30 sq. km may be increased to as much as 260 sq. km when the Jhelum is in flood. A much smaller lake, the *Dal Lake* is close to Srinagar and occupies an area of about 20 sq. km. On its banks are the terraced gardens laid out by the Moghul emperors Jehangir and Shah Jahan. Cascades flow down in a series of steps from the hill side through the gardens between rows of magnificent trees and banks of flowers into the lake.

Most of the Himalayan lakes are thought to be of tectonic origin though some may be due to obstructed drainage through the agency of glacial moraines or ancient landslides. Some of the Kumaon lakes are said to be due to the subsidence of the floor by solution of the underlying calcareous rocks.

LAKES IN BURMA

As in India, there are very few lakes of importance in Burma. These include the Indawgyi and Indaw in the Katha and Myitkyina districts, some crater lakes in the Chindwin district, and ponds of various sizes in the alluvial area.

The *Indawgyi lake* in Myitkyina is 25 km long and 11 km broad with a superficial area of about 210 sq. km. It has an outlet to the north in the Indaw stream, and occupies a depression amidst the hills which may be of tectonic origin. The *Indaw lake* in Katha district is probably also of tectonic origin.

The *Inle lake* in Shan States lies at an altitude of 900 m between two ranges of hills. It has considerably silted up in histroic times, as have several other lakes in the Shan plateau. It is 22 km long and 6 km broad but, even when full, is only about 6 m deep. In the dry season the depth is on an average 2 m, and the bottom is overgrown with weeds. Its margin is marshy and full of vegetation. Its origin is attributed to the sinking of land by the solution of limestone below the bed. Several other areas in the Shan States, now covered with thick alluvial deposits, point to the fact that they were all once lakes.

In the dry zone of Burma, particularly in the Sagaing and Shwebo districts, there are several saline lakes, all of small extent and a few which could yield brine for the manufacture of common salt, sodium sulphate, etc. during the dry season.

There are some 7 or 8 craters or hollows formed by volcanic explosion along both sides of the Chindwin river near Shwezaye. Some of these contain permanent water, while others are dry and may be cultivated for part of the year. In the *Twindaung crater*, 3.2 km N.E. of Shwezaye, the water-level is over 90 m below the rim of the crater and the water is said to be about 21 m deep. The *Taungbyauk crater* is 0.8 km in diameter and 60 m deep, and the bottom is only partly covered with water derived from springs at the sides. The *Twin lake* is the southern-most of the group. The crater is three-quarters of a km in diameter, 45 m deep and has a shallow lake at the bottom, the body of water being about 0.8 km across.

In Lower Burma the lakes are mainly depressions in the alluvium, or obstructed drainage channels. The *Daga lake* in Bassein is an elliptical lake 4.4 km long, 1.6 km wide and 6-13 m deep. It is considered to be an unfilled bend of the Irrawaddy river. The *Imma (Engma) lake* in Prome and the *Htoo* and *Doora* lakes in Henzada occupy abandoned courses of the rivers. These attain their largest extent during the rains and become mere marshes in the dry weather. There are several other similar, but less important lakes in the deltas of the Burmese rivers.

EARTHQUAKES

Earthquakes occur in regions of marked instability of the crust such as mountain belts of geologically recent date. One such region, *par excellence*, is the zone of the Himalayan, Burmese and Baluchistan arcs around the northern borders of the Indian peninsula which have been folded, faulted and over-thrust during the Tertiary era. It is thought that earthquakes originate from places or zones where the accumulated stresses give rise to some movement, mainly by slips along fault planes, by readjustment of material brought about by physico-chemical changes within the crust, or by plastic flowage due to some cause including convection currents. Earthquakes may occur within a few kilometers of the surface in which case they are attributable mainly to slipping along faults. But others originate at intermediate depths (down to 300 km) or at great depths (down to 700 km). Earthquakes at intermediate depths are known in the Pamir region while those of intermediate and deep foci occur in the Japanese and Indonesian island arcs where they are connected with major thrust zones dipping steeply (about 60°) towards the continental mass behind the island arcs. Earthquakes may also be produced in regions where there are active volcanoes, as in the Indonesian Archipelago and the Japanese islands. The major earthquake regions of the earth are in the circum-Pacific mountain belt and also along the Alpine-Himalayan mountain systems. These are generally zones of large negative gravitational anomalies in the crust of the earth.

The Peninsular part of India is a region of high stability as the mountain building movements therein have ceased to be active long ago. It is cut up by a large number of fractures and faults but these are inactive at the present day. Occasionally, however, very feeble shocks are felt in some parts of the Peninsula, particularly around its margins. Sympathetic shocks occasionally arise in the Peninsula at the time of major earthquakes in Northern India. For instance, a minor shock was recorded on the south-west coast of the Peninsula at the time of the great Bihar earthquake of 15th January, 1934. The reason for such shocks may be that the fault along the western coast has not yet completely attained equilibrium. It is also probable that the areas occupied by charnockitic rocks, where occasionally feeble shocks are recorded, were uplifted in the Tertiary times.

The Indo-Gangetic alluvial trough is a region whose origin and structure are closely connected with the formation of the Himalayas. It is a tectonic trough formed in front of the rising Himalayan chains. Changes appear to be still taking place at the bottom of this trough, giving rise to occasional earthquakes. The Sind earthquake of 1819 and the Bihar earthquake of 1934 had their origin in the alluvium-filled trough.

The Himalayan belt is a highly compressed segment of the earth's crust in which the sediments deposited in a large geosyncline as well as the rocks of the northern margin of the Peninsula have been involved. This belt shows the presence of several thrust planes along which slices of the crust have been moved considerable distances in order to adjust themselves to the compressive forces. The compression of the geosyncline took place in several stages extending from the Upper Cretaceous to the Pleistocene, though the mountain building movements have ceased, there are still some adjustments taking place which find expression as earthquakes.

A study of the earthquakes during the last 150 years shows that they are distributed all along the Himalayan, Burmese and Baluchistan mountain belts and also in the Pamir region beyond Kashmir. There is a tendency for a large number of shocks to occur in the regions where the geological formations show sharp changes in strike due to the presence of peninsular wedges underneath. Such regions are north-western Kashmir and Pamir, the north-east corner of Assam and the Quetta region in Baluchistan. Major earthquakes are of frequent occurrence in these areas. The most recent disastrous earthquake was that of August 15, 1950, near Rima (29°N : 97°E) in the Zayul valley on the Assam-China border, in front of the Assam wedge.

Some earthquakes have also occurred in and along the margins of the Assam plateau which is known to be a *horst* uplifted during the Miocene period. The great Assam earthquake of 1897 and the Dhubri earthquake of 1930 occurred in this region.

A large number of earthquakes have been recorded in Northern India and in the Himalayan mountain belt during historic times. A descriptive catalogue of these earthquakes has been published by Dr. T. Oldham in Volume XIX of the *Memoirs of the Geological Survey of India*. Information on Indian earthquakes is also summarized in the monograph by Count Montessus de Ballore in Volume XXXV of the same series. A more recent review of the subject is to be found in the Presidential Address to the Geology Section of the 24th Indian Science Congress by W.D. West in 1937, in which the structure of India is discussed in connection with the seismicity of the different areas. It contains also a list of the important earthquakes in India, Burma and Pakistan during the last 150 years. Individual earthquakes which occurred during the last 60 years or so have been described in various papers published by the Geological Survey of India, a selection of which is given in the bibliography at the end of this Chapter.

VOLCANOES

Though Tertiary volcanism has been fairly wide-spread in the Himalayas, Burma and Baluchistan, recent volcanic activity is known only in the Barren Island and Narcondam in the Burmese arc and in the Nushki desert in Baluchistan.

East of and parallel to the Andaman Arc are the Barren Island and Narcondam which rise from a submerged ridge. That ridge extends north into the volcano zone of Burma and south into the southern border of Sumatra. Barren Island consists of a central symmetrical cone with sides sloping at about 35°, and with a crater some 13 to 15 m deep. It is surrounded by the eroded remnants of a second large cone, whose rim also rises to 180—300 m, while the central cone is 305 m high. The outer cone is breached on the western side by a lava stream which must have flowed in very recent times from four or five parasitic cones on the flanks of the inner cone. Just at this place is a short sandy beach providing entrance to the interior. Vegetation is confined to the outer slopes of the outer cone.

The volcano has been observed in eruption in 1789, 1795, and 1803 and less violently at various times afterwards. It is active at present, 'smoke' being seen occasionally. In the present crater which is elliptic and 90 m in diameter, are seen hot springs and fumaroles which deposit yellow sulphur and some salts. The vulcano is formed of ash beds and lava flows, the latter composed of dolerite and augite andesite.

Narcondam (*Naraka Kundam*, meaning hell pit) is 120 km to the south of Barren Island. It also shows an outer eroded cone inside which is another cone rising to a height of 710 m. This volcano occupies about double the area of Barren Island. It appears to have been active in the Pleistocene. The chief rock types of which it is composed are dacite and hornblende-andesite.

Farther north, in the Central belt of Burma, there is evidence of Pliocene and Pleistocene activity in Mount Popa, Shwebo, Mandalay and Lower Chindwin districts, Jade Mines area and the neighbouring parts of China. It is not known whether any of these were active during historic times.

At the other end of India, in the Baluchistan desert, is Koh-i-Sultan which is also an extinct volcano. Further along the same alignment is the Koh-i-Taftan in Iran which is said to be still active. The lavas are of andesitic type and intercalated with ash-beds. The Solfataras of Koh-i-Sultan deposit sulphur as do those in the Barren Island.

MUD VOLCANOES

Mud volcanoes are not volanoes in the true sense but are merely accumulations of mud in the form of crater and cone due to the eruptive power of hydrocarbon gases in petroliferous strata. They vary in size from small mounds to hillocks 10 to 15 m high, and rarely much larger. Some have the shape of basins with a central vent while others are like volcanic cones. They usually erupt soft liquid mud gently, but in rare cases rather violent eruptions of thick mud and fragments of rocks are known. Mud volcanoes seem to be more active during the rains, perhaps because rain-water helps to soften the mud and thus lessen the pressure on the imprisoned gases. A small difference in temperature between the atmosphere and erupted mud is sometimes recorded. This may be due merely to the depth from which the mud comes or to oxidation of the hydrocarbons escaping to the surface. The temperatures ordinarily recorded are between 85° and 100°F.

Mud vulcanoes occur in Burma on either side of the Arakan Yomas. The eastern group comprises those in Minbu, Prome and Henzada districts and the western group those on the Arakan coast and especially in the Ramri and Cheduba and other islands and near Cape Negrais. They grade from basins on the one hand to cones on the other, the type being apparently controlled by the viscosity of the mud and the energy of eruption. In the basin type the gas escapes as bubbles from a muddy pool, which may have various degrees of permanence. The type which produces cones ejects thick mud and the ejection of gas is much more forceful than in the other type. The cones are often perfect miniatures of volcanic cones and attain heights of 12 to 15 m. Parasitic cones, craters, mud flows, explosions, intermittent activity, rumbling sounds, etc., are all phenomena which have their parallels in true volcanism.

In the Arakan group, the mud volcanoes of Ramri are the best known. The diameter of the cones varies widely, up to 200 or 250 metres. The erupted materials consist of methane and other hydrocarbon gas, petroleum, saline matter (sodium chloride and sodium and calcium sulphates), mud and fragments of rock from the strata underlying the locality. The more violent eruptions tend to be periodical as in the case of geysers. The gases evolved sometimes burn spontaneously. The Cheduba eruption of January 21st, 1904, had a duration of 45 minutes and is said to have been the most violent known in recent times. Submarine eruptions of the same type are known along the Arakan coast, these occasionally producing mud banks.

Mud volcanoes are known to erupt at times of earthquakes if they happen to be in the affected zone. The disturbance in the crust producing earthquakes should naturally be expected to favour the eruption of gases in mud volcanoes.

In and around the oil fields of Burma there are mud volcanoes, which are undoubtedly related to the occurrence of petroliferous strata in anticlinal structures. In some of the oil fields, fissures in the sandstones are noticed to have been filled with clay. These fissure-fillings are of all dimensions, with thickness ranging up to about 25 cm and running in all directions. These are to be explained as due to the liquid mud forced out from beneath, filling up joints and fractures in the dominant sandstone strata of the oil fields.

Mud volcanoes are also seen at the other end of the Himalayan mountain arc, in the Mekran coast of Baluchistan. The region being dry, the cones attain much greater heights (60—75 m) in Mekran than in Burma ; in the latter area they tend to be destroyed by rain.

SELECTED BIBLIOGRAPHY

General References

Medlicott, H.B. and Blanford, W.T. Manual of Geology of India. 4 vols. G.S.I., Calcutta, 1879-1887 ; Second Edition by R.D. Oldham, 1893 ; Third Edition by E. H.Pascoe, Pt. I, 1950 ; Pt. 2, 1960 ; Pt. 3, 1964. Govt. of India Press, Calcutta.
Burrard, S.G., Hayden, H.H. and Heron, A.M. Geography and Geology of the Himalaya Mountains and Tibet (2nd edition), Dehra Dun, 1932.
Chhibber, H.L. Geology of Burma, London, 1934.
Fox, C.S. Physical geography for Indian students. London, 1938.
Spate, O.H.K. 'India and Pakistan'. London, 1960.
Holland, T.H. Geology of India (Imperial Gazetteer of India, Vol. I, Ch. 2). 1904.
Gansser, A. 'Geology of the Himalayas. Interscience, New York, 1964. 290 pp. (Full bibliography).
——Imperial Gazetteers of India for the various Provinces (Introductory chapters on Physical features, etc.). Published by Government of India.

Glaciers

Grinlinton, J.L. Glaciers of the Dhauli and Lissar valleys. *Rec.* **44**, 289-335,1914.
Grinlinton, J.L. Former glaciation of the E. Lidar valley. *Mem.* **49**, Pt. 2, 1928.
Hayden, H.H. *et al.* Notes on certain glaciers in Kashmir, Lahaul and Kumaon. *Rec.* **35**, 123-148, 1907.
La Touche, T.D. Notes on certain glaciers in Sikkim. *Rec.* **40**, 52-62, 1909.

Rivers

Medicott, H.B. Sketch of the Geology of the N.W. Provinces (now U.P.). *Rec.* **6**, 9-17, 1873.

Carless, T.G. Memoir to accompany the survey of the delta of the Indus. *Jour. Roy. Geog. Soc.* **8**, 328-366, 1848.

Cunningham, A. Memorandum on the Irawadi River. *Jour. As. Soc. Beng.* **29**, 175-183, 1860.

Cunningham, A. Ancient geography of India. London, 1871.

Ferguson, J. Delta of the Ganges. *Q.J.G.S.* (London) **19**, 321-354, 1863.

ob, K. and Narayanaswami, S. The Structural and drainage patterns of the area near Palghat Gap. *Proc. Nat. Inst. Sci. India*, **XX**, 104-118, 1954.

Krishnan, M.S. and Aiyengar, N.K.N. Did the Indobrahm or the Siwalik river exist ? *Rec.* **75**, paper 6, 1940.

Oldham, C.F. The Saraswati and the lost river of the Indian desert. *Jour. Roy. As. Soc. N.S.* **25**, 49-76, 1893.

Oldham, C.F. Notes on the lost river of the Indian desert. *Cal. Rev.* **59**, 1-27, 1874.

Oldham, R.D. On probable changes in the geography of the Punjab and its rivers. *J.A.S.B.* **55**, 322-343, 1886.

Pascoe, E.H. The early history of the Indus, Ganges and the Brahmaputra. *Q.J.G.S.* (London) **75**, 138-155, 1919.

Pilgrim, G.E. Suggestions concerning the history of the drainage of Northern India, arising out of a study of the Siwalik Boulder Conglomerate. *J.A.S.B.*, N.S., **15**, 81-101, 1919.

Rennell, J. An account of the Ganges and Burrampooter Rivers. *Phil. Trans. Roy. Soc.* (London) **71**, 87-114, 1781.

Rennell, J. Memoir of a map of Hindoostan. London, 1788 (3rd edn. 1793).

——Physical geography of Bengal, from maps and writings of Major J. Rennell. Calcutta (Bengal Government) 1926.

Lakes

Holland, T.H. and Christie, W.A.K. Origin of the salt deposits of Rajputana. *Rec.* **38**, 154-186, 1909.

La Touche, T.D. Lakes of the Salt Range, Punjab. *Rec.* **40**, 36-51, 1909.

La Touche, T.D. Geology of the Lonar Lake. *Rec.* **41**, 266-285, 1910.

——Symposium on the Rajputana Desert. *Bull. Nat. Inst. Sci. India*, **I**, 1952.

Earthquakes, Volcanoes, Mud Volcanoes

Ballore, M.De. Seismic phenomena in British India and their connection with geology. *Mem.* **35**, Pt. 3, 1904.

Brown, J.C. Burma earthquakes of May, 1912. *Mem.* **42**, Pt. I, 1914.

Brown, J.C. The Pyu earthquake of 3rd and 4th December, 1930. *Mem.* **62**, Pt. I, 1933.

Brown, J.C. et al. Pegu earthquake of May 5th, 1930. *Rec.* **65**, 221-270, 1932.

Brown, J.C. Mud Volcanoes of the Arakan coast. *Rec.* **37**, 264-279, 1909.

Brown, J.C. Submarine mud eruptions of the Arakan coast. *Rec.* **56**, 250-256, 1926.

Dunn, J.A. et al. The Bihar — Nepal earthquake of January 15th, 1934. *Mem.* **73**, 1939.

Gee, E.R. Dhubri earthquake of 3rd July, 1930. *Mem.* **65**, Pt. I. 1934.

Gutenberg, B. and Richter, C.F. Seismicity of the Earth and associated phenomena. Princeton, 1954.

Heron, A.M. Baluchistan earthquake of 21st October, 1909, *Rec.* **41**, 22-35, 1911.

Hobday, J.R. and Mallet, F.R. Volcanoes of Barren Island and Narcondam. *Mem.* **21**, Pt. 4, 1885.

Krishnan, M.S. Volcanic episodes in Indian Geology. *Jour. Madras Univ.* **27B**, 193-209, Jany. 1957.

Mallet, F.R. Mud volcanoes of Ramri and Cheduba. *Rec.* **11**, 188-207, 1878 ; **12**, 70-72, 1879 ; **14**, 196-197, 1881 ; **15**, 141-142, 1882 ; **18**, 124-125, 1885.

Middlemiss, C.S. Kangra earthquake of 4th April, 1905. *Mem.* **38**, 1910.

Noetling, F. Occurrence of petroleum in Burma and its technical exploitation. *Mem.* **27**, 1897.

Oldham, R.D. Assam earthquake of 12th June, 1897. *Mem.* **29**, 1900.
Oldham, R.D. Cutch earthquake of 16th June, 1819, with a revision of the Great Earthquake of 12th June, 1897. *Mem.* **46**, Pt. 2, 1926.
Oldham, T. Catalogue of Indian earthquakes from the earliest times to the end of A.D. 1869. *Mem.* **19**, Pt. 3, 1883.
Pascoe, E.H. The oil fields of Burma. *Mem.* **40**, 211-215, 1912.
Pendse, C.G. Earthquakes in India and neighbourhood. *India Met. Dept., Sci. Notes* **X**, No. 129, 1948.
Poddar, M.C. Preliminary Report on the Assam earthquake of 15th August, 1950. *Bull. G.S.I.*, B. **2**, 1952.
Stuart, M. Srimangal earthquake of 8th July, 1918. *Mem.* **46**, Pt. I, 1920.
West, W.D. Baluchistan earthquakes of August 25th and 27th, 1931. *Mem.* **67**, Pt. I, 1934.
West, W.D. Baluchistan earthquake of May 31st, 1935. *Rec.* **69**, 203-240, 1936.
West, W.D. Presidential address to the section of Geology, 24th (Hyderabad) session of the Indian Science Congress. *Proc. Ind. Sci. Congr.*, 1937.

CHAPTER II
STRUCTURE AND TECTONICS OF INDIA
Peninsular India

The Peninsula of India, as already mentioned, is made up of the Archaean gneisses and schists penetrated profusely by igneous rocks and Pre-Cambrian sediments, the latter including the Dharwarian, Cuddapah and the Vindhyan formations. The earlier rocks were mostly igneous in origin, while the later ones comprise both igneous and sedimentary rocks. More than three periods of granitic intrusion have been recognized and it is possible that others may come to light if detailed studies are undertaken. The earliest igneous suite is a streaky and banded gneissic complex to which the term ' Peninsular Gneiss ' has been applied in South India (Smeeth 1916). The second group comprises porphyritic granite or augen gneiss, while the third is a granite which is practically unaffected by folding and yet of Pre-Cambrian age. Others have received various names in different parts of India — the Closepet granite in Mysore, the Arcot and Hosur granite in Madras, the Erinpura granite in Rajasthan, the Singhbhum granite in Bihar, etc.

Similarly, the highly metamorphosed schistose formation found in various parts of India and which include a large proportion of original sediments, have received different names — the Dharwars in South India, Champaners and Aravallis in Gujarat and Rajasthan, the Sausar and Sakoli Series in Madhya Pradesh, the Shillong Series in Assam, the Bengal Gneiss in Bihar, the Darjeeling and Daling Series in Sikkim, and so on.

Looked at in a broad way, the Archaean rocks reveal certain regional strikes over large stretches of the country as noted below :—

The Aravalli Strike.—The Aravalli mountain belt of Rajasthan is characterised by a N.E.—S.W. trend conspicuously displayed from Champaner in Gujarat at the head of the Gulf of Cambay to near Delhi across Rajasthan. This continues into parts of the Sub-Himalayan zone of Tehri-Garhwal (Auden 1933). In this region, the Tertiary Orogenic movements have not obliterated the original trends of the ancient Aravalli rocks. In the south, in Gujarat, the Aravalli strike splays out so that a part of it continues straight into the Gulf of Cambay and probably into the Laccadives and the Maldives. Another part veers round eastwards and coalesce into the rocks of the Satpura belt in Central India. The southern end of the Aravalli belt near the borders of Rajasthan-Gujarat-Central India is therefore a region requiring critical study. Near the northern end of the Aravalli Range near Delhi, a part of the structure turns sharply towards the northwest and N.N.W. and continues as sub-surface ridge bordering the Punjab plains, and extends as far west as the Kirana and Sangla hills not far from the eastern end of the Punjab Salt Range.

The Dharwar Strike.—The dominant strike in southwestern India (NNW-SSE) is seen in Mysore and adjacent parts of Andhra. How far it continues to the north under the Deccan Traps and whether it connects with the Aravalli strike in Gujarat is not known. In southern Mysore it turns to the south and then southwest, but is overlapped or overthrust by the rocks of the Eastern Ghats province beyond Nanjangud.

The Eastern Ghats Strike.—The major direction of fold axes of the Eastern Ghats is N.E.—S.W. from the Mahanadi valley to the Krishna Valley where it turns to the south and S.S.E. and goes into the sea, to bend back again and emerge near Madras City. From here, rocks characterising this unit proceed in a general W.S.W. direction to the Nilgiri Mountains. In the latter part, as also in the area to its south (in Kerala and Madras), the rocks are thrown into major folds which plunge N.N.W., north or N.N.E. diverging in direction between the two coasts. The major synclinal fold in Ceylon is probably a part of this great fold system. The age relations between the charnockites and the schistose rocks of the region are obscure but it is possible that the latter are much older.

In the northern part of the Eastern Ghats proper in Orissa, the western portion of the belt shows N.—S. strike which terminates at the Mahanadi, while the eastern part turns sharply to the east and E.S.E. and is cut off at the coast. The coastal strip down to Vizagapatam is characterised by this E.S.E. strike.

The Satpura Strike.—Another major trend in Peninsular India is that of the Satpura Range, with an E.N.E.-W.S.W. direction. The Satpura ranges in Western India are composed largely of Deccan Traps, but farther east they include Vindhyan and Gondwana sediments and the Archaean gneisses of Jabalpur, Chota Nagpur and Gaya. The northern part of this belt passes through Jabalpur, Sausar-Chilpi area, Ranchi and the Bihar mica belt. After an interruption in the Rajmahal hills, it continues into the Assam plateau. The southern part passes through the Sakoli tract of Nagpur and through Gangpur-Singhbhum turning gradually towards the southeast in Eastern Singhbhum and Mayurbhanj.

The Mahanadi Strike.—The area of Archaean rocks lying to the west of the Eastern Ghats, between the Godavari and the Mahanadi valleys, exhibits a trend parallel to the course of the Mahanadi river (N.W.-S.E.). The same trend is seen in the area south of the Godavari River, probably extending under the Cuddapah Basin.

There are some other trends in parts of Chattisgarh and in the adjoining areas of Orissa due perhaps to the intervention of rocks of different ages. Systematic mapping of these areas will lead to a proper understanding of the structures affecting them.

It may be pointed out here that there is a rough parallelism between the Aravalli and the Eastern Ghats trends and also between the Dharwarian and Mahanadi trends. Since the strike directions of the rocks in the different belts vary considerably, they meet at sharp angles in some places. The Dharwarian and the Eastern Ghats trends meet in southern Mysore and in the Nilgiris ; the Satpura and Aravalli trends meet in northern Bombay and southern-most

Rajasthan. Three different strike directions meet in a triangle in Bhandara in Madhya Pradesh. These and other regions where the major trends meet each other require to be carefully studied.

MAP III

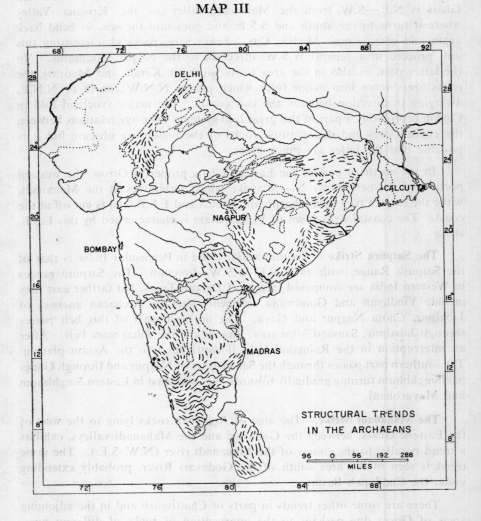

STRUCTURAL TRENDS IN THE ARCHAEANS

Radiometric age determinations of a number of rocks and minerals are now available from different areas, but confirmatory data are necessary for delimiting various orogenic cycles in this region, although it is known that some of these cycles are of world-wide significance. The oldest rocks so far known occur in Mysore (Dharwarian) and in South Singhbhum. As there are three major sedimentary cycles in the Dharwarian, they may possibly extend back to 3000 M.Y. The Singhbhum, Eastern Ghats, Dongargarh, Satpura, Delhi and Tranvacore cycles appear to be successively younger. As

STRUCTURE AND TECTONICS OF INDIA

age data accumulate, it will be possible to define the limits of the sedimentary cycles and igneous activities more precisely. The present knowledge may be summarised as below:

	Million Years
Monazite (Thorianite etc.) bearing pegmatites of Travancore and Ceylon; some granites of Eastern Ghats & Rajasthan, Malani igneous rocks	500- 700
(? Erinpura Granite), Post-Delhi pegmatites	700- 750
Delhi sedimentary cycle	?
Chota Nagpur granite; soda granite and mineralisation of copper belt of Singhbhum; Bihar mica pegmatites	950-1050
Satpura (? Dhanjori) orogeny	950-1050
Dhanjori and Dalma Series with lavas	?
(?) Dongargarh System	?
Amgaon orogeny and granites	1400-1500
Eastern Ghats orogeny	1400-1500
Eastern Ghats sedimentary cycle	1400-1750
Singhbhum orogeny and granite	2000-2100
Iron-ore Series	2000-2400
Post-Dharwar orogeny; pegmatites	2300
Upper Dharwar	? 2200-2500
(? Middle Dharwar) Gold mineralisation	2450
Peninsular gneiss (?) } Bundelkhand granite & gneiss }	2500
Middle Dharwar Sedimentation (?)	2500-2800
Lower Dharwar Sedimentation (?)	2800-3000
Granites and Pegmatites in 'Older Metamorphics'	3000-3100
Older Metamorphics (Singhbhum)	Over 3100

The northern border of the peninsular mass is covered to a large extent by the thick alluvia of the Indo-Gangetic plains which cover a large part of the original Vindhyan basin. The configuration of the mountain belt along the northern, northwestern and northeastern borders of India indicates that there are two important wedge-like masses — one directed northwest from Punjab towards Kashmir and the Pamirs and the other directed northeast from upper Assam towards S.W. China. There are also three minor wedges in the Baluchistan arc, one directed towards Kalabagh at the western end of the Punjab Salt Range, and the other two towards Dera-Ismail Khan and Quetta respectively.

THE ASSAM WEDGE

The Assam plateau, together with the Mikir Hills which is its outlying part, is a horst uplifted during the Mio-Pliocene. It is bordered on all sides by faults, the Brahmaputra flowing along them on the north and west. A wedge of this mass runs up the valley towards the Mishmi hills. The plateau is covered by comparatively thin Cretaceous and Tertiary strata in the southern part. The strata form a monocline dipping fairly steeply into the Bengal plains. Along its axis the monocline is split by the Dauki fault, the fault zone being 5 or 6 km wide. It is nearly vertical or very steep in the west and the northern limb is thrust over and against the southern limb. Near Haflong at its eastern end, it merges into the DISANG THRUST which runs to the northeast for some 400 km. This thrust is the easternmost of the eight to ten thrusts traversing a zone of "schuppen" which is about 20 km wide. The westernmost one of this zone is the NAGA THUST which borders the Brahmaputra valley and presents a truncated fold overturned to the northwest. All the thrusts in the schuppen belt are directed northwest from the Burmese side and show the Tipam or younger strata over-ridden by the Surma and Barail Series. The zone of thrusts ends abruptly in the northeast (Sadiya region) against the MIJU THURST which brings Precambrian gneissic rocks over it, the line of junction running W.N.W.—E.S.E. close to the Dihing River. The MIJU thrust runs to the northwest across the Brahmaputra valley and overrides also the Main Boundary Fault of the Eastern Himalayas.

The DAUKI FAUKI is a dextral transcurrent fault along which the Assam Plateau has been moved to the east over a distance of 250 km. This is inferred from the shelf facies of sediments on the Plateau coming very close to the geosynclinal facies lying to the south of the fault. Moreover, the Sylhet Traps (which are the same as the Rajmahal Traps) are found on the plateau as far east as 92°15′ E. whereas they seem to end along 89°30′ E. longitude in the plains, as proved by boreholes (Evans, 1964).

The effect of the Assam Wedge is felt in the uppermost part of the Burmese Arc and in southwest China where the Himalayan Arc meets the Burmese Arc in a hair-pin bend. This region is highly seismic and is similar to the Pamir region in front of the Kashmir syntaxial bend at the other end of the Himalayas. The rivers beyond the eastern syntaxial region, *viz*., the Irrawaddy, Salween, Mekong and Yangtse, flow along parallel courses close to each other for some distance from north to south, before they diverge widely and finally empty themselves in the Bay of Bengal or the Pacific Ocean. Thus the Assam Wedge plays a very important part in determining the structure of the region of S.W. China.

THE KASHMIR WEDGE

A similar effect is produced by the Punjab-Kashmir wedge in the northwest, but the ancient rocks here are almost entirely buried under the alluvium of the Punjab rivers, except where a few outcrops occur, as in the Kirana and Sangla hills to the south-east of the Punjab Salt Range.

The Potwar plateau of western Punjab is a synclinal trough (Soan Syncline) having its axis along E.N.E.—W.S.W. direction. Along its northern border are the Kala-Chitta and Margala hills which show tightly folded isoclines. The zone to the south of it also contains close-set folds with numerous strike faults. The rocks exposed in the synclinal trough are mainly Murrees and Siwaliks, but along the southern border there are Eocene rocks which are underlain by Mesozoic and Palaeozoic rocks. This southern border, which represents an overthrust limb is the Punjab Salt Range, showing a well marked scarp against the plains of Mianwali. In this part of the Salt Range there appear salt, gypsum and dolomite beds in various stratigraphic positions.

The eastern end of the Salt Range is affected by the presence of the Kashmir wedge and the rocks turn sharply northwards. The course of the Jhelum along the eastern side of the Plateau is apparently controlled by a fault. The most convincing evidence for this wedge is the spectacular hair-pin bend (syntaxis) of the geological formations in north-west Kashmir (Wadia, 1931), On both the arms of this bend the geological formations have a N.W.—S.E. strike but the southwestern side sweeps down in a broad arc into Hazara and the N.W. Frontier Province. It is noteworthy that in this bend the overthrusting of the rocks is directed from both sides towards the central axis. The effect of the wedge is seen as far north as the Karakoram-Hindu Kush mountains and the Pamir region. It is known that in the Alai Range in Russian Turkistan two different periods of folding have been recognised, the earlier one being connected with the Variscan revolution which produced the Tien-shan, Kun Lun and the Karakoram Ranges and the later one connected with the Alpine-Himalayan revolution. The Karakoram ranges have also been affected by the Cimmerian revolution.

THE PURANA FORMATIONS

Lying over the Archaean gneisses are basins of Cuddapah and Vindhyan rocks, which are referable to the upper pre-Cambrian, perhaps extending into the Lower Palaeozoic. Some of the Vindhyan beds have yielded microscopic remains of plants and also some discoidal impressions which are considered to be primitive fossils whose exact nature is still in doubt. These rocks are exposed in three main areas—(1) The Cuddapah basin of Andhra, (2) The Orissa-Chattisgarh region and (3) the great Vindhyan basin extending from Sasaram in Bihar to Agra in U.P. and Chitor in Rajasthan. The Cuddapah basin is conspicuously crescentic in outline, the concave eastern margin being highly disturbed and folded. This basin contains both Cuddapah and Kurnool formations, the former being somewhat disturbed while the latter is practically undisturbed. The Cuddapahs contain also intrusive basic igneous rocks in the form of sills. The eastern margin of this basin has been disturbed along the trend of the Eastern Ghats. This may be due to a rejuvenation of the Eastern Ghats in post-Cuddapah times.

In the exposures of eastern India which have not been studied in any detail, both Cuddapah and Vindhyan rocks are presumably represented. They consist of a number of outcrops now isolated from each other. The more eastern outcrops have been subjected to disturbances probably connected with the Eastern Ghats orogeny. A few isolated patches of Purana rocks also occur along the Godavari, Krishna and Bhima valleys of Hyderabad and Northern Mysore. In the last of these, the Kaladgi formations, which are presumably the equivalents of the Cuddapahs, are said to be intruded by granites. The Pakhals of the Godavari valley near Singareni are also considered to be of Cuddapah age, but Dr. Mahadevan regards them as earlier. It is not improbable that the Cuddapahs and their equivalents have been affected by earth movements and granite intrusions in certain restricted areas.

The Great Vindhyan Basin of Northern India occupies a large semi-circular area to the east of the Aravalli axis but some Vindhyan rocks are found also to the west of that axis. The total area occupied by the Vindhyan basin is of the order of 100,000 sq. km. This basin has also a crescentic outline, the northern margin being concave and enclosing the Bundelkhand granite or being covered over by the Ganges alluvium. Along the western margin of this basin is a large reversed fault (the Great Boundary Fault of Rajasthan) which brings the Aravallis on the western side against the upper Vindhyan Bhander sandstones on the eastern side, and which has been traced over a length of 800 km.

The Vindhyans in this basin consist of two major divisions, the lower being the Semri Series which is intruded by basic lavas and appreciably disturbed by folding and faulting. The southern and southeastern edges of this large basin are more disturbed than the other edges, except for the reversed fault mentioned above. The rocks which originally belonged to the northern part of this basin are now part or the Sub-Himalayan region of U.P., but it would be impossible to identify them and establish their relationship with the Vindhyans because of the fact that Sub-Himalayan area is far removed from the Vindhyan basin and has been involved in the Tertiary mountain building movements. However, according to Auden, the Nagthat beds of the Mussoorie-Garhwal Himalayas show a considerable degree of resemblance to the Vindhyans.

It may be mentioned here that the Delhi formations of Rajasthan were strongly compressed, metamorphosed and intruded by igneous rocks while their equivalents in other areas were affected comparatively mildly. Even the Lower Vindhyans of the Sone valley have been subjected to folding and intrusion of basic lavas because of which it is arguable that they may be the equivalents of parts of the Delhis or Cuddapahs. The upper Vindhyans and the Kurnools are practically unaffected by earth movements.

There are no formations in Peninsular India between the Vindhyans (which are probably early Cambrian in age) and the period of glaciation in the Upper Carboniferous. During this interval the peninsula was generally undergoing denudation and was not affected by any earth movements. If any deposits had been laid down in the Peninsular area during this period, they should have been completely removed by erosion.

THE GONDWANA FORMATIONS

Sedimentation during the Gondwana period was initiated by continental ice sheets which covered large parts of the southern hemisphere. The glacial deposits of the period are the Talchir tillites and their equivalents, which are found not only in various parts of the Peninsula but also in the Salt Range, Hazara and several places in the Sub-Himalayan zone.

The glacial deposits were followed by the Damuda formations which contain all the workable coal-fields in India except those of Assam. The formation of thick coal seams during this period is indicative of the disappearnce of the glacial conditions and an amelioration of climate, which is deduced to have been cold-temperate, from the characters of the flora. The major exposures are found in large faulted troughs along the Damodar-Sone, the Mahanadi and the Godavari valleys. Similar exposures containing coal seams are also found in several places in the Lesser Himalayas of Nepal, Sikkim, Bhutan and Assam. In some cases, as in the Rangit valley in Sikkim, in the Subansiri valley and in the Sela agency in Assam, the plant-bearing strata are associated with marine formations containing *Spirifer, Chonetes, Eurydesma, Conularia*, etc. These therefore indicate the position of the northern border of Gondwanaland in the Sub-Himalayan region. It is likely that the area between the major coal-fields of Eastern India and Himalayan region, which is now occupied by the Ganges and Brahmaputra basins, contains some coal-bearing strata but these would lie at a depth of several hundred metres under alluvial and Tertiary strata.

At the beginning of the Gondwana era there was evidently an arm of the sea extending from Rajasthan and Kutch into the Narmada valley. We find a couple of small patches of marine strata of Permo-Carboniferous age near Umaria. These marine beds overlie Talchir tillites and are in turn overlain by Barakar strata. It is likely that careful search may reveal other patches of marine Permo-Carboniferous rocks, particularly as the Umaria fauna is closely related to that of the Lyons Group in the Carnarvon Basin of West Australia which must have been close to the eastern coast of India.

The glacial conditions during the Upper Carboniferous in India and in other continents followed the great orogenic disturbances known as the Variscan or Hercynian revolution, which affected large parts of Asia and Europe and which was responsible for the rise of great mountain ranges such as the Western Alai, Tien-shan, Kun-lun, Karakoram, Nan-shan and Tsin-lin. This revolution also brought into existence a great Mediterranean sea extending from the region of Atlas mountains and Pyrenees in the west, through the areas now occupied by the Alps, Carpathians, Caucasus, the mountains of Asia Minor and Iran into the Himalayas and farther east into Burma-Arakan-Andamans and the southern border of the East Indian Archipelago. The sediments which were laid down in this great Mediterranean ocean, which was named the *Tethys* by E. Suess, were later compressed and raised up into the mountain ranges just mentioned.

In parts of Kashmir, the Hercynian revolution produced land conditions and there were great volcanic eruptions whose products are now the Panjal Traps. With these volcanics are also associated the Agglomeratic Slates with intercalations of marine strata with Upper Carboniferous fossils. The land conditions in Kashmir continued throughout the Permian and well into the Triassic, permitting a certain amount of inter-migration of plants between India on the one hand and Angaraland and Cathaysia on the other, probably through the Pamirs and Ferghana. It is known that continental conditions prevailed in Ferghana from the Upper Carboniferous to the Jurassic or even up to the Lower Cretaceous; there were also land conditions in parts of northern Iran and Afghanistan.

The chief coal-fields of India owe their preservation to block faulting. The faults appear to be Upper Triassic or Lower Jurassic in age. The coal-bearing rocks are traversed by dykes of dolerite which are similar to the Rajmahal traps and also by dykes and sills of mica-lamprophyre. It is possible that the block faulting was partly contemporaneous with the eruption of Rajmahal lavas and intrusion of basic and ultra-basic dykes in the coal-fields.

THE INDO-GANGETIC BASIN

The Indus Basin in Pakistan stretches south from the Potwar Plateau to the mouths of the Indus. It is filled by sediments extending back in age from the Recent to the Permo-Carboniferous and possibly also by the Vindhyans, remnants of which are found in Western Rajasthan. At the northern end is the Salt Range which is a thrust mass containing a fairly complete succession of strata from the Cambrian to the Siwalik, thrust over the Pleistocene strata of this basin. At that end a promontory of the Indian Peninsula underlies the basin and continues north into Kashmir. Another submerged ridge running from Khairpur to Jacobabad-Sibi has been discovered by geophysical methods and on this ridge are situated some of the natural gas fields of Sind. The basin in Sind is probably about 6000 m deep, for borings in this area to a depth of 3600 m reached only the top of the marine Jurassic formations which are known to be over 1800 m thick in Kutch. The basement in the lower Indus Basin should be formed of rocks of the Indian shield extending west from Rajasthan, being overthrust eastward by the sedimentary pile of the Baluchistan Arc.

The northeastern part of the Indus Basin and the Ganges-Brahmaputra Basin form a continuous strip along the southern border of the Himalayas. This falls structurally into 5 or 6 transverse units, separated by N.E.—S.W. or N.—S. faults or basement 'highs.' The eroded and rather irregular surface of the northern edge of the Peninsular shield underlies the Ganges Basin and the Himalayan border and dips north at a low angle of 1°—3°. The following information is derived mainly from a paper by Mathur and Kohli (1964) and a brochure on Oil in India by Mathur and Evans (1964).

The westernmost unit is the portion in the Punjab lying to the northwest of the basement 'ridge' which is a continuation of the Aravalli Mountain axis into the Dehra Dun area. In this unit the Siwaliks lie directly over most of the basement except near the foot-hill zone where the lower Tertiaries are present. In the Potwar Plateau and Jammu, the marine Eocene is overlain by the brackish-water Murrees with a conglomerate bed in between. This conglomerate indicates a marked unconformity and the absence of the greater part of the Oligocene. The Siwaliks in the basin attain a maximum thickness of 5000 m or more. They are only gently folded, in contrast to the more severe folding and thrusting in the more eastern units. In the eastern part of this unit the Lower Tertiaries are represented by the SUBATHU BEDS which are of shallow marine origin. They are succeeded without discordance by the DAGSHAI and KASAULI Beds (called the DHARMSALA SERIES) which attain a thickness of 4000 m. There is a sharp change from the marine Subathu to the brackish-water Dagshai Beds which are red clays and grey and purple sandstones below and mainly sandstones above. They pass up into the fresh-water Kasauli Beds composed of greenish-grey sandstones with green and brown claystones.

In a borehole at Adampur (31°26′: 75°45′) the Siwaliks rest directly on the basement rocks. Even near the foothills there are no Mesozoic or older rocks between the Tertiaries and the basement. The Lower Siwaliks extend for some distance to the south under the plains; the Middle Siwaliks extend up to about the middle of the plains; while the Upper Siwaliks overlap these and extend well beyond and south of the middle of the plains.

A N.E.-S.W. fault with a southeasterly downthrow of 2000 to 3000 m (revealed by aeromagnetic data) passes through Moradabad which may be taken as the eastern border of the submerged Aravalli ridge. The second unit lies between this on the west and a basement 'high' passing N.E.-S.W. through Faizabad (26°47′ : 82°8′) on the east. The Siwalik belt in this shows north-dipping overturned anticlines thrust toward the south. Other thrust-sheets are present beneath the alluvium wherein the Paleogene rocks are thrust over the Neogene. The Paleogene thins down towards the east and is probably not present to the east of Naini Tal. A borehole at Ujhani (28°1′ : 79°1′) in the northwestern part of this unit passed through 1100 m of Siwalik and 1000 m of Mesozoic and late Palaeozoic rocks, before reaching the basement at 2100 m. The sericite-quartzites in the basement gave an age of 1045 M.Y. by the K-Ar method. The Mesozoic consisted of dolomitic and calcitic limestones, shales, foetid sandy limestones and quartzose sandstones with tuffaceous material. The upper part of the sequence is Mesozoic and probably the equivalent of the Krols, as judged from spores and pollen which are the only fossils present, for the strata are singularly devoid of megascopic and microscopic animal remains. The lowest part of the beds is probably Permo-Carboniferous from the evidence of pollen, and the volcanic material may be connected with activity contemporaneous with the Panjal Volcanic period of Kashmir.

The third unit is between the Faizabad 'high' and another N.E.—S.W. 'high' passing through Patna (25°37′ : 85°15′) and Muzaffarpur (26°7′ : 85°24′). A borehole at Raxaul (near longitude 85°) near the Nepal border passed through the Pre-Tertiary unconformity at 4150 m. The Siwaliks are here thinner than farther west. Pre-Tertiary rocks of 450 m thickness were also penetrated. These consisted of unfossiliferous (except for spores and pollen) sandstones, red, purple and green shales and some basic sills. The shales have a phyllitic lustre. These have some resemblance to the MUSI KHOLA FORMATION and the metamorphosed KISERI FORMATION, both exposed beyond the Main Boundary fault in Nepal and regarded as the equivalents of the Krol formations. In a borehole at Muzaffarpur, the depth to the pre-Tertiary unconformity was 1700 m.

The fourth unit lies between the Patna 'high' and a well-marked fault passing N-S through Kishanganj (26°5′ : 88°0′) and Malda (25°2′ : 88°10′). This fault forms the eastern side of a graben whose western side seems to be another fault passing near Purnea (26°40′ : 87°30′). There is also a conspicuous E-W fault along the border of the Indian shield here, which is thought to be the continuation of the Dauki Fault along the southern side of the Assam Plateau. The Brahmaputra flows along a N-S fault bordering the western foot of the Garo Hills (Assam Plateau) and this appears to continue into a fault which is present in the Himalayan foot hills region near the Torsa valley. The Rajmahal and Garo hills are connected by a submerged basement ridge at a depth of 200 m or less, and the basement slopes down from this ridge both towards the north and south.

The Brahmaputra valley in Assam is similar to the Ganges valley along the Himalayan side. On its southeast are the horst of the Assam Plateau and the thrust zone of Upper Assam which are dealt with on another page.

The Extra-Peninsular Region
THE HIMALAYAS

Kashmir.—In the Himalayan region, at least three major thrust zones are recognised in the rocks of the Siwalik and Lesser Himalayan zones. The southernmost of these thrust zones is generally designated as the *Main Boundary Fault* which usually separates Siwaliks from the earlier Tertiary and older rocks. To the north of the Tertiary belt in Kashmir is the autochthonous zone containing sediments of all ages ranging from the Carboniferous to the Eocene, the sediments having been folded and thrust over the foreland. This thrust is called the *Murree thrust*. Beyond this is the zone of nappes in which two or more important thrusts are known to be present. The nappe zone shows Palaeozoic and Mezosoic rocks which have travelled some distance from their original sites of deposition. Farther to the northeast is the Central Himalayan range consisting of sediments intruded by large masses of granite, presumably of Lower Tertiary age.

Simla-Garhwal.—In the Simla and Garhwal region, which has been mapped by Pilgrim, West and Auden, the Main Boundary Fault separates the Middle and Upper Tertiary beds from the Lower Tertiaries which have been thrust over them. The Upper Tertiaries which occupy a belt over 100 km in width in Kangra, become much narrower in the area south of Simla where the belt is barely 24 km wide. The Tertiaries are separated by pre-Tertiary rocks of the autochthonous belt by the *Krol thrust* which may be considered to be the equivalent of the Murree thrust of Kashmir. North of this is the zone of nappes containing the *Jutogh* and *Giri thrusts*. Farther north is another important thrust called the *Chail thrust*. Granitic intrusives are found in the nappe zones of the Central Himalayan range which exhibit several klippen of highly metamorphosed ancient rocks resting on less metamorphosed and unfossiliferous Palaeozoic and Mesozoic rocks.

East and southeast of Simla, the same structural units are found in Garhwal. Here the Main Boundary Fault separates the Siwaliks from the zone of Simla Slates, overlain by Nummulitic and other Lower Tertiary strata. This zone is thrust over by the Krol nappe which contains rocks of various ages from the Pre-Cambrian to Mesozoic. The Krol thrust unit is overlain by the Giri thrust. Farther north are other nappes which bring highly metamorphosed Pre-Cambrian rocks over the Krol belt and which have their roots in the main Himalayan Range from where they have transported granite. Both the Krol and Garhwal nappe units are folded. These thrusts are probably of Miocene age as they occur over the Nummulitic rocks.

The Eastern Himalayas have yet to be examined in detail but our present knowledge enables us to say that the structures present there are similar to those of Garhwal and Kashmir. In Nepal, the Lower Siwaliks are separated from the Middle and Upper Siwaliks by a thrust, while the Tertiaries are overridden by pre-Tertiary rocks of another thrust, as seen at Sanotar. Still another thrust brings the Pre-Cambrian Darjeeling Gneiss over the above-mentioned zone.

Central Himalayas.—According to Heim and Gansser, there are at least four superimposed thrust sheets in the Lesser Himalayas of Garhwal over the Nummulitic rocks. The synclinal of Nainital is the continuation of the Krol belt. This is thrust over by the gneisses and schists of the Ramgarh region and this again by the crystalline schists of Almora. The northernmost is the Tejam zone which may possibly represent the Krol-Tal succession of rocks. Beyond this again is the crystalline zone of the Central Himalayas and then the Tethyan belt. The last thrust in this zone contains Cretaceous flysch sediments associated with the exotic limestones of Permian and Mesozoic ages and with basic intrusives and lava flows.

Beyond this exotic zone is another sedimentary belt containing rocks assigned to the Chilamkurkur and Rakshas Series which are dominantly shaly and of Mesozoic age. Still farther north is the Darchen zone associated with igneous rocks, but these rocks are thrust in a north-easterly direction (therefore considered to be a counter-thrust), over the Eocene conglomerate of Mount Kailas which occupies a wide belt. Beyond Mount Kailas is the Trans-Himalaya Range containing Mesozoic and earlier granites and metamorphics.

Summarising, we may say that the following units or zones have been recognised in the Himalayas from south to north :—

(1) Main Boundary Fault of the Siwalik zone
(2) Imbricate thrusts of the Himalayan border
(3) Thrusts of the Lesser Himalayan belt
(4) Central Himalayan thrust in which Mount Everest and Kanchanjunga are included
(5) Thrusts of the Tethys Himalayan zone
(6) Thrust of the flysch and exotic zone
(7) Counter-thrust of the Darchen zone

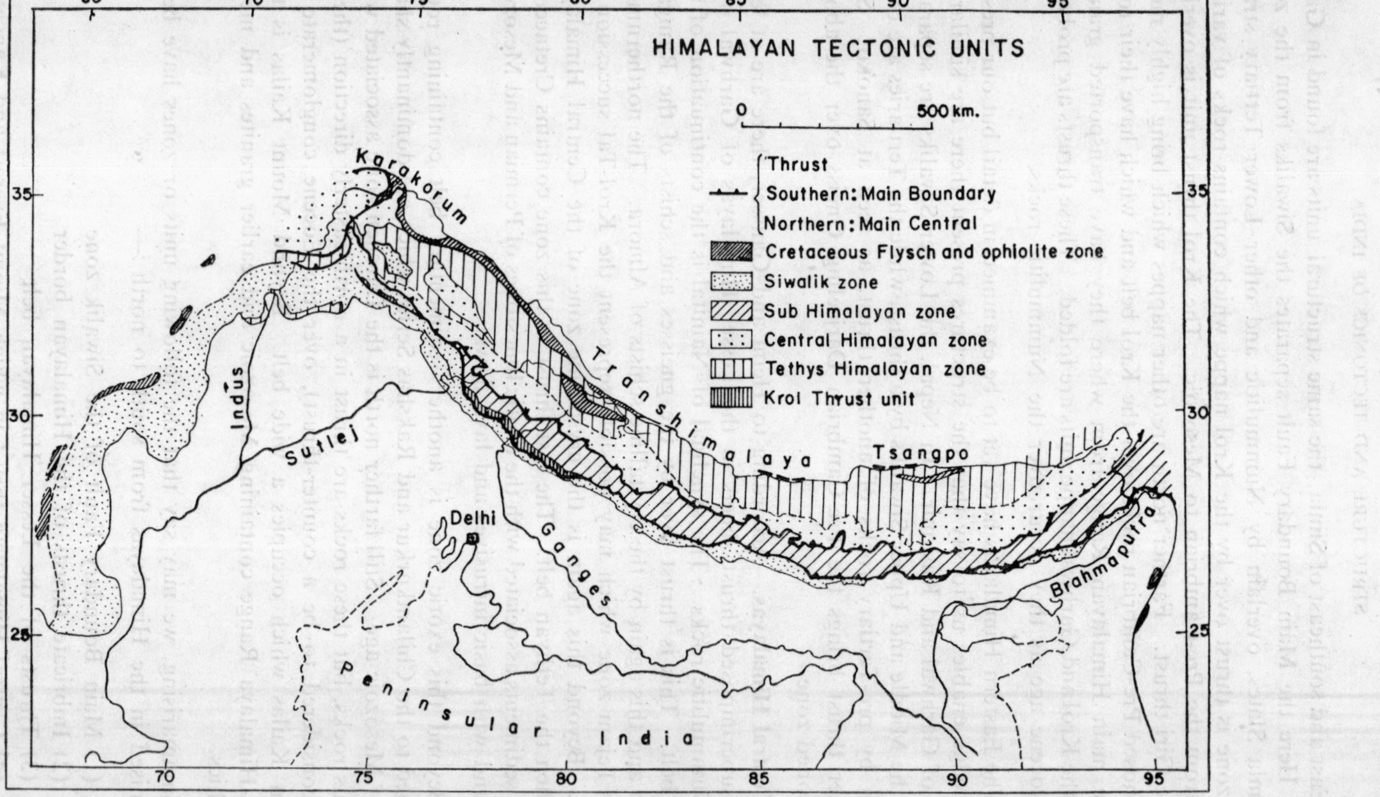
HIMALAYAN TECTONIC UNITS
0
500 km.
Thrust
Southern: Main Boundary
Northern: Main Central
Cretaceous Flysch and ophiolite zone
Siwalik zone
Sub Himalayan zone
Central Himalayan zone
Tethys Himalayan zone
Krol Thrust unit
Karakorum
Indus
Sutlej
Transhimalaya
Tsangpo
Delhi
Ganges
Brahmaputra
Peninsular India
65
70
75
80
85
90
95
35
30
25

MAP VI—HIMALAYAN TECTONIC UNITS

In the east, the Himalayas terminate geographically at Namcha Barwa peak. Though the region beyond Upper Assam is little known, the available knowledge goes to show that, geologically, the Himalayan formations turn sharply to the south and form the mountains of Burma and Arakan which will be referred to here as the Burmese Arc. At the eastern end of the Himalayas the rocks exhibit a north-easterly strike; they then turn sharply to the south-east and finally to the south-west and south. Beyond the Indo-Burma frontier and the central belt of Burma, lies the Shan region which is geologically a part of the Yunnan-Indochina province and which was brought to its present position only in post-Cretaceous times.

THE BURMESE ARC

This Arc commences some distance to the east of the eastern termination of the Himalayas and sweeps in a broad curve through the boundary region between India and Burma, the Arakan region and the Andaman islands into the Indonesian Archipelago. It is convex towards India except for some length in the Arakan region where it is slightly concave. The southern part of this Arc is largely submerged in the Bay of Bengal, the Andamans and Nicobars being really the unsubmerged peaks of a group of ridges. The concavity is due to the northern part of the arc being held back by the Indian shield while the southern part being free to move out into the sea.

Our knowledge of geology of the Assam-Burma border is very sketchy. But it is generally known that it contains Mesozoic rocks intruded by granite and ultra-basic rocks. This mountainous region is composed of a series of over-thrusts directed towards India, including the well-known Haflong-Disang and Naga thrusts. The Digboi Oil field is located in the last anticline cut off by the Naga thrust at the border of the Brahmaputra plains.

Inside this Arc and parallel to it is the main volcanic zone containing the Tertiary and recent volcanoes of the Jade Mines area, Prome, Tharrawaddy, Barren island and Narcondam, continuing into the volcanic zone of the southern edge of Sumatra, Java and other islands of the Indonesian Archipelago. This zone shows faulted junctions both against the eastern border of the Burma-Arakan mountain belt as well as the median Tertiary belt of Burma. Most of the volcanoes in this zone were active in the Upper Tertiary and some in Pleistocene times. The median belt of Burma, composed of moderately folded Tertiary rocks, contains all the important Burmese oil fields. This region was occupied by a gulf during the Tertiary times and was gradually filled up and later gently folded. The rocks are therefore of fresh water to brackish water origin in the north and marine in the south. There are stratigraphical breaks below and above the Miocene. Along their eastern margin, the Tertiary rocks are faulted against the more ancient rocks of the Shan plateau. The zone of junction contains another line of volcanic rocks — e.g., basic and intermediate lavas of Kabwet and Mandalay and the rhyolites of Thaton. The Shan plateau consists of Pre-Cambrian, Palaeozoic and some Mesozoic rocks, intruded into by Pre-Cambrian as well as Jurassic granites. This granite belt runs from Bhamo and Mogok in the north through Yamethin into Tenasserim and the Malay States. The Shan region is geologically allied to Thailand, Malaya and Indo-China.

THE BALUCHISTAN ARC

The Northwestern Arc commences beyond the western end of the Himalayas which culminates geographically at Nanga Parbat. As we have already seen, the geological formations here bend round sharply and proceed first southeastwards and then southwards and south-westwards into Hazara and the Sulaiman-Kirthar Ranges on the border of the Punjab and Sind. Part of it branches off and goes into the Safed Koh mountains of Afghanistan. It will be noticed that in contrast with the smooth flowing curve of the Burmese Arc, the Baluchistan Arc shows three conspicuous festoons because of the gathering up of the strata in the Dehra Ismail Khan and Quetta regions. In this Arc, the over-thrust of the strata is from the northwest and west, *i.e.*, towards India as in all other cases. The festoons are to be attributed to the presence of submerged wedges of ancient rocks under the Indus alluvium projecting towards the northwest in the areas mentioned. On the convex side of the Arc are the Murree and Siwalik sediments deposited in brackish to fresh water environment, while on the concave side, in Baluchistan, are marine strata of various ages.

In the Northern part of the Baluchistan Arc in Hazara, we find two sedimentary facies lying side by side and having a general N.E.-S.W. trend. The northwestern facies is of the Himalayan type in which the rocks resemble those of the Tethys Himalayan zone of Spiti and Kashmir. The southeastern facies continues down into Sind and Baluchistan and is dominantly calcareous, for which reason it is called the *Calcareous zone*. This Calcareous Zone is found along the mountain chains forming the boundary between Sind and Baluchistan and contains rocks extending in age from Permo-Carboniferous to Eocene. This zone strikes into the sea at Cape Monze near Karachi, but turns to the S.S.W. forming a series of ridges underneath the waters of the Arabian Sea and continuing towards the Oman mountains in Arabia.

Two other lithological belts parallel to the Calcareous Zone are recognized in Baluchistan. The zone which lies a few kms west of the Calcareous Zone consists of close-set parallel ridges composed of greenish flysch sandstones and shales of the type of Khojak Shales. This may therefore be termed the *Khojak zone*. It includes the Zhob valley, Khwaja Amran and Sarlat Ranges, the western part of Sarawan and Las Bela and continues southwards into Mekran and the southern border of Iran. The third zone, which may be called after *Chagai*, lies farther west and includes the Nushki area and Ras Koh Range, consisting of ancient igneous and metamorphic rocks as well as Tertiary sedimentary strata intercalated with basic lava flows. There are a few recently extinct volcanoes in this area and one of them, Koh-i-Taftan, is said to be still active.

On the Sind side of the Calcareous Zone there are Upper Eocene, Murree and Siwalik strata, but when followed southwards they become predominantly marine in Sind. This is because there was a gulf in this region which was gradually filled up from the north forming the Nari, Gaj and Manchhar Series.

The Khojak zone and part of the Chagai zone of Baluchistan consist mainly of Tertiary rocks corresponding in position to the median belt of Burma. But the rocks of Baluchistan appear to be much more folded and disturbed than those of Burma, which might explain the general absence of productive oil fields in Baluchistan. The southern part of the Khojak Zone has been faulted down along the Makran coast during the late Pliocene.

THE EASTERN COAST

The Eastern coast is roughly parallel to Eastern Ghats up to the Krishna valley and thereafter it turns south more or less parallel to the outline of the Cuddapah basin. Farther south, the coast bears no relation to the trend of the rocks as it cuts across the strike. The earliest fossiliferous rocks found along the coast are Upper Gondwana estuarine formations which are intercalated in some places with marine beds whose age is Neocomian. An important marine transgression took place during the Middle Cretaceous, both along the southern part of the Assam plateau and in the coastal region of Southern Madras. The earliest fossils found in these beds are of Upper Albian age. Ceylon seems to have been first cut off from India sometime during the Tertiary, the earliest marine formations seen on the north-western coast of Ceylon being of Middle Miocene age (Jaffna beds). The sea between the Indian mainland and Ceylon is very shallow and marked by numerous coral islands.

In the Eastern part of the Bay of Bengal a ridge was formed during the Laramide times along what is now the axis of the Burmese Arc, extending through the Andaman and Nicobar islands and to the south of Sumatra and Java. It was about this time that the Tethyan basin of the Himalayas was cut off from that of the Burmese Arc. The Andaman Sea, *i.e.*, the basin between Burma and the Andamans probably took shape at the end of the Cretaceous. It extended into Upper Burma. East of the Andaman ridge is a volcanic ridge which carries the Barren Island and Narcondam. The Andaman Sea was probably much shallower originally, but was faulted down later to its present depth, reaching over 2,000 fathoms at its deepest. Sewell says that there is another inconspicuous ridge to the west of and close ot the Andaman ridge and this may be part of the Ninety Degree ridge running roughly along 90°E. Longitude in the Indian Ocean. The Andaman ridge is clearly over-folded to the west but the axis of folding is said to have shifted progressively westward during the Tertiary. The western coast of Tenasserim is only moderately faulted, having suffered a slight submergence during geologically recent times. When followed northward, it continues into the Tenasserim-Shan zone whose western margin is faulted and overthrust towards the west in a more pronounced way than is apparent in the Tenasserim region.

THE WESTERN COAST

The edge of the continental shelf of the western coast is remarkably straight as it is a fault line formed in late Pliocene. It runs from south of Cape Comorin to near Karachi, the shelf being broadest opposite Bombay and Kathiawar. The coastal tract in south shows a few small patches of marine Burdigalian strata (The Quilon Limestone) overlain by a narrow but fairly continuous strip of the Mio-Pliocene Varkala Sandstone. Farther north the Deccan Traps occupy the coastal region except in the Surat-Cambay-Ahmedabad area where they are overlain by Tertiary strata.

Available evidence suggests that the southwestern coast of India should have been alongside the eastern coast of Madagascar in pre-drift times. This is supported by the great similarity between the geology of Madagascar and Tavancore-Ceylon. Both the regions contain Charnockite-Khondalite association, cystalline limestones and granulitic metamorphites with garnet, spinel, pyroxene, sillimanite, cordierite, staurolite, etc. The mineral deposits in both include graphite and heavy sands containing monazite and other thorium minerals, ilmenite, garnet, zircon, etc.

India separated from Madagascar in the Cretaceous, as the earliest sedimentary rocks on the eastern coast of the latter are of Maestrichtian age. As no Cretaceous rocks are known on the southwestern coast of India, it is inferred that such deposits as were laid down have been faulted down. The Miocene transgression is wide-spread in India and must have reached farther inland on this coast than the Cretaceous transgression. After the Miocene Varkala Sandstones were laid down, they were raised up and the coast was faulted down simultaneously with the faulting of the Makran coast.

The Deccan Traps occupy the coast and the interior north of Goa up to the border of Rajasthan, but over a good part of the Gujarat coast they are overlain by Tertiary sediments. The Cambay area lies in a trough fault running N.—S. in which the Deccan Traps have been dropped down to a depth of about 2000 m. The sides of the trough appear to be step faults as there is no indication of abrupt downthrow in the gravity and magnetic maps. The trough appears to continue to the N.N.W. from Cambay for some distance. The faulted zone is characterised by well-marked positive gravity anomalies which are noticed from north of Cambay, along the coast, for some distance to the south of Bombay. This was interpreted by Glennie of the Survey of India as indicating the presence of a very thick dyke of dolerite along the Bombay coast. The Deccan Traps dip into the Arabian Sea at an angle of $7°$ to $10°$ as a monocline, the axis of which runs through Panvel and Kalyan to a point south of Surat. This monoclinal feature is called the PANVEL FLEXURE. Along its axis, which is fractured, there are several hot springs. It is not known how far the Deccan Traps originally extended to the west of the Bombay coast. But, from the fact that they are exposed in Kathiawar and in the hills west of Sind, and have been found in boreholes in the Indus delta in Sind, it may be presumed that they may have continued as far west as the meridian of Karachi.

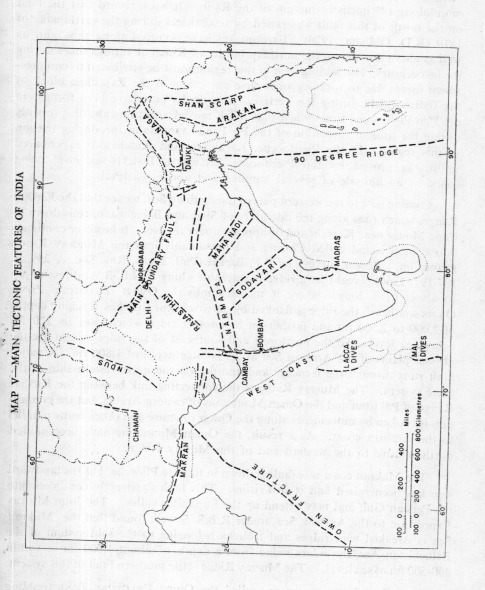

MAP V—MAIN TECTONIC FEATURES OF INDIA

Kutch and the islands in the Rann show some faults and folds trending W.-E. to W.N.W.-E.S.E.. Major E.-W. trending faults have been found bounding Kathiawar on the north and south, the latter one by geophysical methods. There is also a prominent fault seen for a distance of 40 miles or more along the northern margin of the Rann. It is on record that the land on the north of this fault was raised by several feet during the earthquake of 1819 (R.D. Oldham, 1926). Earthquakes have occurred along this fault as well as along the one on the northern border of Kutch at various times during the last century. The folding seen in this region must be attributed to compressional forces due to rotatory or reverse movement of the Rajasthan block of the Indian shield during the Tertiary. The faults may have been formed in the Plio-Pleistocene times when that block moved again northward, probably during the time of formation of the Pir Panjal mountains, producing tension in this region. In the Pleistocene, Kathiawar and Kutch also experienced uplift, as proved by the Pleistocene MILIOLITE LIMESTONES now being found at an altitude of several hundred feet above sea level.

Coming now to the western part of the Arabian Sea, we see that the Kirthar Range, which runs along the boundary of Sind and Baluchistan, runs down to Cape Monze near Karachi and disappears under the sea. It however continues into a small island called Churna and presumably into the MURRAY RIDGE farther on. Recent surveys by P.F. Barker (Phil. Trans. Roy. Soc. A 259, p. 187-197, 1966) reveal that a ridge is traceable along a N.E.-S.W. direction up to about 61°E. long. where it mereges into the OWEN FRACTURE ZONE. On its southeast, the ridge is flanked by narrow linear trenches reaching depths of 2,000 to 2,500 fm and farther on by another ridge which rises to 220 fm below sea level. To the northwest and southeast of the ridge system are the abyssal plains of the Arabian Sea at an average depth of 1,800 fm. A major fault runs through the trenches and another on the northwestern side of the ridge system. The Murray Ridge is the connecting link between the Kirthar Range of Pakistan and the Oman Mountains of eastern Arabia, but the connection is broken by movements along the Owen Fracture and other faults parallel to the Arabian coast. As a result, the Oman Mountains have been shifted north relative to the western end of the Murray Ridge.

The Makran coast was faulted down in the late Pliocene, but the landward side has been raised and is still rising. This fault apparently continues into the Persian Gulf and may extend up the Euphrates valley. The John Murray Expedition to the Arabian Sea, under R.B.S. Sewell, found that the Makran Sea is streaked with ridges and troughs belonging to the fold system of the Baluchistan Arc, the ridges rising from a depth of about 1,500 fm to within 400-500 fm of sea level. The Murray Ridge is the southern limit of this system.

A major fracture system, called the OWEN FRACTURE ZONE trending N.N.E., has been traced over a length of 1,500 miles from east of the Somali coast (3°30′N : 52°20′E.) to the continental slope off Karachi. It comprises ridges and trenches with brecciated rocks, the deepest being the Wheatley Deep near 12°40′N. with a maximum depth of 2,450 fm. The Carlsberg Ridge, which is the Mid-Indian Ocean Ridge in the northwestern part of the

Indian Ocean, is intersected by this fracture, and its western continuation into the Gulf of Aden is shifted northeast by 170 miles in a dextral sense. Near 61°E. longitude it bends and seems to follow the Murray Ridge. It may have a straight westerly branch going into Baluchistan as the Nushki-Chaman Fault, but this has not been established. The fault along the Murray Ridge may continue into the Indus Valley or the zone of dislocation along 66°30′E. longitude. The Owen Fracture and the parallel faults along the Arabian coast must be genetically connected with the drift of Arabia and India. Though a direct connection between the Owen Fracture and the fault along the eastern coast of Madagascar has not been clearly established, it is likely that there are *en echelon* faults in the area between the two.

The Kutch—Gulf of Cambay—Narmada valley region appears to have been faulted down along E.—W. or E.S.E.-W.N.W. faults in the Upper Palaeozoic. Permo-Carboniferous marine strata are found in two or three places in Rewa ; Upper Cretaceous marine strata are found in Bagh in the Narmada valley ; and thick Pleistocene deposits farther west along the same valley. The Narmada and Tapi valleys are in this graben. It is very probable that along this valley may be found Permian, Triassic and Jurassic rocks of marine or estuarine facies. The western coast of India was probably faulted down in the Miocene (Oldham, 1893, p. 493) more or less across the direction of the Narmada valley graben, and this let down the southerly extension of the Aravalli mountains in the Arabian sea. The extent of down-faulting is 1,100 or 1,200 fathoms (about 7,000 ft) for it is known that the Laccadives, which are on this submarine ridge, rise from a platform at this depth. Kathiawar and Kutch, which contain Jurassic and Lower Cretaceous sediments, may have been uplifted in the early Tertiary. Whereas the Laccadives lie in a zone of positive gravity anomaly, the Maldives are in a negative anomaly zone ; it may be that the latter is connected with some structure in South India. There is a gap between the Laccadives and Maldives at 8°N. Lat. and another gap at 2°N. Lat. between the Maldives and the Chagos group, but these two are not on the same alignment. The eastern slope of the Laccadives is steeper than the western, but the sea bottom on the eastern side is only 1,100 fathoms deep while that on the western side is deeper.

OROGENIC BELTS AND ULTRAMAFIC ROCKS

The various occurrences of ultramafic rocks known in India fall into one or the other of the known belts of mountain building. They include rock types such as serpentine, peridotite, dunite, saxonite, lherzolite and pyroxenite which have frequently given rise to magnesite, asbestos, talc, etc., and are associated with epidiorites and amphibolites. The chief groups of occurrences of these rocks are mentioned briefly below.

In Rajasthan numerous lens-like masses occur near Jhiri, Raialo, Maroli, Bhilwara etc. and in the Biana hills of Central Mewar and Ajmer. Talc-bearing serpentine rocks occur along a N.N.E.-S.S.W. zone to the northeast of Udaipur city and also in Sirohi, Dungarpur and Idar. Their distribution generally follows the trend of the axis of folding of the Aravalli and Delhi systems of rocks in this region.

In the Mysore region there are similar ultramafic rocks along a N.N.W.-S.S.E. belt passing through Kadur, Hassan and Mysore districts. They appear in the belts of ancient Dharwarian rocks, and are associated in some places with chromite. At the southern end, near Mysore city, the trend changes its direction to south and then to S.S.W.

Details of the geology of the Eastern Ghats are not well-known, but in some places there are ultramafic rocks, as for instance those which are associated with the magnesite deposits of Salem and with chromite in the Kondapalle hills of the Krishna district and a few places in Orissa. This belt is generally marked by intrusions of large masses of Charnockite.

There are a few occurrences of ultramafic rocks along the Singhbhum Copper belt and its continuation to the south-east into Mayurbhanj. Some of these exposures are associated with chromite or titaniferous magnetite.

In the Extra-Peninsular Region there are ultramafic rocks assciated with chromite in the Zhob and Pishin valleys near Quetta and near Fort Sandeman. Ultramafic rocks are closely associated with orogenic belts. They are intruded into the upper crust through fractures developed during the initial stage of folding, when materials derived from the upper mantle are able to penetrate the crust. Indeed, it would appear that such rocks are almost entirely confined to orogenic belts. The granitic rocks in the root zones of mountains are formed at a later stage, mainly as a result of softening and melting of crustal materials which have been depressed during compression and folding. The roots generally consist of granites and grano-diorites, with inclusions and streaks of schists representing recrystallised earlier rocks. The marginal zones of the batholiths and stocks of the root zone show banded gneisses, migmatites and granitised schists resulting from hybridisation and metasomatism, while the central parts may show true granites crystallised from completely molten materials.

Ultramafic rocks, often associated with radiolarites, basic intrusives and pillow lavas, etc. are found at many places along the older mountain belts as well as along the Tertiary orogenic axis. Several occurrences are known in the Dharwar belt of Mysore, in the Aravalli belt of Rajasthan and in the Eastern Ghats. They are often associated with serpentinites, and deposits of chrysotile, magnesite, steatite and chromite. In the Baluchistan Arc they are found along the main axis between Karachi and Quetta, in the Zhob-Pishin area east of Quetta and near Fort Sandeman farther north. A conspicuous zone associated with Upper Cretaceous flysch occurs along the northern border of the Kashmir syntaxis around Nanga Parbat, continuing to Dras and farther east through Spiti to the neighbourhood of Mt. Kailas. Gansser (Geology of the Himalayas, 1964) believes that this zone may continue along the valley of the Tsangpo to near Namcha Barwa, as the straight course of the river indicates a tectonic line. That the Tsangpo follows a major tectonic dislocation zone has already been shown by the magnetometric traverses made by Wienert in 1938. In the Burmese Arc, there are occurrences along the central axis from the Jade Mines area of Myitkyina, through Naga and Manipur hills, to Prome and Henzada in the Arakan Yomas. This belt continues into the Andaman and Nicobar islands where similar rocks occur.

SEISMIC PHENOMENA

The distribution and significance of seismic phenomena in India in relation to geological structure have been dealt with by various authors including T. Oldham, De. Ballore and W.D. West. It is known that the peninsular part of India is practically immune to all but minor shocks which are occasionally felt in a few places. But earthquakes of great intensity occur along the highly disturbed belt of mountains in the Extra-Peninsula. Occasionally some important shocks originate from the floor of the Gangetic basin, the Bihar earthquake of 1934 being attributed to movements in the floor of the fore-deep now filled up with alluvium.

The regions towards which the wedge-like projections of the peninsular mass are directed are generally known to be suceptible to frequent and severe earthquakes. Such are the Gilgit-Pamir region in the northwest, the Quetta area in Baluchistan and the region of S.W. China adjoining the north-eastern corner of Assam. The Pamir region is indeed visited by earthquakes originating at intermediate depths of 220-230 kilometres, in addition to shallow earthquakes.

ORIGIN OF THE HIMALAYAS AND THE GANGETIC PLAINS

It is generally assumed that the Peninsular mass, which was a part of Gondwanaland, remained passive while the Tethys basin to the north of it was thrust against and over its edges. Along the northern borders of the Peninsular mass we find fragments of older rocks broken off and carried along with the over-thrust sediments. These include the unfossiliferous rocks of Peninsular facies in the Lesser Himalayas, of Cuddapah, Vindhyan and Gondwana ages. In front of this region is a sag or depression which has been formed by buckling down of the crust in obedience to the pressure exerted on the borders of the Peninsula by compressive forces.

The direction of movement deduced from the rocks exposed at the surface is towards the south in the Himalayan Arc, towards the west in the Burmese Arc and towards the east in the Baluchistan Arc. P. Lake who has discussed this question points out the difficulty of explaining how one single continental mass could move in different directions at the same time and suggests that the Pacific and Indian regions have been underthrust towards Asia. It is of interest to note that Central Asia is a region containing excess of matter, which could only be explained as due to the action of compressive forces *towards it*. Though the Baluchistan Arc and the Himalayan Arc appear as if they were compressing the area of north-west Punjab intervening between them, Burrard points out that gravity observations do not indicate the presence of an excess of mass in that region and that indeed the reverse is the case. The explanation appears to be that the mountains here are composed of piles of light sediments pushed over the submerged borders of the Peninsula.

A northerly movement of the Indian shield would thus appear to have given rise to the thrusting of the sediments of the Tethyan basin over the whole of its northern border. This would naturally produce an eastward thrust in the Baluchistan Arc and a westward thrust in the Burmese Arc. The edges of the underthrust mass appear to have buckled down and broken off in places, except where projections have acted like wedges extending far into the adjacent sedimentary basin and producing festoons and syntaxial structures. This would explain the simultaneity of the movements in the different Arcs during the successive periods of mountain building and also the much greater violence of compression and faulting in the Himalayan Arc than in the two lateral Arcs. An important consequence of the great intensity of compression of the Himalayan Arc is the breaking up of most, if not all, of the structures suitable for the accumulation of petroleum in that region, whereas the same sedimentary belt in Iran and Burma contains numerous unbroken oil-bearing structures.

GEODETIC OBSERVATIONS

The making of accurate topographic maps necessitates the assumption of a standard spheroid for the shape of the earth. This would be the sea-level surface continued through the continental portions also. This surface, obtained from the mean sea-level of tidal observations would give the *geoid*, whose shape is an oblate spheroid. The Survey of India has used, for this purpose, Everest's spheroid (H.J. Couchman, 1937) whose equatorial semi-axis is 6377.3 km, with an ellipticity of 1/300.8. The modern value is however slightly different, the major semi-axis being 6,378.4 km and the ellipticity 1/295 (ellipticity being $a-b/a$, where a and b are the equatorial and polar semi-axis respectively). Recent data based on observations of artificial satellites indicate the value to be 1/298.2.

If we have a homogeneous sphere it will have the same force of gravity at every point on its surface, *i.e.*, the geopotential will be the same at every point. But the geopotential will vary according to height (*i.e.*, vertical distance from the level surface of reference) and also according to variation in the distribution of matter. Any extra mass which forms a plateau or mountain will give an extra value of gravity over it which can be measured. As a general rule we can state that if we have a thickness of 1 km of rock of average density 2.5 per unit area (1 square cm.) it increases the gravity by 0.105 cm./sec.^{2} This extra thickness of 1 km therefore gives roughly an extra value of 100 milligals (1 milligal = 0.001 cm/sec.^{2}). This extra mass possesses gravitative attraction, which will be seen as a deflection from the vertical, of a plubline placed at the side of the mass.

The attraction due to gravity is, as we have seen, dependent on the height above the spheroid at which the measurements are taken. It will be different and smaller if we could take it in free air at the same height (*i.e.*, allow for the fact that the place of observation is some distance above the geoid surface and so from the centre of the earth). This gives the 'free air' value. We can calculate this from the observed gravity by allowing for the height above the mean sea-level. The difference is the *free air anomaly*. A correction is also made for the mass present between the point of observation and the sea level as was suggested by Bouguer. This gives the Bouguer anomaly which is quite useful.

Isostasy.—It is also known that gravity does not vary in accordance with the height of solid matter above sea-level or depth below sea-level. In the Alps, for instance, the observed gravity is something like 100 milligals less than what would be the normal. In the 'deep' off the coast of California we should expect gravity anomalies of 300 milligals if the 4-mile depth meant merely replacing normal rock by water. But in many parts of the oceans the anomalies are systematically positive. These observations prove that, in and below mountains, there is matter of lower density than normal, and in ocean basins there is matter of higher density. Thus Nature seems to try to compensate the visible inequalities of matter by density, so that excess of matter is compensated for by lack of density and defect of matter by an excess of density. This relationship between mass and density is called *Isostasy*. The subterranean variation of mass is called *compensation*; if there is too much, it is over-compensation, and if too little, it is under-compensation. If our calculated value of gravity is also corrected for the disturbance due to any type of assumed compensation, we get the *isostatic anomaly*. The free-air anomaly gives us the earth's external gravitational field. The Bouguer anomaly gives us the sum total of all the information about the distribution of density. Isostatic anomaly is only of interest to test any particular theory of compensation.

It is an interesting fact that it was in India that the theory of mountain compensation was first propounded by Archdeacon Pratt of Calcutta, an eminent mathematician to whom Sir Andrew Waugh (Surveyor General of India) referred certain gravitational anomalies for solution. When the deflections of the plumb line were measured at some localities near the foot of the Himalayas, it was found that the observed deflection was much less than the result obtained by calculation from the visible excess of matter of the Himalayas. At Dehra Dun the calculated and observed deflections are 86 seconds of arc and 36 seconds of arc respectively; at Murree they are 45 seconds and 12 seconds respectively.

According to one view of isostasy adopted by Hayford and Bowie of the United States Coast and Geodetic Survey, different vertical sections or blocks of the crust may be thought of as being completely compensated at a certain uniform depth, called the *depth of compensation*. This can be illustrated by an experiment in which blocks of various substances, of equal cross section and of equal weight, are resting on a heavy liquid, say mercury. Here they all sink to the same depth but rise to various heights above the liquid in inverse relation to density. According to another interpretation, blocks of one substance having the same cross section but different weights can be thought of as resting on mercury. Here they sink to various depths and also have their top surface at various heights. The latter view, allowing for varitions in the depths of compensation of different segments of the earth's crust, is perhaps more in consonance with the evidence of seismology. The second view was proposed by Airy and developed by Heiskanen.

It is well known that continental masses are built up, to a large extent, of granite. Granite and large masses of sediments also form the mountain ranges of Tertiary age. In contrast with this, heavier rocks like basalt are found in regions which have been rent by tensional cracks. The ocean basins also have basic rocks at shallow depths. These facts and suppositions are in accord with the principles of isostasy.

A study of the values of gravity in different parts of India and of the deflections of the plumb line has indicated that the Himalayas consist of light materials and are deficient in mass. There are also certain regions with an excess or defect in gravity values, which may indicate the presence of heavy or light sub-crustal rocks. Sir Sidney Burrard of the Survey on India deduced the presence of a sub-surface feature, which can be likened to a ridge of heavy rocks, running roughly east to west across India through Jabalpur, and another parallel zone of light rocks passing through Belgaum and Nellore in South India. The former is referred to as *Burrard's Hidden Range* on the supposition that the heavier layers of the crust had here come nearer the surface than elsewhere, forming a sort of ridge. The other zone is called *Burrard's Hidden Trough*, expressing the idea that the heavier layers formed a trough-like depression here occupied by lighter rocks. To the north of the Hidden Range is another parallel zone passing through the southern border of the Siwalik Zone marked by negative gravity anomalies.

It is an interesting fact that the zones of positive and negative anomalies are more or less parallel to the general trend of the Himalayas and also that the distance between each pair is about 8 to $8\frac{1}{2}$ degrees of latitude. If the Himalayan orogenesis is attributable to the northward or northeastward movement of India, it is possible that this arrangement of 'ridge' and 'trough' may also be due to the same cause.

The alluvium-filled trough through which the Ganges flows in Northern India is of the nature of a 'fore-deep', *viz.*, a slight buckling down of the upper crust in front of the convex mountain arc. It was formerly thought that it was V-shaped in section and that the thickness of the alluvium in it was as much as 40,000—50,000 feet. It is now known, from geophysical data and from borings, that the thickness of the strata overlying the basement is of the order of 25,000 feet.

Some years ago E.A. Glennie of the Survey of India explained the various regional gravity anomalies found in India as due to local up-warps or down-warps of the heavy sub-crustal layers. In some cases there are co-incidences between the supposed crustal warp and the surface geology. For instance, the Archaean ranges of the Aravallis, Dharwars, Satpura belt and the Assam Plateau are regions of positive anomaly which coincide with the up-warps, while the Cuddapah basins of Madras, Chattisgarh and Gwalior which contain thick sedimentary beds are zones of negative anomaly. There are, however, other areas in which there is no such coincidence. A zone of positive anomaly runs through the Gulf of Cambay and along the Bombay Coast, which may probably be due to a large thicness of basalt here. In the area occupied by the Deccan Traps in western and central India the anomalies are very erratic and have no relationship to the surface geology.

In a paper published in 1946, Evans and Crompton have given the results of the extensive gravity survey undertaken by the Burmah Oil Company in the Assam-Burma region. They have computed the anomalies after carrying out corrections for local geology down to a depth of 11.5 km, as data were available on the sub-surface geology from bore-hole records. Their map indicates that there is a zone of negative anomaly along the eastern flank of the Burmese arc continuing into the Andaman-Nicobar Islands and farther on into the well-marked negative anomaly zone just to the south of Sumatra and Java. This marks the zone of uplift during the Tertiary mountain building movement. There is a zone of high positive anomaly along the volcano belt of Burma which goes through Wun-tho, Monywa, Popa, Barren Island and continues into the volcano zone of Sumatra and Java. The Shan Plateau shows only weak positive anomalies as it has apparently nearly attained equilibrium during the long period which has elapsed since it experienced orogenic disturbances. The Upper Assam Valley as well as the Ganges Valley in Bihar are regions of negative anomaly, as should be expected, for they contain a large thickness of light sediments.

From the geological map of Southern Asia it will be noticed that the Tethyan basin of Turkey, Iraq and Iran would, if continued in the direction of its general trend, go into the region of the Indonesian Archipelago, but the portion of the basin between Baluchistan and Sumatra is found to be violently distorted and pushed to the north by the foreign mass of India. It may therefore be suggested that the shifting of the Tethyan belt to the north of India would give us roughly a measure of the drift of India towards the north and of the shortening of the earth's crust by the compressive forces which were responsible for the Himalayan orogeny. The distance between the postulated original position of the southern limit of the Tethyan geosyncline and the present Himalayan arc is of the order of 13° to 14° of latitude or say, 1,300 km or more. This does not however take into account the fact that the Zagros mountain belt of Persia and the Indonesian archipelago are themselves compressed and folded, indicating that the Tethyan basin must originally have been still wider.

Heim and Gansser (1939, p. 226) have made an estimate of the original width of the sedimentary zones of the Himlayas and deduced the shortening of the zones as a result of mountain building. They came to the conclusion that the original width from the northern border of the Indian plains to the zone of Exotic Blocks is at least 460 km, the actual width at present being about 160 km. The difference between the two, about 300 km., would be the minimum shortening of the crust. They also point out that this is only a rough estimate which may be out by 100 to 200 per cent or more, because of the lack of detailed knowledge of the geology of the Himalayas. It should also be noted that the Himalayan orogeny affected a part of the region lying in Tibet, viz., Karakorum, Alai, Kun Lun ranges which took their final shape only in the Tertiary. The estimate of the crustal shortening of about 1,300 km given above would therefore seem to be quite a modest estimate for the numerous folds, overthrusts and nappes present in the Himalaya-Karakorum belt indicate that a region of great width was involved in the Tertiary mountain building movements.

The Indian Peninsular appears to have experienced an upward tilt in the southwest and south, which resulted in producing a general northeasterly slope. The charnockite massifs and the Deccan and Mysore plateaux have also been uplifted during the Tertiary era. Geomorphological studies (Vaidiya nadhan, 1964; Radhakrishna, 1965) have revealed at least three major erosion-surfaces in the southern part of the Peninsula which are of Upper Jurassic, (also Oligocene), Pliocene and Pleistocene ages. These are well seen in the Cuddapah Basin and the Mysore plateau. The post-Miocene uplifts of the Western Ghats and the Mysore and Deccan plateaux are responsible for the precipitous scarps on the west and for the youthful nature of the rivers traversing them, for the same rivers in the plains to the east show mature stages of development.

SELECTED BIBLIOGRAPHY

Aswathanarayana, U. Absolute ages of the Archaean orogenic cycles of India. *Amer. Jour. Sci.* 254, 19-31, 1956.
Auden, J. B. Geology of the Krol Belt. *Rec.* 67(4), 1934.
Auden, J. B. Traverses in the Himalayas. *Rec.* 69, 123-145, 1935.
Auden, J. B. Structure of the Himalayas in Garhwal. *Rec.* 71, 407-433, 1936.
Burrard, S.G. Origin of the Himalaya Mountain — a consideration of the geodetic evidence. *Surv. of Ind. Prof. Paper* 12, 1912.
Burrard, S. G. Presidential Address to the Indian Science Congress. *Proc. A.S.B.*, N.S. 12(2), 1916.
Burrard, S.G. Origin of the Gangetic Trough, commonly called the Himalayan fore-deep. *Proc. Roy. Soc.* London, 91-A, 220-238, 1915.
Burrard, S. G. Attraction of the Himalaya Mountains on the plumb line. *Surv. of Ind. Prof. Paper* 5, 1901.
Burrard, S.G. Hayden, H.H. and Heron, A.M. Geography and geology of the Himalaya Mountains and Tibet. (2nd Ed.) Dehra Dun. 1932.
Crosthwaite, H. L. Investigation of the theory of isostasy in India. *Surv. of Ind. Prof. Paper* 13, 1912.
Couchman, H. J. Progress of Geodesy in India. *Proc. Nat. Inst. Sci. Ind.*, III, 1937.
Du Toit, A. L. Our wandering continents. London, 1937.
Evans, P. The tectonic framework of Assam. *Jour. Geol. Soc. India*, 5, 80-96, 1964.
Glennie, E. A. Gravity anomalies and the structure of the earth's crust. *Surv. of Ind. Prof. Paper* 27, 1932.
Gregory, J. W. *et al.* Structure of Asia. London, 1929.
Gulatee, B. L. Gravity anomalies and the figure of the earth. *Surv. of Ind. Prof. Paper* 30, 1940.
Gulatee, B. L. Gravity in India. *Survey of India, Tech. Paper* 10, 1956.
Gulatee, B. L. Isostasy in India. *Bull. Nat. Inst. Sci. India.*, No. 11, 1958.
Gutenberg, B. and Richter, C.F. Seismicity of the earth and associated phenomena. Princeton, 1954.
Hayden, H. H. Relationship of the Himalaya to the Indo-gangetic plain and the Indian peninsula. *Rec.* 43, 138-157, 1913.
Heim, A. and Gansser, A. Central Himalayas. *Denkschr. Schweiz.Naturf. Gesell.* 73(1), 1939.
Hess, H.H. Gravity anomalies and Island Arc structure. *Proc. Am. Phil. Soc.* 79(1), 71-96, 1937.
Hess, H.H. Serpentines, orogeny and epeirogeny. In "Crust of the Earth". *Geol. Soc. Amer. Spl. Paper* 62, 391-407, 1954.

Holmes, A. The age of uraninite and monazite from post-Delhi pegmatites of Rajputana. *Geol. Mag.* **86**(5), 288-302, 1949.
Holmes, A. Age of uraninite from a pegmatite near Singar ; Gaya district, India. *Amer. Mineral.* **35**, 19-28, 1950.
Holmes, A. Dating the Pre-Cambrian of Peninsular India and Ceylon. *Proc. Geol. Assoc. Canada*, **7**(2), 81-106, 1955.
Jeffreys, H. The Earth. Cambridge, 1929.
Krishnan, M.S. The structural and tectonic history of India. *Mem.* **81**, 1953, 1961.
Krishnan, M.S. Tectonics of India. *Bull. Indian Geophys. Union*, **3**, 1-51 (Also *Bull. Nat. Inst. Sci. India*, 32, 1967)
Lake, P. Geology of South Malabar. *Mem.* **24**, 1890.
Lake, P. Island arcs and Mountain Building. *Geogr. Jour.* **78**, 149-160, 1931.
Mathur, L.P. and Evans, P. Oil in India. *XXII Int. Geol. Cong.* (New Delhi), *Special Brochure*, 86, 1964.
Mathur, L.P. and Kohli, G. Exploration and development of the oil resources of India. *6th World Petroleum Cong.* Sec. i, 1963.
Oldham, R.D. Structure of the Himalayas and the Gangetic plain. *Mem.* **42**(2), 1917.
Oldham, R.D. Support of the mountains of Central Asia. *Rec.* **49**, 117-135, 1918.
Oldham, R.D. Geological interpretation of some recent geodetic investigations. *Rec.* **55** 78-94, 1923.
Sewell, R.B.S. Geographic and oceanographic researches in Indian waters. *Mem. A.S.B.* IX, parts 1-8, 1925-38.
Sewell, R.B.S. The John Murray expedition to the Arabian Sea. *Nature*, **133**, 669-672, **134**, 685-688, 1934.
de Terra, H. Himalayan and Alpine orogenies, *16th Int. Geol. Cong. Proc.* **2**, 859-871, 1936.
Umbgrove, J.H.F. Structural history of the East Indies. Cambridge, 1949.
Wadia, D.N. Syntaxis of the N.W. Himalayas — their rocks, tectonics and orogeny. *Rec.* **65**, 189-220, 1932.
Wadia, D.N. Structure of the Himalayas and the North Indian foreland. *Proc. 25th Ind. Sci. Cong.* Pt. II, 91-118, 1938.
West, W.D. Recent advances in Indian geology. Geology of the Himalayas. *Carr. Sci.* III (6), 286-291, 1934.
West, W.D. Structure of the Shali Window near Simla. *Rec.* LXXIV, 1939.
Wilson, J.T. Major structures in the Canadian Shield. *Canad. Min. Met. Bull.* **42** (No. 450), 543-554, 1949.
Wiseman, J.D.H. and Sewell, R.B.S. The floor of the Arabian Sea. *Geol. Mag.* **74**, 219-230, 1937; **75**, 143-144, 239-240, 1938.
John Murray Expedition, Scientific Reports I, 1936.
Symposium on the Northwestern Indian Ocean. *Phil. Trans. Roy. Soc.*, London, A 259, 133-298, 1966 (Several papers).

CHAPTER III

REVIEW OF INDIAN STRATIGRAPHY

Principles of Stratigraphy

Stratigraphical or historical geology has, as its aim, the description and classification of rocks with a view to arranging them in the chronological order in which they were laid down on the surface of the earth. Of the three groups of pocks, sedimentary, igneous and metamorphic, only the sedimentary rocks are easily amenable to such an arrangement, since they have been deposited bed by bed and contain the remains of organisms which flourished while they were formed. The lithological characters of the units or formations and particularly their fossil content have been invaluable for determining the chronology of the materials of the earth's crust, as will be explained below.

Lithology.—The lithological characters of the different formations are persistent over the area in which they are exposed, though there may be variations when followed over some distance. Each lithological unit may comprise a number of individual beds having more or less the same characters, when it is spoken of as a *formation* and given a local or specific name to distinguish it from a similar formation of different age or belonging to another area. We have thus the Barakar Sandstone, Kamthi Sandstone, Bhander Sandstone; the Attock Slate and Cumbum Slate; the Vempalle Limestone and Megalodon Limestone, etc. The lithology is often of help in correlation as in the case of the Spiti Shales of various parts of the Himalayas or the carbonaceous Barakar Sandstones in various parts of the Peninsula and in the Himalayan foothills.

Fossil content.—Each formation has not only distinct petrological characters but also encloses a fossil assemblage which is characteristic and different from that of the underlying and overlying formations. Animal and vegetable organisms of each geological age bear special characters not found in those of other ages. Though some species are long-lived and have a long range in time, there are others which have a very short range, and each assemblage contains a mixture of many different species and groups of animals. Some species, for example of graptolites and ammonites, are so highly specialized in morphological characters and so restricted in range of time that they are highly valuable indicators of very small sub-divisions of geological time.

Fossil assemblages of the same age are not necessarily identical, for the species in them will depend on the conditions of environment and development in each area of sedimentation. If the environment was the same or very similar, the species may be identical or closely allied, as is often the case with marine fauna; if the environment were different, as in the case of estuarine and lacustrine deposits, the elements of the fauna will be different but will show the same stage of evolution or development in relation to each other and in comparison with the parallel faunas of other areas. The conditions which control sedimentation and life give rise, therefore, to different facies, such as the deep sea, coastal, estuarine, fluviatile, etc. ; and also, depending on lithology, to shale, limestone or sandstone, or other facies. Hence, in comparing the faunas or floras of two areas, the lithological as well as environmental facies will have to be fully taken into account.

Order of superposition—Every geological formation rests on another and is superposed by a third. The formation at the bottom is naturally older than the one at the top, and when we deal with several, the upper ones are successively younger than those below. The sequence is the same wherever the same formations are met with.

If the formations have been laid down continuously each of them grades perfectly into the succeeding one. They are then said to be conformable. The gradation is not only lithological but also faunistic. It often happens however that, owing to local upheavals, some formations are locally missing. In this case, the transition from the underlying to the overlying beds will be abrupt, such a break in continuity being called an unconformity. If there is no visible break, and the formations show complete parallellsm in spite a gap in sedimentation, it is called a disconformity. The unconformity may be marked by a change in rock type, by the different disposition of the overlying beds, by the intervention of a horizon of conglomerate containing pebbles from the underlying formation, or by other features. The overlying formation may sperad over and transgress the limits of the lower one, thereby showing the phenomenon of *overlap*. Or, there may be regression, producing a *gap*. Yet these phenomena do not affect the order of superposition of the strata.

The earth's crust is the scene of constant changes and the rocks are affected by them in various ways. They may be tilted, folded, and faulted. They may be intruded by igneous rocks, or metamorphosed as a result of earth movements. The final result of these changes, as seen at the present day, is often very complex but the geologist should observe all the facts carefully and unravel the history of the formations after weighing all the available evidence.

After careful study, the geological formations have been arranged into a few major groups. These are shown in Table 3 in the order of increasing antiquity. The latter terms in these groups indicate the stage or development of the organisms. The Azoic is entirely devoid of organisms while the Proterozoic shows traces of the most primitive life ; the Palaeozoic contains the remains of ancient animals and plants, and so on to Recent times.

TABLE 3.—Geological Time Scale

(Figures without brackets show the total duration of the Group or System in millions of years, while those within brackets show the lapse of time from the beginning of the particular era or period to the present).

Group	System	Chief Fossils
Quaternary 1	Recent (.01) Pleistocene 1(1)	Living animals. Man appears. Many mammals die off during glacial periods.
Tertiary or Kainozoic 65	Pliocene 7 (8) Miocene 17 (25) Oligocene 13 (38) Eocene 27 (65)	Mammals, mollusca and flowering plants dominant. Division largely based on proportion of living to extinct species of mollusca and the presence of mammal species.
Secondary or Mesozoic 180	Cretaceous 75 (140)	Giant reptiles and ammonites disappear at the end. Flowering plants become numerous.
	Jurassic 60 (200)	Ammonites abundant. First birds, flowering plants and sea urchins.
	Triassic 40 (240)	Ammonites, reptiles and amphibia abundant. Arid climate.
Primary or Palaeozoic 370	Permian 50 (290)	Trilobites disappear at the end.
	Carboniferous 60 (350)	Many non-flowering plants; first reptiles appear.
	Devonian 60 (410)	Abundance of Corals, Brachiopoda; first amphibians and lung-fishes.
	Silurian 35 (544)	Graptolites disappear at the end; first fishes; probably first land plants.
	Ordovician 60 (505)	Abundance of Trilobites, and Graptolites.
	Cambrian 100 (605)	Abundance of Trilobites.
Pre-Cambrian or Proterozoic	Pre-Cambrian (2,500)	Soft-bodied animals and plants.
Archaean or Azoic	Archaean (3,600)	Lifeless.

Note :—The latest results from lead ratios in meteorites are said to indicate an age of about 4,600 million years for the planets in' the Solar System. The earliest radiometrically dated rocks are 3,600 million years old.

REVIEW OF INDIAN STRATIGRAPHY

The major groups are divided into Systems; each System into Series; each Series into Stages; each Stage into Zones. Corresponding to these divisions of formations there are divisions of geological time, as shown below:

Formations	Time
Group (*e.g.* Mesozoic)	Era
System (*e.g.* Triassic)	Period
Series (*e.g.* Upper Triassic)	Epoch
Stage (*e.g.* Carnic)	Age
Zone (*e.g.* Tropites)	

The American Commission on stratigraphic nomenclature has recommended the use of the terms Bed, Member, Formation and Group for units of increasing magnitude. It is necessary to standardise the nomenclature and usage of stratigraphic terms in India.

As the geological formations were first studied in Western Europe, the names of formations in the European region are now universally used as standards of reference to facilitate the correlation and comparison of formations of all parts of the world. Table 4 gives the names of the chief divisions in usage, and many of them will be frequently referred to in the following pages.

TABLE 4.—Standard Geological Divisions.

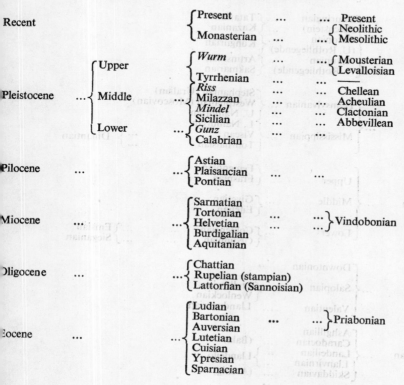

Era/Period			Stage		
Paleocene			Thanetian / Montian		
Cretaceous			Danian		
			Maestrichtian		
			Campanian		Senonian
			Santonian		
			Coniacian		
			Turonian		
			Cenomanian		
			Albian		Gault and Up. Greensand (L. Greensand)
			Aptian		
			Barremian		
			Hauterivian		Neocomian
			Valanginian		
Jurassic		Upper (Malm)	Purbeckian / Portlandian		Tithonian
			Kimmeridgian		
			Argovian		Oxfordian
			Divesian		
			Callovian		
		Middle (Dogger)	Bathonian		Great Oolite
			Bajocian		Inferior Oolite
		Lower (Lias)	Toarcian		
			Charmouthian		Pliensbachian
			Sinemurian		
			Hettangian		
Triassic		Upper	Rhaetic		Keuper
			Noric		
			Carnic		
		Middle	Ladinic		Muschelkalk
			Anisic (Virgloric)		
		Lower	Scythic (Werfenic)		Bunter
Permian		Thuringian (Zechstein)	Tatarian		
			Kazanian		
		Saxonian (U. Rothliegende)	Kungurian		
		Autunian (L. Rothliegende)	Artinskian		
			Sakmarian		
Carboniferous		Pennsylvanian	Stephanian (Uralian)		
			Westphalian (Moscovian)		
			U. Namurian		
		Mississippian	L. Namurian		
			Visean		Dinantian
			Tournaisian		
Devonian		Upper	Femennian		
			Frasnian		
		Middle	Givetian		
			Eifelian		
		Lower	Coblenzian		Emsian
			Gedinnian		Sieganian
Silurian		Downtonian			
		Salopian	Ludlowian		
			Wenlockian		
		Valentian	Llandoverian		
Ordovician		Ashgillian			
		Caradocian	(Bala)		
		Landeilian	Llandeilo		
		Llanvirnian			
		Skiddavian	(Arenig)		

Cambrian	...	Tremadoc Olenus	...⎫ ...⎬ Potsdamian
		Paradoxides	... (Acadian)
		Olenellus	... (Georgian)

General Review of Indian Stratigraphy

Before commencing the description of the stratigraphical units of India, a general summary might prove useful so that the subject can be viewed in the roughest outline.

More than half of the Peninsula is occupied by gneissic and schistose rocks of the Archaean and Proterozoic ages. The Cuddapahs, Vindhyans, the Gondwanas and the Deccan Traps occupy the rest of the area, except parts of the coastal regions. In the Extra-Peninsula, marine sedimentary systems predominate, though parts of the sub-Himalaya and the main axis of the Himalaya are occupied by ancient metamorphic rocks and intrusive igneous rocks. A full succession of fossiliferous sedimentary systems, extending from the Cambrian to Eocene, is met with on the Tibetan side of the Himalaya, while the southern or Sub-Himalayan zone contains a different facies which is practically unfossiliferous. The fossiliferous facies transgresses to the south of the central Himalayan zone in Kashmir.

Beyond the sharp syntaxial bend of the northwestern Himalaya near the Nanga Parbat, the Hazara area shows unfossiliferous Palaeozoics and fossiliferous later formations; but farther south, the Baluchistan arc is built up mainly of post-Carboniferous systems which sweep down in a broad arc to the Makran region. The eastern part of this, forming the mountain chains of Sulaiman-Kirthar-Laki, shows a calcareous facies while the western part is largely a shale facies with flysch-like lower Tertiaries. To the east and south of the Syntaxis of northeastern Assam, in the Burmese arc, Tertiary rocks form a broad zone with a core of Mesozoic rocks which constitute the mountains of the Assam-Burma border and the Arakan Yomas. To their east is the Shan-Tenasserim belt of pre-Tertiary rocks which belong to the S.E. Asian geological province.

The major stratigraphical divisions of India are shown in Table 5 together with their standard European equivalents, these being arranged, as usual, in the order of increasing antiquity.

TABLE 5.—Major Geological Formations of India

Recent	... ...	Recent Alluvia, Sand dunes, Soils.
Pleistocene	... ...	Older Alluvia, Karewas of Kashmir, and Pleistocene river terraces etc.
Mio-Pliocene	... ...	Siwalik, Irrawaddy and Manchhar Systems; Cuddalore, Warkilli and Rajamahendri Sandstones.
Oligo-Miocene	... ...	Murree and Pegu Systems; Nari and Gaj Series.
Eocene	... ...	Ranikot-Laki-Kirthar-Chharat Series; Eocene of Burma.
L. Eocene : Up. Cretaceous		Deccan Traps and Inter-trappeans.
Cretaceous	... ...	Cretaceous of Trichinopoly, Assam and Narmada Valley; Giumal and Chikkim Series; Umia beds,

Jurassic	... ...	Kioto Limestone and Spiti Shales; Kota-Rajmahal and Jabalpur Series.
Triassic	... ...	Lilang System including Kioto Limestone; Mahadeva and Panchet Series.
Permian	... ...	Kuling System; Damuda System.
Carboniferous	... ...	Lipak and Po Series; Talchir Series.
Devonian	... ...	Muth Quartzite.
Silurian	... ...	Silurian of Burma and Himalayas.
Ordovician	... ...	Ordovician of Burma and Himalayas.
Cambrian	... ...	Haimanta System; Garbyang Series.
Pre-Cambrian and	... ...	Cuddapah and Vindhyan Systems; Dogra and Simla Slates; Martoli Series.
Archaean	... ...	Dharwar and Aravalli Systems; Salkhala, Jutogh and Daling Series, various gneisses, etc.

In 1904 in an article in the Imperial Gazetteer of India, Sir T.H. Holland proposed a new classification of the Indian strata in which the Cuddapah and Vindhyan systems were grouped together under the name of *Purana Group*, corresponding to the Algonkian of America, which is now referred to as the Proterozoic. The strata from the base of the Cambrian to Middle Carboniferous were put together under the *Dravidian Group*. At the top of this group and below the Talchir boulder-bed there is a well marked and universal unconformity. All the rocks from the Talchir boulder-beds upwards were placed under the *Aryan Group* which therefore includes everything from the Upper Carboniferous to the Pleistocene. Of these, only the term 'Purana' is sometimes used in Indian geological writings and the other two have not gained any currency.

The main divisions shown in the Table above have representatives in different areas, varying in facies and lithology. Besides deep-sea and coastal facies, we have also estuarine, fluviatile and continental facies of different ages. It is fairly easy to correlate the marine systems of the Extra-Peninsula with those of the coastal regions of the Peninsula because of the contained fauna. But in the case of the fluviatile and continental deposits, the faunas are of local distribution and have special characters for which there are not exact equivalents elsewhere. Their age can be settled with some confidence if they are in some way connected with marine beds or if the age of any similar formations in other areas has been worked out.

There are several stratigraphical problems in this sub-continent which await solution. The regional peculiarities of strata have necessitated the growth of a considerable number of local names which have only a limited application. Geological work was originally done in a series of detached areas which compelled the adoption of local nomenclature. As these areas were connected up by the mapping of the intervening tracts, some of the local formation names have become superfluous. But, in a large number of cases, even though the general equivalence or homotaxis could be recognised, the local nomenclature persists because of the lack of identity of characters.

By far the greater part of India has been mapped in a general way but there are still some blanks in the Assam-Burma border and in the Himalaya. These are gradually being filled up while several of the more important areas have been undergoing revision. Hence the stratigraphic information available on different parts of the country is of varying degrees of reliability, detail and precision.

A generalised picture of the geological succession in different areas is presented in Table 6 which may be useful for reference purposes. Further details about individual areas will be found in the relevant chapters in which each geological system is described in detail.

TABLE 6.—GENERAL GEOLOGICAL SUCCESSION IN DIFFERENT PARTS OF INDIA AND BURMA

(Sy. = System ; S. = Series ; B. = Beds ; Sst. : = Sandstone ; Lst = Limestone)

Standard scale	Northern Himalaya	Baluchistan arc	Salt Range & Potwar	Kashmir-Hazara	Simla-Garhwal	Assam	Burma	Peninsula	Coastal areas
RECENT	Alluvium	Sands, Loess	Alluvium, Loess	Alluvium &c.	Alluvium &c.	Alluvium, River gravels	Alluvium, River gravels	Alluvium Khadar	Alluvium sands
PLEISTOCENE	Older alluvium, gravels, moraines	Older alluvium, sands	Older alluvium, etc.	Older alluvium, moraines Upper Karewa	Older alluvium, moraines	Older alluvium, river terraces	Older alluvium, river terraces	Older alluvium Bhangar, cave deposits, river terraces	Raised beaches, Porbander stone etc.
PLIOCENE UPPER MIOCENE	...	Siwalik Sy. Manchhar Sy.	Siwalik Sy	Lower Karewa Siwalik Sy.	Siwalik Sy.	Dihing S.	Irrawaddy Sy.	...	Siwalik Sy. Karaikal B. Quilon B. Cuddalore Sst. etc.
MIDDLE MIOCENE LOWER MIOCENE	...	Mekran Sy. Flysch, Bugti B. Gaj S.	Murree S.	Murree, S. Fatehjang Zone	Kasauli B. Dagshai B.	Tipam S. Sarma S.	Upper Pegu S.	...	Gaj S.
OLIGOCENE	...	Nari S., jak shales	...	...	...	Barail S. (in part).	Lower Pegu S.	...	Nari S. Dwarka B.

REVIEW OF INDIAN STRATIGRAPHY

EOCENE	...	Eocene	Kirthar S. Laki S. Ranikot S.	Chharat S. Kirthar S. Laki S. Ranikot S.	Chharat S. Hill Lst. Ranikot S.	Subathu B.	Barail S. Jaintia S. Sylhet Lst. Disang S. Cherra Sst	Yaw and Pondaung Stages, Tabyin, Tilin, and Laungshe stages	...	Kirthar S. Laki S. Ranikot S.
UP. CRETACEOUS		Up. Cretaceous Chikkim S.	Cardita Beaumonti B., Pab. Sst.	Up. Cretaceous	Up. Cretaceous	...	Disang S. (in part) Up. Cretaceous	Axial S. (in part) Negrais S.	⎱ Deccan Trap and Intertrappeans, Lametas	Bagh B. Trichinopoly Cretaceous
L. CRETACEOUS		Giumal Sst.	Parh Lst.	L. Cretaceous	Giumal S. Orbitolina Lst.!	...	...	L. Cretaceous	⎰ ... Umia B.	Umia B.
JURASSIC	...	Spiti Shales Kioto Lst.	Massive Lst. Crinoidal Lst.	Jurassic	Spiti shale Jurassic	?Tal.S.	...	Namyau B. ?Loi-an	Jubbulpore Kota Rajmahal	Katrol Chari Patcham
TRIASSIC	...	Triassic	Triassic	Triassic	Triassic	?Shali Lst. ?Krol S.	...	Napeng B.	Maleri Pachmarhi Panchet	...
PERMIAN	...	Products B. Syringothyris Lst.	Permo-Carboniferous	Productus Lst. Speckled Sst.	Agglomeratic Slates, Zewan B.	!Infra-krol	...	IUp. Plateau Lst. Moulmein Lst.	Raniganj S. Barren Measures, Barakar S.	...
		...	...	Conularia and Eurydesma B. Boulder bed	Agglomeraticfish. Gangamopteris B. Tanakki Boulder B.	...	...	?	Umaria B.	...
UP-CARBONIFEROUS		...	...			?Blaini boulder Bed	Subansiri B.		Talchir boulder B.	

TABLE 6—Contd.

Standard scale	Northern Himalaya	Baluchistan arc	Salt Range & Potwar	Kashmir-Hazara	Simla-Garhwal	Assam	Burma	Peninsula
MIDDLE & L. CARBONIFEROUS	Po series Fenestella sh.	...	...	Fenestella sh.	...	...	L. Plateau Lst.	...
DEVONIAN	Muth Qtzt.	...	...	Muth Qtzt. Devonian (Chitral)	? Jaunsar S.	...	Padaukpin Lst. Wetwin sh.	...
SILURIAN	Silurian	...	...	Silurian	...	...	Zebingy S. Namshim S.	...
ORDOVICIAN	Ordovician	...	...	Ordovician	...	...	Naugkangyi S. Mawson S.	...
CAMBRIAN	Cambrian	...	Cambrian	Cambrian	? Deoban Lst.	...	Bawdin volcanics ? Mergui S.	Vindhyan Sy
UPPER PRE-CAMBRIAN	Haimanta	...	Attock Slates	Dogra Slates	Simla Slates	Buxa S. Daling S. Darjeeling S.	Chaung. Magyi S.	Cuddapah Sy. Delhi Sy.
L. PRE-CAMBRIAN AND ARCHAEAN	Vaikrita Sy.	...	?	Salkhala S. Gneisses	Chail S. Jotogh S.	Shillong S. Gneisses	Megogk S. Gneisses	Dharwar Sy. Aravalli Sy. etc. Gneisses

CHAPTER IV
THE ARCHAEAN GROUP—PENINSULA
INTRODUCTION

The term Archaean was introduced by J. D. Dana in 1872 (*Amer. Jour. Sci.* 3, p. 253) to designate the geological formations older than the Cambrian. It represents an enormous span of time and can be divided into two major groups, the Archaean and Proterozoic, the latter being applicable to those formations which contain remains of primitive organisms. But there is always much difficulty in identifying the very primitive organic structures because of the possibility of confusing them with inorganic ones. During the last two decades much work has been done through the use of radiometric techniques to determine the age of minerals and rocks. The data have revealed that the earth was formed about 4600 M.Y. ago and that the oldest rocks from different areas in the world have ages near about 3600 M.Y. These areas are occupied by the most ancient gneisses and schists and are referred to as SHIELDS as they have remained virtually undisturbed and unaffected by changes except for vertical movements. The well-known shield areas are Eastern Canada, the Guianas, Brazil, West Africa, Central and South Africa, Arabia, the Baltic region, Ukraine, parts of Siberia, Peninsular India, West Australia and East Antarctica.

The Archaean rocks were formed during the very early period when there was no life on the earth. They are mostly of igneous origin, comprising metamorphosed granitic and basaltic rocks together with a subordinate amount of sediments. They consist of greenstone, amphibolites, amphibole-schists, garnetiferous, micaceous and other schists, granodioritic gneisses and granites, etc. Because of the fact that these rocks form the basements of all other formations they are referred to a BASEMENT COMPLEX or FUNDAMENTAL GNEISS.

There is as yet no agreement amongst geologists about the duration of the Archaean era. In Canada, the beginning of the Proterozoic is placed at 2500 M.Y., while in the U.S.S.R. it is at 2000 M.Y. It is possible that life actually began on this planet 2500 M.Y. ago and that, after many vicissitudes, it gradually evolved from unicellular organisms like bacteria into multicellular ones like algae, fungi and Archaeocyathinae. Such organic structures have been identified in recent years in early Proterozoic rocks 1600 to 2000 M.Y. old, such as the Gunflint Formation of the Lake Superior region in N. America, from the Saksagam and Krivoi Rog belts of European Russia and doubtfully from the Bulawayan of Rhodesia. A long period of time elapsed before protective covers of chitinous, siliceous or calcareous nature, capable of being preserved, were evolved and the organisms could be identified by such shells or skeletons.

The almost sudden apperance of a variety of megascopic life with hard parts, at the base of the Cambrian System 600 M.Y. ago, indicates that conditions on the earth's surface had become favourable by then for such explosive development. The factors involved in the stabilisation of such conditions were the temperature on the earth's surface, the composition of the atmosphere, a suitable range of climate, protection from harmful radiations from space and formation of large basins of water which later became the oceans.

It is not yet possible in India to divide the various eras in the Pre-Cambrian into Archaean and Proterozoic units for lack of well established and acceptable age data. Only in the case of a few limited areas, sufficiently reliable, radiometric age data are available. For other areas the data are meagre and sporadic and require confirmation by different methods (U-Pb, Th-Pb, K-Ar and Rb-Sr.) and for different horizons. Some years must elapse before we have enough massive and confirmatory data on which to base a reliable classification. Until then, it would be best to continue the present practice of referring to the Pre-Delhi and Pre-Cuddapah formations as ARCHAEAN and to the others as PURANA or Proterozoic.

DISTRIBUTION OF THE ARCHAEAN

The Archaean formations in the Indian shield occupy most of southern and eastern India and parts of Assam, Bihar, Madhya Pradesh and Rajasthan. They occur also in the Sub-Himalayas and central Himalayas and in the Shan-Tenasserim belt of Burma. Only a few regions have been mapped in detail, so that the information on different areas is of very uneven character. Pakistan has been mapped on modern aerial maps during the last decade though the more inaccessible areas in the Himalayas, N.W.F.P. and Chagai will require further attention. The more inaccessible areas in the Himalyas and elsewhere in India are not likely to be mapped in detail in the very near future, not only because of lack of facilities, but also because the present emphasis is on the development of mineral resources.

MYSORE

This constitutes the type region of the Dharwar System studied by R. Bruce Foote in the eighteen-eighties. Since then a considerable amount of work has been done by the Geologists of the Mysore Geological Department. This region is occupied by gneisses and granites which are traversed by a number of bands of schistose rocks of the Dharwar System containing the oldest rocks so far known in South India. Amongst the dozen bands exposed, those of Shimoga and Chitaldrug are best developed.

Besides these there are several small strips scattered over this and neighbouring regions. These are all thought to be remnants of a great formation which formerly covered a large part of Southern India and which have escaped denudation because they form synclinal strips folded in with the gneisses. The larger ones are evidently closely folded synclinoria in which some members may be repeated by folding. For example, Bruce Foote noted about 36 beds across a section in the Sandur band which he believed to form a simple syncline with an overturned easterly limb; since the total thickness, on this interpretation, would amount to 9.6 km it is very likely that this is a synclinorium in which some part of the section is repeated by folding (Fermor, *Mem.* LXX, p. 62, 1936).

The Dharwarian rocks have a regional strike of N.N.W.-S.S.E. which becomes N.-S. in the southern part of Mysore and veers to a N.E.-S.W. direction near the southern border.

The Archaean succession of Mysore was described by W.F. Smeeth in 1915. This has since been revised by B. Rama Rao, whose latest ideas have been given in Bulletin 17 of the Mysore Geological Department and in his Handbook of the Geology of Mysore (1962). These two classifications are given in Table 7 for comparison.

In Smeeth's classification, the Dharwars, the oldest formations in Mysore, were held to be entirely of igneous origin and divisible into a lower *Hornblendic division* and an upper *Chloritic division*. This lithological classification has been found to be untenable for it depends more on the regional metamorphic grade than on the age and stratigraphical relationships of the constituent members. In the northern parts of the State, where the Dharwars form wide bands, the chloritic types predominate in the form of greenstones and chlorite-schists, while normal limestones, argillites and quartzites are also seen. In the central region, amphibolites begin to develop while the sedimentary types become schistose, recrystallised and silicified. The southern tract contains only comparatively small lenses and strips amidst the gneisses and they are conspicuously hornblendic, coarsely recrystallised, granulitic and fresh-looking because of the higher grade of metamorphism. Here the argillaceous sediments have been reconstituted into schists and granulites with garnet, cordierite, staurolite, kyanite, sillimanite, etc., while the ferruginous quartzites have become dominantly magnetitic. The Dharwarians in Mysore exhibit northerly plunging folds; the southern region exposes the deeper parts of the folds with their highly metamorphosed rocks while the northern region shows only the shallower parts of the folds with rocks of low grade metamorphism.

Formerly, all the Dharwarian types were regarded as of igneous origin; the conglomerates as autoclastic and derived from felsites and porphyrites; the haematite-quartzites as altered and silicified amphibole rocks, and so on. This view has changed in recent years and a considerable part of the Dharwarians is now recongnised as of sedimentary origin. The conglomerates and grits are found on closer examination to exhibit ripple-marks and current-bedding. The banded ferruginous rocks are recognised as original sediments with alternating bands of cherty or chalcedonic silica and iron-rich jasper or haematite, the nature of the iron oxide as well as the presence or absence of amphibole and garnet being determined by the degree of metamorphism. The limestones and calc-granulites, which were regarded as products of metasomatism of igneous rocks, are in part at least sedimentary. Some of the conglomerates which were first described as autoclastic *i.e*, as pseudo-conglomerates whose

TABLE 7.—THE ARCHAEAN SUCCESSION IN MYSORE

Smeeth (1915) Original formations	B. Rama Rao (1940) Original formations	Probable alterations
Pre-Cambrian — Basic dykes	Basic dykes, chiefly dolerites	
Eparchaean interval		
Felsite and porphyry dykes	Felsites and Porphyry dykes	
Closepet Granite (coarse pink or grey biotite-granite rarely slightly foliated)	Closepet granite.	slightly foliated
Charnockite massifs and later dykes	Recrystallisation and reconstitution of older rocks into complex types of the charnockite series	
Hornblende and pyroxene granulite dykes	Norite dykes	
	Hornblendic dykes	
Peninsular Gneiss (Biotite-granite and gneiss with inclusions of schists)	Peninsular Gneiss : Complex granite gneisses.	Slightly crushed and graulitic
Champion Gneiss (crushed granitic gneiss with zones of autoclastic conglomerate)		
Eruptive unconformity	*Eruptive unconformity*	
Dharwar system — Upper Dharwars (Chloritic Division), Chloritic schists and greenstones, mica-schists, conglomerates, quartzites, crystalline limestones and banded ferruginous quartzites. Also schists with kyanite, staurolite, etc.	**Dharwar system — Upper:** Some cherty and ferruginous silts, clays, calcareous silts and clays, impure quartzites and conglomerates forming in part the G.R. formation (Local).	Somewhat altered but easily recognisable.
	Middle: Granite porphyry and granitic rocks, fine and coarse.	Micaceous granitic gneisses and crushed and foliated gneissic granites.
	Basic and ultrabasic intrusives, Ironstones, limestones, argillites, quartzites and conglomerates; also ashes, tuffs and other volcanic products.	Banded ironstones with amphibole, etc., granular crystalline limestones, micaceous gneisses with cordierite, sillimanite, etc schistose conglomerates — all highly crushed and crystalline.
Lower Dharwars (Hornblendic division) schistose hornblendic rocks with subordinate magnetite — and haematite-quartzites, some calc-granulites etc.	**Lower:** Rhyolites, felsites and quartz-porphyry and other acid volcanics with opalescent quartz.	Quartz-schists, micaceous quartz-schists and gneisses with opalescent quartz – highly crushed.
	Basic volcanic dykes and flows.	Greenstones, hornblende-schists, etc.
	Original basement not recognized	

structures were produced by shearing and rolling, are now regarded as sedimentary in origin. There are, of course, also metamorphosed igneous rocks such as amphibolites, amphibole-schists, talc-tremolite-schists, serpentinites, sheared porphyries, felsites, etc. Amongst the mixed types are various gneisses containing biotite, hornblende and garnet, pyroxene-granulites and calc-granulites.

Rama Rao has divided the Mysore occurrences into five geographical groups from west to east. The westernmost contains mainly hornblendic schists and thin bands of haematite-quartzites. The west-central group, comprising the Shimoga and Bababudan belts, shows a fairly full succession, banded ferruginous rocks and manganiferous rocks occurring in force in the Bababudan hills. The central group, comprising the Chitaldrug, Chiknayakanhalli and Nagamangala belts, shows much igneous material and also banded ferruginous rocks and limestones. The east-central group includes various small occurrences in which both regional and thermal metamorphic effects are discernible which have produced several interesting rock types —quartz-magnetite-granulites, garnetiferous quartzites, garnet-quartz-pyroxene-granulites, sillimanite-cordierite-gneisses, sillimanite-quartzites, cordierite-hypersthene-gneisses, cordierite-mica-gneisses, cummingtonite-schists, pyroxene-gneisses, etc. The easternmost is the Kolar schist-belt which is of great economic importance because of the workable gold-bearing quartz lodes in it. It is 65 km long and 6 km broad at its widest and is composed of hornblendic rocks believed to be of igneous origin, with a band of autoclastic conglomerate at its eastern border.

According to Rama Rao, the Shimoga schist-belt in the west-central area exposes one of the best developed sections of Dharwars. Typical current-bedding and ripple marks, indicating undoubted features of sedimentary origin, have been found at several places in this belt in recent years. The oldest rocks are basic volcanics overlain by rhyolitic flows and tuffs and intruded by sills of felsite and porphyry. With these are intercalated bands of chert and halleflinta. They are succeeded by thick beds of conglomerate containing pebbles of felstite, quartz-porphyry and quartzite. Above these are micaceous quartzites showing current-bedding in the upper layers. These are the earliest undoubted sediments in the Dharwars. The quartzites are succeeded by slaty schists, limestones and banded ferruginous quartzites, intruded by bosses of granite-porphyry. This series is followed by a bed of conglomerate indicating a period of uplift and denudation. Further sedimentation laid down a series of silts and ferruginous quartzites above these rocks. The succession in the Shimoga belt as worked out from several sections, is shown in Table 8.

It will be seen that the Middle Dharwars contain conglomerates, limestones and dolomites and banded ferruginous rocks. Several of the conglomerates which were originally thought to be sheared igneous rocks like porphyries, have since been recognised as sedimentary conglomerates. They are seen in several exposures in the northeastern parts of the State. Many of the limestones are really magnesian or dolomitic to a varying extent and these are particularly abundant in the northern part of the State. The banded ferruginous quartzites occur both in the north and in the south. In the north they are of the nature of hematite-quartzites which have given rise to bodies of rich hematite ore like those of the Bababudan hills which are now being worked. In the southern parts of Mysore they are composed of magnetite and quartz and sometimes also amphibole (cummingtonite and bababudanite) and garnet, as a result of the metamorphism.

TABLE 8—DHARWARIAN SUCCESSION IN THE SHIMOGA BELT

After B. Rama Rao, *Mysore Geol. Dept. Bull.* 17, p. 36).

Upper Dharwars (Sulekere series)	(e)	Ferruginous quartzites and cherty ferruginous slates with thin intercalations of argillitic layers and probably of ash beds. (Rain-prints and sun-cracks in some sections).
	(d)	Friable ferruginous silts and micaceous ferruginous grits intercalated with thin bands of limestone towards the top. Basic hornblendic sills.
	(c)	Argillitic and calcareous silts and fine grained quartzites with minute grains of opalescent quartz.
	(b)	Quartzites.
	(a)	Jandimatti and Kaldurga conglomerates containing pebbles of granitic rocks, ferruginous quartzites, schists, etc.

Granite-porphyry masses of Rangadurga, Balekal and probably granites of Honnali, Shimoga and adjacent parts.

Middle Dharwars (Hosur series)	(e)	Band haematite-quartzites (Chandigudda outcrops.)
	(d)	Limestones, dolomites and siliceous limestones.
	(c)	Phyllitic and chloritic schists, grey or greenish.
	(c)	Phyllitic and chloritic schists, grey or greenish.
	(b)	Sericitic grits and quartzites with coarse grains of opalescent quartz.
	(a)	Conglomerates (showing pebbles of "quartzites" and quartz-porphyries), felspathic grits and greywackes.

Lower Dharwars (Igneous Complex)	(c)	Sills of quartz-porphyry, felsite and other types of acid intrusives and their schistose phases.
	(b)	Acid and intermediate flows — rhyolites, keratophyres etc., with intercalated tuffs and ash-beds now seen e dark grey or bluish argillitic layers and beds, altered in places into compact hornstones in contact with (e
	(d)	Compact greenstone and greenstone-schists, micaceou or calciferous chloriteschists, etc. (Basic and Intermediat lava flows probably with admixed ash beds).

It is not known definitely whether there are any Pre-Dharwarian formations in the region. P.Lake (1890) in his paper on the geology of Malabar, has noted the presence of gneisses with E.-W. fold axes in south Malabar which are worth investigating. The oldest conglomerates in the Dharwarian, assigned to the base of the Middle Dharwarian, contain pebbles of granite, gneiss, pegmatite, etc. which appear to have been derived from pre-Dharwarian rocks as such rocks are not present in the Lower Dharwarian. A group of rocks composed of manganiferous calc-rocks, calc-gneisses, sillimanite-quartzites and granulites (the SAKARSANITE SERIES of Rama Rao) occurring at Sakarsanhalli, S.S.W. of Bisanattam in the Kolar District, may possibly be pre-Dharwar. Detailed field examination and radiometric age determination are necessary for deciding whether formations older than the Dharwars are present in the region.

Champion Gneiss.—Presumably younger than the Dharwars is the Champion gneiss which is a sheared, grey, micaceous gneiss whose type area is the eastern edge of the Kolar schist belt. It contains blebs of opalescent, grey-coloured quartz. Some other acid igneous rocks like keratophyres, rhyolites, quartz-porphyries and some granites, which also contain the same opalescent quartz, are included with the Champion gneiss group, but it is not always easy to distinguish some of them from the types belonging to the Peninsular gneiss group. According to Rama Rao it is desirable to restrict the use of the name Champion gneiss to the stocks and bosses of granite and granite-porphyry which are older than the Peninsular gneiss.

Peninsular Gneiss.—Gneissic rocks belonging to this are the most widespread group of rocks in Mysore and in many parts of Southern India. They consist of a very heterogeneous mixture of different types of granites intrusive into the schistose rocks after the latter were folded, crumpled and metamorphosed. They include granites, granodiorites, gneissic granites and banded or composite gneisses, the granitic constituents of which show distinct signs of intrusion. The banded gneisses consist of white bands of quartz-feldspar alternating with dark bands containing hornblende, biotite and minor accessories. The gneissic types are due to granitisation of older schistose rocks and show streaky and contorted bands some of which are granitoid to porphyritic and others granulitic. The granitic group ranges in composition from granite, through granodiorite to adamellite, augite-diorite, monzonite, etc, and contains inclusions of hornblendic rocks. To what extent they represent intrusives of different ages is yet to be determined, but their very complex nature is unquestionable since they include composite gneisses, migmatites, granitised older crystalline rocks and true granites with their aplitic and quartz vein systems.

Charnockites.—The next important group of rocks in Peninsular India is the Charnockite Series. The rocks of this series were described by earlier workers as Nilgiri gneiss as they formed the principal rock type in the Nilgiri massif. The occurrence of these rocks with a granular texture and the invariable presence of orthopyroxene in them was first recognised by Judd (1885) who described a collection of rocks from the Nilgiri hills. Sir Thomas Holland collected and described similar rocks from the Salem district in the year 1891, which was reported in the tri-monthly notes by the Director, Geological Survey of India in 1892. It was noticed about this time that the tombstone of Job Charnock, the founder of the city of Calcutta, at St. John's Churchyard at Calcutta was made of hypersthene-granite composed of blue quartz and feldspars besides hypersthene. In view of its distinct mineralogical and petrographic character, Holland named it *Charnockite* in honour of Job Charnock. In 1900, Holland published his famous memoir on the Charnockite Series, the term being used for designating a whole group of rocks varying in composition from acid to ultramafic, characterised by the invariable presence of orthopyroxene and by a general megascopic similarity in colour and texture. In this Memoir he had described rocks from the Magazine hill at St. Thomas Mount, and the adjacent hills around Pallavaram near the city of Madras, as also similar rocks from the southern districts of Madras including those of the Nilgiri and Shevaroy massifs. For convenience of description Holland classified the rocks of his charnockite series into three divisions; an acid division represented by charnockite with a sp. gr. of 2.67 and silica percentage of about 75; an intermediate division comprising rocks characterised by the presence of all the minerals of the series and having a sp. gr. of 2.77 and a silica percentage of about 54; a basic division of norites with a sp. gr. of 3.03; and an ultrabasic division having a sp. gr. of 3.73 and silica percentage of 47-50.

As Holland found all the above types in close association in one area he regarded them as a genetically related suite of rocks belonging to a petrographic province and gave the name Charnockite Series to them. Since the publication of Holland's Memoir, rocks with charnockitic affinities have been described from other places in India and from many parts of the world —e.g. Burma; Western Australia; Kenya, Uganda, Tanganyika, Madagascar, Central and South Africa, Adelie, Enderby and Queen Mary Lands in Antarctica; Bahia in Brazil; and also in Finland, Scandinavia, Scotland, Greenland and New York State. Some of these occurrences from outside India have been investigated in much greater detail than those of the type area. Excellent summaries of the present status of our knowledge on charnockitic rocks have been presented by Quensel (1951), Pichamuthu (1953), Wilson (1955) and Howie (1964).

The type area of charnockites near Madras city has been the subject of detailed field and laboratory studies by Subramaniam and Howie, who have published some of the results of their investigations. Howie (1954) has contributed a wealth of geochemical data on the rocks and minerals of the type area near Madras based on which he considers the charnockites to represent a plutonic igneous rock series which has undergone recrystallization in the solid state as a result of plutonic metamorphism. Subramaniam (1959) on the basis of his field and laboratory studies has redefined the term 'Charnockite' and 'Charnockite Series'. The *acid division* of the charnockite series, composed of alaskites, birkremites, enderbites and hypersthene-quartz-syenites, is considered an igneous suite which has undergone metamorphic reconstitution and recrytallisation with concomitant changes in mineralogy, such as unmixing of perthites and formation of garnet. In fact, re-examination of Charnock's tombstone has revealed the presence of garnet. The rocks of the *basic division* of Holland are essentially pyroxene-granulites and variants which have no genetic relationship to the charnockites *sensu stricto*. Sporadic exposures of norite with pyroxenite layers and lenses are thought to be syntectonic lenses), unrelated to the charnockite suite. The rocks of the *intermediate division* of Holland grade from homogeneous hypersthene-diorites to pyroxene granulites and granulitic migmatites, and are considered hybrids resulting from partial assimilation and incorporation of pyroxene-granulite by the charnockite magma. Garnetiferous sillimanite-gneisses (khondalites) are also developed in force in the area; Holland's *leptynites* are inferred to be a thoroughly reconsitituted and recrystalised facies of khondalite. It would be well to recall here that the term leptynite was introduced by Hauy in 1822 for highly acid granulites which he defined as 'quartz-feldspathic metamorphic rocks poor in mica but often rich in garnet'. Leptynite is akin to very acid khondalite and may sometimes be associated with bands of migmatite. This should be taken note of, as it is sometimes used loosely and incorrectly in India, although scarcely used nowadays outside India. The association of charnockites, leptynites, khondalites and other granulitic rocks is prominently noticed in the Eastern Ghats proper and in the charnokite areas on Nilgiri, Anaimalai, Shevaroy, Palni and Travancore.

The above suggestions appear to fit in with the geological setting of charnockite rocks in India and elsewhere. The chemical composition of charnockites and associated rocks of the type area near Madras and some of the minerals in them are presented in the accompanying tables 9 and 10.

The charnockites show a combination of characters of igneous and metamorphic rocks. They form bosses and lenses, along the margins of which crush-zones are sometimes seen. They send out apophyses and veins into the surrounding rocks. Dark *schlieren* or segregations are found as inclusions; these may consist of augite, hypersthene and biotite or corundum and sillimanite. Tongues of charnockite cutting across the foliation of the country rocks are found in Salem and North Arcot. Phenomena of partial assimilation and hybridism are noted near Kondapalle (Krishna District), Polavaram (West Godavari) and other places in the Eastern Ghats. Banding is fairly common (*e.g.*, in the Shevaroys, Nilgiris and in parts of the Western and Eastern Ghats), but the bands are neither so conspicuous as in gneisses nor persistent, for they can be followed only for a few yards and are generally distinct only on weathered surfaces. The bands consist of light feldspathic and dark hornblendic or pyroxenic materials. Rarely are lit-par-lit injections of charnockite into other rocks seen. Contact phenomena and fine grained selvages along margins are not common, but are occasionally seen. Bands of mylonite have been noted along the margins of lenses in Salem and Mysore (these were originally called trap-shotten gneiss). The banding or crude foliation which is generally parallel to the regional structure seems to have been imparted during emplacement and there are no marked evidences of post-consolidation folding or compression.

TABLE 9.—CHEMICAL COMPOSITION OF CHARNOCKITES AND ASSOCIATED ROCKS
(DATA FROM HOWIE (1955) HOWIE AND SUBRAMANIAM (1957) AND SUBRAMANIAM (1959))

	1	2	3	4	5	6	7	8	9	10	11	12
SiO_2	77.32	71.97	64.73	64.24	65.95	51.55	50.73	47.71	46.30	48.50	75.88	77.93
TiO_2	0.56	0.40	0.40	1.51	0.99	1.12	1.38	0.97	0.83	0.12	0.58	0.31
Al_2O_3	9.81	13.30	15.16	13.57	15.27	13.86	13.33	15.57	16.24	4.87	11.03	10.65
Fe O	0.41	1.29	1.36	0.97	1.12	1.51	4.99	2.31	2.54	3.49	2.46	0.99
FeO	3.33	2.30	2.99	7.10	4.86	11.63	11.80	10.85	9.42	12.08	6.27	2.50
MnO	0.03	0.04	0.12	0.24	0.22	0.23	0.44	0.38	0.33	0.22	0.11	0.04
MgO	1.21	0.58	2.14	1.36	2.70	5.80	4.50	7.24	7.83	19.98	1.51	0.18
CaO	1.14	2.10	1.92	5.16	2.95	10.28	8.79	11.52	12.24	9.32	0.18	0.40
Na_2O	1.94	2.78	3.67	4.00	3.74	3.08	2.94	2.07	2.50	0.17	0.59	2.19
K_2O	3.87	4.24	6.97	1.27	2.06	0.54	0.51	0.42	0.15	0.01	0.11	4.54
H_2O+	0.25	...	0.58	0.43	0.25	0.25	0.72	1.11	1.26	0.16	0.85	0.08
H_2O-	0.07	0.27	0.30	0.13	0.20	0.12	0.15	0.07	0.07	0.10	0.17	0.16
P_2O_5	0.11	0.14	0.19	0.17	Tr	0.23	0.17	0.12	0.10	0.06	0.03	Tr
S	...	0.05 0.03	...	...	...	...	...	...	...	...	...	...
Totals	100.05	99.43	100.53	100.15	100.31	100.20	100.45	100.34	99.81	100.17	99.77	99.77
Density gm/cm^3	2.70	2.72	2.58	2.83	2.84	3.10	3.10	3.10	3.02	3.42	2.87	2.67

C.I.P.W. Norms

	1	2	3	4	5	6	7	8	9	10	11	12
Quartz	45.06	33.60	8.82	20.64	22.92	...	4.14	...	...	...	65.28	45.48
Corundum	0.55	0.61	...	...	1.53	...	...	...	...	...	9.38	1.43
Nepheline	...	...	...	...	...	...	...	...	0.85	...	...	...
Albite	16.40	23.58	31.44	34.06	31.44	26.00	24.63	17.82	19.39	1.41	5.24	18.34
Anorthite	4.92	9.45	4.17	15.01	14.73	22.41	21.68	31.97	33.08	12.19	1.11	1.95
Orthoclase	22.90	25.02	41.14	7.78	12.23	3.02	2.78	2.22	0.56	0.61	0.56	26.69
Diopside	...	...	3.37	8.11	...	22.72	17.45	21.00	21.80	26.97	...	...
Hypersthene	7.90	3.91	7.60	9.40	13.53	17.94	18.52	8.89	...	43.80	12.38	4.28
Olivine	...	...	...	...	...	2.92	...	12.89	17.13	7.90	...	...
Magnetite	0.46	1.86	2.09	1.39	1.62	2.09	7.19	3.25	3.71	5.10	3.77	1.39
Apatite	0.27	0.34	0.44	0.40	...	0.54	0.40	0.29	0.34	0.13	0.07	...
Pyrite	...	0.12	...	...	...	...	...	...	...	...	...	...

(1) Charnockite 4639, Pallavaram
(2) Charnockite 6436, Tirusulam hill, Minambakkam
(3) Charnockite (Hypersthene Quartz Syenite) Ch 112, West of Hasanapuram south of Pallavaram
(4) Enderbite Ch 29, northern slopes of Vandalur hill .563
(5) Garnetiferous enderbite Ch 113, Hasanapuram quarry
(6) Hypersthene diorite 4642A, Pallavaram
(7) Hypersthene diorite (hybrid rock) Ch 207, Ridge west of Muvarasampattu, near Pallavaram
(8) Basic granulite (garnetiferous) Ch 199, summit of Paravatta hill south of Mosque hill, Pallavaram
(9) Norite Ch 171, south eastern flank of .265, Pammal hill
(10) Pyroxenite 4645, Pammal hill
(11) Khondalite Ch 119, Pachchai Mala hill .360 Tambaram
(12) Garnetiferous leptynite 3708, Pallavaram

Table 10—Chemical composition of minerals from charnockite and associated rocks of type area near Madras city (Data from Howie (1955), and Howie and Subramaniam (1957)

	(1)	(2)	(3)	(4)	(5)	(6)	(7)	(8)
SiO_2	65.53	60.07	47.98	46.17	50.09	38.02	37.65	37.20
TiO_2	0.02	0.16	0.94	1.78	0.30	0.03	0.04	0.07
Al_2O_3	18.90	24.84	3.39	3.22	2.86	21.02	20.87	21.75
Fe_2O_3	0.13	0.35	1.54	1.16	1.32	1.98	1.87	0.71
FeO	0.09	0.21	32.16	33.17	14.09	28.12	27.65	33.30
MnO	Tr	Tr	0.26	0.75	0.24	0.64	1.10	0.74
MgO	0.33	0.02	13.16	12.24	10.51	7.87	3.74	4.82
CaO	0.54	6.65	0.03	0.04	20.11	2.25	7.16	1.58
Na_2O	2.96	7.54	0.06	0.52	0.39	0.11	0.09	n.d.
K_2O	11.59	0.34	0.00	0.34	0.18	0.01	0.02	n.d.
H_2O+	0.03	0.04	0.31	n.d.	0.01	nd.	n.d.	n.d.
H_2O-	0.14	0.05	0.42	0.31	0.09	0.09	0.02	0.09
Total	100.06	100.27	100.20	99.70	100.19	100.14	100.21	100.26

(1) Microcline perthite from charnockite 4639, Pallavaram
(2) Plagiclase from charnockite 6446, Minambakkam
(3) Orthopyroxene (Ferrohypersthene) from charnockite 4639, Pallavaram
(4) Orthopyroxene (Ferrohypersthene) from Charnockite 6436, Minambakkam
(5) Clinopyroxene from hypersthene diorite 4642A
(6) Garnet from Enderbite Ch 113
(7) Garnet from Basic granulite Ch 199
(8) Garnet from garnetiferous leptynite 3708

Subramaniam has stated that the rocks of the type area near Madras city have a N.N.E.-S.S.W. strike of foliation occasionally veering to N.E.-S.W, with steep easterly dips varying from 60° to 80°. The charnockites occasionally display a distinct linear element due to the stretching of the quartz grains and to a lesser extent, due to the dimensional orientation of the mafic minerals. This lineation is parallel or sub-parallel to the foliation, occasionally plunging at low angles to the N.E. The lineation in the charnockites is conformable to the lineation in the pyroxene-granulites and garnetiferous-sillimanite gneisses with which they are closely associated in the field, and is inferred to indicate the general trend of the regional fold axes. The horizontal lineation and the generally uniform easterly dip of all the rock units are interpreted to indicate an isoclinally folded group, overturned to the west. It is considered by Subramaniam that the rocks of the type area were tightly folded isoclinally, after the emplacement of the charnockite suite as interstratified sheets in a basement of khondalites and pyroxene granulites. Howie now fully agrees with Subramaniam after his visit to the Madras area in 1964.

The rocks were studied first by Sir T.H. Holland (1900) and described in a monograph. He thought that they represent an igneous suite as they possess a complete range in composition from acid to ultrabasic types and show the phenomena of intrusion, segregation and assimilation. Several years later, the study of similar rocks from the Adelie Land in Antarctica led F. L. Stillwell to describe them as products of plutonic metamorphism, though exhibiting phenomena characteristic of igneous masses. Vredenburg (1918) believed that they represented the metamorphosed members of the Dharwarian system. A.W. Groves (1935) from a study of the charnockites of Uganda advocated the view that they were more or less normal igneous rocks subjected to metamorphism at great depths and characterised by 'dry' minerals with a low proportion of hydroxyl molecules, absence of carbon dioxide, abundant myrmekitisation and containing feldspars with exsolved perthite or anti-perthite, etc. C.E. Tilley (1936) described a granodioritic rock with quartz, acid plagioclase (antiperthitic), hypersthene, magnetite and zircon, which he named *Enderbite*. This is therefore an acid charnockite containing high silica, low potash and high lime, with the soda-lime feldspars dominating over the potash feldspars.

B. Rama Rao (1945), from his extensive acquaintance of the charnockites of Mysore, came to the conclusion that they were formed by different ways from a variety of pre-existing rocks: these included recrystallised siliceous, argillaceous and ferruginous sediments which would produce hypersthene-granulites, norites etc; rocks which would give rise to pyroxenites, rocks rich in magnesia and iron which would be converted into intermediate or acid charnockite by the action of acid magmas. He found no clear evidence of intrusive phenomena and believed that much of the charnockite was older than the Peninsular gneiss. It may be added here that the descriptions and discussions of Rama Rao refer not to typical charnockites but to charnockitic rocks and granulites included in the Charnockite Series.

In his study of similar rocks from West Greenland, H Ramberg (1951) found enderbitic gneisses in association with khondalite (garnet-sillimanite-graphite gneisses) and kinzigite (cordierite-sillimanite-garnet gneisses). He came to the conclusion that enderbitic gneiss was formed at low levels in the crust, the conditions there favouring the upward migration of hydroxyl molecules and elements of low density and large atomic diameter (*e.g.* potassium). Titanium and iron would tend to be liberated from the silicate lattices and form rutile and ilmenite. The minerals stable under those conditions are hypersthene, garnet and sillimanite which would develop at the expense of hornblende, biotite, muscovite, epidote, etc. At higher levels, nearer the surface, granitic and granodioritic rocks would develop and also high-alumina rocks like khondalites.

F. J. Turner (1948), in his treatise on metamorphic rocks, shows that under the conditions of formation of the granulite facies, the acid charnockite would be produced from acid magmas and basic charnockite from rocks of noritic and similar composition. Hornblende in these rocks may be in equilibrium or may be formed from diopside; it is generally rich in alumina and poor in hydroxyl. The garnet of this facies is pyrope-almandite. Feldspars hold much Na or K in solution which are later exsolved and appear as microperthite or antiperthite as the case may be. Corundum, spinel and olivine appear in silica-poor rocks.

It will therefore be seen that rocks with the characters of charnockites may be formed under deep-seated (kata-zone) conditions and may behave as igneous rocks. They were intruded into the country rocks at considerable depths so that the contact phenomena are not as prominent as in the case of intrusions of hot magmas at shallow depths. In course of time some mineralogical changes have occurred, such as the formation of amphibole from pyroxene, appearance of coronas of garnet around hypersthene, and separation of potash from the soda-lime feldspar and soda from the potash feldspar. Their presence at the surface as prominent hill masses may be due to uplift in post-Archaean times, which may to some extent explain the general freshness of their appearance and the comparatively light weathering they exhibit.

Closepet Granite.—This name is applied to a granite suite much younger than the Peninsular Gneiss. These rocks form a N.-S. band (N.N.W.-S.S.E. in the north) right through Mysore, with a width of about 15-25 km. The outcrops are marked by tors and rounded hummocky blocks. B. P. Radhakrishna has shown that the granites are of metasomatic origin, except for a few small patches along the centre which may have crystallised from a melt. Gneissic structure is seen almost everywhere and the inclusions and marginal rocks are related to the country rocks on either side. The common rocks are grey and pink porphyritic gneisses with large feldspars. The feldspars have gradually been changed from oligoclase to microcline while the petrographic types show variation from trondjhemite through granodiorite to granite. Myrmekite is sometimes seen, while fine-grained aplite is common.

The granite band cuts through the regional metamorphic grades which vary from granulitic facies in the south, through amphibolites in the middle green-schists in the north. It has been emplaced after the period of generaly metamorphism and regional tilting which has brought up charnockites and granulitic rocks to the surface in the south.

There are a few small linear patches in the Peninsular Gneiss which should be classed with the newer granites, while those in the schist country are probably to be assigned to the Peninsular Gneiss suite which may itself contain granites of different ages.

ANDHRA

The Dharwars are well displayed in the south-western parts of the former Hyderabad State which have now been politically merged into Mysore. A few outlying bands are seen farther east in Karimnagar and Waranal districts. They consist, as in Mysore, of hornblende- talc-, chlorite- and mica-schists, quartzites, ferruginous quartzites, etc., having the same N.N.W.-S.S.E.) direction of strike. The rest of the country is occupied by gneisses of which there are two types, *Grey gneiss* and *Pink gneiss*. The Gray gneiss, which corresponds to the Peninsular gneiss, is conspicuously banded, the light bands being rich in quartz and feldspar and the dark bands in mica and hornblende. The Pink gneiss is granitoid, though occasionally gneissic, and intrusive into the schists and into the Grey gneiss, and therefore similar to the Bellary gneiss. It consists of quartz, microcline, orthoclase, acid plagioclase, some hornblende, mica and epidote. Phases of the Pink gneiss are red syenitic rocks and porphyritic pink granites. They are cut up by later elsite and porphyry dykes. Gold-bearing quartz veins occur in shear zones in the schistose Dharwars and near the margin of the Dharwars and gneisses.

In the Khammamet district there are some nepheline-corundum and zircon-syenites which are intrusive into the Dharwarian schists and Khondates. The zircon crystals are short and stumpy, measuring upto $1\frac{1}{4}$ cm. across.

W. King distinguished four types of gneisses in the Nellore region, two being schistose and two massive. The schistose gneisses include quartz-, mica-, hornblende- and talc-schists, and quartz-magnetite rocks. The quartz-magnetite rocks are banded and form several hills in the Guntur district near Ongole and in the valley of the Gundlakamma river. The massive gneisses include a grey, sometimes porphyritic gneiss and a red, granitoid gneiss. The grey gneiss is banded and very variable in composition and contains streaks and bands of micaceous gneiss and charnockites; it is called by King the *Carnatic gneiss*, and belongs undoubtedly to the Peninsular gneiss group. The red gneiss is mainly granitic and younger. The granitic gneisses and schists are intruded by pegmatites and quartz veins. In northern Nellore and adjoining parts of Guntur, the granite is in places quite rich in fluorine-bearing minerals, such as fluorite, apatite and topaz. The north-western portion of Nellore is practically devoid of pegmatite intrusions but the south-eastern portion (Gudur and Rapur area) is rich in them including numerous lenses and veins containing workable deposits of muscovite mica. The schistose rocks and the gneisses have here roughly the same strike—(N.N.W varying to N.N.E.) as in Hyderabad and Mysore. In the western part of this district are quartzites and mica schists associated with sills (and flows) of basic volcanics, now converted int epidiorites and amphibolites. They are called *Kandra volcanics* and belon to the earlier schistose group, as they have also been involved in post-Dharwaria diastrophism and granitisation.

MADRAS

Only some parts of the State have been mapped in recent years so that complete picture is yet to emerge. There appear to be three or four majo bands of charnockites with intervening schistose rocks and Peninsular Gneisse the whole assemblage having been thrown into folds more than once. Th fold axes run in a N.N.W. direction ou the western side and in a N.N.E direction on the eastern side, with a general plunge to the north. The relativ ages of the rock groups and their mutual structural relationships will becom clear when the geological mapping, now in progress, is completed.

In the districts of Salem and Coimbatore there are a few bands of magnetite quartzite, derived from banded hematite-quartzite. There are also severa schistose bands widely distributed. Older grey gneisses and younger pin gneisses are found in Arcot, Tiruchirapalli and Madurai. Important band of crystalline limestone occur amidst the gneisses at many places. Some gra phite deposits are known in Travancore and in Tirunelveli district. Th assemblage, together with the monazite-bearing younger granites (one abou 500 M.Y. old and two others probably 700 and 1000 M.Y. old) emphasize the similarity of the Travancore-Tirunelveli region to Ceylon and Madagasca which are believed to have been parts of one land mass in Pre-Cretaceous time

The areas described above contain a few interesting suites of rocks whic are mentioned below:—

Magnesia-rich rocks, Salem.—Olivine-rocks, in which magnesite has bee developed, are found in several places in Salem and Southern Mysore, th best known being the "Chalk Hills" between the foot of the Shevaroys an Salem town. The ultrabasic rocks are highly altered and replaced by abundar veins of magnesite and sometimes by steatite and asbestos. A few pockets an segregations of chromite have also been found in them. The chromite deposit of Southern Mysore also occur in similar rocks. At Neyyoor in South Travan core as well as near Punalur on the Madras-Travancore border, there ar magnesia-rich pyroxenites which have given rise to books of phlogopite mic especially near the contact with pegmatites. The phlogopite has been worke intermittently.

In the district of North Arcot, an occurrence of ultramafic igneous rock has been found near Ten Mudiyanur (12° 7′ : 78° 57′) about 20 km. southwes of Tiruvannamalai. The regional trend of the charnockites which are th country rocks here is N.E.-S.W., dipping to the S.W. or nearly vertical. Th ultramafics comprising serpentinised dunites and peridotites with veins of magnesite, are flanked on the one side by pyroxenites and on the other by corundum-feldspar rocks associated with granite and pegmatite. These rocks appear to be intrusive into charnockites according to V. Gopal (private communication).

Anorthosites, Salem.—A layered sequence of meta-anorthositic gneisses and meta-gabbros, with the former containing layers of chromitite and perknite, is exposed as a well defined band extending W.S.W-E.N.E. from near Pattalur on the Cauvery River, through Sittampundi (11° 14′: 77°54′) to Suryapatti and beyond, for a distance of over 32 km. The entire sequence is arcuate and convex to the south. Of special interest the very calcic plagioclase (An 80-100) in anorthosites, the mineral in some types approaching pure anorthite in composition. It is of interest to note that this anorthite was described by Count de Bournon in 1802 and named INDIANITE by him. Based on comprehensive petrographic, mineralogical and chemical studies, this sequence of rocks has been interpreted by A.P. Subramaniam (1956) to be a metamorphosed gravity-stratified layered complex. Deformation and mineral reconstitution are inferred to have taken place during two periods of Archaean orogeny following primary crystallisation from a basic magma, with consequent changes in mineralogy. The development of new minerals such as hornblende, anthophyllite, pyralmandite, epidote-clinozoisite, grossularite, porphyroblastic corundum (in part with calcite rims) and rutile are attributed to metamorphic reconstitution and recrystallisation. The corundum in the meta-anorthosites appears to have resulted by the breakdown of anorthite during mineral reconstitution, when other minerals such as clinozoisite, lime garnet, scapolite, etc. were also formed. The chromitite layers carry aluminian chromite in varying proportions, besides green amphiboles, spinels, red corundum and rutile and are sufficiently extensive to be worked for chromite. This layered sequence is called SITTAMPUNDI COMPLEX after the village near its type exposure. This is one of the very few known occurrences of highly metamorphosed stratiform anorthosites with such mineral associations, but it is not unlikely that complexes with similar mineralogy may be discovered in other Archaean terrains of the world.

In the northern part of the Salem district there are bands of syenites and syenite-pegmatites containing corundum and hercynite crystals which are found strewn over the weathered outcrops. They occur as lenses in the granite gneisses which include also biotite-hypersthene granulitic gneisses.

Anorthosite, Tiruchirapalli District.—An area composed of quartzites amphibolites and crystalline limestones occurs in and around Kadavur (10°35′: 8°10′) in this district. In this area is found a circular group of ridges around a central basin. The country rocks dip steeply away from this basin. The peripheral portions of the circular ridges are occupied by well foliated mafic rocks which gradually become massive, structure-less and felsic in the centre. The outer gabbroic rocks show inward dips of foliation, varying from 90° to 30° or less as one proceeds towards the centre. The outer margins consist of chilled granular facies of gabbro which give place successively to gabbroic anorthosite, mottled anorthosite and anorthosite without dark minerals, when going towards the centre. According to A.P. Subramaniam, this mass appears to be a funnel-shaped intrusion which has been tilted towards the S.W. The principal minerals of the rocks are medium plagioclase, green hornblende clino-pyroxene and ortho-pyroxene, while apatite, scapolite, ilmenite and magnetite are accessories, the last two being often abundant in the peripheral zones. The feldspars show antiperthitic textures with rods and spindles of potash feldspar separated out. They contain about 45 per cent anorthite near the borders but are more calcic (60 per cent An.) in the centre. The hornblende is secondary while the ortho-pyroxene varies from bronzite to ferro hypersthene, exhibiting lamellar structure of the Bushveld type.

Alkali Rocks of Sivamalai.—At and around Sivamalai (11° 2′ : 77° 33′) are exposed a group of alkali rocks in about half a dozen hills which were originally described by Sir Thomas Holland as the Sivamalai series (*Mem.* 30, 164-217, 1901). The rocks include nepheline-syenite and its varieties containing hornblende and biotite, aegirine-augite, feldspar rock and feldspar-corundum rock. The whole suite is characterised by soda-rich feldspars and pyroxene, the former being generally micro-perthitic and containing 15 to 18 per cent of the anorthite molecule. The pyroxene in the more basic type is an aegirine augite often intimately associated with some olivine and hypersthene. Occasionally also hornblende and titanaugite may be present. The accessory minerals are calcite, graphite, zircon, spinel, corundum, sphene, magnetite and ilmenite. The corundum and spinel are associated particularly with feldspar-rich rock. Both calcite and graphite are primary. Some banding is seen in a few places expecially near the contact with the surrounding gneissic rocks and parallel to the strike of foliation which is W.N.W.-E.S.E., and this banding may be an effect of later regional metamorphism.

Cordierite-Sillimanite rocks, Tiruchirapalli and Madurai. — A very interesting suite of rocks found near Kiranur (south of Kulitalai) in the Tiruchirappalli district contains cordierite, sillimanite, sapphirine, kornerupine, etc. associated with garnetiferous gneisses, amphibolites, crystalline limestones with calc-silicates and corundum. These have been described by M. N. Balasubramanian (unpublished doctoral thesis, Madras University 1965). Similar rocks occur also at the foot of the Palni hills at Ganguvarpatti near Kodaikanal Road, the cordierite in which is optically positive and has been described by the author (Mineral Mag., London, 1924, **20**, p. 248-251). Garnetiferous cordierite-sillimanite gneisses, sometimes with sapphirine, occur in the khondalite-charnockite suite in Tranvancore and in the Vizagapatam Ganjam hill tracts of Andhra.

CEYLON

The island of Ceylon is geologically continuous with the adjacent part of Southern India. Physiographically it is said to consist of three peneplains, the lowest forming the general plain country, the highest the mountainous uplands and the middle one a unit of intermediate altitude. According to Wadia, the two upper units are the result of uplift.

The intermediate peneplain is said to be of post-Jurassic age and the upper one probably post-Miocene. The entire island, except for a small coastal fringe in the north and northwest, where Upper Gondwana and Miocene strata occur, is occupied by Pre-Cambrian gneisses and schists. The central N.-S. belt appears to form a northward plunging synclinorium exposing khondalites, crystalline limestones and charnockites, with granitic gneisses on the flanks. The folds trend N. by E. to N.E. in the north, gradually becoming N.-S. and N.N.W.-S.S.E. in the central and south-central parts. In the southwest the rocks trend N.W.-S.E. and appear to be a continuation of the rocks of Travancore-Tirunelveli area in the southernmost part of India.

J.S. Coates (1935) and Fernando (1949) gave the following sequence of the formations:-

>Pegmatites and Basic dykes
>Wanni Gneiss (in the north and northwest)
>Charnockites
>Khondalites and associated schists
>Vijayan Series (Bintenne and other Gneisses)

Cooray (1962) has renamed the khondalite-charnockite suite of the central belt as the HIGHLAND SERIES. It is flanked on both sides by the banded gneisses of the VIJAYAN SERIES which are mainly microcline-bearing biotite- hornblende-gneisses of granitic to granodioritic composition, partly assimilating earlier schistose rocks. On the west they are intruded by the TONIGALA GRANITE which is probably a much younger pink microcline-granite associated with much migmatite. Between the Highland Series and the Vijayan Series is a transition zone containing "biotite-gneisses with a charnockitic aspect" which are apparently charnockitised gneisses. They contain amphibolite bands with hornblende, uralite and occasional pyroxene.

The Highland Series belt is cut off, on the south, by another suite of rocks along a line running southeast from a little north of Colombo to a place 65 km east of Galle. This belt contains biotitic charnockite-gneiss with wide bands of leucocratic granite-gneiss and minor bands of calc-granulites, sillimanite-cordierite-gneisses and amphibolites. It is apparently closely related to the rocks of South Tranvancore and Tirunelveli. The granitic material in this is apparently of young age and it yields some heavy minerals including monazite, thorianite, etc.

The southeastern part of the island contains the BINTENNE GNEISS which appears to dip under the Highland Series. The WANNI GNEISS occupies the north-western part along a N.E.-S.W. belt, a short distance north of Colombo and Trincomalee. This is pink to buff and contains bands of hornblende granite.

These VIJAYAN SERIES is presumably the equivalent of the Peninsular Gneiss of South India, while the HIGHLAND SERIES is the counterpart of the rocks of the Eastern Ghats province. The well known graphite deposits of Ceylon are associated with crystalline limestones of the Highland Series of the central and southern parts of the island.

MAP VII

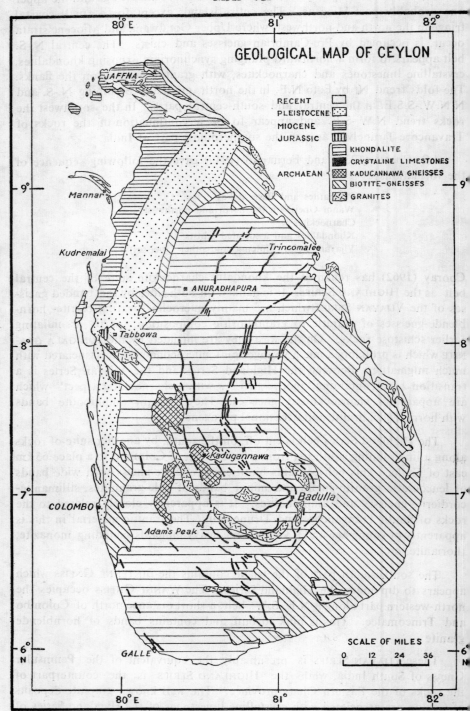

(By kind permission of the Director, Dept. of Mineralogy, Ceylon)

The gneisses as well as the khondalites are intruded profusely by veins of granite and pegmatite. But it is only the khondalite suite that contains useful mineral deposits — e.g., graphite, phlogopite, sillimanite, apatite. The pegmatite intrusives contain several interesting minerals among which thorianite, thorite, fergusonite, monazite, zircon, xenotime, columbite-tantalite, etc. may be mentioned. The feldspars in some of the pegmatites have given rise to moonstone.

The latest rocks of the igneous suite are dykes of dolerite and lenses and dykes of ultrabasic rocks whose age is not known.

EASTERN GHATS

The Eastern Ghats region between Bezwada and Cuttack, which attains the greatest width in the Ganjam-Cuttack tract, is composed of ridges trending in a N.E.-S.W. direction which is also the regional strike of the rocks. The hills are made up of gneisses, charnockites and khondalites. To their west lies a gneissic tract overlain by a basin of Proterozoic rocks which is highly disturbed on its eastern side, and which might have originally extended farther to the east.

The Eastern Ghats is a belt of high grade metamorphism as evidenced by the abundance of garnet and sillimanite. The charnockite shows intrusive relationship towards the khondalite and has itself undergone post-magmatic changes such as albitisation and myrmekitisation as described by H. Crookshank. The charnockites have also formed hybrid gneisses with the khondalites and sometimes with other ancient gneisses. There are, in parts of this area, some metamorphosed manganiferous rocks which have formed hybrid mixtures with acid igneous rocks. The *kodurites*, as these have been called by Fermor, consist of spessartite-andradite (contracted to spandite), orthoclase, apatite and manganese-pyroxene in varying proportions. Associated with these rocks here are also crystalline limestones, cordierite-gneisses, sapphirine bearing rocks, nepheline-syenites, etc., the last being intrusive into them.

In introducing the term "Khondalite" T.L. Walkar (1902, p. 11) wrote:—
> Instead of the awkward descriptive name garnet-sillimanite-graphite-schist, I propose to call these rocks KHONDALITE in honour of the fine hillmen the khonds in whose mountain jungles khondalite is better developed than in any region hitherto described. The rocks are not new to petrology neither is the selection of a special name imperative, but for the sake of brevity of nomenclature and as a local convenience I trust the new designation will be of service".

The typical khondalite is a rock of the granulite facies and is often accompanied by charnockite. Eduard Wenk (Indian Mineralogist, **6** p. 21-26, 1965) has called attention to the fact that *Stronalite* was similarly introduced by Artini and Melzi (1900, p. 284-295) to describe rocks of identical composition and origin found at Strona in the Italian Alps, in their classic paper on Alpine Geology. The rock has been described also by Bertolani (Rend. Soc. Mineral

Ital. **20**, p. 31-69, 1964). Eskola has compared the two types in his paper on mineral facies of rocks (1957) and shown that they are virtually the same.

The khondalites have been derived from rather high-alumina clays which were also fairly rich in iron. Some varieties rich in feldspar, such as those seen in the Eastern Ghats near Vizianagaram, owe their feldspar to granitisation. In some cases the feldspars form even large phenocrysts. The khondalites have been described under such names as BEZWADA GNEISS, KAILASA GNEISS and other names. At Bezwada they contain the variety of feldspar known as cleavelandite. On weathering, these rocks give rise to laterite and bauxite.

Fermor was of the opinion that the khondalites and charnockites were formed at some depth and that the whole Eastern Ghats belt was uplifted at a later date, bringing these high grade metamorphic rocks to the surface. The margins of the Eastern Ghats must therefore show zones of shearing and faulting. In his paper on Southern Jeypore, Crookshank (1938) stated that such faulting is not identifiable. But it is known that parts of the margin have been subjected to migmatisation. It is also known that the Machkund River follows a straight course for several km southwestwards from the Dudma Falls (18° 32' : 82° 30') along the junction of khondalite and charnockite, which must be a faulted one. The charnockite shows often a zone of crushing along its junction with hornblendic gneiss, so that there seem to be more than one parallel fault zone bordering the Eastern Ghats here.

Alkali Rocks and other types.—Near Koraput town in southern Orissa an igneous complex of alkali-gabbro, calc-alkali syenites and nepheline-syenites occurs concordant to the foliation of the gneissic rocks of the region. The nepheline-syenite is miaskitic in composition but perthite and albite occur separately only in some parts of the mass. The main part of the mass consists of nepheline, microperthite, iron-rich biotite, alkali- ferrohastingsite and titanomagnetite. The albite constituent forms strings parallel to (100) crystallographic direction. The nepheline is slightly higher in silica than the Morozewicz composition and the feldspar is a soda-rich orthoclase. Some crushing is developed at the contact of the mass with the country rock. The syenite and associated rocks are considered to be intrusive (M. K. Bose, 1964, Geol. Mag. **101**, p. 556-561).

Another rock type found near Paderu in Visakhapatnam is a sapphirine-bearing hypersthene-biotite rock. The sapphirine encloses grains and streaks of green spinel. At the border of basic charnockite and khondalite are developed cordierite-sillimanite gneisses and types containing biotite sapphirine spinel and magnetite.

In the Bison Hill area of the Eastern Ghats near the Godavari gorge, there are intrusives of corundum-syenite containing large zircon crystals. The country rocks are the khondalite suite consisting of garnetiferous sillimanite schists, biotite schists and charnockites. Nepheline, corundum and zircon syenites are found in the Paloncha area of Khammamet where they are intrusive into Khondalites and occupy a tract over 65 km long.

JEYPORE-BASTAR

The Jeypore-Baster tract behind the Eastern Ghats of Visakhapatnam (Vizagapatam) has been examined by H. Crookshank and P...K. Ghosh. The succession worked out by Crookshank (1963) for a part of this area is shown in Table 11. There is no comprehensive modern description available for the rest of the region.

TABLE 11.— Geological Succession in Jeypore-Bastar

PURANA	Upper :	Limestones, purple shales and slates
	Lower :	Pale sandstones and shales, purple shales quartzites, grits, conglomerates
— — — —	Unconformity	— — — —
IGNEOUS ROCKS		Dolerite dykes Granite and Pegmatite Charnockites Greenstones and Granite-gneiss
— — — —	Unconformity	— — — —
KHONDALITES		(Position uncertain)
— — — —	Unconformity	— — — —
BAILADILA IRON-ORE SERIES		Banded hematite-quartzites, grunerite-quartzites and white quartzites etc.
— — — —	Unconformity	— — — —
BENGPAL SERIES		Ferruginous schists, schistose conglomerates, biotite-hornblende-quartzites, shales, slates Slates, schists, phyllites, grunerite-garnet-schists, magnetite-quartizites, garnet-biotite-gneiss, with baslatic flows and tuffs Sericite-quartzites, andalustite-gneiss, banded magnetite-quartzites, grunerite-schists and quartzites with intercalated basalt flows
— — — —	Line of division uncertain	— — — —
SUKMA SERIES		Sillimanite-quartzites, grunerite-schists magnetite and diopside-quartzites, hornblende-schists, biotite-cordierite gneiss, etc.

No definite line of division can be drawn between the SUKMA SERIES and BENGPAL SERIES, but the former shows a higher grade of metamorphism being characterised by the development of sillimanite-and pyroxene-gneiss which are absent from the latter. The latter contains andalusite. Otherwise, both contain a variety of quartzites (magnetite, hematite, garnet or grunerite bearing) ferruginous quartzites, with contemporaneous basaltic flows which have generally been converted into greenstones. These two series, with associated granite-gneisses, have a general W.N.W. to N.W. strike of foliation and appear to form a synclinorium in the west where the BAILADILA IRON-ORE SERIES overlie them. The gneisses seem to be older than the SUKMA and BENGPAL SERIES. As Crookshank sometimes refers to the metamorphism of these two series by the gneissic rocks present in them, it would seem that they contain some later granites which are younger than the two series and have metamorphosed them at the contact.

The BAILADILA IRON-ORE SERIES forms a synclinorium with N.-S. axis, with a central eroded anticline flanked by two synclines which form ridges with prominent bands of iron ore. This Series is younger than the other rocks mentioned above and has been folded in a different direction.

SAMBALPUR

In Sambalpur, in the Mahanadi drainage, there are biotite- and hornblende gneisses, schists and granites. The town of Sambalpur is situated on a ridge of quartzite and quartz-schists of Pre-Cambrian age.

Sonakhan Beds.—To the west and north of the Sambalpur area there is a group of rocks, typically developed in the Sonakhan hills and called the *Sonakhan Beds*[1] by F.H. Smith, strikingly similar to the Chilpi Ghat and Sakoli Series. They are steep-dipping, highly crushed and schistose rocks which pass under the Cuddapahs to the north. They comprise quartzites, conglomerates, slates, phyllites, hornfelses, quartz-magnetite-schists, garnetiferous gneiss, etc. The regional strike of foliation is N.E.-S.W. There are also interbedded traps and hornblende-schists, and intrusive basic dykes of a late age. These schistose rocks are intruded by a coarse porphyritic pink granite with a little biotite and hornblende.

RAIPUR—DRUG

Immediately to the south and southwest of the Proterozoic basin of Raipur there is a large tract occupied by granites and gneisses. Schistose rocks occur in the adjoining parts of Kanker and BASTAR their strike being N.W.-S.E. The rock types include quartzites, phyllites, mica-schists and banded haematite-quartzites (*e.g.*, those of the Dhalli-Rajhara ridges). These schistose rocks in the Drug district are overlain by volcanic agglomerates and epidiorites and intruded by granites which occasionally show quartz-porphyries along their margin.

BILASPUR—BALAGHAT

Chilpi Ghat Series.—The schistose rocks of the Chhattisgarh[2] basin have been designated the *Chilpi Beds* by W. King and the *Chilpi Ghat Series* by R. C. Burton. The rocks wedge in at the eastern end between the Vindhyans and granitic gneisses but expand westwards into two strips, the northern one going into Nagpur and Chhindwara and the southern one into Nagpur and Bhandara, over a distance of some 65 km.

The rocks of the Chilpi Ghat Series strike roughly N.E. and have a straight northern margin and a sinuous and irregular southern margin. The northern

---

[1] General Report of the G.S.I. for 1898-99, pp. 39-42, 1899.

[2] The Raipur-Bilaspur-Raigarh area is called Chhattisgarh.

band of exposures comprises quartzites, felspathic grits, shales and slates with intercalations of trap. The basal conglomerate contains pebbles of rocks which appear to have been originally sedimentary. These rocks are now known to be part of the Sausar Series.

Sonawani Series.—In northern Balaghat there is another group of rocks to which Burton has given the name of Sonawani Series. They contain quartz-magnetite-schists, calc-silicate rocks, phyllites and muscovite- schists with some manganese-ore beds. It has now been proved by continouous mapping that the Sonawanis are also part of the Sausars.

NAGPUR—BHANDARA

Sakoli Series.—The Chilpi rocks continue westwards and bifurcate, the southern strip occupying parts of the Nagpur and Bhandara districts, and the northern strip going into Nagpur and Chhindwara. They are called the Sakoli (southern) Series and Sausar (northern) Series respectively. The rocks of the northern belt dip generally to the S.S.E. and S. and those of the southern to the N.N.W., while the middle or axial region may be a zone of faulting or overthrusting. The southern belt (Sakoli Series) contains chlorite- and sericite-schists and hematitic iron ore of a low grade of metamorphism, in contrast with the northern belt (Sausar Series) which is characterised by calc-granulites, marbles, garnetiferous schists and manganese-silicates and gondites.

TABLE 12.—The Sakoli Series

Sausar Series		Quartz-dolerite Tourmaline-muscovite granite and pegmatite
	Sakoli Series	Crushed albite-microcline-quartzite Phyllite and slate Hematite-sericite-quartzite Chlorite-muscovite schist with chloritoid, epidote-chlorite-schist, jaspilite, phyllite, chloritic horn-blende-schist Amphibolite and garnet-amphibolite Dolomites, crystalline limestones, calciphyre and chlorite-tremolite-schist Microcline-muscovite-quartzite

An area of critical importance near the eastern end of the Sausar-Sakoli belt constitutes what D.S. Bhattacharjee called the Bhandara triangle (Rec.G.S.I 71, p. 90, 1936). It is bordered on the north by the Sausar belt, on the east by the Purana basin of Raipur-Drug and on the soutwest by the continuation of the Godavari trough a little beyond Nagpur. Bhattacharjee worked only on a part of this area. The only published work on the whole of this triangle is that of S.N. Sarkar (1957-1958) whose modified table of succession (Table 13) is given below.

TABLE 13.— Pre-Cambrian Succession in the Bhandara Triangle

Khairagarh Orogeny (Ca. 600 M.Y.)

KAIRAGARH SERIES	Mangikhuta Andesite Karutola Sandstone Sitagota Andesite (637 M.Y.) Bortalao Sandstone

Dongargarh Granite (710-2000 M.Y.)

NANDGAON SERIES	Pitepani Andesite : Nandgaon Orogeny (Ca 1200 M.Y.) Bijli Rhyolite (1260 M.Y.)

Sakoli Orogeny (1335 M.Y.)

SAKOLI SERIES

Amgaon Orogeny (1434 - 1630 M.Y.)

AMGAON SERIES

Sarkar has deciphered three series here, viz., AMGAON, NANDGAON and KHAIRAGARH, the latter two of which overlie the SAKOLI SERIES. The last, which is seen in the middle of the triangle, opens out to the southwest, being bordered on the north by the Sausar belt and on the east by the Nandgaon-Amgaon formations. The KHAIRAGARH SERIES appears near the north-east apex of the triangle, apparently overlying the Nandgaon-Amgaon belt of rocks. The BIJLI RHYOLITE occupies a tract 110 km. long and 60 km. broad, intercalated with inter-trappean rhyolitic conglomerate and sandstone. The thickness has been estimated to be over 15000 ft. (4500 m.). Bhattacharjee however considered this formation to be mainly of mylonitic nature. Three series of andesites occur in the succession, one above the Bijli Rhyolite and thought to be connected with Nandgaon orogeny (1200 M.Y.) and the other two in the Khairagarh Series. The DONGARGARH GRANITE occupies an area of 6500 sq km. The chronology has apparently been deduced from age data (mainly K-Ar) of a variety of materials coming from scattered localities. Obviously, much more work needs to be done in this area on the stratigraphy, structure and chronology of the formations.

NAGPUR—CHHINDWARA

Sausar Series.—The manganese belt of Madhya Pradesh (part of which now goes into the newly constituted Maharashtra State) was first studied systematically by Dr. L. L. Fermor in the first few years of the present century and his observations were published in his well known monograph (*Mem.* XXXVII, 1909). Though the occurrence of manganese in this region has been known since 1829, it recieved the attention of the Geological Survey only from 1879 onwards when Mallet (*Rec.* XII) published an account of braunite and rhodonite from Nagpur. P. N. Dutta studied some of the deposits in 1893-94 but his report was not published. King mapped the Bilaspur area in 1888 and designated the rocks the Chilpi Series (*Rec.*XXII, p. 5). P.N. Bose worked in Balaghat in 1888-89 when he distinguished the Chilpi Ghat

Series, Baihar Gneiss (composite genisses) and Chauria Gneiss (mainly intrusive granite). Much work was done by Fermor and other officers of the Survey (Burton, Cotter, Bhattacharjee, West, S.K. Chatterjee and others) in this belt between 1911 and 1926. A comprehensive account of the geology of the region was published by Fermor (*Mem.* LXX, Part 2, 1940) when he revised his earlier work and gave a full account of the succession in the *Sausar* and *Sakoli Series.* The Sonawani Series of Burton and the Chilpi Series of King were recognised by Fermor as parts of his Sausar Series, showing different metamorphic grades, and perhaps containing some beds not represented in the Sausars. West (1936) made detailed studies of the rocks in the Deolapar part of the belt and deciphered the existence of a *nappe* structure. A party of the Geological Survey of India remapped the manganese belt between 1951 and 1955 under the supervision of John A. Straczek of the U.S. Geological Survey. A summary of the work has been published in the Manganese ore symposium of the XX International Geological Congress held at Mexico City in 1956. A full account of the work has since been written up by S. Narayanaswami, N.A. Vemban, S.C Chkarvarti K. D. Shukla and other officers who participated in it but it is not yet published. The following account is based on the above mentioned publication. The general succession now adopted for the Sausar Series which forms the northren band (mentioned under Bilaspur—Balaghat) is a modification of that given by Fermor and West.

TABLE 14—The Sausar Series

Minor intrusions	— Leucocratic granite, granite-pegmatite and quartz veins.
Granitic intrusives	— Gneissic granite and ortho-gneiss.
SAUSAR SERIES :	
BICHUA STAGE	— Dolomitic marble, serpentine marble, diopsidites, actinolite-schists, calc-silicate granulites with tremoltie, anthophyllite, wollastonite and grossularite. Occurs in all areas except in the south and east.
JUNEWANI STAGE	— Muscovite-biotite-schists, quartz-biotite-granulite, locally biotite-gneiss, often garnetiferous; in places staurolite, kyanite and sillimanite; sometimes interdigitating with Bichua stage. Widespread but lenticular.
CHORBAOLI STAGE	— Quartzites, quartz-muscovite and felspathic quartz-schists, occasionally garnetiferous ; sillimanite and kyanite in places ; also autoclastic quartz conglomerates. Widespread except in the Central part of the belt.

Manganese ore and Gondite Horizon :

MANSAR STAGE	— Muscovite and biotite schists, phyllite, often garnetiferous, become gneissic where feldspathised. Generally highly argillaceous. *Two or Three Manganese horizons in the schists.* Most widespread stage of the Series.

Manganese ore and Gondite Horizon :

LOHANGI STAGE :—	
(*a*) LOHANGI	— Pink and white calcitic marbles (locally dolomitic) and calciphyres, etc. Occurs mainly in the northern part of the belt.

TABLE 14—The Sausar Series—*Contd.*

(*b*) UTEKATA	—	Calc-granulite and calc-gneiss with silicates; contains microcline-bearing bands. Extensively distributed but thin and impersistent.
(*c*) KADBIKHERA	—	Quartz-biotite granulite with epidote and magnetite intercalated with quartz-biotite-gneiss. Occurs only in the western part but thin and lenticular.
SITASAONGI STAGE		—Quartz-muscovite-schist, feldspar-muscovite-schist and intercalated quartzites; locally a gneiss with kynite and garnet. Well developed in Bhandara, extending to the east as a schistose feldspathic grit. Passes laterally into the Lohangi Stage to the west.

Unconformity

TIRODI GNEISS	—	Biotite gneiss with subordinate intercalations of amphibolite, hornblende-schist, calc-gneiss, feldspar-muscovite-schist, biotite-granulite; commonly garnetiferous; locally porphyroblastic; varies from arkose-grit to gneiss according to metamorphic grade. Occupies the middle strip of the belt mainly but found also in other parts extensively.

Unconformity

METAMORPHIC GROUP	—	Hypersthene-granite gneiss, biotite-gneiss, hornblende-gneiss, amphibolite, etc.

(*Note* — The members of the substages of Lohangi tend to grade into each other laterally. They pinch out in the eastern and south-eastern part of the belt where they are replaced by the Sitasaongi Stage).

The manganese belt is structurally very complex. It forms an arc, broadly convex to the south. It extends from Balaghat in the east to Chhindwara in the west, between longitudes 78° 49' and 80° 30' E. and latitudes 21° 21' and 22° 05' N.., having a length of about 210 km and a width of 24-30 km. The country has an average elevation of 300 m and is undulating. The general strike of the rocks is N.E.-S.W. in the east, E.-W. in the middle and W.N.W.-E.S.E. in the west. The southern part of the belt constitutes a zone of isoclinal folds with steep (50°-80°) dips to the south; in the middle strip the folds are recumbent, with 30° to 60° dip to south; the northern strip shows thrust sheets (nappes). In Balaghat, in the east the folds lean to the south with the axial plane dipping to the north at low angles of 10° to 30°. There are numerous steep dipping strike-faults in different parts of the belt and they are generally thrust-faults. The folds as well as the thrust-faults tend to be arranged *en echelon*, these being shifted progressively to north as we proceed east or west from the central region. The fold axes plunge to the S.E. in Chhindwara, E. in Nagpur and E.N.E. or N.E. in Balaghat. The main folds have been cross-folded at a later date, the cores of the cross-folds being often occupied by granite of post-folding age. The axes of the cross-folds plunge at low angles to the west or south-west. Four nappe units have been recognised, there being the Sapghota, Ambajhari, Deolapar and Sonawani nappes from west to east, but they have yet to be connected together after mapping the area further north. All the nappes have low southern dip, the dip increasing gradually southwards. In these nappes the younger stages of Bichua-Chorbaoli are thrust over the older Lohangi-Chorbaoli and over the Tirodi gneiss, and the succession is inverted in some areas.

Continuous mapping from the Balaghat area along the belt has shown that the Chilpi and Sonawani Series are merely parts of the Sausar Series, allowing for lateral variation and difference in metamorphism. The Chilpis are arkosic and argillaceous while the Sonawanis contain calcareous rocks, quartz-muscovite schists and granulites.

The sediments are largely argillaceous and arenaceous in the south and east, becoming more calcareous in the north and west. Thus the Lohangi and Bichua stages are better developed in the north and west, while Mansar and Chorbaoli stages are better seen in the south and east. The manganese horizons occur associated with the argillaceous materials rather than with the calcareous.

The grade of metamorphism increases gradually westwards and northwards. Near the eastern end of the belt the rocks are of epi-grade, with phyllites containing sericite, chlorite, ottrelite, etc. Garnet, staurolite, biotite and kyanite come in farther west, while in the north-west and north, sillimanite, pyroxene and iron-garnets are well developed. The middle strip in which the Tirodi gneiss is prominently developed represents probably the root zone of the former mountain belt which has since been largely planed off. The sillimanite zone is characterised by the development of tabloids or thin lenticular plates of sillimanite and quartz.

As shown in the table, manganese—ores are associated with three horizons, one each at the bottom, middle and top of the Mansar Stage. Of these the one at the top of the Mansar (and bottom of the Chorbaoli) is not important, as only a few deposits occur there. Within the Mansar Stage occur several important deposits but the stratigraphic position is not constant and has been conditioned by the local supply of sediments at different times. The zone at the bottom of the Mansar (and top of the Lohangi) contains the best and largest deposits. They sometimes replace the calcareous rocks. In the eastern area this horizon is sometimes developed between the Sitasaongi and Mansar Stages, or between the Sitasongi and Tirodi gneiss where the Mansar Stage is absent. The rocks exposed in the synclinal structure of Ramtek (21° 24′ : 79° 18′) on the convex side of the belt would appear to form an additional (Ramtek) stage at the top of the Sausar Series.

The manganiferous sediments have been converted into ore or gondite, as the case may be, depending on purity and grade of metamorphism. In the east the ore is fine grained braunite, with hollandite and cryptomelane and thin bands of chert and quartzite. The sediments are similar to the banded iron formations but the rhythmic precipitation has not been so fine or regualr as in the banded hematite-jaspers. In the zone of mesograde metamorphism gondite has developed from the impure ores (quartz, spessartite, manganese, amphiboles, piedmontite, manganese mica) while the ore becomes a coarsley crystallised braunite with some sitaparite and jacobsite-hausmanite. In the still higher zone of metamorphism (sillimanite zone) manganese pyroxenes are also developed with braunite as the chief ore mineral. The gondite as well as the ores have been subjected to weathering and alteration; while the silicates have given rise to oxides like pyrolusite, psilomelane and wad, the ores may have given rise to psilomelane, hollandite and pyrolusite. In some cases the weathering and supergene alteration have proceeded to considerable depths as a result perhaps of special conditions of topography and drainage. For a discussion of the mineralogy and petrology of there formations, the work of Supriya Roy (1966) may be consulted.

Ages have been determined on five mica specimens from rocks of different horizons in the Sausar series (Sarkar et al, 1964). The ages obtained are between 864 and 984 M.Y. which would indicate the period of final folding and metamorphism. The Sausar Series itself may therefore be 200 to 300 million years older.

The Sakoli Series lying to the south of the Sausars widen out in a S.W. direction. Their structure is interpreted as a synclinorium plunging to the northeast. According to Sarkar et al (1964) two very different groups of ages have been obtained for materials from the two flanks. The northwestern flank near the Sausar Series suffered metamorphism between 860 and 950 M.Y. (more or less contemporaneously with the Sausars) while the southeastern flank was affected between 1330 and 1340 M.Y. As there is a difference of over 300 M.Y. between the two, it would be necessary to examine wheter the rocks on the two flanks belong to the same sedimentary series or whether they are different. In any case it is necessary to reexamine and map the Sakoli and Khairagarh-Nandgaon areas and assess the structure and chronology of the formations present therein.

JABALPUR

This region is separated from the Nagpur-Bhandara-Balaghat region by a stretch of Decean Traps which cover all the earlier formations. The formations here comprise conglomerates, phyllites, mica-schists, calcitic and dolomitic marbles, banded ferruginous rocks associated with manganese and iron-ores, with sills of altered basic igneous rocks, all these having a foliation strike of E.N.E.-W.S.W. In the neighbourhood of Sleemanabad, the schistose rocks are traversed by veins containg copper-ores. The conglomerates are auriferous but are too low in grade to be workable, as detailed prospecting indicated less than two dwt. of gold per ton.

The above series of rocks was first referred to the Bijawar series (Cuddapah system) by C.A Hacket and divided into four stages, but these are not definite stratigraphic units. Marbles and calc-rocks occur in the lower-beds while ferruginous formations and iron-ores occur in the upper beds. The rocks bear an extraordinary resemblance to the Pre-Cambrians of Singhbhum and Gangpur (Fermor, 1909 p. 805; Krishnan, 1935).

GUJARAT

The Champaner Series:- Outcrops of Pre-Cambrian schists associated with gneisses are found in Narukot 65 km. E.N.E. of Baroda in Gujarat. They were named the CHAMPANER SERIES by W. T. BLANFORD (Rec. G. S. I. 2, 1869). Their relationship to other Pre-Cambrian rocks was obscure until Heron continued the mapping of Rajasthan into Gujarat and found that the Champaners were the same as the Aravallis. The Champaners comprise quartzites, conglomerates, slates and limestones. Near Jothwad and Jambughoda they contain also manganiferous quartzites. At Champaner (22° 26′: 73° 37′) near Shivrajpur in Panch Mahals, the Champaners consist of quartzites at the base, followed by slates, conglomerates and grits, manganiferous quartzites and phyllites, limestones and calc-schists. The quartzites show cross-bedding and ripple marks. The rocks show a low grade of metamorphism and are folded into a synclinorium with W.N.W.-E.S.E. to W.-E. strike of fold axes. On the west, the Deccan Traps occur at Pavagad hill while in the north are granites which are presumably the same as the Erinpura Granites. There are two ore horizons in the phyllite. The Aravallis which occur to the north are separated by a distance of over 20 km. of granite country. Detailed work should be done in this region to establish that the Champaners are identical with the Aravallis.

RAJASTHAN

Parts of Rajasthan wree mapped originally by C. A. Hackett and later by La Touche and Middlemiss. During the present century the whole region has been mapped by A. M. Heron, assisted by A. L. Coulson, B.C. Gupta, P. K. Ghosh and others. The geology has been described in a series of papers and an excellent summary of the work has been given by Heron (1935).

The characteristic feature of the country is the presence of several groups of rocks belonging to the Archaean and Pre-Cambrian, forming a folded mountain system running across it from the north of Delhi in the north-east to the Gulf of Cambay in the south-west. This mountain system was formed in Pre-Cambrian times, folded again in post-Delhi (? Cambrian) and affected by uplift and faulting probably in Mesozoic times. The central part of the Aravalli ranges is occupied by a great synclinorium composed of Delhi and Aravalli rocks. Trending roughly in the same direction as the ranges is the *Great Boundary Fault* of Rajasthan along which the Aravallis lie against the Vindhyans on the east. Because of the semi-arid nature of the country the rock exposures are good but in the west and south-west they are often engulfed in sandy alluvium and desert sands. The geology is of great complexity and at present all we can say is that the preliminary work, of which an account is given in the following pages, is liable to drastic revision in coming years.

The major formations of pre-Vindhyan age which have been classified in Rajasthan are shown below :—

5. Malani suite of igneous rocks
4. Delhi System
3. Raialo Series
2. Aravalli System
1. Banded Gneissic Complex and Bundelkhand Gneiss

The Archaeans consist of the Bundelkhand Gneiss and the Banded Gneissic Complex, the latter forming composite gneisses which include much undoubtedly original sediments. The Aravallis which are an enormously thick series of mainly argillaceous rocks are probably the equivalents of the Dharwars of South India. The Raialos which may probably be more than 600 m thick, are considered to be intermediate in age between the Aravallis and the Delhis. The Delhis, consisting of sandstones and shales, resemble the Cuddapahs and are probably about 6,000 m thick. After the deposition and folding of the Delhis there occurred a series of igneous intrusions which include the Erinpura, Jalor-Siwana and Idar granites as well as the Malani suite of volcanic and plutonic rocks. The Vindhyans which are considered to be partly of Cambrian age are younger than the succession represented by the rocks mentioned above. The Pre-Vindhyan succession as given by Heron is reproduced in Table 16. Since the Delhi System is now regarded as the equivalent of the Cuddapah System which is considered post-Archaean, this section will deal only with the Pre-Delhi rocks.

The Bundelkhand Gneiss.—The main exposure of this group occurs in the Berach Valley between Chitor and Bhilwara and is over 70 miles long. It is overlain by the Vindhyans in the south and elsewhere by the Banded Gneissic Complex. In its typical form, it is a pink to reddish, medium grained, non-foliated, non-porphyritic granite. The chief minerals are quartz, orthoclase, subordinate microcline, and a little of ferromagnesian minerals. The quartz has a violetish opalescence while the feldspars are usually somewhat altered. The rather sparsely occurring ferromagnesian minerals, biotite and green hornblende, are more or less altered to epidote and calcite. Accessory minerals are generally scarce. Veins of pegmatite are infrequent but those of microgranite and aplite are common. Here, as in Bundelkhand, the rock is traversed by prominent quartz reefs and numerous dolerite dykes. The Bundelkhand Gneiss of Rajasthan is believed to be identical with that of Rajasthan, though the two are separated by over 250 miles of younger rocks. One or two age determinations indicated an age of 1500 M.Y. for this rock while that of the rock in Bundelkhand (near Jhansi) was 2500 M.Y. If this is confirmed by further work, the Bundelkhand Gneiss or Granite of Rajasthan will have to be called BERACH GRANITE as suggested by Saxena.

Towards the west, near the junction of Berach and Began rivers, the Bundelkhand gneiss gradually becomes well foliated and grey coloured, with knots of quartz and feldspar and small quantities of sericite and chlorite. The gradation is probably due to the partial assimilation of the schistose country rocks.

The relationship of this to the Banded Gneiss is not known as the junction is covered by the Aravallis. It is possible, according to Heron, that it may really represent the granitic constituent of the Banded Gneiss. Near the northeastern edge of the great mass in Bundelkhand there are numerous xenoliths of older age, consisting of quartzites, argillites, hornblende-biotite-schists, etc., which have undoubtedly been intruded by them and torn off from the original schistose basement.

TABLE 15.—PRE-VINDHYAN FORMATIONS OF RAJASTHAN (After A.M. Heron)

System	Jodhpur	Mewar; Ajmer-Merwara (Main syncline)	Jaipur	Alwar
Vindhyans	Vindhyans	Chitor; Nimbahera; Sadri — Upper Vindhyans (Semri series) / Lower Vindhyans (unmetamorphosed)	Jiran Sandstone; Binota shales	
— Great Boundary Fault —				
Delhi system	Malani Volcanic series	Calc-gneisses; Calc-schists; Phyllites & biotite-schists; Quartzites; Basal arkose-grits	Ajabgarh series; Alwar series	Ajabgarh series; Hornstone breccia; Kushalgarh limestones; Alwar Series
Raialo	Raialo (Makrana) Marble; Limestones of Ras.	Garnetiferous biotite-schists; Raialo (Rajnagar) marble; Basal grit — local	Railao (Bhagwanpura) limestone; Sawa grits	Raialo limestone; Raialo quartzite
Aravalli system	Shales (Sojat) Schists of Godwar	Phyllites, cherty limestones, quartzites and composite gneisses; Basal quartzites, grits and local conglomerates; Thick volcanics (local)	Khardeola grits; Badesar quartzites; Ranthambhor quartzites; Shales and cherty limestones; Basal quartzites and grits; Quartzites and schists of Baonli-Awan ridge and Bechum, Biana and Lalsot hills, Volcanics of Basi, Schists of Rajmahal	Limestones and schists of Rajgarh; Conglomerates and quartzites of Rewasa
Basement	Grey, homogeneous gneiss	Banded gneissic complex	Bundelkhand Gneiss	Gneissic granite of Karela and Ganor

The Banded Gneissic Complex.—The rocks belonging to this group consist of alternating bands of biotite-gneiss and granite. Biotite and chlorite schists, which may represent early sediments, are found as constituents of these in Southern Mewar. In places they grade into a granite-gneiss or even into an unfoliated granite. They contain also some hornblende-schists and epidiorites, representing interbedded altered basic igneous rocks. The Gneissic Complex is traversed by pegmatite and aplite veins, apparently derived from granitic rocks of different ages.

Banded Gneisses also occur in Central and North Mewar and in Ajmer, comprising dark-coloured schists and garnetiferous granulites intruded by biotite-granite. Another type of gneiss which may belong to the same group occurs west of the synclinorium and consists of a fine-grained and somewhat foliated porphyritic granite. Another variety of this gneiss is the grey fine-grained slightly foliated granitic gneiss along the foot of the Arvalli Range on the northwestern side of the Delhi synclinorium.

The Banded Gneissic Complex is also intimately associated with crystalline limestone near Ras (26° 19′ : 74° 11′), which is probably the same as the Raialo limestone. Northwest and west of Amet (25° 9′ : 73° 56′) the gneisses are found to surround some exposures of quartzites with which they have conformable dips. In the southwestern area the gneisses are cut up and penetrated by the Erinpura Granite and is ultimately completely replaced by it.

With regard to the age of the Banded Gneissic Complex, it may be mentioned that Crookshank (1948) states that their schistose components in northern Mewar are essentially the same as Aravalli schists and that they represent granitised Aravalli schists. N. L. Sharma (1953) states that K. L. Bhola, who has studied the Aravallis on either flank of the synclinorium in Ajmer and Jodhpur, is also of the opinion that the Banded Gneisses are the granitised representatives of the Aravallis. According to Heron there is a distinct erosion unconformity between the Banded Gneisses and the overlying Aravallis and the former are nowhere exposed in juxtaposition to the Bundelkhand Gneiss. The ages of these different gneissic rocks can be settled only after careful dating of materials collected from the different bands.

The Aravalli System.—The Aravalli System is dominantly argillaceous in composition and of great thickness. The rocks show increasing metamorphism as they are followed from east to west into the highly folded region.

The basal beds, which rest on Bundelkhand Gneiss or the Gneissic Complex, are arkose and gritty quartzites. Above these come shales and phyllites with which are associated some altered basic volcanics in places. Impure argillaceous and ferruginous limestones occur in two facies, one being a lenticular ferruginous limestone as in Bundi and Mewar and the other black massive limestone as near Udaipur city. In some places there are quartzites instead of limestones. The whole series of rocks is well foliated and injected lit-par-lit by granitic rock, resulting in mica-schists and composite gneisses as in Mewar and Idar. Recent work by the Geological Survey seems to indicate that the Aravallis are younger than the Banded Gneissic Complex, though the latter may contain some later granitic constituents.

An unmetamorphosed facies of the Aravallis occurs in Eastern Mewar east of the Great Boundary Fault of Rajasthan. This has been named the *Binota Shales* and consists of low-dipping, brown and olive shales with ferruginous and clay concretions. In the east, the Binota Shales are succeeded by the Jiran Sandstones, Vindhyans or the Deccan trap. To the west of the Boundary Fault, the Aravallis are represented by steep dipping slates and impure limestones intruded by dolerite. Followed westwards, these shales become first distinctly slaty or phyllitic, and later schistose, with the development of staurolite, garnet and kyanite.

The youngest members of the Aravallis are the reddish sandstones and quartzites seen near Ranthambhor and Sawai Madhopur in Jaipur State. The unmetamorphosed Aravallis of Chitor can be followed northeastwards along the strike through Bundi into southeastern Jaipur, where they are associated with the RANTHAMBHOR QUARTZITES. These rocks have resemblance to the shales of the Gwalior Series but there seems to be no doubt that they are of Aravalli age, according to Heron.

In the Ranthambhor area the rocks form a syncline composed chiefly of quartzites 1,100 m thick, interbedded with subordinate shales and sills and flows of dolerite. The dip of the rocks is 25° to 30′ at the edges of the syncline, but practically horizontal at the centre. The edges form fairly well marked scarps often 150 to 180 m high. The quartzites are reddish to pink and are much more compact than Vindhyan quartzites. They are well jointed and contain intercalations of shale beds. The shales are more compact than similar Vindhyan shales and are purplish to black and sometimes spotted. They are not cleaved but break into tilelike fragments. Sometimes ripple marks are seen in both the shales and quartzites.

There are at least five horizons of trap in these rocks, which appear to be intrusive. The traps are uniform in grain and of varying thickness. The topmost trap sill in the Ranthambhor syncline is at least 20 m thick, while the one below is about 60 m thick. The other sills are much thinner. They weather at the surface into a ferruginous gravelly material known as *moorum*. The Traps consist of feldspars and pale green augite showing a little alteration to chlorite. Olivine may or may not be present. Grains of ilmenite and pyrite are common. The traps of the Ranthambhor area are identical in characters with those of the neighbourhood of Gwalior.

The quartzites are underlain by a large thickness of shales which have been gently folded during the compression of the whole series. There is no unconformity between the two types of rocks, though the shales have yielded more to compression than the sandstones.

Near Hindaun (26° 44′ : 77° 6′), beds of a similar character (and similarly resembling those of the Gwalior System) are found forming a series of outcrops 32 km long along the edge of the plateau of Upper-Vindhyans. They strike parallel to the Aravalli ranges. The characteristic formation consists of a dark siliceous hematite, banded white chert and red jasper. The bands are much contorted and the strata dip at high angles (60° to 70°). There are also quartzites, black slates, trap and impure limestones in the sequence. The hematite beds have given rise to iron-ore and red and yellow ochres. These are undoubtedly much older than the lower Vindhyans which rest on their upturned and denuded edges with a profound unconformity. The Hinduan rocks resemble those of the Morar Stage of the Gwaliors very closely, though the latter are 130 km away, but whereas the Gwalior Series of Gwalior are practically horizontal, those of Hindaun have been subjected to folding because they are near the Aravalli fold axis.

The Gwalior System of Gwalior city and neighbourhood are separated from the Aravallis by a belt of Vindhyans having a width of about 130 km, but they resemble the unmetamorphosed Aravallis. Though they lie distinctly to the east of the continuation of the strike of the Aravallis, they may possibly belong to the Aravalli system.

Near the Mewar-Partabgarh border, there are several exposures of an amygdaloid, associated with ferruginous sandstones and cherts and overlain by the *Khardeola Grits*. These last consist of conglomerates, grits, grey-wackes and slates intercalated with slates of Aravalli aspect. There may be a slight unconformity between the Aravallis and these rocks.

The Aravallis have been intruded by fine-grained aplo-granite which is found as bosses and also as lit-par-lit intrusives, *e.g.*, near Udaipur, continuing there into Dungarpur. There are also ultrabasic rocks, now seen as talc-chlorite-serpentine rocks. Others, including granite, epidiorite and post-Delhi dolerite, are also met with.

Intrusive into the Banded Gneisses and the Aravallis, are the *Soda-syenites of Kishengarh* (26° 34' : 74° 52') which have been described by Heron (*Mem.* 65, Pt. 2, p. 152, 1934). They include nepheline-syenite, theralite and camptonite dykes. They are found in a series of small hills near the junction of Aravallis and the overlying Delhi quartzites. They are presumed to be older than the Delhis as the Delhi Quartzites overlie them, though they do not contain any fragments derived from the syenites. A new interpretation has been advanced that the Delhis have been thrust over the Aravallis and that the soda-syenites have been intruded into junction zone and that therefore they are post-Delhi. It is also probable that they may be correlated with the soda-syenites associated with pyroxenite, gabbro and picrite described in Sirohi by Coulson (1933) which are known to be post-Delhi but pre-Malani.

These syenites may be granitoid, or banded. They consist of nepheline, sodalite, cancrinite, orthoclase, microcline, some albite and a pleochroic amphibole (blue green to greyish yellow). The weathered surface is pitted and greasy looking when rich in feldspathoids and present a lustrous pearly surface when highly feldspathic. They are associated with pegmatites consisting of coarse crystals of nepheline, blue to colourless sodalite, white to yellow cancrinite, and dark amphibole. Some of the pegmatites have a pale carmine colour when freshly broken, but the colour fades very soon.

Raialo Series.—The Raialo Series overlies the Aravallis and is overlain by the Delhi system, the junctions in both cases being markedly unconformable.

They consist in the main of limestones, about 600 m thick, with thin basal sandstone and conglomerate. The limestones often rest directly on the older rocks without the intervention of the basal sandstones. In the Rajsamand area in Mewar, the limestones are overlain by micaceous quartzites and mica-schists which have been highly metamorphosed and partly converted into gneisses.

The main exposures are those of Alwar-Jaipur around Raialo, the type locality; in Ajmer and Mewar on the north-western flank of the synclinorium at Makrana, Ras and Godwar; and on the southeastern side of the synclinorium, north of Udaipur city and through Nathdwara, Rajsamand and Kankroli into Par-Benara and Jahazpur hills. The exposure of the unmetamorphosed limestone of Bhagawanpura, east of Udaipur is also considered to be Raialo.

In the type area in Alwar (*Mem.* 45) there are basal quartzites resting on granite passing upwards into thin-bedded slaty and micaceous quartzites. They are overlain by 600 m of pure white, hard saccharoidal dolomite marble with obscure stratification. There are also yellow pink and brown marbles, the impure bands showing tremolite and actinolite. The top beds are cherty and contain some patchy deposits of hematite. The celebrated Makrana marble deposits consist of a series of ridges trending N.N.E.-S.S.W. for over five miles, the intervening hollows being filled by sand. The rocks show steep easterly dip and considerable variation in thickness. Much of the greater part of the formation is a white, medium to coarse grained, calc-marble with cloudy grey patches. There are also beds of rose-pink and blue-grey colours. A few veins of pegmatite related to Erinpura granite have penetrated the marble here though in most other exposures the marbles are free from intrusions.

South-west of Makrana, near Ras (26° 19' : 74° 11') fairly extensive outcrops of marble are found over a strip 80 km long and 1.6 km wide. The composition of the material is very varied, from limestone to calc-gneiss, the usual type being a coarse saccharoidal calcitic marble with diopside and a little white mica. There are inter-bedded bands of blue grey limestone with knots of quartz, feldspars and calc-silicates which stand out on weathered surfaces.

At Godwar in Jodhpur, 145 km S.W. of Ras, they are exposed as crystalline marbles and calcareous schists. In the Sarangwa quarries, where the marble band is 1.6 km long, they are surrounded by the Erinpura Granite which has been responsible for marmorization.

In the Udaipur, Nathdwara, Rajsamand and Kankroli areas, the marble (Rajnagar marble) is underlain by 9 m of conglomerate and thin quartzite. The marble is a pure white dolomite. In the Rajsamand syncline the beds above the marbles have been converted into mica-schists intruded by granite, sometimes forming banded gneiss.

In the Jahazpur and Sabalpura Hills in N.E. Mewar, they form two ridges exposing both the basal quartzites and dolomitic marbles. The quartzites contain numerous quartz and granite veins. The passage beds between the two formations consist of a ferruginous breccia.

The Bhagwanpura Limestone in Mewar, 300 m thick, is an unmetamorphosed dolomite forming a broad outcrop parallel to the Great Boundary Fault from Chitorgarh southwards. White, grey, pink, crimson and brown colours are seen but the material is fine-grained, hard and not visibly crystalline. It contains disseminated silica, iron oxide and small clots of jasper. Stromatolites have been found in it near Chitor.

Though frequently closely associated with the Erinpura Granite the Raialo Limestone is not ordinarily penetrated by it. The metamorphism to marble and calciphyre with calc-silicates has been mainly due to the regional folding to which the formations were subjected.

There are several areas in the Aravalli belt containing mafic and ultramafic intrusives. As instances may be cited the steatitic rocks derived from olivine-bearing intrusives in Biana-Lalsot and Nawai hills. Talc-chlorite rocks associated with serpentine and magnesite occur near Beawar, at a few localities northeast of Udaipur, near Rikhab Deo and in Dungarpur. The occurrences near Dev Mori and a few other places in Idar are serpentinites containing some asbestos and talc. Most of these may be pre-Delhi and some even pre-Aravalli.

Two interesting occurrences, both probably ring-complexes, are found at Toa and Mer (Mundwara) in the Sirohi area. They have been described briefly by Coulson (1933). In both the hills, the central part is occupied by coarse picritic rock surrounded or flanked by olivine-gabbro and dolerite, olivine- and hornblende-basalt, and pyroxenite. On the south flank of Mer there are also alkali syenites with sodalite and aegirine. Coulson thought that these complexes may be of about the same age as the post-Delhi Erinpura Granites.

BUNDELKHAND

Bundelkhand Gneiss :—The Pre-Cambrians of Bundelkhand are separated by the Deccan Traps and Vindhyan formations from those of the neighbouring areas. In Bundelkhand there is a large semicircular outcrop of gneissic granite 320 km long (E.-W.) and 200 km broad (N.-S.) at the maximum. It was described by T. Oldham (1859) and General McMahon (1875) and later by Heron (*Mem. G.S.I.* 68, 1936). The typical rock is a pink, medium grained granite with obscure foliation, but coarser and finer varieties as well as porphyritic ones are known. Recent studies by Saxena (*Res. Bull. Punjab Univ.* June, 1961) have shown that inclusions and xenoliths are common in the granite, which should really be called a gneiss. It is now considered to be mainly a metasomatic granite which has assimilated older rocks and contains inclusions of quartzite, mica-schist, talc-schist, amphibolite, etc. (Jhingran, 1958). These inclusions are generally elongated in a N.E.—S.W. direction in the main mass and E.N.E.-W.S.E. near the margins. The rock is composed of quartz, pink orthoclase, perthite, plagioclase, some biotite and hornblende. Quartz often shows a violetish opalescence and strain shadows under the microscope. It is traversed by prominent reefs of quartz, some of which are 60 m high, for several miles at a stretch. Some of the reefs contain veins of pyrophyllite along the joint planes and along their junction with the granite. There are also basic dykes, these being younger than the quartz veins. Sarkar *et al* (1964) give the K-Ar age of biotite inclusions in the gneiss as 2500 and 2510 M.Y. The age will need confirmation by additional data.

SINGHBHUM

Singhbhum in Southern Bihar is one of the regions which has been mapped in some detail and information about which is of modern character. The rocks show two facies, an unmetamorphosed one in the south and a metamorphosed one in the north, separated by a major thrust zone.

This thrust zone extends from Porahat in western Singhbhum through Chakradharpur, Amda, Rakha Mines, Mosaboni and Sunrgi into Mayurbhanj, over a distance of 160 km. It has an E.-W. course in the western part and turns to the S.E. in the eastern part. The thrust zone marks the overfolded limb of a geanticline. Two lesser zones of thrust are found farther north, one along the northern border of the Dalma lavas in Southern Manbhum and Midnapur, and the other still farther north along the boundary of the granitic and schistose rocks. The three zones are parallel to each other and to the Satpura strike which prevails over Southern Bihar. They converge in the neighbourhood of Goilkera (between Manoharpur and Chakradharpur) where the rocks are seen to be tightly folded.

South of the main thrust zone, the rocks are little metamorphosed, though they have been thrown into folds whose axes are parallel to N.E.-S.W. (or N.N.E.-S.S.W.). The following succession was recognised by H.C. Jones in South Singhbhum, Keonjhar and Bonai.

	Newer dolerite
	Granite
	Ultrabasic rocks
Iron-ore Series ...	Basic lavas Upper shales Banded haematite-quartzite Lower shales Purple and grey limestones (local) Basal conglomerate and sandstone
	Older Metamorphics — Quartzites, quartz-, mica-, hornblende- and chlorite-schists.

The Older Metamorphics consist of a group of metamorphic rocks including hornblende-schists, quartzites, quartz-schists, micaceous and chloritic schists which have been highly folded and eroded before the deposition of the Iron-ore Series. They are found as a series of small exposures isolated by the Singhbhum Granite. Jones states that there is a profound unconformity between them and the overlying Iron-ore Series. These are overlain by purple, somewhat satiny, phyllitic shales and beds of banded haematite-quartzites. The shales contain some deposits of manganese-ore, mainly pyrolusite and psilomelane, derived from the shales themselves by a process of concentration by meteoric waters. They are worked in the Koira valley in Keonjhar south of Jamda and in several other places farther south in Keonjhar and Bonai. The haematite-quartzites, which have a thickness of well over 300 m form prominent, isoclinally folded ridges capped by beds of very high grade haematite. They are composed of alternating layers of cherty silica, jasper and haematite, the individual layers varying in colour from white, through grey and brown to nearly black. The layers are from a tenth to a quarter of an inch in thickness but may sometimes be much thicker. The haematite-quartzites give evidence in some places of the presence of oolitic structures and of siderite which has since been replaced by hematite. They are generally free from clastic debris, indicating deposition in quiet waters far from the shore. They are intricately folded and contorted, the minor structures being apparently attributable to local adjustments after deposition, mostly while still in a plastic condition, and also during replacement and slumping. The iron-ore is thought to have been derived from the enrichment of the haematite-quartzites through solution and replacement of the silica by ferric oxide. In addition to the rich, compact, massive ore forming the outcrops, there are also slightly porous biscuit-like ores and finely crystalline powdery ores known as *blue dust*. Lateral variation from one type of ore to another and to haematite-quartzite, the presence of partly enriched fragments of haematite-quartzite in the ore and other features point to the replacement origin of the ore. Other modes of origin and derivation from tuffs have been attributed to some of the less important ore-bodies of Northern and Eastern Singhbhum.

The Copper Belt Shear Zone.—The Singhbhum thrust zone is from 2 to 5 km wide and is bordered on its north and south by well marked shearing. There is also a major shear zone along the middle, separating two parallel folds both of which are over-turned to the south, dipping at 20° to 40° to N.N.E. There are two sets of cross-folds, one gentle and widely spaced and the other sharper and close-set. One set of joints is parallel to the strike and two others transverse (one roughly parallel to the dip of the folds and the other oblique). The rocks within the thrust zone are metamorphosed shales and sandy shales in the main, with subordinate metavolcanics, *viz.*, chlorite-schists, amphibole-schists and some talc-sepentine-chlorite-magnesite schists with titanomagnetite. The pelitic materials have been extensively migmatised, resulting in feldspathic mica-schists, quartz-muscovite-biotite-schists and gneisses and highly feldspathic granophyric rocks which have been described previously as Soda-Granite or Arkasani Granophyre. Some of these are banded, with alternating bands of quartz-feldspar and biotite-muscovite. Near the northern border of the zone are bands of quartz-mica-schists with tourmaline and kyanite, while garnet, staurolite and kyanite are developed close to the northern shear-border

According to Banerji (1962), mineralization took place in three stages. The earliest was the formation of apatite-magnetite lenses, the individual lenses usually upto 10 m long and 20 cm thick. These closely follow the bands of chlorite- and amphibole-schists. Uranium mineralization is noted in the form of disseminated uraninite, torbernite and autunite along the chlorite-schist as well as in the migmatised bands. Copper sulphide mineralization took place last, but is associated in space closely with the uranium bearing veins. Workable grade of uranium bearing ore, amounting to a few million tons, has been proved in the section near Tatanagar. The only producing copper mine in India is at Mosaboni northwest of Tatanagar, the ores being composed mainly of chalcopyrite, pyrrhotite and pentlandite, with about 2 to 2.5 per cent copper. Other parts of the thrust belt also contain copper ores but the deposits remain to be proved. To the north of the thrust zone, there is a geanticlinal structure exposing mica-, hornblende-, and chlorite-schists, which generally show fairly high grade metamorphism. Immediately adjoining the thrust is a zone containing lenses of kyanite and rocks composed of kyanite, topaz, dumortierite, tourmaline and mica, closely associated with pegmatite. Large deposits of coarsely crystallised kyanite occur here in the Lapsa Buru and its neighbourhood. Just north of the abovementioned geanticline there is a syncline occupied mainly by the Dalma Traps, composed of epidiorites, amphibolites, talcose and chloritic schists, the basal portion often showing a quartzite.

Farther north is a large area of granite-gneiss, known as the CHOTA NAGPUR GRANITE-GNEISS (*Rec.* 65, p. 490, 1932). This is a batholithic mass, part of which is a tourmaline-granite, as in southern Ranchi. Near its borders the granite becomes gneissic and has granitised the schistose rocks as seen in northeastern part of Gangpur (now called Sundargarh district).

J. A. Dunn remapped parts of the Iron-ore Series of South Singhbhum and re-interpreted the succession (*Rec.* 74, p. 28, 1939; *Mem.* 62, Pt. 3, 1940; and *Mem.* 69, Pt. 2, 1942). He thought that the Older Metamorphics were part of the IRON ORE SERIES isolated and metamorphosed by the SINGHBHUM GRANITE. The conglomerates and sandstones south of Chaibasa were assigned to the lower part of the KOLHAN SERIES, which include also limestones and shales. These limestones are worked near Chaibasa for cement manufacture. This series lies partly on the Singhbhum Granite, but is much younger than the Iron-ore Series.

Based on the presently available knowledge, the succession in Singhbhum will be as follows :

Newer Dolerite

Soda granite and Cu-U mineralisation

Chota Nagpur Granite-Gneiss

Dhanjori Orogeny (Singhbhum Orogeny of Sarkar and Saha)

Dalma and Dhanjori lavas

Dhanjori Stage (? Kolhan Series)

Singhbhum granite
Singhbhum Orogeny (Iron-ore Orogeny of Sarkar and Saha)
Iron-ore Series

Older Metamorphics

The geology of the area north of the Singhbhum thrust zone and east of the large Singhbhum Granite exposure has been described by Sarkar and Saha (1962) and by Iyengar and Alwar (1965). According to the latter authors, a geosyncline developed to the north and east of the Singhbhum Granite craton, in the Dhanjori and Dalma areas. The basal quartzite-conglomerates in this basin are followed by black carbonaceous phyllites and a turbidite sequence which is overlain and intercalated with submarine volcanic flows and tuffs. Marginal fractures in the Mayurbhanj and Dhanjori areas were intruded by gabbro-anorthosites accompanied by vanadiferous magnetite lenses and pyroxene-granulite granophyre. A quartzite conglomerate horizon occurs above the carbonaceous phyllites. This geosyncline was then subjected to folding and intrusion of granites, *viz.*, the ROMAPAHARI GRANITE in Mayurbhanj, the CHAKRADHARPUR GRANITE in the Ghatsila region, and the Soda-Granite in the Singhbhum Shear zone. The Dhanjori geosyncline contains some 9,000 m of sediments but the thickness is about 4,500 m to the north of the shear zone. The cratonic region also experienced intrusion of ultrabasic lenses and dykes of Newer Dolerite. These authors also believe that the Mayurbhanj iron formation is older than the Iron-ore Series of Keonjhar-Bonai and that the sequence found to the north of the Singhbhum Copper Belt shear is younger than the Iron-Ore Series as it is the same as, and continuous with, that in the Dhanjori syncline.

The sequence in North Singhbhum and in Dhanjori and Mayurbhanj may now be given as below :

Romapahari (Mayurbhanj), Kuilipal (Dhanjori) and Arkasani Soda-Granite (shear zone), all syntectonic
" Dhanjori orogeny " (Singhbhum orogeny of Sarkar and Saha)
Pyroxene granite-granophyre
Gabbro-anorthosite (V magnetite in places)
Spilitic lavas and quartzite (Dhanjori and Dalma lavas)
Turbidite sequence with quartz-conglomerate
Black phyllite
Conglomerate
———————————— Unconformity ————————————
Singhbhum Granite, mostly syntectonic
" Singhbhum orogeny " and folding of Iron-ore Series (— Iron-ore Orogeny of Sarkar and Saha)
Iron-ore Series of S. Singhbhum-Keonjhar-Bonai

Sarkar and Saha called the mountain building movements which followed the deposition of Iron-ore Series and which was accompanied by the intrusion of the Singhbum Granite as ' Iron-Ore Orogeny. ' It would most appropriately be called the SINGHBHUM OROGENY, particularly as Iyengar and Alwar have now designated the one following the Dhanjori-Dalma assemblage of rocks as DHANJORI OROGENY.

Intrusive into the Iron-ore Series are ultramafic rocks, the Chota Nagpur Granite-Gneiss, Singhbhum Granite, the Arkasani Soda-granite and dyke of Newer Dolerite.

Ultramafic Rocks.—In South Sighbhum there is a patch of altered peridotiti rock near Nurda which may be pre-Singhbhum in age. To the same age may belong the chromite bearing rocks (saxonite, lherzolite etc.) found near Chaibasa. The chromite bodies in them have been worked for nearly 50 years.

The gabbro-anorthosites associated with lenses of vanadiferous magnetite in East Singhbhum and the pyroxene-riebeckite-granite-granophyre of Simlipal in Mayurbhanj belong to the period of development of marginal fractures along the Dhanjori geosyncline, prior to the Dhanjori Orogeny and the emplacement of the Romapahari and Kuilipal Granites. These granites appear to be of a later age, probably contemporaneous with the formation of the shear zone and its mineralisation by copper and uranium bearing fluids.

The Chota Nagpur Granite-gneiss occupies an immense tract to the north of the Dharwarian rocks of Singhbhum and Gangpur. The northern belt of this extends from Santhal Parganas through Hazaribagh to Palamau and the southern one from Bankura to Ranchi and Jashpur and farther west. It is distinctly intrusive into the Iron-ore Series and assumes a banded and composite aspect near the margins of the schistose rocks, as for example along the Gangpur — Ranchi border. It is generally coarse and prophyritic and contains quartz, microcline, orthoclase, oligoclase, biotite, a little apatite (and occasionally some green hornblende, in Singhbhum). Tourmaline is frequently seen but especially abundant in the pegmatite phase as in Southern Ranchi. In parts of Manbhum, Ranchi and Hazaribagh, it weathers into tors and is called the *Dome Gneiss*. The composite form of the rock used to be referred to as the *Bengal Gneiss* in earlier geological literature. Its later phases are pegmatites, aplites and quartz veins, the last being often auriferous in Chota Nagpur.

Singhbhum Granite.—This great batholithic mass cocupies an area of several hundred sq. km. in Singhbhum, Keonjhar, and Mayurbhanj, from Chaibasa in the north and to beyond Keonjhargarh in the south. According to A. K. Saha, who has made a detailed study of the northern part of this mass, it consists of a series of domed up intrusions varying in composition from biotite-granodiorite to adamallite, biotite-trondjhemite and leucogranite. The margins are hornblendic, chloritic or epidotic granodiorites and pyroxene-diorites derived by the granitisation of the country rocks. The main mass shows distinct N.-S. or N.N.W.-S.S.E. foliation which tends to become parallel to the margin of the country rocks. A few patches of unassimilated older basic rocks and quartzites occur in the granite. The common granodiorite consists of perthitic microcline (20 to 30 per cent Ab.) with oligoclase-andesine (12 to 28 per cent An.), quartz, biotite, subordinate muscovite and chlorite. The accessory minerals are epidote, sphene, zircon, apatite and rutile. The marginal structures have been controlled by the country rocks, while the central portions show evidences of forceful intrusion. The granite is traversed by numerous dykes of NEWER DOLERITE. Iyengar and Alwar believe that the granite, especially near its eastern margin, may contain younger intrusive units belonging to the Dhanjori orogeny which appears to be contemporaneous with the Satpura orogeny of the Bihar Mica Belt (950-1000 M.Y.) associated with intrusion of granite and mica bearing pegmatites.

The Arkasani Soda-granite.—This rock, which is granophyric in part, is found along the thrust zone of the Copper Belt in several places. It varies from a feldspathic schist to a fairly coarse-grained granite, the former often showing thin bands and streaks of mafic minerals. Granophyric texture is fairly common in some exposures. The typical rock contains partly altered plagioclase, quartz, muscovite, biotite, and grains of magnetite and apatite. Sericitisation and epidotisation of the feldspars is common. The first two minerals make up the bulk of the rock. It is regarded as having been formed from the granitization and replacement of muscovite-quartz-schists which originally occupied much of the thrust zone. Radiometric dating seems to show that it is approximately of the same age as the mica bearing pegmatites of Bihar (about 950 M.Y.).

Newer Dolerite.—The Shingbhum Granite is traversed by very numerous basic dykes which are often several km. long and several metres thick, with a maximum thickness of 700 m. They are mainly doleritic to basaltic in composition, but differentiation is indicated by the presence of some dykes of picrite, perknite, norite, granophyre and epidote-bearing rock. The majority run N.N.E.-S.S.W. while a subsidiary direction of N.N.W.-S.S.E. is also seen. A dyke in Seraikela described by Dunn and Dey is a chilled Mg-rich rock with an Fe/Mg ratio of 1 : 1. There is indication that the fine grained dykes represent the original magma with Fe/Mg ratio of 2 : 1 (Iyer, 1932 ; Krishnan, 1936 ; Dunn and Dey, 1941). A well differentiated sill, called the Amjori sill, of the same age, has been described by Iyengar et al (1964) from the Simplipal hills in Mayurbhanj, on the eastern margin of the Singhbhum Granite. It is a disc-shaped sill, 400-metres thick, intruded into the upper Dhanjori beds, the covering layers having been removed by erosion. According to them the bottom of the sill shows a chilled rock which represents the original magma, consisting of plagioclase, pyroxene, some iron ore and no glass. This is succeeded, in order, by dunite, peridotite (with coarse pyroxene), picrite, and gabbro at the top. The peridotite contains orthopyroxene and some augite with a little biotite in the upper portion. The dunite consists of olivine, ortho- and clino-pyroxene, rare plagioclase, some chromite and magnetite. The picrite with pyroxene and calcic plagioclase passes into gabbro. A fault traverses the centre of the basin and it is occupied by a feldspar-rich hyperstene-gabbro with a little quartz, which is altered in places to hornblende-gabbro. The altered phase contains andesine, microcline, chlorite, epidote, and quartz-oligoclase intergrowth. The last phase is a rock with hornblende, actinolite, biotite, perthite, microcline and occasional quartz. The trend of differentiation indicates that the original Mg-rich magma became Fe-rich in the later stages.

A few radiometric age data are now available on some of the Singhbhum rocks but they should be treated with caution until they are confirmed by more data by using different methods. Muscovite from pegmatite in the Older Metamorphics gave an average age of 3035 M.Y. showing that these formations are the oldest in the area. Muscovite and feldspar from Singhbum granite show that the granite was emplaced 2000 M.Y. ago. Inclusions of Iron-ore Series in the granite indicate that these sediments may be 2200 to 2500 M.Y. old. Phyllitic shales from the Kolhan Series gave discordant ages of 1035 and 1584 M.Y. Rocks and minerals from the shear zone of the copper belt have ages in the range 900 to 1000 M.Y., showing that the metamorphism and mineralisation in this belt are of the same age as the mica-pegmatites of Bihar.

GANGPUR

To the west of the Singhbum district, in Gangpur (now called Sundargarh district), there is an anticlinorium or geanticline which has an E.N.E.-W.S.W. axial direction. The structure is closed towards the east but is cut up and obscured by granitic intrusives to the west. The strike becomes W.N.W.-E.S.E. in western Gangpur, in conformity with and influenced by the structural trends of the Mahanadi and Brahmani valleys just to the south and southwest of this region.

The anticlinorium shows the following succession of rocks, as worked out by Krishnan (Table 18) and named by him the Gangpur Series.

TABLE 16—The Gangpur Series

Iron-ore Series (?)	Phyllites, slates and lavas
	Raghunathpali conglomerate
— — — — —	— shear zone — — — — —
Gangpur Series	Phyllites and mica-schists
	Upper carbonaceous phyllites
	Calcitic marbles
	Dolomitic marbles
	Mica-schists and phyllites
	Lower carbonaceous quartzites and phyllites
	Gondites with associated phyllites (Base not seen)

There is a general increase in the grade of metamoprhism when the rocks are followed from the Singhbum border on the east to the centre of the anticlinorium on the west. It should, however, be noted that some of the rocks, which have phyllitic appearance and characters, are really products of retrogressive metamorphism, containing relics of garnet, staurolite, biotite, etc. The Satpura strike (E.N.E.-W.S.W.) is found to be superimposed on an earlier, presumably Eastern Ghats, strike which is prominent.

The oldest rocks are gondites, found in the central or axial region of the anticlinorium. They contain, besides quartz-spessartite rocks, also those with rhodonite, blanfordite, winchite, etc., associated with workable bodies of manganese-ore. They are succeeded by carbonaceous quartzites and phyllites, dolomitic and calcitic marbles and carbonaceous phyllites, these being intercalated with phyllites and mica-schists. The carbonaceous phyllites are flaggy or slaty in certain places while the marbles contain very large reserves of good limestone and dolomite which are now being used as fluxes in the iron-smelting furnaces of Bengal and Bihar. Large qunatities of the limestone are also burnt into quick-lime, well-known in the Calcutta market as Bisra lime, named after Bisra which is a railway station near the Singhbhum-Gangpur border. At the top of the succession is a shear zone in which the Raghunathpali conglomerate is involved. It is a sedimentary conglomerate which has suffered intense shearing as a result of which an autoclastic character has been imposed on it. The overlying beds are phyllites and mica-schists belonging to some part of the Iron-ore Series. The Gangpur Series is intruded by basic sills (Dalma traps) and by bosses of the Chota Nagpur granite. The basic rocks have been converted into schistose amphibolites and epidiorites containing amphibole, clinozoisite, ilmenite and magnetite.

Radiometric dating of biotites etc. from three different horizons in the Gangpur Series has given ages of 910 to 990 M.Y. indicating that the rocks were folded and metamorphosed at the time of folding of the rocks north of the Copper Belt and the intrusion of Chota Nagpur granite-gneiss. This needs confirmation by further work.

THE SONE VALLEY AND ADJOINING AREAS

North and northwest of Ranchi and Hazaribagh, schistose rocks are found in the Mirzapur district, in the drainage basin of the Sone river which is a tributary of the Ganges. Mallet recognised two series in this area, separated by an unconformity. The lower division, called the *Agori Stage*, includes slates, chloritic schists, schistose quartzites, jaspers, thin limestones and basic igneous rocks. There are also some slates and porcellanoids which are thought to constitute the upper division. No marbles or gonditic rocks seem to have been found amongst these rocks. It is intruded by gneissic granite which is evidently the same as the Chota Nagpur Granite-gneiss. In the Palamau district there are crystalline limestones, graphitic schists, epidiorites and quartz-magnetite rocks.

In parts of the Gaya and Hazaribagh districts there are various types of schistose rocks — biotite-schists, sillimanite-gneisses, calc-granulites and epidiorites, which have been extensively granitised by the Dome Gneiss intrusive into them. The general strike of foliation of the rocks is E.N.E.-W.S.W. The Dome Gneiss is here contaminated by the absorption of schists and is characterised by the presence of quartz, microcline, oligoclase, biotite and occasional hornblende, with fluorite as an important accessory. Numerous mica-bearing pegmatites traverse the gneiss and schists and contain rich deposits of mica together with beryl, monazite, pitchblende, triplite and other minerals. The uraninite indicates an age of about 950 M.Y. for the pegmatites.

Farther east, in the Rajgir, Kharakpur, Gidhaur, Shaikpura and other hills, there are quartzites, crush-conglomerates, jaspery quartzites, slates, phyllites and mica-schists, having a general E.-W. or E.N.E.-W.S.W. strike. The quartzites are generally the most prominent members and form scarps. The schistose rocks have been highly disturbed and have an irregular boundary with the gneisses which are intrusive into them. The age of biotites in some of these rocks indicates that the rocks suffered metamorphism about 360-420 M.Y. ago. (Sarkar et al., 1964).

BENGAL

The Midnapur area of Bengal is continuous with Dhalbhum (Eastern Singhbhum) and contains gneisses and schists similar to those found in the latter. The gneissic rocks are of the type formerly known as the Bengal Gneiss, akin to the Peninsular Gneiss of South India.

A few miles to the south of the Raniganj coalfield, south of the Damodar River, there is an interesting area of anorthosites, varying from almost pure plagioclase rocks to anorthosite-gabbro and norite with minor patches of olivine norite and with dykes of dark granulitic norite. The mass extends in an eastwest direction for about 32 km with a maximum width of 10 Km. in the eastern part where its northern boundary is faulted against the Gondwanas and is more or less marked by the Damodar River. In the south it shows intrusive relationship towards the Bengal Gneiss. Near the junction, there are inclusions of gneiss in the anorthosite and veins of anortho ite in the gneiss. The average composition of the feldspar of the ancrthosite is An 65. The rocks show marked parallelism of minerals, protoclastic marginal granulation of the plagioclase and the presence of Carlsbad twinning, these features indicating primary origin. The feldspars have been altered to scapolite and epidote. The rocks have undergone subsequent regional metamorphism with the development of garnet, clouding of the feldspars and corona structure with garnet around olivine. Dr. S.C. Chatterjee considers these anorthosites to have been formed by the differentiation of a gabbroic magma.

ASSAM

The Assam plateau lies along the continuation of the Archaeans of Bihar but is separated from the latter by the Ganges-Brahmaputra valley. The plateau comprises the Garo, Khasi and Jaintia hills and to its northeast is the detached area of the Mikir hills.

The Archaeans are represented here by gneisses, schists and granites, having a general N.E.-S.W. direction of strike of foliation. There is naturally much local variation, and in parts of the Garo hills the Satpura strike may be seen.

Extensive tracts of the ancient rocks are found in the Khasi and Jaintia hills. The oldest seem to be banded, composite, biotite-granite-gneisses. The granitic constituent is sometimes porphyritic and sometimes fine grained and aplitic, and consists of quartz, microperthite, some microcline, oligoclase and biotite, with garnet, apatite, zircon and rare sphene as accessories. The gneisses are associated with garnet-quartzites with or without sillimanite. In the area west of Shillong, there are hornblende-biotite-gneisses and biotite-cordierite gneisses with N.E.-S.W. strike of foliation.

In the granite-gneiss of the Nongstoin area, Khasi Hills (Assam) occur lenses of quartz-biotite-sillimanite-cordierite rocks and quartz-sillimanite rocks with sillimanite-corundum masses. These occupy a belt one Km. wide, with general E.-W. strike. The economically important lenses consist of massive sillimanite with a little corundum, some showing sillimanite only and a few corundum only. To the east, these rocks are cut up by granite, in which lenses of sillimanite rock may be found. The massive sillimanite rock is sometimes traversed by veins of coarse fibrous sillimanite. On weathering the sillimanite alters to kaolin or to a micaceous material.

The gneissic complex is apparently overlain by the Shillong Series which is regarded as younger. The Shillong Series is mainly of sedimentary origin. It is composed of quartzites, conglomerates, phyllites, sericite-, chlorite-, mica-, and hornblende-schists, with occasional carbonaceous slates and banded ferruginous rocks. Some of the schists are garnetiferous. The assemblage bears some resemblance to the Iron-ore Series of Chota Nagpur. Similar rocks are exposed in the Simsang valley in the Garo hills and also in parts of the Mikir hills.

The Shillong Series was first intruded by the KHASI GREENSTONES — epidiorites, amphibolites and amphibole-schists — which have been folded up with them and are therefore presumably of Pre-Camb4ian age. It shows some gradation towards gneisses near the junction with granitic rocks. Distinctly of later age is the MYLLIEM GRANITE which forms bosses and also thin interfoliar veins in the schists. It is a homogeneous, fairly coarse biotite-granite containing porphyritic pink microcline, orthoclase, some acid plagioclase, biotite and hornblende, with apatite, zircon, magnetite and sphene as accessories. The potash feldspars are often perthitic. Over most of its area the granite is fairly massive and non-foliated but is occasionally streaky and may show xenoliths of quartzite and basic segregations. A grey granite also occurs in these areas which is thought to be a variety of the Mylliem granite. The granite is later than the Khasi greenstone.

In the granite and gneisses are found certain lenses and patches of intermediate to basic pyroxene-granulites which greatly resemble the charnockites. Similar occurrences, thought to be metamorphosed mixed rocks, have been found in recent years in Mysore, Bastar, and other areas.

The granite, and to some extent the gneisses, are traversed by dykes of dolerite; in the more southern areas there are flows of the same rock which are frequently vesicular and amygdaloidal, with intercalated ash-beds. These are the *Sylhet Traps*, which are pre-Upper Cretaceous in age and resemble the Rajmahal Traps. The rock is a dolerite or basalt, sometimes with olivine which is generally more or less serpentinised.

CORRELATION OF THE PENINSULAR ARCHAEANS

The study of the Archaean rocks is beset with many difficulties which do not crop up in the case of later sedimentary systems. There is, in the Archaeans, a complete absence of fossils which are of invaluable help in determining the geological age. They include a great variety of formations, both igneous sedimentary, but the original characters have been obliterated by repeated changes. Metamorphism has produced not only mineralogical and structural changes but has also removed or introduced materials resulting in marked changes in composition. In addition to the effects of temperature and pressure, there are also those of igneous contact, magmatic stoping, assimilation and hybridism. The cumulative effect of these factors is the production of a bewildering variety of petrological types with complex characters which must necessarily be confusing and difficult to unravel. It is common experience that similar rock types may originate from very diverse original materials and that quite dissimilar types may be evolved from one type of original rock.

In the correlation of sedimentary formations it is usual to rely on the lithology, stratigraphical superposition and fossil contents for purposes of age determination and correlation, since these are available for settling questions of inter-relationship of strata. These criteria are, however, either absent or comparatively of little help in the case of Pre-Cambrian formations, and especially of metamorphic complexes, for they are devoid of fossils, their lithology is transformed by metamorphism and their stratigraphical relationships confused by intricate folding, inversion, disturbance and dislocation. Sir L.L. Fermor has discussed this question in the introductory part of his memoir on the 'Ancient schistose formations of Peninsular India' (*Mem. G.S.I.* LXX, Pt. I, p. 9-25).

In recent years, the constant rate at which radioactive disintegration of certain elements takes place has been employed to measure the absolute ages of minerals and rocks. By suitable techniques the present amounts of the isotopes of such elements are measured and calculations are made to find out the time taken to produce the quantities of the derived isotopes which gives the age of the rock or mineral containing the elements. In using these methods, precautions have to be taken to ensure that the materials used have not been altered by geological processes and no appreciable loss or gain of isotopes has occurred. By taking suitable precautions and allowing for uncertainties, the age can be found. U-Pb, Th-Pb, K-Ar and Rb-Sr methods are in vogue at present. If all of them could be used and if concordant results are obtained, it may be taken that the results obtained are accurate.

CORRELATION

The schistose rocks are now recognised as the oldest in most of the areas. The oldest rocks are probably of igneous origin, the sedimentary material gradually increasing in the younger strata. In Rajasthan, however, the Balanced Gneissic Complex and the Bundelkhand gneiss have been regarded by Heron as older than the Aravallis.

The Older Metamophics of Singhbhum appear to be the oldest rocks in Northern India. Radiometric ages are now becoming available, but it will be some time before enough data are available to unravel the Pre-Cambrian history of the Indian shield.

TABLE 17 — ROUGH CORRELATION OF THE PENINSULAR ARCHAENS

Mysore	Madras	Ceylon	Eastern Ghats and Bastar	Chota Nagpur	Central Provinces	Rajasthan	Assam
Closepet granite	Bellary, Hosur Arcot and other granites	Wanni gneiss	Garnite	Singbhum granite Dome gneiss ?	Amla granite	(Bundelkhand Gneiss ?) Alkali-syenite ?	Mylliem Granite
Charnockite	Charnockite	Charnockite	Charnockite	...	...	...	...
Peninsular gneiss	Peninsular gneiss	Bintenne gneiss	Granite-gneiss	Chota Nagpur or Bengal gneiss	Granite-gneiss	?	Granite-gneiss
Upper Dharwar (Clays, silts, grits, etc.)	?	?	?	Kolhan Series ?	?	Raialo Series ?	Shillong series
Middle Dharwar (Banded ferruginous rocks quartzites, etc.)	Magnetite W and haematite quartz schist, etc.	Khondalites, Calc-gneisses, etc.	Bailadila Iron-ore series	Iron-ore series	Sakoli series	Aravalli system & Champaner series	Granulites, calc-gneiss, etc.
Lower Dharwar (Chloritic, hornblendic and micaceous schists, etc.)	Chloritic schists, etc.	?	Bengal series, Kohndalites, Kodurites, et.	?	Sausar series Somawani series (Gondites, marbles, etc.)		
Oldest Archaeans	?	?	?	Older metamorphics	Older schists ?	Bundelkhand gneiss ? Banded gneiss ?	?

CHAPTER V

THE ARCHAEAN GROUP—EXTRA-PENINSULA

Pre-Cambrian formations occur throughout the length of the Himalaya but only a few regions have so far been studied — Kashmir-Hazara, Simla-Garhwal, Sikkin-Bhutan and parts of Burma. Information on other parts is either very meagre or wanting.

There is also a special difficulty in this region, *viz.*, that there is not always sufficiently detailed information for separating the Archaean from Proterozoic formations. We have therefore to deal with all the Pre-Cambrian rocks of the Himalayas here. Formations regarded as of Purana age will be dealt with in a later chapter.

NORTH-WEST HIMALAYA

Pre-Cambrian rocks are developed in this region in Chilas, Gilgit, Baltistan, Northern Kashmir, Ladakh and Zanskar and continue through Kumaon into Nepal and Sikkim. In Kashmir and Hazara they are called the SALKHALA SERIES and comprise slates, phyllites, quartzites, mica-schists, carbonaceous and graphitic schists, crystalline limestones, dolomites and biotite-gneisses. They are highly folded and compressed and have been involved in the movements which brought the Himalaya mountains into being. The Salkhalas are well seen in the Nanga Parbat, in the mountains north of the Kishenganga and in the Pir Panjal where they are highly metamorphosed and subjected to granitisation. They are often found thrust over rocks of Permo-Carboniferous or later ages. As is to be expected, the grade of metamorphism varies from place to place, some of the slates in the less affected areas being scarcely distinguishable from the Dogra slates of a later (Purana) age.

The Salkhalas are associated with a gneissic complex, parts of the constituents of which might possibly be older. The gneisses include granulites and biotite-gneisses containing quartz, orthoclase, acid plagioclase and biotite, sometimes with prophyritic structure and prominent gneissic banding. There are also some hornblendic gneisses in the complex. The bands of the gneisses comprise schists of various descriptions — biotitic, muscovitic, hornblendic, talcose and chloritic. The gneissic rocks are well displayed in the region of the Zanskar Range and to its north. Exposures are also seen in the Dhauladhar Range, Pir Panjal, and other areas.

The Salkhalas are, according to D. N. Wadia, comparable to, and probably homotaxial with, the Jutogh Series of the Simla area. Much of the intervening area remains yet to be mapped.

The Salkhalas and the gneisses are traversed by later igneous intrusives including gabbro, pyroxenite, dolerite, hornblende-granite, tourmaline-granite and pegmatite. The "Central Gneiss" of the Himalayan is apparently a mixture of rocks of various ages, mainly of granitic composition, some being Tertiary and some pre-Tertiary. The hornblende-granite is presumably of

Tertiary age as it is seen to be intrusive into Cretaceous rocks at the head of the Burzil valley.

The Salkahalas and the gneisses are succeeded by the DOGRA SLATES which are mainly argillaceous with minor layers of quartzites, quartzitic slates and flags. They are unfossiliferous and are overlain by the fossiliferous Cambrians of Kashmir. Since they form a thick series, their range in age is not known, but they would appear to extend downward from Lower Cambrian to Pre-Cambrian and may be the equivalents of the Vindhyans and possibly also part of the Cuddapahs.

The Dogra slates are also found in the Pir Panjal and parts of the Kishenganga valley (Muzaffarabad district). Similar rocks are present in Hazara and in the Attock district of the Punjab where they are called the Attock Slates.

SPITI VALLEY AND KUMAON

The region between the Central Himalayan range and the Zanskar range is of great interest as it contains a complete set of formations from the Pre-Cambrian to the Cretaceous. The region of the Spiti Valley has been studied by Griesbach, Hayden, Diener and Von Krafft. In this region, there are highly folded mica-schists, slates and phyllites which constitute the VAIKRITA SYSTEM named by Griesbach. (*Mem.* XXIII, p. 10, 1891). They rest upon gneisses but do not show any clear boundary with the overlying rocks. Recent work indicates that the Vaikritas are merely the highly metamorphosed and granitised Haimantas and their equivalents. The term is vague and may therefore be discontinued.

In the same area there are large thicknesses of unfossiliferous slates, phyllites and conglomerates which were named the HAIMANTA SYSTEM by Griesbach (*Mem.* XXIII, 1891). The thick conglomerates forming the Lower Haimantas have since been proved to be of Ordovician age. The original Middle and Upper Haimantas are now regarded as of Lower Cambrian to late Pre-Cambrian age. They will be described in the chapter dealing with the Cambrian formations.

SIMLA — GARHWAL

Jutogh Series — In the Sub-Himalayas of Simla there are rocks similar to the Salkhalas of Kashmir, assigned to the JUTOGH and CHAIL SERIES. The Jutogh Series is composed of quartzites (Boileauganj Quartzites) carbonaceous slates, limestones and mica-schists with intercalations of hornblende-schists which are metamorphosed basic igneous rocks. Two large exposures are present, one around Jutogh near Simla and the other around the Chor Mountain (30° 52' : 77° 29'), besides a couple of small patches. The rocks show a high degree of metamorphism and have been thrust over the less metamorphosed and younger Chail Series. They are regarded as forming a highly compressed recumbent double anticline, the Chor granite occupying the core of the intervening syncline. Near its contacts, the Chor granite has produced a metamorphic aureole in which different minerals have been developed at different distances from the contact — chlorite, muscovite, chloritoid, ottrelite, staurolite, kyanite, garnet. The granite has been intruded after the folding of the Jutoghs, for the contact zone of the granite cuts across the fold structures. The granite is probably of late Palaeozoic age.

The CHAL SERIES are mainly quartzites, limestones and schistose slates. They are thrust over the Simla Slates and the Blainis They resemble part of the Jutoghs. They are represented in the Chakrata area where they have been thrust over the Deoban Limestone and the Mandhali Series which are probably of late Pre-Cambrian age.

In Garhwal there are metamorphosed sandstones, shales, calcareous shales and limestones which may be the equivalents of the Jutoghs. They have been thrust over unmetamorphosed sediments. They are well seen for some miles northwards of Vishnuganga where they comprise para-gneisses, quartzites, garnetiferous mica-schists and granulites, having an estimated thickness of 9,600 m. The arrenaceous members of this group have preserved current-bedding structures in spite of the general metamorphism (*Rec*. LXIX, p. 134, 1935). The calcareous members of the suite, which are exposed near Badrinath, resemble the Salkhalas and they have been invaded profusely by tourmaline-pegmatites. The slaty and schistose rocks of this series in Kumaon and Garhwal contain copper-ores in places. They thus resemble the Dalings of Sikkim.

NEPAL—SIKKIM

The Archaeans are here represented by the DALING and DARJEELING SERIES. The Daling Series is a schistose group; it grades through a transition zone into the dominantly gneissic Darjeeling Series.

The Dalings are typically slates and phyllites in the lower part and sericitic and chloritic phyllites in the upper part. The Dalings of Sikkim contain lodes of copper-ore in some places. In the transition zone the phyllites carry porphyroblasts of chlorite and biotite with occasional zones containing tiny garnets. These pass into garnet-biotite-schists, granulites and schists containing staurolite, kyanite and sillimanite. These schistose rocks are interbanded with granite-gneiss. In Northern Sikkim and adjacent parts of Nepal, there are also such rock types as marbles, calciphyres, quartzites and pyroxene-granulites amidst the gneisses.

Since the Dalings practically always underlie the Darjeelings and show a different grade of metamorphism, the two were formerly considered to be distinct series separated by a hypothetical thrust zone. J. B. Auden regards the presence of a thrust zone improbable (*Rec. G.S.I.* LXIX, p. 123-167, 1935). The Darjeelings seem to be merely the granite-injected and highly metamorphosed upper part of the Dalings. Auden has also suggested that the granitic rocks of Dudatoli and Lansdowne in Garhwal may be the same as the granitic constituent of the Darjeelings, and that a similar granite has given rise, by disintegration, to the felspathic sandstones of the Middle Siwaliks of the Darjeeling-Nepal foot-hills.

The DALING and DARJEELING SERIES have been thrust over the Gondwanas with which they are sometimes intercalated as thrust slices. They form a large recumbent fold, thrust to the south. In the core of the fold are the Kanchenjunga, Paunri and some other massifs, while Mount Everest is in its upper limb. The Dalings stretch eastward into Bhutan where, beyond Mo Chu, the Buxa Series intervenes between them and the Tertiaries. In the Everest region the representatives of both the Dalings and Darjeelings are seen, the former attaining a thickness of about 1000 m. They seem to have been intruded by earlier (? Palaeozoic) as well as later (Tertiary) granites. A belt of the Dalings passes through the Tamur Kosi and Arun valleys and through Katmandu plains extending a few miles to the north. They comprise some metamorphosed and granite-injected sandy limestones, granulites, chlorite-biotite-schists and phyllites in the lower beds, followed by quartzites, phyllites, and carbonaceous phyllites in the upper beds. Near Chandragiri Pass, they are overlain by purple and green phyllites, banded shales and sandy limestones, some of the beds showing ripple marks and resembling the Semris of the Sone valley. The associated thin marble beds are reminiscent of the Rohtas limestone of the same series. Near the top of these beds some Ordovician fossils have been obtained, so that it would appear that the succession in the Chandragiri-Katmandu section contains Dalings, Semris and Lower Palaeozoic rocks.

The Dalings resemble the Salkhalas. The non-calcareous part is similar to the Chandpurs which have also a close resemblance to the Cambrians of Northern Kashmir. The calcareous part, on the other hand, resembles the Jutoghs of Simla. On the whole they are considered as belonging to the Pre-Cambrian.

BHUTAN AND EASTERN HIMALAYA

The Darjeeling and Daling Series continue into the sub-Himalayan area of Bhutan and also farther to the east. A summary of the work of Gansser in Bhutan is given in the chapter on the Cuddapah System. Argillaceous schists and gneisses are known in the Aka hills, these having been correlated with the Darjeelings by La Touche. Godwin-Austen found similar rocks in the Daphla hills. The Abor hills near the Sadiya frontier tract contain quartzites, phyllites, slates, mica-schists, limestones and dolomites, according to J. C. Brown. The Mishmi hills show a large variety of schistose and gneissic rocks including quartzites, mica-schists, chlorite-schists, amphitolites, granulites and kyanite and garnet bearing rocks. A. M. N. Ghosh has recorded that these are associated with composite gneisses, intrusive granites and pegmatites. The rocks therefore correspond to the Dalings and Darjeelings. The Miju hills and the Daphabhum similarly show quartzites, limestones, schists, garnetiferous gneisses, etc.

The Himalayan Archaeans have not been studied in detail, but a rough indication of the correlation may be given here. The Salkhalas of Kashmir-Hazara, the Jutoghs and Chails of the Simla region, some of the gneisses of the Central Himalayas of Garhwal and Kumaon, the Dalings and Darjeelings of Nepal-Sikkim, and the gneisses and schists of the Assam Himalaya seem to be generally the equivalents of the Dharwarian rocks of the Peninsula. In addition to the folding, metamorphism and igneous intrusions suffered by the Himalayan rocks in Archaean times, they have been subjected to the mountain-building movements of Tertiary times, being thereby sheared, overthrust and often inverted. They present therefore extremely complicated structures which can be unravelled only by prolonged and careful study.

BURMA

The western part of Burma consists of comparatively young strata ranging in age from Cretaceous to Tertiary. The older rocks — Archaeans, Palaeozoic and Mesozoic — occur in the belt which includes the Shan Plateau in the north and Tenasserim in the south.

MYITKYINA

The northern end of the Shan belt is the Myitkyina region where the occurrence of ortho-gneisses and schists of Archaeans age is known. They extend northwards into the adjoining parts of China.

MOGOK STONE TRACT

The Mogok tract, which lies between the Irrawaddy and the Shan Plateau some distance north of Mandalay, is an exceedingly interesting area since it contains a variety of rock types of yields several gemstones. The crystalline complex here, called the MOGOK SERIES, consists of a group of gneisses and schists of mixed origin — biotite-gneisses, cordierite-gneisses, garnet-biotite-granulites, garnet-sillimanite-schists, crystalline limestones, calciphyres calc-granulites, quartzites, pyroxene-scapolite-gneisses, etc. The garnet-sillimanite-schists resemble the khondalites of the Eastern Ghats. There is a gradation from the crystalline limestones to calciphyres and calc-granulites, and these are interbanded and occur also as inclusions in the later intrusives. They are folded with the gneisses and have a general strike of E. 30° N. and very steep dips. The minerals in the calciphyres and calc-granulites are calcite, dolomite, spinel, diopside, phlogopite, forsterite, scapolite, sphene and graphite. At their contact with acid intrusives are found rocks containing diopside, feldspar, nepheline and calcite. The excess alumina in the intrusive and in the calcareous rocks has crystallised out as ruby, sapphire or ordinary corundum. The quartzites of this region apparently represent arenaceous bands associated with the calcareous sediments.

The sedimentary and metamorphic series are intruded by syenites and granites and by a series of minor intrusions including basic and ultrabasic rocks. The syenites are found mainly as sheets and lenses and have been involved in the folding of the region. They consist of dominant orthoclase and microperthite with moonstone schiller, some quartz, aegirine, hypersthene or titanaugite. Iron-ore, zircon and apatite are the chief accessories. There are also alkali rocks in close association with the limestones ; indeed, the alkali rocks seem to be absent away from the calcareous rocks. This lends support to the view that the alkali rocks and syenites are products of desilication of a granite by the limestone, in accordance with Daly's hypothesis of origin of alkali rocks. But, as the syenites are disturbed and folded, they may represent an earlier phase of granitic intrusion than the KABAING GRANITE. The Kabaing granite and associated pegmatites are unaffected by the folding, the former often containing inclusions of limestones and other rocks in its peripheral portions. The Kabaing granite consists of abundant feldspar, quartz, biotite and minor accessories, the feldspar being frequently partly kaolinised.

Closely associated with the calcareous rocks there are hornblende-augite rocks, amphibole-pyroxene-feldspar rocks, hornblende-nepheline and aegirine-nepheline rocks and different types of nepheline syenites, which occur as dykes and sills. There are also utlrabasic types including peridotites, pictites, norites and garnet-pyroxene rocks resembling eclogites. Amongst these minor groups of intrusives are rocks closely resembling intermediate and basic members of the charnockite suite.

A large variety of gemstones is found in this region, of which the ruby is the best known. Ruby and other gem mining has been carried on here for many years and still continues. The other gemstones are sapphire, colourless corundum, spinel, scapolite, apatite, nepheline, garnet, peridot, lapis-lazuli, zircon, topaz, tourmaline, beryl (acquamarine), rock crystal, amethyst and moonstone.

The rock types of this area bear the impress of high grade metamorphism and have a typical Archaean aspect. All recent workers agree that the crystalline limestones and associated schists are metamorphosed sedimentaries. Amongst the gneissic complex of this region there are igneous constituents of different ages. Some of the basic and ultrabasic intrusives and the Kabaing granite represent later intrusives.

L. A. N. Iyer (*Mem.* 82, 1953) who mapped a large part of the area and L. L. Fermor appear to regard all the Mogok rocks as of Archaean age. But E. L. G. Clegg has suggested (*Mem.* 74, Pt. 1, p. 9, 1941), that some of the rocks, and especially the limestones, may be of a later age. The Mogok limestones are said to be continuous with those farther north and the limestones in the second defile of the Irrawaddy have been proved to be as young as the Cretaceous.

SHAN STATES

Between Mogok and the fossiliferous formations of the Northern Shan States there is an extensive area of the Archaean rocks, well developed in the Tawng-Peng State and called the TAWNG PENG STYSTEM by La Touche. This includes the biotite-schists of Mong Long, the Chaung Magyi series and the Bawdwin volcanic series. Of these only the MONG LONG SCHISTS are probably definitely referable to the Archaean. The biotite-schists are intruded by granites containing tourmaline and garnet and traversed by veins of quartz. There are also mixed gneisses, clac-granulites and other rocks.

ARCHAEAN GROUP — EXTRA-PENINSULA

The Chung Magyi Series overlies the Mong Long schists with a transitional zone. It consists of slaty shales, phyllites, quartzites, greywackes and also carbonaceous slates. The series is dominantly argillaceous and non-calcareous and is developed in the hilly parts of the Shan Plateau, extending into the Southern Shan States and the Yamethin district on the one hand and into Northern Shan States and Yunnan on the other. The Chaung Magyis are intruded by granite bosses and basic dykes, the former having produced contact alterations in the argillaceous rocks. The age of this Series is not clear — it may be Archaean or Purana.

TENASSERIM

The Mergui Series.— In the Mergui, Tavoy and Amherst districts of Lower Burma there is developed a group of rocks called the MERGUI SERIES. It includes quartzites, conglomerates, limestones, argillites, greywackes and agglomerates, the argillites being sometimes carbonaceous, and also the most important by volume. The greywackes and agglomerates are next in importance and apparently represent pyroclastics. Dark and white, fine-grained to saccharoidal limestones occur sparingly. The series is intruded by granite which has produced contact metamorphism, with the development of hornstones and schists containing biotite, andalusite, sillimanite, garnet, etc. The rocks have a N.N.W.-S.S.E. strike parallel to the general trend of the mountains. They are much disturbed and folded and are overlain by the MOULMEIN LIMESTONE which is regarded as of Permo-Carboniferous age.

The age of the Mergui Series is in doubt, some taking it to be Pre-Cambrian and others Upper Palaeozoic. In support of the latter view is the fact that the Moulmein limestone lies conformably over it in some places. The Mergui rocks are, however, entirely unfossiliferous.

Another series of rocks, called the TAUNGNYO SERIES, occurs in the same region and contains similar rock types. This series is said to overlie the Mergui Series, but it is not always posisble to distinguish the two and map them separately. They may both be of the same age.

CORRELATION OF THE BURMESE ROCKS

The Chaung Magyis are Pre-Cambrian in age, as they are overlain by the Bawdwin volcanics which intervene between them and the graptolite-bearing Ordovician strata. In the present state of our knowledge it is not possible to say whether they are Archaean or Purana or both. They are unfossiliferous and have suffered regional and contact metamorphism.

There is even greater uncertainty about the Merguis. They are also unfossiliferous and much disturbed and may be of any age older than the Permo-Carboniferous. If they are Palaeozoic in age, we can find their parallel amidst the rocks of the Sub-Himalayan region.

The gneisses and crystalline limestones of the Mogok tract have a typical Archaean aspect. A suggestion has however been made that the limestones may belong to the Plateau Limestone age or later, but this can be proved by connecting up this area with the neighbouring tracts by continuous mapping.

SELECTED BIBLIOGRAPHY

Adams, F. D. The Geology of Ceylon. *Canadian Jour. Res.* **1**, 467-486, 1929.
Artini, E. and Melzi, G. 'Richerche petrographiche e geologiche nulla Valsasia.' Milan, 1900.
Aswathanarayana, U. Age determinations of rocks and geochronology of India. XXII *Int. Geol, Cong.* (New Delhi) *Special Brochure*, 25. pp 1964.

Ball, V. Geology of the Mahanadi basin. *Rec.* **10**, 167-185, 1877.
Banerji, A.K. Cross folding, migmatization and ore localisation along part of Singhbhum shear zone south of Tatanagar, Bihar. *Econ. Geol.* **57**, 50-71, 1962.
Banerji, A.K. Structure and stratigraphy of part of northern Singhbhum south of Tatanagar. *Proc. Nat. Inst. Sci. India*, **30A** (4), 486-510, 1964.
Brown, J. C. and Heron, A.M. Geology and ore deposits of Tavoy. *Mem.* **44**(2), 1923.

Chatterjee, S. C. Anorthosites of Bengal, Calcutta Univ. Press, 1937.
Chatterjee, S. C. Gabbro rocks near Gorumahisani Pahar. *Proc. Nat. Inst. India* XI, 255-281, 1945.
Coates, J. S. Geology of Ceylon. *Ceylon Jour. Sci. B.* **19**(2), 101-187, 1935.
Cooray, P. G. Charnockites and their associated gneisses in the Precambrian of Ceylon. *Q.J.G.S.* **118**(3), 239-273, 1962.
Coulson, A. L. Geology of Sirohi State, Rajputana, *Mem.* **63** (1), 1933.
Crookshank, H. Western margin of the Eastern Ghats in Southern Jeypore, *Rec.* **73**, 398-434, 1938.
Crookshank, H. Geology of Southern Bastar and Jeypore from the Bailadila Range to the Eastern Ghats. *Mem. G.S.I.* **87**, 150 pp. 1963.

Dunn, J. A. Geology of Northern Singhbhum. *Mem.* **54**, 1929.
Dunn, J. A. Stratigraphy of South Singhbhum. *Mem.* **63**(3), 1940.
Dunn, J. A. and Dey, A. K. The Geology and petrology of Eastern Singhbhum and surrounding areas. *Mem.* **69** (2), 1942.

Eskola, P. The mineral facies of Charnockites. *Jour. Madras Univ.*, **27 B**, 101-109, 1957.

Fermor, L. L. The manganese ore deposits of India. *Mem.* **37**, 1909.
Fermor, L. L. The age of the Aravalli Range. *Rec.* **62**, 391-409, 1930.
Fermor, L. L. An attempt at correlation of the ancient schistose rocks of Peninsular India. *Mem.* **70**, 1936-40.
Fernando L. J. D. The Geology and Mineral Resources of Ceylon. *Bull. Imp.Inst.* XLVI, 303-325, 1948.
Foote, R. B. Dharwar system, the chief auriferous series in South India. *Rec.* **21**, 40-56, 1888 ; *Rec.* **22**, 17-39, 1889.
Foote, R. B. Geology of the Southern Mahratta country. *Mem.* **12**, 1876.
Foote, R. B. Geology of the Bellary district. *Mem.* **25**, 1895.
Foote, R. B. Geology of Madura and Tinnevelly districts. *Mem.* **20** (1), 1883.

Ghosh, P.K. Charnockite series of Bastar State *Rec.* **75**, paper 15, 1941.
Griesbach, C.L. Geology of the Central Himalayas. *Mem.* **23**, 1891.
Groves, A. W. The charnockite series of Uganda, *Q.J.G.S.* **91**, 1935.
Gupta, B.C. Geology of Central Mewar. *Mem.* **65**(2), 1934.
Gupta, B.C. and Mukerjee, P.N. Geology of Gujarat and S. Rajputana. *Rec.* **73**, 164-205, 1938.

Hayden, H. H. Geology of Spiti, *Mem.* **36**(1), 1904.
Heim, A. and Gansser, A. Central Himalaya. *Denkschr. Schweiz. Naturf. Gesell.* LXXIII (I), 1939.
Heron, A. M. Geology of N. E. Rajputana. *Mem.* **45** (1), 1917.
Heron, A. M. Soda rocks of Kishengarh. *Rec.* **56**, 179-197, 1924.
Heron, A.M. Synopsis of the pre-Vindhyan geology of Rajputana. *Trans. Nat. Inst. Sci.*, **1** (2), 1935.
Heron, A. M. Geology of South-eastern Mewar, Rajputana, *Mem.* **68** (1), 1936.
Heron, A. M. Geology of Central Rajputana. *Mem.* **79**, 1953.
Holland, T. H. The Charnockite series. *Mem.* **28**(2), 1900.

Holland, T. H.   The Sivamahlai series of elaeolite syenites and corundum syenites, *Mem.* **30**(3), 1901.

Howie, R. A.   The Geolochemistry of the Charnockite Series of Madras, India. *Trans. Roy. Soc. Edin.* **62** (No. 18), 1955.

Howie, R. A.   Charnockites.   *Sci. Progress*, **52**, 628-644, 1964.

Howie, R. A.   The geochemistry and mineralogy of charnockites from Baffin Land, Brazil, British Guiana, Congo, Madras and Uganda.   *Indian Mineralogist*, **6**, 67-76, 1965.

Iyengar, P. Sampat.   The acid rocks of Mysore.   *M.G.D. Bull.* **9**, 1920.

Iyengar, S. V.P. and Alwar, A.   The Dhanjori geosyncline and its bearing on the stratigraphy of Mayurbhanj and Singhbhum.   ' Wadia Commemorative Volume ' (*Min. Geol. Met. Inst. India*) 138-162, 1965.

Iyengar, S. V.P., Venkatraman, P.K. and Banerjee, S.   The Amjori Sill — a differentiated dolerite still in the Simplipal Hills, Orissa State, India.   In ' Research Papers in Petrology' (G.S.I.) 1-26, 1964.

Iyer, L. A. N.   Granitic ntrusions and associated rocks in Ranchi and Singhbhum.   *Rec.* **65**, 490-533, 1932.

Iyer, L. A.N.   The Geology and Gemstones of the Mogok stone tract, Burma.   *Mem.* **82**, 1953.

Jhingran, A.G.   The Bundelkhand Gneiss.   *Proc. Ind. Sci. Cong.* (Pres. Addr. Geology Section), 1958.

Jones, H. C.   Iron ores of Bihar and Orissa.   *Mem.* **63**(2), 1934.

King, W. and Foote, R. B.   Geology of Trichinopoly, Salem, etc.   *Mem.* **4**(2), 1864.

King, W.   The gneiss and transition rocks of Nellore.   *Mem.* **16**(2), 1880.

King, W.   Sketch of the progress of geological work in the Chattisgarh division of the C.P.   *Rec.* **18**, 169-200, 1885.

King, W.   Geology of Travancore State.   *Rec.* **15**, 87-92, 1882.

Krishnan, M. S.   Dharwars of Chota Nagpur, etc.   *Proc. 22nd Ind. Sci. Cong.* 1935 (Pres. Address to the Geology Section).

Krishnan, M. S.   The dyke rocks of Keonjhar.   *Rec.* **69**, 102-120, 1935.

Krishnan, M.S.   Geology of Gangpur State.   *Mem.* **71**, 1937.

Mallet, F. R.   Geology of Darjeeling and Western Duars.   *Mem.* **11** (1), 1874.

Pichamuthu, C.S.   Some aspects of Dharwar geology with special reference to Mysore State.   (Presidential address to Geology Section)   *Proc. 34th Indiam Science Congress*, Delhi session, 1947.

Pichamuthu, C. S.   The Charnockite Problem.   Mysore Geologists Association, Bangalore, 1953.

Pilgrim, G. E. and West, W. D.   The structure and correlation of the Simla rocks.   *Mem.* **53**, 1928.

Pilgrim, G.E.   Geology of a portion of Bhutan.   *Rec.* **34**, 22-30, 1906.

Quensel, P.   The Charnockite Series of the Varberg district on the South-western coast of Sweden.   *Arkiv for Min. Geol.* Bd I, No. 10, 229-332, 1951.

Radhakrishna, B. P.   The Closepet granites of Mysore.   *Mysore Geol. Assn. Spl. Pub.* 1956, 110 pp.

Ramberg, H.   The facies classification of rocks.   *Jour. Geol.* **57**, 1951.

Rao, B. Rama.   Recent investigations on the Archaean Complex of Mysore.   *Proc. Ind. Sci. Cong.* 1936, 215-244.

Rao, B. Rama.   The Charnockite rocks of Mysore.   *M.G.D. Bull.* **18**, 1945.

Rao, B. Rama.   The Archaean Complex of Mysore.   *M.G.D. Bull.* **17**, 1940.

Rao, B. Rama ' Handbook of the Geology of Mysore.'   Bangalore Press, Bangalore, 1964.

Rao, M.B. Ramachandra.   Petrography of hornblende schists, Kolar.   *M.G.D. Bull.* **16**, 1937.

Rao, M. B. Ramachandra and Rao. K. Sripada.   Origin and correlation of the metamorphic rocks of Sakarsanhalli area.   *M.G.D. Bull.* **14**, 1934.

Rau, S. Sethurama.   The Geology of Mergui.   *Mem.* **55** (1), 1930.

Roy, Supriya.   ' Syngenetic Manganese Formations of India.'   Jadavpur University, Calcutta 220 pp.   1966.

Saxena, M.N.   The Bundelkhand granites and associated rocks from Kabrai and Mau-Ranipur areas of Hamirpur and Jhamsi Districts U.P.   *Punjab Univ. Res. Bull.* **12**, 1961

Sarkar, S. N., Polkanov, A., Gerling, E.K., and Chukrov, F.V.   Geochronology of the Pre-Cambrians of Peninsular India — a synopsis.   *Science & Culture*. **30**, 527-537, 1964.

Sarkar, S. N. and Saha, A.K.   A revision of the Pre-Cambrian stratigraphy and tectonics of Singhbhum and adjacent regins. Q.J.G.M.M.S.I, **31**, 97-136, 1962.

Sharma, N. L.   Problems in the correlation of the Pre-Vindhyan igneous rocks of Rajasthan, (Presidential address, Geology Section)   *Proc. 40th Ind. Sci. Cong.* Lunckow Session. 1953.

Smeeth, W. F.   Outlines of the Geology of Mysore. *M.G.D. Bull.* **6**, 1916.
Stillwell, F. L.   The metamorphic rocks of Adelie Land. *Australian Antarctic Expedition*, 1911-14. *Sci. Rep.* A, III (1), 1918.
Subramaniam, A. P.   Mineralogy and Petrology of the Sittampundi Complex, Salem district, Madras. *Bull. Geol. Soc. Amer.* **67**, 317-389, 1956.
Subramaniam, A. P.   Charnockites and associated rocks of the type area near Madras — a reinterpretation. *Amer. Jour. Sci.* **257**, 321-353, 1959.
Subramaniam, A. P.   Petrology of the anorthosite-gabbro mass at Kadavur, Madras. *Geol. Mag.* **93** (4), 287-300, 1956.

Tilley, C. E.   Enderbite, a new member of the Charnockite series. *Geol. Mag.* **73**, 312-316, 1936.
Turner, F. J.   Mineralogical and structural evolution of metamorphic rocks. *Bull. Geol. Soc. Amer.* **30**, 194.

Vredenburg, E. W.   Considerations regarding a possible relationship between the charnockites and the Dharwars. *J.A.S.B.* N.S. **15**, 433-448, 1918.

Walker, T.L.   Geology of Kalahandi State, Central Provinces, *Mem. G.S.I.* **33** (3), 1902.
Washington, H.S.   The charnockite series. *Amer. Jour. Sci.* **41**, 323-337, 1916.
Wemk, E.   Khondalite and Stronalite. *Inidan Mineralogist*, **6**, 21-26, 1966.
Wilson, A. F.   Charnockitic rock in Australia — a review. *Proc. Pan-Indian Ocean Science Congress.* Sec. C. Perth (1954), p. 10-17, 1955.
Wilson, J. T.   Some major structures in the Canadian shield. *Canada. Min. Met. Bull.* **42** No. 450, 543-554. 1949.

CHAPTER VI
MINERAL RICHES OF THE ARCHAEANS

The Dharwarian schists and some of the igneous intrusives associated with them constitute the most important mineral bearing formations of India. They contain a large variety of metallic ores, industrial minerals and rocks such as the gold lodes in Mysore, Madras, Hyderabad, Bihar and Orissa; the copper ores of Sikkim, Rajasthan, Bihar and Andhra; manganese ores of Madhya Pradesh, Bombay and the Eastern Ghats; the chromite of Bihar, Orissa and Mysore; the banded iron ores of Bihar, Orissa, Madhya Pradesh, Mysore and Madras; the lead-zinc ores of Rajsthan and many non-metallic minerals used as abrasives, refractories, ceramic materials, gems, building stones, etc. The occurrences of the chief minerals will be described briefly in the following pages. For more details the original papers published by the Geological Survey of India, including the Quinquennial Reviews of Mineral Production, may be consulted.

METALLIC ORES
ANTIMONY

The antimony sulphide, stibnite, occurs in gneisses near the Shigri glacier in Lahaul. The deposit is said to be large enough for regular production but the locality is very inaccessible. The stibnite is associated with galena and blende and also some gold.

Several small deposits of stibnite have been recorded in the Southern Shan States, in the Amherst district of Burma, in the Jhelum district of the Punjab, in the Jabalpur district of Madhya Pradesh and in Mysore State. The Burmese deposits, however, belong to post-Archaean age.

ARSENIC

Minerals of arsenic occur sporadically in the mica belt of Hazaribagh and near Darjeeling. The orpiment deposits of Chitral on the Afghan border in which good veins of orpiment and realgar occur together, are in calcareous shales and marbles of Palaeozoic (?) age in close association with a basic intrusive. There are six individual occurrences of which four are on the same strike. They have been worked intermittently during the last five or six decades.

CHROMITE

Chromite deposits are found in various places in India, Pakistan and Burma. They occur in association with some types of ultrabasic rocks — dunites, saxonite, lherzolites, pyroxenites, enstatitites, norites, etc., mainly in the root zone of ancient mountains. The chief deposits are those of Keonjhar,

Cuttack and Dhenkanal in Orissa ; near Chaibasa in Singhbhum ; near Kondapalle in the Krishna District, Andhra; Sittampundi and Chalk Hills in Salem ; Shinduvalli and several other places in Southern Mysore ; Ratnagiri in Bombay. These are all of Archaean age.

Deposits of Cretaceous age are found at Hindubagh near Quetta in Pakistan; in Dras, Bembat and Tashgam in Ladakh ; near Hanle in the Spiti valley ; in the Manipur hills on the Manipur-Burma border ; and in the Arakan Yomas and Minbu district of Burma. Other deposits may be found in the Sutlej and Indus valleys in the Great Himalayas and along the Burma-Assam border.

Chromite consists mainly of iron and chromium oxides but may also show varying quantities of magnesium, ferric iron and alumium. Those with a high chromium content are useful for making ferrochrome and chromium chemicals while those with high magnesia and aulmina are useful for refractory purposes.

COLUMBITE — TANTALITE

These minerals are found in pegmatites traversing Archaean rocks in the Singar area in Gaya, at Pananoa hill of Monghyr, in the Kadavur hills of Trichinopoly and occasionally in the pegmatites of Nellore and Rajasthan.

COPPER

Copper ores are found in several localities — in the Singhbhum and Manbhum districts of Bihar, in the Rangpo area of Sikkim, in the Guntur and Nellore districts of Andhra, in the Jabalpur district of M.P., in the Chitaldrug district in Mysore and a few places in Garhwal and Almora in U.P. and in some places in Nepal. The copper ores of Khetri and Singhana in Jaipur, of Daribo and Kho in Alwar are probably of Delhi age, white those of Andhra are of Cuddapah or Kurnool age.

Ores composed of chalcopyrite, pyrrhotite and pentlandite, together with a few other minerals, occur as important lodes in the main shear zone in the Iron-Ore Series of Singhbhum, extending from Seraikela and Kharsawan to the border of Mayurbhanj. They are thought to be related to the sheared soda-granophyre which also occurs in the same zone. Some lodes occur in the adjacent country rocks such as epidiorite. They are now worked at Mosaboni where the average ore contains about 2.5 percent copper and a little nickel. The mines have now reached a vertical depth of 600 m. The ore is smelted and refined locally, the annual output of copper being around 9,000 tons. Most of this is converted into brass for which there is a ready market in India.

Copper-ores occur at Bhotang, Dikchu and other places as lodes and stringers in the Daling schists which are intruded by the Sikkim gneiss. The ores contain chalcopyrite, pyrrhotite, galena and blende with small quantities of bismuth and antimony sulphides. The average grade is rather poor but if all the lodes are developed they may be workable.

Copper-ores are also known to occur in the Cuddapahs at Agnigundala in the Guntur district and at Garimenapenta in the Nellore district of Andhra. They are connected with post-Cuddapah mineralisation. Some veins are found consisting of chalcopyrite, tetrahedrite, galena, pyrite, barite, quartz and calcite in the Dharwarian dolomites and schists near Sleemanabad in the Jabalpur district. Other occurrences, some of which have occasionally produced small quantities of ore, are:— Chota Udepur in Gujarat; Tamakhum in Manbhum, Bihar; Dhanpur, Pokhri and Askot in Almora and Garhwal. Several other deposits have also been found and they are under investigation.

GOLD

Kolar Gold Field.—The Kolar schists belt, in which this gold field occurs, is 65 km long and 4 to 6 km broad. According to S. Narayanaswamy it is occupied by metamorphosed and folded mafic rocks and sills of associated basic intrusives. Some gneissic and ferruginous rocks occur along the margins. The eastern side of the field is bordered by the Champion gneiss and by a band of autoclastic conglomerate. The field is surrounded by the Peninsular gneisses which are traversed by two sets of basic dykes — one set trending E.-W. and the other N.-S. The amphibolites are composed mainly of the amphibole hastingsite and calcic plagioclase. They may be subdivided into four main types: 1. massive; 2. schistose and spotted; 3. granular and 4. fibrous. They are all apparently concordant.

The amphibolite series has been folded into a major syncline or synclinorium the folds being isoclinal along N.-S. axis and dipping steeply (60°—80°) to the west. They have been refolded along N.N.W.-S.S.E. axis plunging 30° — 40° northward. The refolding, involving shearing movements, has led to drag and cross-folding and to a westerly shift of the marginal folds as we proceed southwards.

There are more than 25 lodes, all of which are gold-bearing and some rich in sulphides. Only 5 of the lodes have been worked. The Champion and Mundy lodes form the eastern group while the McTaggart East, McTaggart West, Oriental and other lodes belong to the western group. The lodes are localised in the schistose amphibolites at their stratigraphical contacts with the other types of amphibolites. The Champion lode has been worked most, its average width being perhaps 1.2 — 1.5 m. It dips at about 50° near the surface but is practically vertical at depth. The main mineralised zone is the Mysore North Fault, in the cross-faulted area of the field. The mineralisation fades both to the north and to the south, beyond the limits of this field. The lodes are composed of a series of parallel echelon veins and stringers of bluish grey to white vitreous quartz. The individual veins show echelon structure. They attain prominence in the synclinal troughs and noses of the drag folds. They show a northerly pitch corresponding to the northerly plunge of the cross folds. The shearing and dragfolding have produced a north and upward movement in the western bolck relatively to a south and downward movement in the adjoining eastern block. Occasionally there are local reversals of the drag folds.

The wall rocks show alteration and zoning. Thye have been altered to diopside, hornblende, epidote and biotite. The associated sulphides are pyrrhotite, pyrite, arsenopyrite, chalcopyrite, galena and sphalerite. The western lodes contain much more sulphides (10 — 15 per cent of the lode matter) than the eastern. Gold occurs free in the quartz, silicates or sulphides. The wall rock alteration consists mainly in the formation of brown micaceous material and extends to a maximum distance of 15 m from the lode on either side. This usually shows three zones: — 1. a thin shell of diopside immediately bordering the vein ; 2. a fairly wide zone of hornblende and epidote ; and 3, a thick zone of biotite rich rock. The introduction of the vein quartz seems to have been accompanied by the alteration of wall rock, followed by replacement of quartz by sulphides to some extent and finally by the introduction of gold. The mineralisation may be classed under hypothermal metasomatism.

All the reefs contain quartz, tourmaline, albite, carbonate, brownmica, epidote, graphite and scheelite. The Champion reef is very poor in sulphides but the Oriental and McTaggart's lodes contain usually about 1 to 2 per cent, but occasionally up to 3 or 5 per cent, these including pyrite, subordinate pyrrhotite, chalcopyrite, galena and arsenopyrite. Some telluride is found in the Oriental lode.

The gold is generally not visible in the ore to the naked eye. The ore is crushed in stamp mills and washed into tanks containing fibre mats or blankets to recover the coarse particles. The material is then treated wtih sodium cyanide which is passed through tanks containing zinc shavings which help to precipitate the gold.

The chief mines are Champion, Mysore, Nundydroog, Ooregum and Balaghat, but the last ceased working some years ago. The average grade of ore at present worked contains more than 5.5 dwt. (penny-weights) per ton. The field used to produce between a half and one-third of a million ounces per year obtained from crushing about 700,000 tons of ore, but in recent years the output has dropped to around 0.2 million ounces. The total production from 1882 upto 1950 was a little over 22 million ounces, valued at 131 million pounds sterling.

In the extension of the Kolar Field to the south there are three major folds and the mineralization is largely confined to the middle fold.

Wainad.—In the Wainad region of the Nilgiris district, numerous gold bearing quartz lodes were opened up and worked between 1880 and 1900. Most of them are said to contain an average of less than three penny-weights per ton. The lodes have a N.N.W. strike, almost at right angles to the foliation of the country rocks which is N.E.-S.W. The country rock is hornblende or biotite gneiss and the gold is associated with pyrite and ferruginous matter.

Hatti.—Gold veins are worked also in the Hatti Gold Fields in the Maski band of Dharwarian schists, in the Raichur district. They were worked between 1903 to 1920 producing a total of 256,500 oz. of gold valued at £ 1,000,000. The mines have been reopened in recent years and 3 or 4 lodes are being worked at shallow depths. There is evidence that the lodes would continue to several thousand feet depth and that this field is of some importance.

In the Anantapur district of Andhra, near Ramagiri, some quartz reefs were worked between 1908 and 1924, producing a total of 136,700 ounces. Workable lodes are also said to exist in the Gadag band of Dharwar schists running through Dharwar and Sangli districts and in the Gooty taluk, Anantapur district, about 55 km north of the Ramagiri mines. Other areas where gold-quartz veins are known are in Jashpur and Dhalbhum in Bihar ; Gangpur, Bamra, Singhbhum, Sambalpur, Koraput and other districts in Orissa.

IRON-ORE

The Dharwarian sediments contain some of the richest and largest iron-ore deposits in the world. The ore bodies are associated with and derived from banded hematite-quartzites. The original rocks consist of alternating this ribbons of quartz, jasper, and hematite, from which the silica has been leached out leaving rich ore bodies. At the surface the ores are massive, compact and rich, containing 60 to 69 per cent iron. They grade partly into shaly and powdery ores whose composition is variable. The largest concentration of deposits is in southern Singhbhum and the adjoining Keonjhar and Bonai districts of Orissa, where the rocks form isoclinally folded series. Other deposits occur in Bastar, Chanda and Drug districts of Madhya Pradesh, in the Ratnagiri district of Bombay, and in Mysore, Madras and Andhra. When metamorphosed, these banded ferruginous quartzites have given rise to quartz-magnetite ores such as are developed in Southern Mysore, Salem and Guntur. Many of the ore bodies occur on top of hill ranges and are of considerable size, containing millions of tons. The total resources of Pre-Cambrian iron-ores of comparatively high grade are probably of the order of 20,000 million tons. The magnetite quartizites are of low grade containing usually 35 to 40 percent iron, but they are amenable to magnetic concentration. The banded hematite-quartzites themselves contain 25 to 30 per cent iron, and as they occupy large areas, they should amount to many thousand millions of tons.

There are also some ore bodies consisting of titaniferous magnetite (with some vanadium or chromium) associated with basic igneous intrusives traversing Dharwarian rocks. Such ores are found in Eastern Singhbhum, Mayurbhunj and Southern Mysore. They may contain a few million tons of ore.

LEAD-ZINC ORES

Deposits of lead and zinc ores occur in Jaipur and Mewar districts of Rajasthan. The deposits at Zawar (24°21' : 73°44') near Udaipur city have been found to be extensive and of large size. They occur as replacement veins and fissure fillings occupying fault and fracture zones in dolomites of Aravalli age associated with phyllites and quartzites. The rocks in the Mochia Magra hill dip steeply towards the north while the lodes which traverse them have a steep southerly dip. In addition to massive lodes and veins, there are also fairly rich zones of dissemination in the dolomite, which may be workable. The ores consist mainly of a mixture of galena and sphalerite (with a little silver sulphide and native silver) the proportion of which varies greatly. It would appear that the ore is richer in lead in the upper horizons and in zinc in the lower horizons. The average grade is 4 to 5 per cent of the combined metals but richer ores occur which can yield 10 to 15 per cent of the two metals together. There are three or four other hills in the neighbourhood which are also similarly mineralised. The deposits are expected to contain extensive resources running into several million tons of rather low grade.

There are also several small occurrences of lead-zinc ore veins in the schistose rocks of Bihar, Orissa and Madhya Pradesh. Lead ore (galena) occurs also as veins and disseminations in the Cuddapah formations of Cuddapah and Kurnool districts and some of these appear to have been worked formerly, many decades ago. A few other deposits occur in Gujarat and Madras.

MANGANESE-ORE

Manganese ores are found in India in three types of association, *viz.*, in metamorphosed manganiferous sediments, called GONDITES, in similar sediments intruded by igneous rocks called KODURITES, and as lateritic concentrations derived from Dharwarian schists and phyllites.

The gondite type of deposit is found in Jhabua, Shivrajpur, Chhindwara, Nagpur, Bhandara, Balaghat and Gangpur, the most important being the manganese belt of the former Madhya Pradesh in the Sausar Series. The ores occur in three geological horizons, one within the Mansar Stage and the other two at the top and bottom of the same Stage. Of these, the uppermost horizon is unimportant, and the other two contain good deposits. The magnaniferous beds resemble the Pre-Cambrian iron formations in that they show alternating bands of manganese oxides and silica. The layering is not clear in the thick beds of ore because of high compression. The ore bands are ordinarily up to 1.8 m in thickness but in exceptional cases attain 12-15 m as a result of tight folding. The ore and the manganiferous bands (composed of gondite and manganese silicates) are intercalated with biotite-quartz-schists and quartz-mica-feldspar rocks which were originally part of the normal sediments laid down at the time. The ore minerals are braunite, psilomelane, pyrolusite, sitaparite (bixbyite), vredenburgite (jacobsite and hausmanite). The silicate bands were formed as a result of metamorphism of the impure materials. These have been reconverted, in the zone of weathering and oxidadation, to secondary oxides. The ores are of varying grades. Phosphorus is high in the highly metamorphosed bands, amounting to 0.1 to 0.3 per cent. P, but less than 0.1 per cent, in the bands showing low metamorphism. The primary ore is expected to persist to the depths to which the original manganese sediments extend. Taking the manganese belt as a whole, it has been calculated that, for a depth of every 30 m along the dip, there would be roughly 12 million tons of workable ore. Of this about half would be first grade (48% Mn and over). It may be expected that in most areas the deposits would extend to a depth of at least 300 m. The total reserves of the whole belt may be of the order of 80 to 100 million tons.

There are two important manganese horizons associated with the Lohangi and Mansar stages of the Sausar Series the latter being the more important and sometimes attaining a thickness of 6 m or more of solid manganese-ore. These manganese bearing horizons form part of the sequence and are folded with the associated Archaean rocks. The Balaghat-Nagpur belt contains the richest ores and the largest reserves in India, the manganese content of the ore exceeding 52 per cent in some cases.

The second type of deposits occurs mainly in the Eastern Ghats associated with garnet-sillimanite schists (Khondalites) and crystalline limestones of Archaean age, intruded by granite, producing a hybrid rock called KODURITE, which ordinarily consists of quartz, orthoclase, garnet (spessartite-andradite, contracted to spandite), manganese-pyroxene and apatite. The regional strike of the folds is N.W.-S.E. with S.E. pitch at about 30 degrees. Cross folds parallel to the direction of the Eastern Ghats are also found. The ore bodies occur in the axial parts of the main folds and in some cross folds. The ores are worked in a few places in the Ganjam, Koraput, and Visakhapatnam districts of Orissa and Andhra. When altered, the kodurite gives rise to lithomargic clay and wad. The ore bodies are generally irregular and occasionally attain large dimensions as at Garbham and Kodur. The ores consist of psilomelane with some pyrolusite, braunite and mangan-magnetite. They are high in iron and phosphorus and are generally of second and third grades.

The lateritoid deposits are fairly widely distributed in Mysore, Sandur, Southern Bombay, Singhbhum, Keonjhar and Bonai and a few other areas. They are due to the concentration near the surface, by meteoric waters, of the manganese contained in the Dharwarian schists and phyllites. The ores generally form irregular bodies confined to the zone of weathering. The minerals found in these ores are psilomelane, pyrolusite wad and limonite. They are high in iron and mostly of rather low grade, though occasionally bands consisting of pyrolusite yield very rich ore, which is in demand for making dry-cell batteries.

MOLYBDENITE

There is occasionally a small production of molybdenite from Tavoy. The mineral is known to occur in the Palni hills, in the Godavari district, in Chota Nagpur and in Kishengarh. Occasionally it also occurs in association with graphite deposits. No workable deposit has so far come to light.

MONAZITE

This mineral, a phosphate of cerium earths, is prized mainly for its thorium content. It is a constituent of the beach sands of Travancore, Madras, Andhra and Orissa but has been worked only in Travancore. The sands are derived from the pegmatites and gneisses of the interior. The mineral generally contains 7 to 10 per cent thorium oxide (ThO_2) and 0.3 per cent uranium oxide (U_3O_8). The thorium content of the sands is estimated at several hundred tons. It was in demand in the earlier part of this century as a source of thorium which was used as thorium nitrate in the gas mantle industry But the demand fell off in the twenties owing to the rapid development of electric lighting. Now again it has assumed importance as a possible source of atomic energy. The cerium and rare earths are also utilised in the manufacture of alloys and chemicals.

NICKEL AND COBALT

The presence of nickel in the copper ores of Bihar has already been mentioned. Cobalt ores (cobaltite and danaite) are found as thin veins and disseminations in the Aravalli schists and slates in the Khetri and Babai areas of Rajasthan. The cobalt ores have been used for the manufacture of a blue enamel employed in jewellery. Cobalt and nickel sulphide and sulpharsenide ores are known to occur and to have been worked near Tamgas in Western Nepal but no details are available regarding their geology and reserves.

TIN

A few small deposits of cassiterite are found as thin veins and disseminations in granite and pegmatite in the districts of Hazaribagh, Gaya and Ranchi in Bihar. None of them seem to be large enough for steady commercial exploitation, though some prospecting was carried out at various times in some of these deposits.

Tin ores occur in the Mergui and Tavoy districts of Lower Burma and also along the same granite belt continuing to the north and south. The great majority of the deposits are now worked as eluvial gravels and weathered materials over the granite outcrops. The granite is characterised by the presence of tourmaline, topaz, lepidolite, etc. The tin bearing mineral is cassiterite which is accompanied by wolfram, molybdenite, arsenopyrite, etc. The granite which gave rise to the deposits are, however, thought to be of Jurassic age. The cassiterite is concentrated by dredging or by the use of water jets and gravel pumps.

TITANIUM

Titaniferous magnetite and ilmenite are of common occurrence in many rocks, but workable deposits of ilmenite are few. Ilmenite is found as small veins and aggregates in the mica pegmatites of Bihar and as veins associated with quartz veins traversing granite gneiss in Kishengarh and also in association with the wolfram veins of Degana in Rajasthan. There are some fairly important masses of titaniferous and vanadiferous magnetites in Southeast Singhbhum and the adjoining parts of Mayurbhanj associated with gabbros, anorthosites and ultrabasic rocks. They contain upto 28 per cent of TiO_2 and 7 per cent of V_2O_5. Similar bodies have also been noted to be associated with the basic and ultrabasic rocks of the Nuggihalli schist belt of Mysore where they also contain some chromium.

The most important sources of titanium in India are the beach sands of Travancore, Ratnagiri in Bombay, the southern districts of Madras and of the Vishakapatnam and Orissa coast. The ilmenite is derived from the garnetiferous gneisses and charnockites of these regions. The beach sands of Travancore are often quite rich and are credited to hold reserves of the order of 250 to 300 million tons. They are associated with monazite, zircon, garnet, rutile, sillimanite and other resistant minerals. Some of the Tranvacore ilmenite deposits contain over 62 per cent of titanium dioxide ; while the others contain between 50 and 55 per cent. The beach deposits have been formed by the concentration, by sea waves, of the products of denudation of the counry rocks brought to the coast by the numerous streams and rivers draining the area.

TUNGSTEN

Tungsten ores generally occur associated with tin, the host rocks being either the granites or the country rocks immediately in contact with them. Small deposits are known in India in the Bankura and Singhbhum districts in Bengal and Bihar, in the Nagpur district of Maharashtra, and at Degana in Rajasthan. Here the wolfram occurs in quartz veins traversing Archaean granite and phyllite and in the weathered detrital material derived from them. The wolfram in the granite is associated with muscovite, fluorite and some pyrite and chalcopyrite, and forms veins and stockworks. The deposits have been worked intermittently and are likely to contain a few thousand tons of the mineral. At Agargaon in Nagpur district, wolfram and scheelite occur in tourmalinised phyllites and quartz veins. These may amount to a few tons.

Important deposits occur in the granite and pegmatite of the Shan-Tenasserim belt of Burma where numerous mines have been, and are being, worked. In some cases scheelite and cassiterite accompany the wolfram, as also subordinate amounts of arsenopyrite, molybdenite and bismuthinite. These are, however, of Jurassic age.

URANIUM

Pitchblende and uranium-ochre have been recorded as occurring in the pegmatites of Gaya though no workable deposits have yet been discovered. Disseminations of uraninite, torbernite and uranium ochre have been recorded in the rocks of the shear zone in the copper belt of Singhbhum. Though low grade, these deposits are being developed and exploitated, the reserves being of the order of a few million tons. Parts of these deposits contain recoverable amounts of nickel and molybdenum. Other uranium bearing minerals like allanite, columbite-tantalite, samarskite, triplite, etc. are occasionally found in pegmatites in various parts of India, particularly those which have been worked for mica in Bihar, Rajasthan, Andhra and Madras. Workable deposits of uranium bearing minerals are however more likely to be found in mineralized areas than in pegmatites.

VANADIUM

Vanadiferous titaniferous magnetite ore bodies occur associated with ultrabasic intrusives of South-east Singhbhum and Mayurbhanj. The deposits in Mayurbhanj are said to be of large size but have not yet been investigated in detail. The vanadium pentoxide content in the magnetite varies from 1 to 7 per cent, the average being around 1.5 to 2 per cent.

NON-METALLIC MINERALS

APATITE

The only important deposits of apatite known in India are the apatite-magnetite rock occurring in the shear-zone of Singhbhum and Mayurbhanj. The mineral has been worked intermittently. The rock has to be crushed fairly fine in order to separate the apatite from the magnetite. Apatite is also a constituent of the kodurite rocks of the Eastern Ghats and to some extent of the pegmatites in various parts of India, but the separation and concentration of the mineral from the associated rock would be rather difficult, because of the very small proportion of it in the rock.

ASBESTOS

Amphibole asbestos occurs in Archaean rocks in various parts of India — in the Hassan and Bangalore districts of Mysore State, in the Salem district of Madras, in the Seraikela area of Bihar, and in Idar and Ajmer-Merwara in Rajasthan. The material is usually associated with basic and ultrabasic rocks. The fibres are of poor strength and not useful for making asbestos-cloth, though they may be suitable for other purposes.

BERYL

Beryl deposits of some size have been worked in Mewar, Jodhpur and Ajmer in Rajasthan. Small quantities of beryl occur in the pegmatites of Rajasthan, Bihar and Andhra and other areas and in some years an output of 1,200 to 1,500 tons has been attained. The beryl crystals are mostly 2 to 15 cm. long but, occasionally, huge crystals several m long are found. The beryllium oxide content of the beryl varies between 9 and 12 per cent. Because of its sporadic occurrence in pegmatites, the mining of beryl for itself is generally uneconomical. Most of the output is therefore in conjunction with the mining of mica.

BUILDING AND ORNAMENTAL STONES

The Archaean gneisses, granites, charnockites, slates, quartzites and crystalline limestones provide an inexhaustible storehouse of excellent stones for building and decoration. Gneisses and granites are widespread throughout the Archaean terrain. The banded gneisses and Bundelkhand granite of Rajsthan, the various gneissic rocks and granites of Madhya Pradesh, Bihar, Hyderabad, Andhra, Mysore and Madras have provided excellent stones for temples, forts, palaces, bridges and other large structures as well as for local buildings. The ruins of the city of Hampi, which was the capital of Vijayanagar kingdom, a few km from Hospet, contains many old structures built of the excellent grey and pink gneisses found in the neighbourhood. The charnockites of Madras, Mysore and Andhra are amongst the strongest and most durable stones found anywhere in the world. These and the granites of South India have been tested and found superior to some of the famous granites of the British Isles and have been used in harbour construction in Bombay and Madras. The temples and the architectural monuments of Mahabalipuram near Madras have been hewn out of solid charnockite. In many places in Madras and Hyderabad the granites are capable of yielding beams, pillars and slabs, the last 8-9 m long and 5-10 m wide. Such material has been extensively used for pillars, beams and flooring and roofing slabs in many great buildings, particularly temples and palaces in the south. Pink, white and light grey granites and finely banded gneisses of Archaean age in Rajasthan are extensively quarried in Jodhpur, Mewar, Idar, Ajmer and other areas for local use. The granites and gneisses of other parts of India such as Ranchi, Hazaribagh, Gaya and other districts of Bihar ; of Sambalpur and neighbouring regions of Orissa ; of Balaghat, Bhandara and Chhindwara in Madhya Pradesh — these have all been used in important local buildings from time immemorial.

The khondalites of the Eastern Ghats are not particularly durable stones as they weather into a sandy rock and give rise to red and brown streaks and patches in a buff ground-mass. Yet they have been used extensively in many buildings and temples in Andhra and Orissa, for instance in the Puri and Konarak temples. Most of the finest statues in the Konarak and Bhuvaneshwar temples have been carved out of the khondalites but many of them have weathered rather badly and show unsightly spots and streaks and have developed irregular surfaces.

Talc and talc-chlorite schists are mainly used for carving doorposts, windows and statues, etc. as they are suitable for delicate carving and stand weathering exceedingly well. The statues of the Sun God in the Konarak temple (13th Century A.D.) have been made of such rocks and they have preserved the exquisite delicacy and beauty of the original finish remarkably well even after seven centuries of neglect since they were made.

The crystalline limestones and marbles of Rajasthan have been extensively used in that region as building and ornamental stones. The Raialo marbles are worked at Raialo and other places in Alwar and Jaipur ; at Rajnagar, Nathdwara and Kankroli in Mewar ; and Makrana in Jodhpur. The usual type of Makrana marble is white with vague cloudy streaks and patches of grey, but white marbles are also available. The quarries in Mewar yield also pink, salmon and grey coloured marbles. White sachcharoidal Raialo marble is worked in Jhiri, Kho and Baldeogarh in Alwar. The Kharwa stone is of beautiful pink and green colours. The Makrana marble has been used in the Taj Mahal at Agra and some of the famous mosques in Agra, and Delhi, in the Victoria Memorial at Calcutta and in many buildings in Rajasthan, U.P. and Punjab. Several fine marbles of Archaean age are worked in Madhya

Pradesh. Amongst these may be mentioned the marble rocks of Jabalpur, the white marbles of Betul, and the various coloured marbles of Narasinghpur, Chhindwara and Nagpur districts. The Motipura marble in Baroda is a green serpentinous marble mottled with rose and pink which has been used for architectural purposes in Baroda and the neighbouring areas in Gujarat. A hard limestone of varying colours occurs at Sandara in Baroda, while dolomitic and serpentinous marbles are found in Dharwar and Idar districts. Excellent stones of various colours are also found in Visakhapatnam, Coimbatore and Madura in Madras, in Chitaldrug and Mysore in Mysore and in Koraput and Gangpur in Orissa.

The Aravallis contain good limestones and flagstones. The limestone worked at Umrar in Bundi is grey, massive, flaggy and capable of taking a good polish and also of being carved for architectural purposes. Good flagstones are quarried at Raghunathgarh in Jaipur and at Datunda in Bundi, for use as paving and building stones. Well cleaved but not very fine slaty rocks are woked in Ajmer and used for roofing and paving purposes.

CLAYS

As the Archaean and Pre-Cambrian rocks are generally strongly metamorphosed and they do not yield clays except where they are highly weathered in outcrops. Some feldspathic gneisses and granites in Mysore, Travancore, Malabar, Kanara and South Arcot in the south, and similar rocks in Singhbhum and Manbhum in Bihar and some parts of Orissa have given rise to good china clay. The gneisses of Salem, Malabar and Travancore-Cochin and the khondalites of the Eastern Ghats have given rise to bauxitic clays.

CORUNDUM

This mineral occurs as a constituent of ultrabasic rocks, syenites and highly metamorphosed aluminous sediments. It is found in the anorthite-gneiss of Sittampundi in the Salem district ; in the gabbroid and associated ultrabasics of souttern Mysore ; in the corundum-syenites of the northern part of Salem ; in the South Kanara district in Madras ; in Anantapur in Andhra and neighbouring districts of Mysore and in an area in Hyderabad adjoining the Godavari gorge. Corundum is a constituent also of the sillimanite schists in the Khasi hills, and of similar rocks in the Pipra area of Rewa and in the Bhandra district of Madhya Pradesh.

FELDSPAR

The mica pegmatites in various parts of India contain large quantities of potash and soda feldspar, much of it being oligoclase with perthitic structure and microcline. Orthoclase occurs as a constituent of non-mica bearing pegmatites and as veins in granite and gneisses. The bluish green variety of microcline called *amazonite* is found in a few of the mica mines of Nellore and Bihar. Potash-soda feldspar is worked generally in conjunction with mica in several States for meeting the demand from the ceramic industries.

Highly sodic feldspar and nepheline are found in the nepheline (and sodatite) syenites and associated pegmatites in the Coimbatore district of Madras, in the Eastern Ghats of Koraput and Visakhapatnam and in Kishengarh and Jaipur in Rajasthan. They may become useful for the ceramic industries in future.

GEMSTONES

A large number of precious and semi-precious stones are found in the Archaean rocks; moonstone, zircon, topaz, garnet and other gemstones from Ceylon; ruby and sapphire, spinel, garnet, scapolite, lapislazuli, olivine, tourmaline, zircon, topaz and moonstone from the Mogok Stone Tract in Burma; sapphire from the Ladakh region of Kashmir; aquamarine from Kishengarh in Rajasthan, Kashmir and Madras; chrysoberyl from Coimbatore in Madras and Kishengarh in Rajasthan; garnet from Kishengarh, Jaipur and Mewar in Rajasthan, Nellore in Andhra, Tirunelveli and South Kanara in Madras, and several other places in other parts of India; tourmaline from the Shan States, Nepal and Kashmir; rock crystals from various parts of India. Emeralds are found in biotite- and amphibole-schists associated with ultrabasic rocks and intruded by pegmatite, at Kaliguman and a few places in Mewar and Ajmer.

GRAPHITE

Workable deposits of graphite occur in the Khondalite and associated gneisses of Kalahandi, Bolangir, Ganjam and Koraput in Orissa, and in the Visakhapatnam and Godavari districts of Andhra. Some deposits are also known in Travancore, in the Tirunelveli district of Madras and in the Warangal district of Hyderabad State. In some cases good pockets occur in pegmatites. Graphitic schists are known to form part of the Archaean succession in several areas in India but they are too poor in graphite content and too finely and intimately mixed with the other constituents to be workable as sources of graphite. Some have been used as cheap paint materials.

KYANITE AND SILLIMANITE

Though sillimanite is widely distributed as a constituent of certain Archaean schists and gneisses, workable deposits occur only in the Khasi hills of Assam, in the Pipra area in Rewa, and in the Bhandara district of Madhya Pradesh. The sillimanite content of the Khondalite is generally not rich enough for being economically exploitable. The Khasi hills deposits are in the form of massive sillimanite bodies associated with gneisses of the Shillong Series. The sillimanite does not change much in volume after firing at $1,500° - 1,600°$ C., when it is changed to mullite and gives excellent service when employed as linings to tanks used for glass making.

Kyanite is rather more widespread than sillimanite and is found in gneisses as well as in association with veins containing quartz. The best known occurrences are those of Lapsa Buru and a few other places in mica schists and hornblende schists close to the shear zone of Singhbhum. The Lapsa Buru deposits have produced 20,000 to 30,000 tons per year over a period of years. Other deposits occur in Madhya Pradesh, Rajasthan, Andhra and Mysore and are being worked. In contrast with sillimanite, kyanite shows much shrinkage when calcined. It is, therefore, crushed, calcined to change it to mullite, bonded with fireclay, bauxite or other suitable material, and made into refractory materials.

MAGNESITE

The ultrabasic rocks in the Salem district of Madras and the Hassan and Mysore districts of Mysore have, in some places, been altered to magnesite. The magnesite has probably been derived from the action of magmatic waters containing carbon-dioxide on the magnesia-rich ultrabasics. The best known occurrences are those of the Chalk hills near Salem and of Kadakola in Mysore which are under active exploitation. The reserves of these deposits are probably of the order of 80 to 100 million tons.

MICA

Both muscovite and phlogopite occur in commercially workable deposits in India but the muscovite deposits are by far the most important. Indeed India is one of the chief sources of block mica and splittings in the world.

The chief muscovite occurrences are those of the mica belt of Bihar, Rajasthan and Andhra but there are a number of other occurrences in different districts containing Archaean gneiss. The Bihar mica belt traverses the districts of Hazaribagh, Gaya and Monghyr, being about 100 km long and 20 km wide. The pregmatites are probably related to the Chota Nagpur granite gneiss. They traverse the planes of schistosity and joints of the mica schists of the region and are found to be productive only when they are in the mica schists and not in the hornblende schists or granite. The explanation seems to be that the residual liquids from granites, which ultimately crystallised to form the pegmatites, absorbed some of the mica-schists and produced the coarse mica crystals. The larger pegmatites are up to 600 m long. They have a quartz core, the mica sheets being found along one or both of their contacts with the country rocks. The feldspars in the productive pegmatites are oligoclase and microcline, but when orthoclase occurs there is practically no commercial mica. The mica is generally of the quality known to the trade as " ruby mica " as it has a coppery red colour in thick sheets. The other minerals found in the pegmatites are garnet, tourmaline, beryl, columbite, tantalite, allanite, triplite, etc.

In Rajasthan, mica-pegmatites occur in Jaipur, Ajmer-Merwara and Mewar, the Bhilwara region of Mewar being the richest. The mica is generally colourless, but light ruby and light green varieties are also obtained. In some cases the pegmatites have produced large quantities of beryl.

In the Nellore district of Andhra a large number of pegmatites occur in a belt some 65 km long and 12 km wide. The muscovite is generally of a light greenish colour, though the light ruby variety is also obtained from a few mines. Similar but comparatively small occurrences of mica are found in many districts in Madras, Mysore, Andhra, Orissa, Bihar, Madhya Pradesh and Rajasthan.

The only occurrences of phlogopite in India are those of Travancore where there are small deposits near Neyyoor in South Travancore, and near Punalur in the Western Ghats in east-central Travancore. The mica occurs in association with basic pegmatites traversing pyroxenic rocks. Compared to muscovite, the phlogopite sheets are small in size and generally of a deep amber colour.

There are also a few occurrences of a semi-phlogopite mica (which has been named MAHADEVITF) in Travancore, in the Tinnevelly district of Madras and in the Visakhapatnam-Ganjam hill tracts. The mica is of a pale yellow colour.

STEATITE

Steatite and talc-schists are widely distributed amidst Archaean rocks. Deposits are found in the Jabalpur district of Madhya Pradesh, in Jaipur district of Rajasthan, in the Singhbhum district of Bihar, in the Idar district of Bombay, in the Bellary district of Mysore, in the Nellore district of Andhra, in the Salem district of Madras, and several other places. A few are of high quality and are used in the manufacture of insulators and in the cosmetic industry, while the impure talc-schists are used in making stone utensils, for carved work, etc.

MINERAL RESOURCES

Note.—This is a general list of papers on mineral deposits irrespective of their geological age.

Quinquennial Reviews of Mineral Production : *Rec.* **39**, 1910 ; **46**, 1915 ; **52**, 1921 ; **57**, 1925 ; **64**, 1930 ; **70**, 1936 ; **80**, 1954. Minerals Year Book (I.B.M.) 1956 onwards.

Bulletins of Economic Minerals. Rec. Vol. **76** : No. 2—Chromite ; 3—Strontium Minerals ; 4—Phosphates ; 5—Titanium ; 6—Indian Precious stones ; 7—Magnesite ; 8—Clay ; 9—Manganese ore ; 10—M ca ; 11—Copper ; 12—Abrasives and grinding materials ; 13—Beryllium ; 14—Vanadium ; 15—Tin and Wolfram ; 16—Coal.
Bulletins of the G.S.I. Series A : No. 1—Coal for synthetic Petroleum ; 2—Limestones in the Sone Valley ; 3—Glass-making sands in V.P., U.P. and Parts of Rajasthan ; 4—Cement ; 5—Asbestos and Barytes in Pulivendla taluk ; 6—Magnesite ; 7—Chromite ; 8—Iron-ore deposits of Salem and Trichinopoly ; 9—Iron -ore, iron and steel ; 10—Mineral Resources of Madhya Pradesh, and later ones.
Bichan, J. Structural control of ore deposition in Kolar Gold Field, *Econ. Geol.* **42**, (1947).
Brown, J.C. Geology and ore deposits of the Bawdwin mines. *Rec.* **48**, 122-155, 1917.
Brown, J.C. Cassiterite deposits of Tavoy, *Rec.* **49**, 23-33, 1918.
Brown, J.C. A geographical classification of the mineral deposits of Burma. *Rec.* **56**, 65-108, 1924.
Brown, J.C. and Dey, A.K. India's mineral wealth. Oxford Univ. Press, 1956.
Brown, J.C. and Heron, A.M. Tungsten and Tin in Burma. *Rec.* **50**, 101-121, 1919.

Chhibber, H.L. Mineral Resources of Burma, MacMillan, London, 1934.
Clegg, E.L.G. A note on the Bawdwin mines. *Rec.* **75**, Paper 13, 1941.
Clegg, E.L.G. Mineral Resources of Burma, Government Press, Rangoon, 1939.
Coulson, A.L. Mineral Resources of N.W. Frontier Province, *Rec.* **75**, Paper 2, 1940.

Dunn, J.A. Aluminous refractory minerals — Kyanite, Sillimanite and Corundum. *Mem.* **52** (2), 1929.

Dunn, J.A. Mineral deposits of Eastern Singhbhum. *Mem.* **69,** 1937.
Dunn, J.A. Sulphide mineralisation in Singhbhum, India. *T.M.G.I.I.* XXIX, 163-172, 1935.
Dunn, J.A. et al. Mineral Resources of Orissa (Orissa Government Press, Cuttack), (1949).
Dunn, J.A. and Dey, A.K. Vanadium bearing titaniferous iron ores of Singhbhum and Mayurbhanj. *T.M.G.I.I.* XXXI (3), 117-183, 1937.
Dunn, J.A. Mineral Resources of Bihar. *Mem.* **78,** 1941.

Fermor, L.L. Manganese-ore deposits of India. *Mem.* **37,** 1909.
Fox, C.S. Bauxite and aluminous laterite occurrences of India. *Mem,* **49** (1), 1923.

Gee, E.R. Economic Geology of Northern Punjab. *T.M.G.I.I.* **33** (3), 1937.

Hatch, F.H. The Kolar Gold Field. *Mem.* **33** (1), 1900.
Hayden, H.H. and Hatch, F.H. Gold fields of Wainad. *Mem.* **33** (2), 1901.
Heron, A.M. Mineral Resources of Rajputana. *T.M.G.I.I.* **29,** 1935.

Iyer, L.A.N. The Geology and gemstones of the Mogok Stone Tract. *Mem.* **82,** 1953.

Jones, H.C. Iron ores of Bihar and Orissa. *Mem.* 63(2), 1934.

Krishnan, M.S. Mineral Resources of Central Provinces and Berar. *Rec.* **74,** 386-429, 1939.
Krishnan, M.S. Mineral Resources of Madras. *Mem.* **80,** 1952.

LaTouche, T.D. Bibliography of Indian Geology. Part I-B, Mineral Index. Government of India Press, Calcutta, 1918.

Middlemiss, C.S. Corundum localities in Salem and Coimbatore. *Rec.* **29,** 39-50 ; 1896 ; *Rec.* **30,** 118, 1897.
Middlemiss, C.S. Mineral Resources of Jammu and Kashmir. (Govt. of Kashmir).
Mirza, K. Mineral Resources of Hyderabad State. *Hyd. Geol. Surv.* 1937.

Pryor, T. Underground geology of Kolar Gold Field. *Trans. Inst. Min. Met.* **33,** 159, 1923.

Roy, B.C. Mineral Resources of Bombay. Bombay Government Press, 1953.
Roy, B.C. Mineral Resources of Saurashtra. Saurashtra Government, 1954.
Roy Choudhury, M.K. Bauxite in Bihar, Madhya Pradesh, etc. *Mem.* **85,** 1959.

Straczek, J.A. et al. Managnese ore deposits of Madhya Pradesh. XX *Internat. Geol. Cong.* (Mexico) Symposium on Manganese Deposits. Vol. IV, Asia, p. 63-96 ; 1956.
Smeeth, W.F. and Iyengar, P.S. Mineral Resources of Mysore. *M.G.D. Bull.* **7,** 1916 (Also *Bull.* **11,** 1937).

Tipper, G.H. Monazite sands of Travancore. *Rec.* **44,** 186-195, 1914....

CHAPTER VII
THE CUDDAPAH (KADAPA) SYSTEM
Introduction

The Cuddapah and Kurnool systems of South India as well as Delhi and Vindhyan Systems of North India have long been recognised as belonging to the Proterozoic. It is, however, obvious that certain formations in Madhya Pradesh and Bihar, which have hitherto been described as Archaean, will have to be included in the Proterozoic. But reliable age data would be required for distinguishing these formations. The Cuddapah and Delhi Systems, which are believed to be roughly contemporaneous, consist of sedimentary rocks which have been subjected to varying grades of metamorphism, but the Delhi System has undergone more severe folding. The available age data for the rocks of the Cuddapah System indicate that the oldest of them may be about 1500 m.y., which would place them in the Middle Proterozoic.

THE CUDDAPAH BASIN, ANDHRA

The rocks of the Cuddapah and Kurnool Systems are exposed in a large basin within the limits of Andhra Pradesh (formerly included in the Madras Presidency). The basin is crescent shaped with the concave side facing the east. It is 340 km long, between the Singareni coalfield in the north and the Nagari Hills near Madras City in the south, with a maximum width of 145 km about the middle, and an area of 42,000 sq. km. The greatest part of the basin is occupied by strata of the Cuddapah System, the younger Kurnool System being confined mainly to the northern and northwestern parts. The Cuddapah sediments were deposited on the upturned edges of the gneissic and schistose rocks of Archaean age, with a profound unconformity which is generally referred to in previous literature as the *eparchaean unconformity*. As the earlier Proterozoic rocks are missing here, the unconformity represents a long period of time during which deposition may have taken place elsewhere. The Cuddapah sediments have been derived from the area to the west and southwest of the basin, as indicated by the presence of basal conglomerates along the western margin. The basin is deepest near the eastern concave margin where the strata may attain a vertical thickness of 3,000 to 4,000 m. The rocks in the basin are undisturbed in the western part but show folding and increasing metamorphism as the eastern margin is approached. Along this margin the gneisses on the east are thrust over the sedimentary rocks at a steep angle. The fold axes are parallel to the margin, but the basement under the basin appears to be irregular. There seem to be also some faults parallel to the fold-axes and along one such zone a few km west of the margin, high magnetic anomalies are encountered, which may be attributed to the presence of magnetic material along the zone. Folding and shearing along the margin is responsible for the high grade metamorphism exhibited by the rocks of the Nallamalai Series present there. It appears likely that the basin extended originally far to the east of the present limits. This is supported not only by the large depth of the basin near the eastern margin, but also by the presence, amidst the gneisses of the Nellore plains, of schistose rocks resembling metamorphosed representatives of the Nallamalai Series.

Near the southwestern margin, within the basin, there are several thick sills of basaltic rock associated with lava flows and tuffs, whose outcrops are elliptical. This configuration of the sills and the large magnetic anomaly in the central part of the area occupied by them may be due to their arrangement in the form of cone-sheets or to the presence of a very thick sill.

Near the eastern margin of the basin, especially in the northern part, there are a few deposits of copper, lead and zinc sulphides. These appear to be associated with granitic intrusives of post-Cuddapah age.

Another sedimentary basin of the same age, which may or may not have been connected with the Cuddapah basin, appears to have occupied parts of Orissa and Madhya Pradesh. The large exposures of the Purana or Proterozoic rocks in the Raipur and Bastar areas appear to belong to the Kurnools, but there may be equivalents of the Cuddapah System in other parts of the area. This northern basin should also have extended to the east originally, before the uplift of the Eastern Ghats in Post-Cuddapah times. This northern basin is now separated from the Cuddapah basin by the trough-faulted Godavari valley which must have been formed some time during the Palaeozoic.

The Cuddapah basin was studied and described by W. King nearly a century ago (1872). Parts of the basin have been remapped during the last two decades. But, except for a few papers on the Kurnool System, the Cuddapah Traps and some mineral deposits, no comprehensive account has been published on the stratigraphy. It is therefore still necessary to depend largely on King's monograph for much of our information. Table 18 shows the stratigraphical succession established by King:

TABLE 18.—The Cuddapah System

Kurnool System	Various Sedimentary Rocks
Kistna Series (600 m)	{ Srisailam Quartzites { Kolamnala Shales { Irlakonda Quartzites
Unconformity	
Nallamalai Series (1,000 m)	{ Cumbum Shales { Bairenkonda Quartzites
Unconformity	
Cheyair (Cheyyeru) Series (3,300 m)	{ Tadpatri (Pullampet) Shales { Cumbum Shales
Unconformity	
Papaghni Series (1,400 m)	{ Vempalle Shales and Limestones { Gulcheru Quartzites
Unconformity	
Archaeans —	Gneisses and schists

Each of the major divisions of the Cuddapah System is makred by an unconformity but the largest occurs at the base of the Nallamalai Series. Each series overlaps the previous ones while the topmost Kistna Series overlaps all of them on to the gneiss. The actual extent of the different series is variable.

probably because the shape and depth of the basin varied from time to time. The sediments seem to have been derived mainly from the west and south-west. It is also to be noticed that each series begins with a quartzite and is later followed by slates and limestones indicating that at the beginning of each series the basin became shallow and gradually became deeper.

Papaghni Series.—The lowest series is named after the Papaghni river, a tributary of the Penner. It is found only in the west. It consists of two stages, the lower one being GULCHERU (Guvvalacheruvu) STAGE consisting of conglomerates, grits and sandstones, resting unconformably on the Archean basement. These sandstones present a fine scarp on the southern margin of the basin, south of Cuddapah. The sandstones are generally quartzitic and contain pebbles of jasper and of vein quartz, derived from the Dharwarians. They become finer as they are followed from west to east.

The upper stage of this series is called the VEMPALLE (VAIMPALLI) STAGE. It consists mainly of fine grained flaggy limestone with bands and nests of chert and chalcedony, with subordinate shales. The limestones are grey coloured, weathering to buff. The weathered surfaces occasionally show what may be interpreted as algal structures. They show intrusive trap sills in the upper portion, along the contacts of which chrysotile asbestos has been developed in some places. The traps are considered to be responsible for the deposition of barytes which occurs mainly in the limestone and sometimes near the contact of the trap. They are considered to have been intruded during the Nallamalai times or probably even somewhat later. In some places the traps are contemporaneous flows associated with some tuffs.

Cheyair Series.—This series, named after the Cheyyeru river, is well developed in two areas, the north-western one in the Penner valley and the southeastern in the Cheyyeru valley. In the northwestern area the lower division is the PULIVENDLA STAGE consisting of quartzites, conglomerates, sandstones and flags, the lower beds often showing ripple marks. The pebbles and conglomerates are to some extent derived from the chert bands of the Vempalles. Their equivalents in the southern area are the NAGARI QUARTZITES which rest directly on the gneisses and form the hills of Nagari, Kalahasti and Tiripati. At Kalahasti, on the eastern margin, the quartzites are much crushed and apparently faulted.

The upper stage in the northwestern area is the TADPATRI STAGE, consisting of slaty shales with thin beds of siliceous limestone, chert, jasper and intrusive basic sills. This stage is well exposed in the Penner and Krishna valleys. The shales are soft and not well cleaved and have a tendency to break into long thin fragments. Their equivalents in the southern area are the PULLAMPET SHALES which also contain some limestone bands.

Nallamalai Series.— These occupy the largest area of any of the subdivisions of the Cuddapahs and take their name from the Nallamalai hills. They are particularly well developed in the eastern part of the Cuddapah basin. The lower beds are the BAIRENKONDA QUARTZITES which rest with a slight unconformity on Cheyairs. They are highly folded and contorted in the Nallamalai hills, the succeeding CUMBUM SHALES forming the cores of synclinal folds. The CUMBUM SHALES comprises shales and slates of varying shades of

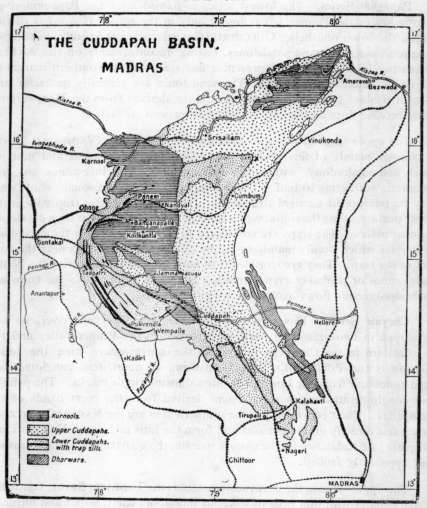

MAP VIII — THE CUDDAPAH BASIN, MADRAS

colour and degrees of hardness, intercalated with thin bands of quartzite and limestone. Quite well cleaved slates of this formation are worked near Markapur and Cumbum in the Kurnool district. The quartzite bands are thin-bedded and fine grained and suitable for use as sharpening stones. At places the slates have been converted into phyllites, while there are also softer varieties which are more shaly though cleaved. The limestone bands are grey, finely crystalline and sometimes micaceous. The lead ores of Nandialampet north of Cuddapah and the Cu-Pb-Zn ores of Agnigundala are found in these limestones. The slates become schistose at and near the disturbed eastern border.

Kistna (Krishna) Series.—These lie unconformably on the Nallamalais and are exposed on the plateau overlooking the Krishna river towards the north. They lie over the gneisses or dip under the Kurnools of the Palnad tract. The lower beds, called IRLAKONDA QUARTZITES, form a plateau. The KOLAMNALA SHALES which form the middle division are found in the valley of the stream of that name. The topmost stage is the SRISAILAM QURATZITES forming the plateau on the north and named after the well-known Srisailam temple on the Krishna river.

IGNEOUS ROCKS IN THE CUDDAPAHS

The Vempalle limestone belt has been intruded by sills of basic igneous rocks in many places, the sills varying in thickness from a few feet to a few hundred feet. They occur in the southern part of the western margin. Some contemporaneous flows and tuffs are also known. They are generally doleritic and basaltic in composition, occasionally showing chilled margins. The sills may sometimes split and again coalese but in general they maintain their thickness and horizon for long distances. Some of the sills are of the nature of quartz-dolerite as they contain a fair amount of silica or granophyric inclusions. There are also a few olivine-bearing types. The chief minerals are plagioclase and augite showing ophitic, intergranular and intersertal textures. Biotite, hornblende, magnetite, apatite and pyrite as well as occasionally a little micropegmatite are present. The rocks are similar to the Newer Dolerites of Singhbhum and Keonjhar and differ from the Deccan Traps in containing no pigeonite. They belong to the tholeiite type and show a differentiation trend similar to that of the calc-alkaline suite of rocks (Vemban, 1946).

KALADGI SERIES

Between Kaladgi and Belgaum in southern Bombay are exposed a group of formations bearing resemblance to the Cuddapahs. They show little or no metamorphism. They are divisible into a lower and upper series. The Lower Kaladgis lie with a profound unconformity on the gneisses. The Kaladgi basin is over 250 km long from east to west and in many places is covered by the Deccan Trap, amidst which several inliers appear. The Lower Kaladgis comprise basal conglomerates and quartzites with banded jasper pebbles, cherty siliceous limestones, shales and more limestones. The cherty limestones bear much resemblance to the Vempalle Stage of the Cuddapahs. The lower quartzites are well exposed at Gokak, over which the Gokak falls drop to a depth of 54m. The Malprabha river has cut a gorge through these rocks, 90m deep and only 45 m broad at the narrowest part. The limestones are well seen near Kaladgi, being of different colours and capable of yielding fine ornamental stones. The Lower Kaladgis have a total thickness of nearly 3,300 m.

			Metres
Upper	6.	Shales, limestones, hematite schists	(600)
	5.	Quartzite, conglomerates	360— 550
Lower	4.	Limestones, clays and shales	1,500—1,800
	3.	Sandstones and shales	
	2.	Siliceous limestones, hornstones	900—1,500
	1.	Quartzites, sandstones, conglomerates	

The Upper Kaladgis, consisting of quartzites, conglomerates, shales, limestones and hematite-schists, are about 1,100 m thick, the hematite-schists being sometimes rich enough to be used as iron-ore. The Upper Kaladgis are of restricted distribution, being found in synclinal folds in the northeastern part of the area, the axis of folding having a trend of W.N.W.-E.S.E.

THE PAKHAL SERIES

A large tract of rocks, presumably of Cuddapah age, extends for 150 km up the Godavari valley in Andhra in a N.W.-S.E. direction. This is divided into two by a band of younger rocks (Sullavais) lying in their middle with the same strike. The rocks are crushed and altered at their southeastern end in the Singareni area, and are folded rather sharply around the northeastern margins of each of the outcrops. The southwestern outcrop which forms the Pakhal hills shows the beds dipping towards the northeast. The rocks are divided into two stages, the lower being the PAKHAL STAGE, and the upper the ALBAKA STAGE. The first occupies the greater part of the southwestern outcrop while the second forms the north-eastern outcrop. Each of the stages is about 1,500 m thick. The Pakhal Stage comprises basal conglomerate resting on the gneisses, overlain by siliceous limestones and thick slates. The slates are not well cleaved but are flaggy, except in the north-east. The limestones have been metamorphosed to a tremolitic marble. The Albaka Stage comprises sandstones and quartzites with a few slate bands.

The Pakhal Series were regarded by King (*Mem.* XVIII, p. 209, 1880). as the equivalents of the Cuddapahs. According to Dr. C. Mahadevan (1949) the crystalline limestones, banded ferruginous rocks and the schistose rocks of this series recall the Archaean succession in Singhbhum and Gangpur. The rocks have been intruded into by granites which have formed composite gneisses as well as hybrid types. The phyllites have been converted into andalusite-mica-schists and garnet-kyanite-staurolite mica-schists. Copper mineralisation is found in the Singareni area, as indicated by old workings. Pegmatites and quartz veins traverse the Pakhals. All these go to prove their pre-Cuddapah age, according to Dr. Mahadevan, but others do not agree with this view.

PENGANGA BEDS

Rocks similar to the Pakhals, comprising a lower limestone group and upper shaly group, occur in the Pranhita valley to the west of the Wardha valley coalfield. The rocks are undisturbed and the limestones are said to contain bands of ribbon-jasper. These rocks generally lie directly over the gneisses but in Adilabad district, Hyderabad, they are said to show basal quartzites.

KOLHAN SERIES

Lying upon the Iron-ore Series of South Singhbhum and Keonjhar with a well marked unconformity, there is a series consisting of basal conglomerates, purple sandstones, shales and limestones, which are practically unmetamorphosed but subjected to folding in the western parts of their outcrops. The conglomerates contain pebbles derived from Singhbhum granite, banded hematite-jasper and iron-ore. The limestone rests conformably on the basal sandstone but is not extensive, being lenticular and about 13 m thick. It grades into phyllitic shale. The overlying shale is purplish grey or buff and somewhat phyllitic. The rocks are folded and cleaved in some places. The maximum thickness of the Kolhan Series is only 75 m, but there is no doubt that it is later than the Iron-ore Series.

GWALIOR SYSTEM

The rocks of the Gwalior System form hill ranges extending east to west along the northern fringe of a narrow belt of Bundelkhand gneiss at and near Gwalior. They have been regarded as the equivalents of the Bijawar Series which lie 190 km to the east. There is still some doubt whether they are of Cuddapah or Aravalli age.

The outcrops of the Gwaliors occupy an area 80 km long and 25 to 30 km wide, with a very gentle northerly dip. They comprise sandstones, ferruginous jaspers, limestones and interbedded traps. They lie with an unconformity on the denuded surface of the Bundelkhand granite. They have a great resemblance to the unmetamorphosed rocks near Hindaun and the jaspers and shales of Ranthombhor (Jaipur) which are regarded as of Aravalli age by Heron.

The Gwaliors have been divided into a lower PAR SERIES upto 60 m thick, and an upper MORAR SERIES which is 600 m thick (C.A. Hacket : *Rec*. III, p. 34, 1870). The Par Series consists of thin-bedded sandstones at the base resting on an irregular denuded surface and some sandstones and shales into which sometimes Bundelkhand quartz reefs project. The sandstones form well marked scarps. The Morar Series consists of siliceous and ferruginous shales with bands of bright red jasper. There are also ochreous shales and limestones with chert concretions. There are at least two chief horizons of traps, the lower 21m and the upper more than 150 m thick. The ferruginous shales have been used as iron-ore.

The correlation of the Gwaliors is a matter of some difficulty. They lie on the Bundelkhand gneiss and are practically unaltered sedimentaries. Though they resemble the unmetamorphosed Aravallis of Hindaun and Ranthambhor in lithology, they are probably of Cuddapah age, since they have a horizontal disposition and are unaltered. Their immunity to metamorphism and disturbance may, however, be due to their isolation from the main area of orogenic activity in Rajasthan. The age of the traps from the upper beds, determined by the helium method, was 500 million years according to Dubey, who equates them to the Cuddapahs.

BIJAWAR SERIES

This Series, first recognised in the Bijawar State in Bundelkhand, occurs in a series of outcrops extending from Bundelkhand to the south of the Narmada and has a thickness of less than 240 m in the type area (H.B. Medlicott : *Mem*. II, p. 35, 1860). Qurtzites and sandstones, sometimes conglomeratic, form the basal beds resting on gneisses. Siliceous limestones and hornstone-breccia are found with the quartzites. These are rather irregularly distributed and are of less than 60 m thickness. These are overlain by ferruginous sandstone containing pockets of hematite. The rocks are either horizontal or have a low southeasterly dip, though in a few places in the south they have been subjected to crushing and disturbance before the Vindhyans were deposited over them.

The Bijawars are associated with lavas, tuffs, sills and dykes of basic composition with micro-pegmatitic patches, but there are also olivine bearing rocks. The dykes of the Bijawar igneous suite were formerly supposed to be the parent rocks of the diamonds found in the conglomerates of the Vindhyan and Kurnool Systems, though so far as known, none of these basic rocks has the composition of kimberlite. Recently, however, kimberlite pipes have been found near Panna penetrating the Vindhyans.

A similar succession of rocks is found in the Dhar Forest, Jabalpur and Rewa on the one hand, and in the Sone valley of Bihar on the other.

The rocks near Bag (22° 22' : 74° 50') and Jobat (22° 25' : 74° 37') were originally correlated with the Bijawars. Those of Jobat were assigned to the Archaeans by P.N. Bose. There is some doubt whether the slates and phyllites of the Bag area are not also of Archaen age. The dolomites and quartzites of the Nimanpur area here are also probably of Archaean age according to M.K. Roy Chowdhury.

The succession in Jabalpur consists of phyllites, mica-schists, calcitic and dolomitic marbles, banded ferruginous quartzites with which are associated iron and manganese ores, and basic sills (and flows?), the whole assemblage bearing a remarkable similarity to the Iron-ore Series and part of the Gangpur Series of Chota Nagpur. There is little doubt that these " Bijawars " of Jabalpur are to be assigned the same age as the Singhbhum-Gangpur rocks (*Mem.* XXXVII, pp. 803-806, 1909).

The same remarks may perhaps apply also the ' Bijawars' of the Sone valley, whose easternmost extension is found in the Rajgir, Kharakpur and Sheikhpura hills in the Gaya and Monghyr districts. They comprise sandstones and conglomerates, slates, limestones, jasper and porcellanoid beds, chlorite-schists and basic lavas. In the Mirzapur district, Mallet regarded these (*Mem.* VII, Art. I, pp. 22-23, 1871) as comprising two series separated by an unconformity. At the contact with intrusive granitic rocks, the mica-schists and phyllites have been converted into composite gneisses, while the basic rocks have been epidioritised. The opinion has been expressed that these 'Bijawars' are also to be classed with the Dharwars (*Mem.* LXII, Pt. II, p. 145), (footnote), 1933), though they may, with equal justification, be classed with the Cuddapahs.

THE DELHI SYSTEM

This System extends right along the main axis of folding of the Aravalli mountains from near Delhi in the north, through Ajmer and Mewar to Idar and Palanpur in the south. In the north it consists of patchy exposures interrupted by alluvium. The exposures are fuller and much broader in the main synclinorium in Ajmer-Merwara and Mewar. Here they consist of two major synclines separated by a tongue of pre-Aravalli gneisses, the junction zones on both sides of the genisses being marked by shearing. The western margin of this tongue of gneisses is in places marked by a thrust fault and the western syncline is profusely intruded by igneous rocks which gradually increase in a S.W. direction until they almost obliterate and assimilate the Delhi sediments. The synclines coalese in the south. A few strips of the Delhis are also seen to the east of the main synclinorium. The exposures between Mewar and Jodhpur are only 9 km wide though exposed for a distance of 65 km along the strike. Farther south, the syncline widens and is also evidently buried deeper, enabling higher stratigraphical zones to appear in the centre.

The Delhi System lies over the gneisses and the Raialos with a great unconformity and is in turn overlain unconformably by the Vindhyans. It is generally correlated with the Cuddapahs, both consisting of about equal thickness of sediments, namely 6,000 m ; but the Delhis differ from the Cuddapahs in that they have been subjected to mountain building forces and to extensive folding, faulting and igneous intrusions. The Cuddapahs are comparatively less distrubed, and even that only in the eastern part of their exposures where the argillaceous strata have developed cleavage but have not been converted into schists.

The general succession in the main synclinorium is shown in the accompanying table.

TABLE 19 — The Delhi System

Jodhpur	Main Synclinorium, Mewar & Ajmer-Merwara	Chitor and Nimbahera	Jaipur	Alwar
Vindhyans of W. Rajasthan. Malani igneous suite. Delhi System (not present).	'Calc-gneisses' 'Calc-schists' Phyllites and biotite-schists Quartzites Arkose-grits	Lower Vindhyans. Sawa shales and grits.	Boundary Fault Jiran sandstones.	Ajabgarh Series Alwar Series.	Ajabgarh Series. Hornstone breccia. Kushalgarh limestone Alwar Series.
Raialo Series (Makrana marbles).	Raialo Series	Raialo Series.	...	...	Railao Series.

The succession is fully developed in Alwar, where two extra horizons namely KUSHALGARH LIMESTONE (500 m), and HORNSTONE BRECCIA (small but variable thickness), intervene between the Alwar and Ajabgarh Series. The ALWAR SERIES are rather unevenly developed in the various exposures. They die out in the middle portion of the syncline but reappear again in the south where faulted outliers appear in an inverted position under the Aravallis. The AJABGARH SERIES often show injections of pegmatites and aplites which have produced composite gneisses in different places. The CALC-SCHISTS are thinly bedded flaggy biotitic limestones which, when metamorphosed, have become schistose with the development of tremolite, diopside, biotite and feldspar and show dark and white bands. In the extreme south they are found in an unmetamorphosed condition. The CALC-GNEISSES, on the other hand, are coarse and contain much carbonate in the less metamorphosed areas such as those east of Beawar where they are 1,000 m thick, and also in Merwara. They are mainly darkbanded siliceous limestones. When followed along the strike they become more metamorphosed and gneissose. They are generally more massive and harder than the Calc-Schists. Their southeastern border in the south is a shear zone along which Erinpura Granite has been intruded. Near the batholith of Mt. Abu, the Calc-Gneisses are very profusely injected with granitic materials.

Kirana, Chiniot, Sangla and Shahkot Hills.—A low range of hills rises from the alluvial valley of Western Punjab about 70 km. south of the Salt Range. The hill range is about 100 km. long and consists of Delhi rocks such as grey and greenish slates and shales and reddish quartzites, together with Malani rhyolites and tuffs, possibly underlain by the intrusive phase of the Malanis. These hills as well as the area of Sirohi-Danta in Rajasthan are believed to have contributed the materials of the tillite in the Salt Range. In the western part of the range, NW-SE strike is seen in the formations, being a continuation of the strike of the southern part of the basement in Rajsathan. A northerly continuation of this submerged basement is believed to extend into Kashmir towards the Pamirs, producing the Kashmir syntaxis.

North-Eastern Area.—The Delhis in the northeastern area consist of narrow strike-ridges in which the rocks show moderate folding. The general strike is N.N.E.-S.S.W. but the rocks are overfolded to the southeast as a series of isoclines. The Alwars are 3000-4000 m thick, consisting of compact quartzites, conglomerates and grits. Some of the quartzites consist of almost pure quartz grains. A few subordinate shaly and calcareous bands are present and also sills of epidiorite and metadolerite. In the upper beds there is an admixture of argillaceous materials and the quartzites become mottled and streaked with brown colour. The Alwar Quartzites contain some deposits of iron and copper ores. The quartzite ridges prominently exposed at and near Delhi belong to this formation.

The Kushalgarh Limestones overlie the Alwars. They are dolomitic limestones with dust-like inclusions, showing a banded structure, with practically no jointing. Their maximum thickness is 450 m. The Hornstone Breccia appears at some horizon in the Kushalgarh Limestone, more often near the top. It is developed only in the Alwar State. It is considered to have been derived from the shattering of the alternating thin beds of quartzites and shales during the intense folding of the strata. The rock consists of angular pieces of quartz in a fine grained dark matrix of ferruginous and siliceous matter which is sometimes rich in iron. Being quite hard, it forms hillocks, but its outcrops are irregular and sinuous. Its autoclastic origin explains its varying stratigraphical horizon.

The Ajabgarhs are mainly an argillaceous series, forming synclinal valleys because of their comparative softness. They contain subordinate siliceous limestones, calcareous silts and ferruginous quartzites. The shales are sometimes phyllitic, and when metamorphosed have developed chiastolite, staurolite and small garnets. They are thin-bedded and show minor crumpling as well as abundant irregular joints.

In the lower part of the Alwars there are several contemporaneous basic lava flows and sills. Thin and numerous parallel sills are sometimes noted in the quartzites in some places and these may represent metamorphosed volcanic ash interbedded with the quartzites. When the basic intrusions are found in limestone bands, they have produced zones of actinolite, tremolite, epidote and other calc-silicates.

Western Jaipur.—Being close to the Aravalli axis, the Delhis in this area are highly folded, metamorphosed and intruded by granitic and basic igneous rocks. The Alwar quartzites are well developed. The Kushalgarh Limestone and Hornstone Breccia are missing. The Ajabgarhs consist of mica schists, micaceous quartzites and calciphyres. The copper ores of Khetri and Singhana and the cobalt ores of Babai are found in quartzites and slates associated with brown mica schists, and impure limestones of Delhi age.

Main Synclinorium.—The Delhis are exposed in the main synclinorium from Ajmer through Mewar into Sirohi and Idar, over a distance of more than 320 km. The Kushalgarh Limestones and Hornstone Breccia are not developed in this region. The basal beds seem to show a distinct relation to the rocks on which they lie; arkose and grit occur where they lie on gneisses and pegmatite, but only quartzites are found where they lie on Aravalli phyllites. Thick conglomerates (300 m) occur at Barr and can be traced for 50 km. They contain large quartz pebbles which are much drawn out in the direction of dip. Near Todgarh in Ajmer an excellent section of the Ajabgarh Series is developed. Here the Calc-Schists are profusely injected by the Erinpura Granite.

The SAWA GRITS which unconformably lie on the Bhagawanpura limestone are considered to represent the Delhis. The grits pass upwards into the Sawa Shales, the thickness of the whole formations being about 60 m. There are also the JIRAN SANDSTONES which lie below the Vindhyans or the Deccan Traps, separated by an unconformity. They consist of 60 m of compact, hard, pale grey quartzites, sometimes ferruginous and mottled with purple stains. They may also represent the Delhis. They resemble the Kaimur Sandstones to some extent and were originally described as Delhi Quartzites.

Erinpura Granite.—The Delhi System is intruded by the Erinpura Granite which shows a great deal of variety in its form, size, texture and degree of foliation. It has no effusive representatives. It is generally a biotite-granite but its pegmatites contain muscovite and tourmaline. It occurs in two forms, a massive granite which occurs as bosses, and a sheeted type intercalated with other rocks when it shows variation in grain size and occasional development of porphyritic types. In the type area, in eastern Mewar, it shows gneissic foliation especially near the border of schistose rocks. This granite forms the chief intrusive igneous material in the Delhi rocks. It occupies a large area on the northwestern side of the Aravalli range, obliterating part of the western portion of the Delhi synclinorium in the south. It is exposed also in Palanpur, Idar, Sirohi, Beawar, Jaipur and Alwar. Mt. Abu is composed of a large batholith of this granite. It is here a grey biotitic granite with hornblende, being composed mainly of quartz, microcline, orthoclase, some plagiclase, sphene, iron-ore and fluorite. It becomes somewhat micaceous when it approaches the schistose rocks. It shows xenoliths of amphibolites which represent partly assimilated invaded rocks, and also dykes of pegmatite and dolerite.

Malani Igneous Suite.—Over a large area in Jodhpur are found exposures of the MALANI RHYOLITES occypying a tract 200 km by 240 km. The outcrops are partly covered by desert sands and in places by younger formations. Some parts must also have been eroded away, so that the original extent must have been large. They are named after Malani (Barmer) near Jodhpur town. Rocks resembling them are seen in the Kirana-Sangla hills in West Punjab, about 70 km south of the Salt Range. The Talchir Boulder-beds in the Salt Range contain pebbles of the rhyolites and other associated formations in Rajasthan. They consist of beds or layers of rhyolite and ash beds some of which may be ignimbrites, and subordinate porphyries and felsites. The rhyolites are reddish brown, with pink orthoclase, occasional sanidine, some oligoclase and small idiomorphic but corroded quartz crystals. Laths of hornblende and microlites of magnetite are present. The groundmass is crypto-crystalline to micrographic, showing flow structure and devitrification of glassy material. They overlie the Aravallis and Vindhyan sandstones and are of early Vindhyan age. Their intrusive phase is now believed to be represented by the IDAR, JALOR and SIWANA GRANITES. The Idar granite shows granitic, microgranitic and granophyric types. The Jalor granite is a hornblende-biotite granite white the Siwana granite is mainly a hornblende granite. It is believed that the granites in Tusham hills and Sangla hills in West Punjab plains belong to this same suite.

AGE OF IGNEOUS ROCKS OF RAJASTHAN

Dr. Heron considers the Bundelkhand Granite as the earliest igneous rocks in this region. The igneous components of the Banded Gneissic Complex may have been contributed by the Bundelkhand Granite or by some other unknown igneous rocks. The post-Aravalli granites are, according to him, of very limited extent as he has mapped two types of granites in northeastern

Rajasthan, one pre-Delhi and the other post-Delhi. There are, in addition the Erinpura Granite and the Malani igneous rocks which are respectively, post-Delhi and Lower Vindhyan.

In his Presidential Address to the Geology Section of the 40th Session of the Indian Science Congress in 1953, N.L. Sharma has discussed the age relationship of the various igneous rocks in Rajasthan. According to him there are no definitely proved pre-Aravalli granites. The Bundelkhand Gneiss is considered most likely to be post-Aravalli as it has apparently contributed the granitic material of the Banded Gneissic Complex as well as of the Aravalli schists. In support he quotes the views of Crookshank who states that the Banded Gneissic Complex is essentially the same as the Aravalli schists which have been granitised by the Bundelkhand Granite. Sharma agrees that the Erinpura Granite which is so widepread in Rajasthan is post-Delhi and pre-Vindhyan. He is of the opinion that there are only three proved periods of granitic intrusions — the first post-Aravalli, the second post-Delhi and the third Malani.

Regarding the basic intrusives, three periods of activity are postulated by Sharma, as shown below :—

3. Olivine dolerite and basalt (post-Erinpura Granite)
2. Meta-gabbro and meta-dolerite (pre-Erinpura Granite but post-granitoid gneiss)
1. Epidiorite, pyroxene-granulite and hornblend-schists (post-Aravalli but pre-granitoid gneiss).

EXTRA-PENINSULAR AREAS

The equivalents of the Cuddapahs and Delhis in the Himalayan areas are the Dogra Slates of Kashmir, Attock Slates of Punjab and N.W. Frontier Province, the Chails and Simla Slates of Simla Hills, the Chandpurs of Garhwal and Chakrata, and part of the Haimanta System of the Central Himalayas of Kumaon. In Burma the Chaung Magyi Series may be partly of Cuddapah age.

HIMALAYAS

Chail Series.—Named after Chail (30° 58′ : 77° 12′) which is situated on a small Jutogh-Chail outcrop south-east of Simla, this series is exposed around Jutogh and the Chor mountain. The latter exposure extends towards Chakrata to the east and over the Shali mountain to the north. A large area is also seen in the Chamba district where the Chails are thrust over the Krol belt. The series consists of slates, phyllites, limestones, talc-schists, flaggy sandstones and schistose quartzites. They are separated by the Chail thrust from the underlying Shali Series of the Shali window (*Rec.* LXXIV, p. 133-163, 1939), or from the Jaunsars, Blainis and Simla Slates elsewhere. They are also traversed by basic igneous dykes.

The Chails may be merely a local facies of the Jutoghs ; or probably younger. The Chandpurs of Garhwal and the Ajabgarhs of the Delhi System may also be of the same age. Oldham's Jaunsar Series is mainly the equivalent of the Lower Chails while the Bawar Series will represent the Upper Chails.

Martoli Series.—Heim and Gansser have described, in the northern part of Kumaon, a series of phyllites, calcareous phyllites and quartzites which are injected by pegmatites and quartz veins in some places. This assemblage has been called the MARTOLI SERIES, after the village of Martoli on the Gori Ganga below Milam. The rocks are unfossiliferous and flyschilke, and may have accumulated under geosynclinal conditions. They are about 16,500 ft. (5,000 metres) thick and may be of Algonkian age. They resemble some of the Haimanta rocks and form the peaks Nampa, Nanda Kot and Nanda Devi. In the valley of the Kali river they are represented by the BUDHI SCHISTS which show meso-grade metamorphism.

Chandpur Series.—Two suites of rocks, known as the MANDHALI and CHANDPUR SERIES, are found in the Chakrata district, continuing into Garhwal. The stratigraphic position of the Mandhalis is uncertain but the Chandpurs are taken to be the equivalents of the Chails.

South of the Chor Mountain there is a syncline in which the Jaunsars, Blainis, Infra-Krols, Krols and Tals are exposed. The synclinal axis rises to the west, bringing up the Jaunsars. In the east the Mandhalis and Chandpurs are exposed beneath the Nagthat Series. Another syncline is found south of Chakrata and it is bounded on the south by the Krol thrust and on the north by the Tons thrust. The succession in the Chakrata area is :—

```
            NAGTHAT SERIES
         ———— Uaconformity ————
        CHANDPUR (? CHAIL) SERIES
         ———— Thrust plane ————
            MANDHALI SERIES
```

The Chandpurs are thin-bedded (1 to 3 mm) alternations of quartzite and phyllite, giving a fine banded effect, associated with abundant green chloritised tuffs and volcanics. The banded rock may also be mainly tuffaceous. Current bedding and ripple marking are frequent. The chloritic tuffs and slates exhibit polygonal jointing. Some amygdaloidal basalts and intrusive dolerite sills have also been noticed in them. South of Chakrata, the Nagthats are seen to lie over the folded and disturbed Chandpur phyllites.

In the Mussoorie area, the Chandpurs show different grades of metamorphism, the beds in the northern areas being more metamorphosed than those elsewhere. The *Inner Schistose Series*, described by Middlemiss in Garhwal, are mainly the equivalents of the Chandpurs. The rocks of this series are metamorphosed to schists at Lansdowne, Dutatoli, Ranikhet and Almora. The Chandpur Series resemble the Chails to some extent. The Simla Slates, Chails and Chandpurs may be taken as roughly the equivalents of the Cuddapahs.

Simla Slates.—A great and important slate series, presumably of Purana age, is developed in many parts of the western Himalayas. The rocks of this series always occur beneath the Talchir tillite and its equivalents. They have been given local names like Simla, Attock, Hazara and Dogra Slates. They are all unfossiliferous.

There are no slates of similar character either below or above the Cambrians of the Salt Range, so that it is not possible to fix their age with any precision. Their nearest equivalents in the Peninsula are the argillaceous Ajabgarh Series in Rajasthan.

The SIMLA SLATES of the Simla area appear to be separated from the Blaini Series which contain boulder beds, by the Jaunsars which also contain some slates and beds of pebbly quartzites. The Simla Slates are well seen in the spurs between Simla and the Sutlej and along the Mashobra road. They swing round to the west of the Jutogh area and form a band along the Gambhar, Ashni and Giri rivers in a N.W.-S.E. direction. The succession exposed in the latter river valleys is normal, as proved by the attitude of the current bedding. Here the series can be divided into two. The lower part, having a thickness of 420 m, is an alternation of greywackes and slates, in massive beds. The slate bands contain numerous flat-lying, pillow and kidney shaped concretions with sizes up to 45 cm in length. The concretions are of clay slate or quartzite, set in a clay-slate matrix. Current bedding is common in the strata. The upper part, which attains a thickness of about 360 m, is seen only west of Kandaghat. It consists of soft green micaceous and gritty slates and cindery micaceous sandstones which often exhibit ripple marks on bedding planes. The clay-slates are dark grey, carbonaceous and sometimes purple. Occasionally they are phyllitic.

The Simla Slates are seen also alongside the Tons river east of Chandpur and in the Chakrata ridge from Deoban hill to Kailana where they adjoin the Mandhali beds whose age is not known. In Garhwal they are seen along the Nayar river. They are known to occur in Nepal.

Hazara Slates.—The Hazara Slate Series is seen forming a broad belt about 16 to 20 km wide in the Hazara plains. They are separated from the Nummulitics as well as the Salkhalas by faults. They continue to the northeast into Kaghan and to the south into Attock and Peshawar districts. They are dark grey, occasionally green or purple, with interbedded bands of fine grit. The general strike is N.W.-S.W. (*Mem.* XXVI, p. 10, 1896). They contain some limestone bands which appear to be the equivalents of the Kakarhatti limestones occurring in the Simla Slates of Simla. These limestones contain the same type of stromatoporoid markings as in the Deoban limestone of Chakrata. The slates are irregularly cleaved and folded. They underlie the Infra-Trias whose lowest member is the Tanakki conglomerate which is regarded the equivalent of the Talchir boulder bed, and which is well exposed in the Tanakki glen in Mount Sirban near Abbottabad.

Attock Slates.—The Hazara Slates continue into the Attock district and thence into the Peshawar district where they are known as the Attock Slates. They are blue-black, well cleaved slates but the cleavage is irregular. They contain layers of gritty sandstone and are generally steep-dipping in the Attock district (*Mem.* LV, p. 75, 1933).

Dogra Slates.—The Dogra Slates of Kashmir are similar to the above mentioned slates in constitution and stratigraphical position. They occur on the southwestern flanks of the Pir Panjal, intimately folded with the Salkhalas.

They are black to green, highly cleaved, flaggy slates with interbedded green chloritised traps which appear to be contemporaneous. They are highly jointed and wavy and therefore of little use. Here also they lie below the Tanawal conglomerates. The same formations are seen at the western end of the Kashmir valley where they are overthrust by the Salkhalas (*Rec.* LXVI, p. 123, 1932), and in the Wardwan valley below the Permian Kuling Shales.

BHUTAN

We owe most of our knowledge of the interior of Bhutan to Gansser (1964) as no good summary of the work done by the G.S.I. there during the last 6 or 7 years is available. The Main Boundary Fault is a steep-dipping thrust limiting the Siwalik zone. On its north there is a narrow zone 70-200 m wide, of coal-bearing Lower Gondwana rocks forming an anticlinal structure and exposing a reversed succession. Its northern border is another steep thrust bringing up the BAXA SERIES which are exposed over a width of 2000 m. The lowest beds are highly sheared slates, followed by massive quartzites, which become thin-bedded and ripple-marked above. This forms the SINCHULA STAGE. It is marked off from the succeeding JAINTI QUARTZITE STAGE by thin green steatite schists and a band of epidiorite. Bands of hematite-quartzites, hematite-schists and jasper conglomerate are intercalated in this. The succeeding BAXA DOLOMITE STAGE forms prominent ridges in western Bhutan but becomes less important in the east, where the Sinchula quartzites become dominant.

Farther to the north, the Baxa Series is succeeded by the DALING SERIES with the intervention of a thrust. The Dalings are slates, phyllites and mica-schists with subordinate quartzites, which are gradually invaded by bands and veins of augen-gneiss and migmatites. These banded gneisses form a broad zone and have been named the CHASILAHA GNEISS. They continue eastward but become narrow east of the Mochu (Sankosh) valley. There is very little difference between the Baxa and Daling Series except that the former contain much dolomite which is rare in the Dalings. The Baxa Series may therefore be a local development (? corresponding to the Krols) which is apparently absent east of Bhutan.

North of the gneiss zone is the broad PARO belt of metamorphosed rocks which continues east and is quite prominent in the Tongsachu valley. It exposes slates, clacareous quartzites and marbles, with two prominent zones of disturbance. The general direction of fold axes in the Paro schists in NW-S.E. On its north we come into the TACHTSUNG GNEISSES which mark the Central Himalayan zone. This zone also continues east and forms the broad dome of the TONGSA GNEISS near and to the north of Tongsa. The northern part of this gneiss belt shows increasingly numerous intrusions by the young tourmaline granite which is prominently displayed along the Central Himalayan zone and is called the CHOMOLHARI GRANITE. In Pelela area ESE of Punakha, there is a zone of unmetamorphosed Silurian-Devonian TANGCHU SERIES, which may correspond to the Paleozoic rocks of Nepal. North fo this comes the crystalline zone of the Higher Himalayas.

The Chomolhari mountains in N.W. Bhutan are separated from the Yelela area on their south and east by a prominent thrust. The zone on the south shows Palaezoic beds, conglomerates and sandstones containing crushed crinoids, recalling the Blaini Series and limestones and shales with Triassic and Jurassic fossils. In the Lingshi area north of Yelela, Lower Cretaceous beds have also been found, but these are different in facies from the sediments of the Kampa area in Tibet.

The Chomolhari mountains are formed of bands of marbles with prominent sills of the white tourmaline granite. The top is almost entirely of granite. This granite is post-tectonic and corresponds to the one found in Badrinath peak in U.P. It covers a large area in NE Bhutan (Gangkerpunzum). The major mountains in this region, show calc-schists and marbles intruded by the tourmaline granite. The calc-rocks forming the line of high peaks appear to be the equivalents of the Everest Limestone, and are overlain, on the north, by sediments of the Tethyan basin containing a full Mesozoic succession.

ASSAM HIMALAYA

This comprises the Aka, Daphla, Miri, Abor and Mishmi hills but the last are to be placed in the Burmese arc as the Himalayas terminate at the valley of the Tsangpo (Dihang or Brahmaputra). The geology of the eastern extremity of the Himalayas is similar to that of the Sikkim-Bhutan portion. The narrow Siwalik belt is overthrust by the Lower Gondwanas. In the Subansiri tract there are fossiliferous marine Permo-Carboniferous rocks which must mark a marine zone to the north of the terrestrial Gondwanas. The Gondwana thrust is usually followed by one containing the Dalings with their granitised equivalents, the Darjeeling gneisses. In the Aka hills there are two belts volcanic rocks. As the Gondwanas appear to pass into the volcanic sequence transitionally, it is likely that the latter is of the same age as the Panjal Volcanics and not of Jurassic age as suggested by Coggin Brown. The northern belt of volcanics, in the Abor Hills is 20 km. wide, containing frequent intercalations of reddish quartzites.

The Daling belt of slightly metamorphosed slates and thick argillites with quartzite intercalations succeeds the volcanic belt. The metamorphism increases towards the north. Some dolomites appear in the section and there are indications of gneissic intercalations. The NE-SW course of the Dihang River just below the highest part of the Himalayas marks a major thrust zone, to the north of which is the Central Himalayan belt of crystalline rocks with a high grade of metamorphism. Nothing is known of the High Himalayan zone of this part of the Himalayas.

The Mishmi Hills occupy the northeast corner of Assam where the NE strike of the Himalayas turns to the east and then sharply to the southwest and south. The Siwaliks can be followed over a distance of 30 km east of Dibong river, after which they completely disappear and their place is taken by the Tertiaries of the Burmese Arc. Northeast of the Siwalik zone is a thrust bringing up a series of pink and green quartzites which gradually show higher metamorphism. The rocks dip to the northeast. Beyond this appear hornblende gneiss, amphibole schist, graphitic garnetiferous schists etc. There is no trace of Gondwana rocks or of the volcanics met with in the Abor Hills and there must be a major line of separation between the two groups of rocks.

Beyond the zone of metamorphosed sediments there are gneisses and granites. These are thrust over the folded series of the Indo-Burma border in a south-westerly direction along the MIJU THRUST, which overrides the N.E.-trending zone of Naga and Haflong-Disang thrusts.

Only a little information is available on the north-eastern end of the Himalayas. In Tibet the Tsangpo flows parallel to the main Himalayan range in a N.E. direction over a wide bed which rapidly narrows down to a hundred metres as it reaches Namcha Barwa (7755 m), near Timpu. It goes round this peak and to the south and west of Gyala Peri (7150 m). Here it is joined by a tributary called the Yigrong River from beyond Gyala Peri. It then turns sharply south to cut across the mighty Himalayan ranges and reach the Assam plains over a series of precipitous cataracts and falls. The border ranges also turn round to the south and then southwest, finally taking a southward trend along the Indo-Burma border.

BURMA

The Chaung Magyi Series of N. Shan States which are an argillaceous Series with some quartzites, are overlain by fossiliferous Ordovician rocks. They have been folded and there is an unconformity separating them from the younger rocks. It is considered most probable that they are of Purana (Cuddapah) age. They are said to have been traced continuously into Yunnan where they merge into the Kao-Liang Series of Pre-Cambrian age.

ECONOMIC MINERAL DEPOSITS

Copper.—The copper ores of Khetri and Singhana in Jaipur occur as stringers of chalcopyrite, pyrite and other minerals in black slates associated with amphibolites, presumably of Delhi age. Old workings are seen over a length of several miles and the mineralised zone extends for some 25 km in a N.E.-S.W. direction.

At Daribo and Kho in Alwar, thin lodes of copper ore containing chalcopyrite, pyrrhotite, arsenopyrite etc. are found in slates near the base of the Alwar Series. These occurrences are considered promising.

Cobalt.—At Babai (27° 53' - 75° 49') in Jaipur, copper-cobalt ores comprising chalcopyrite, cobaltite, danaite etc. are found as stringers and disseminations in slates (? of Alwar age). They used to be worked in a small way of making cobalt glazes on metals.

Asbestos.—The trap sills traversing the dolomitic limestones of the Vempalle Stage have produced serpentinisation in several places in a zone about 1 m thick, mainly near the upper contact. In this zone, as well as in the traps near the contact, chrysotile asbestos has been developed as cross-fibre veins of good quality. The best deposits are found about 5-10 km west of Pulivendla in the Cuddapah district. The length of the fibres in the veins varies from a small fraction of an inch to about 6 inches, the average being around a quarter to half an inch.

Barytes.—The Cuddapah traps have also been responsible for the formation of veins of barytes in the Vempalle limestone. Some of the veins are several feet thick and a few hundred feet long. The most important deposits occur near Vempalle, Pulivendla and Kotapalle in the Cuddapah district, Nerijumapalle and Mutssukota in the Anantapur district and near Balapalapalle and other places in the Kurnool district. The Alwar Quartzites of the Delhi System contain fissure-veins of barytes at Sainpuri and Bhakhera near Alwar. The deposits in Andhra State and in Alwar produce an average of 25,000 tons of barytes per year.

Steatite.—Steatite and talc of excellent quality is found developed in the dolomitic limestone of Vempalle Series near the contact with basic igneous sills at Muddavaram in the Kurnool district and Tadpatri in the Anantapur district. 'Lava' grade has been obtained from these deposits. Talc-schists are also found to have been derived from ultrabasic rocks intrusive into the Delhi System near Beawar and Ajmer in Rajasthan.

Building Stones.—The basement arkose-conglomerate of Alwar Series at Srinagar (Ajmer) is a fairly well cleaved rock yielding slabs and blocks. A similar conglomerate-grit at Barr, in which the pebbles are much flattened and elongated, yields good slabs up to 4.5 m long, 1 m broad and 7 to 10 cm thick. The Alwar Quartzite of Ghat and Maundla and the micaceous grits of Ajmer and Nasirabad yield thick slabs and blocks. All these are hard and durable building stones.

The slates worked at Kund in Alwar are of good quality, yielding school slates as well as thin slabs useful for paving and roofing.

Ajabgarh limestones are also worked at various places. The grey and black slabby limestone of Jhak and Sanodia (Kishangarh) yields good slabs. The Bhainslana stone is a hard compact finely crystalline limestone used for building and carving. The Tonkra stone is a coarsely granular, dull white crystalline dolomite, while the pink Narwar marble is a beautiful stone of ornamental quality.

In the Cuddapah basin of Andhra some of the quartzites — *e.g.*, Pulivendla and Nagari — are used as building stones where they are well bedded and yield rectangular blocks and slabs. The Cumbum Shales contain some good well-cleaved slates worked near Cumbum and Markapur in Kurnool district. They are associated with thin bands of sandstone which can be used as sharpening stones.

The Cuddapah shales in several places have yielded refractory clays. There are various types of clays derived from the weathering of shales, which can be used in the ceramic industry.

CHAPTER VIII

THE VINDHYAN SYSTEM

The Cuddapahs were succeeded by the rocks of the Vindhyan System after a time interval marked by earth movements and erosion. The Cuddapahs were then folded and metamorphosed to some extent though the intensity of the forces at play was much feebler than that at the close of the Dharwarian era. In Rajasthan, however, the post-Delhi movements were of great intensity along the Aravalli axis and were followed by the granitic intrusions on a large scale.

The Vindhyan System derives its name from the great Vindhya Mountains, a part of which is found to form the prominent palateau-like range of sandstones to the north of the Narmada valley, particularly in Bundelkhand and Malwa. It occupies a large basin extending from Dehri-on-Sone to Hoshangabad and from Chitorgarh to Agra and Gwalior, surrounding the batholithic mass of Bundelkhand Granite. Oldham estimates the area of the exposures as about 100,000 sq. km with a further 65,000 sq. km lying underneath the Deccan Traps.

Over the greater part of the area, only the upper portion of the Vindhyans is developed, usually resting on the Cuddapahs or older rocks with a very pronounced unconformity. In the Sone valley, where the Lower Vindhyans are well developed, an unconformity is seen between them and the upper divisions. The Vindhyans are distinctly less disturbed than the Cuddapahs but the lines of disturbance tend to be common in both. Within the Vindhyan System itself there are distinct unconformities, often marked by conglomerates, separating the different Series.

In recent years the Suket Shales of Rampura (24° 28' : 75° 26') in Central India have yielded small discoid impressions considered to be organic remains and assigned to the genus *Fermoria* related to the primitive brachiopod *Acrothele*. There is, however, a difference of opinion as to whether they are inorganic or organic, and if the latter, whether they are plant or animal remains (Chapman, F. *Rec. G.S.I.* LXIX, pp. 109-120, 1935 ; Sahni, M.R., *Ibid.*, p. 458). A new genus of similar nature, named *Krishnania*, has been described by Sahni (*Curr. Sci.* Feb. 1954). The Suket shales, as also the shales and limestones of the Kheinjua and Rohtas Stages in the Mirzapur district, have yielded coaly matter and spores and tracheids of vascular plants, algal thalus, carbonised casts of algae, fungal spores etc. described by a number of workers. Stromatoliths have also been described from the Fawn Limestones and Bhander Limestones of the Sone Valley. Howell (1956) has reviewed the evidence of fossils of different types from the Semris and the Kaimurs and come to the conclusion that as trilobites are absent, these strata should be assigned mainly to the Pre-Cambrian, though the upper part may come into the Cambrian.

Coaly matter with streaks of vitrain have been reported from the Lower Vindhyans at various times. Fox reported vitrain from the limestone at Japla several years ago. Fragments and streaks of coaly material are known o occur in the Bijaigarh Shales. A seam of carbonaceous matter 60 m long and 1.6 m thick occurs between two clay layers in the basal beds of the Semris below the Kajrahat limestone. The richer part yielded 23 to 31% fixed carbon and 9 to 13% volatile matter on analysis. It consists of algal spores and cuticles and tracheids of vascular plants. Thin carbonaceous layers were also seen in the foundations of the dam across the Rihand river near Obra. Most of the localities are near Dehri-on-Sone and in the Mirzapur district of Uttar Pradesh (Mathur, 1964).

The Vindhyans consist of four main series, named as follows :—

				Metres
Upper	Bhander Series	...	arenaceous and calcareous ...	450
	Rewa Series	...	mainly arenaceous ...	150—300
	Kaimur Series	...	mainly arenaceous ...	150—300
Lower	... Semri Series	...	mainly calcareous ...	300—900

LOWER VINDHYANS : SEMRI SERIES

From Sasaram westwards to the watershed between the Sone and the Narmada, the Lower Vindhyans are exposed underneath the prominent scarp of Kaimur quartzites for a length of some 400 km. Here the maximum width of this series is about 25 km, but farther east it narrows down to a width of less than 3 km. This is the type area of the Semri Series.

The lowermost beds of this series in the Sone valley, called the *Basal Stage*, are 600 m thick and consist of basal conglomerates and the *Kajrahat limestone* beds. They are followed by shales, sandstones and tuffs which have been silicified and converted to porcellanites (the *Porcellanite Stage*) attaining a thickness of about 100 m. The *Kheinjua Stage* overlying this is about 180 m thick and consists of olive shales, fawn limestone and glauconitic sandstones which show ripple marks and other characters pointing to shallow water and sub-aerial deposition. Above this comes the *Rohtas Stage*, 120 to 210 m thick, consisting of alternating beds of limestones and shales which support a flourishing lime and cement industry in Bihar. The limestone varies in quality from bed to bed, much of it being of high grade and containing over 80 per cent calcium carbonate, less than 3 per cent magnesium carbonate and about 10 per cent silica. In the upper part there are large stone nodules in shales, while still further up siliceous limestones occur.

The Semri Series is intruded by dykes of dolerite and basalt in a few places in the Sone valley. The basic rocks contain both augite and rhombic pyroxene, zoned plagioclase, ilmenite and pyrite, with patches of micrographic quartz and feldspar and occasional glass.

The Semri Series is found also in the Karauli area of Rajasthan where the Aravalli phyllites are overlain by sandstones and conglomerates and these in turn by the TIROHAN LIMESTONE. Above the Tirohan limestone is a zone of breccia (TIROHAN BRECCIA) which is due to the partial removal of lime by solution from the beds and the consequent collapse. An unconformity intervenes between these and the overlying Kaimurs. The Tirohan limestone is apparently the equivalent of the Rohtas limestone and both are underlain by beds containing glauconite.

On the southern side, in the Chitor-Jhalrapatan area, shales of probable Aravalli age are overlain successively by grits and conglomerates, NIMBAHERA SHALES, NIMBAHERA LIMESTONES and SURET SHALES, the thickness of the group of Vindhyan beds being about 300 m.

TABLE 20.—THE SEMRI SERIES AND ITS EQUIVALENTS

	Sone Valley	Karauli	Chitor
Rohtas Stage	Alternating limestones and shales	Tirohan Breccia ... Tirohan Limestone ...	Suket Shales Nimbahera Limestone
Kheinjua Stage	Glauconite beds Fawn limestone Olive shales	Glauconite-bearing beds	Nimbahera Shales
Porcellanite Stage	Porcellanites and silicified rocks	Sandstones and conglomerates	Grits and conglomerates
Basal Stage ...	Kajrahat Limestone Basal Conglomerate		

UPPER VINDHYANS

The Upper Vindhyans are exposed in the great Vindhyan basin. They consist largely of sandstones and shales with subordinate limestones, the sandstones forming extensive plateaux around and to the south of the Bundelkhand granite mass. The sub-divisions are shown in the accoumpanying Table.

TABLE 21.—UPPER VINDHYAN SUCCESSION

Bhander Series	Upper Bhander Sandstones Sirbu Shales Lower Bhander Sanstones Bhander Limestone (Nagode) Ganurgarh Shales
—Diamond-bearing Conglomerate—	
Rewa Series	Upper Rewa Sandstones Jhiri Shales Lower Rewa Sandstones Panna Shales
—Diamond-bearing Conglomerate—	
Kaimur Series	Upper { Dhandraul Quartzite Scarp Sandstone & Conglomerate Bijaigarh Shales Lower { Upper Quartzites and Sandstones Susnai Breccia Lower Quartzites and Shales

KAIMUR SERIES

In the Sone Valley the Kaimur Series contains two bands of quartzit in the lower division which may be gritty and even conglomeratic and show current bedding. The lower quartzite passes upward into flagstones and shales showing ripple-marks and sun-cracks, and these into thin bedded micaceous and carbonaceous shales with sideritic bands. Interbedded with these are banded and jointed porcellanites, fragments of which are found in the next succeeding gritty bed called the Susnai Breccia. This breccia is undoubtedly of epi-clastic origin and marks a break in sedimentation, though the base of the Kaimurs is to be put at the base of the lower quartzite.

The SUSNAI BRECCIA is overlain by the upper silicified quartzite (Lower Kaimur) with marked current and lenticular bedding and ripple-marks, which forms a very conspicuous scarp, 15 m high, in the Sone Valley. Above this are other quartzites and also sandstones and mudstones which show extensive replacement by iron and have the characteristics of shallow water deposits. These pass upwards into the BIJAIGARH SHALES which are carbonaceous, pyritiferous and micaceous and generally bleached or yellow in colour. Lenticles of bright coal (vitrain) are found in these and some beds are fairly rich in carbonaceous matter.

It is in these shales that a bed of pyrite about 1 m thick occurs below the scarp of the quartzite at Amjhor, Banjari and other places south of Dehri-on-Sone. The pyrite bed is generally rich enough in sulphur (around 40 per cent) to be worked.

The Upper Kaimurs, overlying the Bijaigarh Shales, consist of greenish flagstones and sandy silt stones (generally showing current-bedding and ripple-marks) which crop out along the Kaimur scarp and are exemplified in the Mangesar hill. The green material apparently includes chamosite, chlorite and green mica. Above these are the DHANDRAUL QUARTZITES which are white to purplish in colour. The Upper Kaimurs have a thickness varying from 150 to 300 m.

In Bundelkhand the Kaimurs show a basal conglomerate containing pebbles of jasper, the main formation being a fine-grained quartzite of greyish or brownish colour with conspicuous current-bedding.

REWA SERIES

The Kaimurs are succeeded by the Rewa Series composed of somewhat coarser sandstones than those of the Kaimurs, and current-bedded flagstones. The two series are separated by a zone of diamond-bearing conglomerate. The divisions recognised in Central India in the Rewa and the overlying Bhander series are shown in the Table given above.

The existence, in Bundelkhand, of the Lower Rewa Sandstones and Panna Shales is questioned by Vredenburg who states that the diamond-bearing conglomerate occurs at the base of the Jhiri Shales. In Gwalior, however, there are two shale bands separated by a sandstone, between the Kaimurs and the main Rewa Sandstone.

BHANDER SERIES

The uppermost division of the Vindhyans is the Bhander Series, which is separated from the Rewa series by a horizon of diamond-bearing conglomerate. The Bhander Sandstones are fine-grained and soft, usually of a red colour with white specks. When light-coloured they often show red streaks. They are fairly thick-bedded and yielded large blocks which are used in building. The Upper Bhanders frequently show ripple-marks. The Bhander Limestone is of variable thickness and quality, passing from a good limestone to a calcareous shale. The Bhanders contain veins and beds of gypsum.

The Sirbu Shales of this Series bear a good resemblance to the Salt Pseudomorph Shale of the Salt Range and contain salt pseudomorphs at a few places in the Maihar and Jodhpur areas. A greenish type of these shales occasionally shows black rounded discoidal bodies similar to *Fermoria*.

The scarp near Sonia (26° 44′ : 73° 44′) in Rajasthan exposes 40 m of basal conglomerates and grits, succeeded by 8 m of purple shales with salt pseudomorphs, 30 m of variegated shales and sandstones and 30 m of pink to brown sandstones. This section is overlain by the BILARA FORMATION (same as Phalodi Limestones) consisting of light to dark grey cherty limestones which are 40 m thick. The whole section is the equivalent of a part of the Cambrian of the Salt Range and of the upper part of the Upper Vindhyan.

In the great Vindhyan basin the sandstones and quartzites form a series of well-marked scarps while the intervening strata being soft, give rise to sloping talus. The chief members persist over large areas with fairly uniform characters. Taken as a whole, the structure of the Vindhyan area is that of a basin, the sandstones forming plateaux. Over the greater part of the area the beds are nearly horizontal, but they show evidence of disturbance near the northwest and southeast margins. In the Dhar Forest and near Jhalrapatan, the Vindhyans are folded and show steep dips.

The Vindhyans are thickest in the southern and southwestern areas. The Upper Vindhyans are 3,300 m thick in the southwest, 1,300 m in the northwest and about 1,200 m in Bundi State. The Lower Vindhyans are either thin or absent in the northwest, the Kaimurs overlapping them and coming to rest directly on the gneisses or on the Bijawars.

The margins of the Vindhyan basin show a good development of sandstones, while the shales are best developed in the centre and east, passing gradually laterally into sandstone. The prevalence of current-bedding and ripple marks in the strata is indicative of shallow water origin ; while the red sandstones, of the Kaimurs and Bhanders indicate semi-arid and continental conditions.

The Vindhyans have been deposited on peneplaned older rocks under stable shelf conditions, the sediments having come from the southeast and south. There are evidences of semi-contemporaneous earth movements. In the Sone-Narmada Valley the compression seems to have come from the south or southwest, while in the area between Chitor and Hoshangabad the compresive forces have acted from the southwest and west. In Rajasthan they have been affected by overthrusts from the west, along the Main Boundary Fault which has a throw of some 1,500 m and which brings the undisturbed Bhanders against the highly folded Aravallis. This fault can be traced for a distance of about 800 km, parts of it coinciding with the course of the Chambal river. There are, however, some strips and outliers of Vindhyans to the west of this fault, *e.g.*, the Kaimurs from Bundi to Indargarh.

The Vindhyans probably continue to the north under the Gangetic alluvium of Bihar, perhaps buckled down to form the basement of the Himalayan foredeep. It is not known whether they have equivalents in some of the unfossiliferous rocks of the sub-Himalayan region in U.P. and Nepal, which are now found broken up and thrust southward over the Permo-Carboniferous and Tertiary rocks.

Rocks resembling the Vindhyans occur as scattered outcrops in Western Rajasthan northwest of the Aravalli mountains, some 150-200 km from the western margin of the main Vindhyan basin. They rest on the Malani rhyolites or the Delhis and consist of reddish brown sandstones with conglomerate beds, with limestone layers near the top. The conglomerates are only coarse grits and are devoid of large pebbles and boulders. They appear to have been deposited by currents coming from the east. These rocks may be Vindhyan or Lower Palaeozoic, but much older than the Bap beds (correlated with the Talchir tillite) which contain derived pebbles from the Malani igneous suite and from the Vindhyan limestones.

KURNOOL SYSTEM

The Cuddapah basin in the Andhra State contains two areas of younger rocks resting unconformably on the Cuddapahs — one in the Kundair Valley stretching up to the Krishna, and the other in the Palnad tract. This younger group of rocks, constituting the Kurnool System, is about 400 m thick in the west but much thicker in the Palnad area and has been affected by disturbances in the eastern part but by forces which were less intense than those which acted on the Cuddapahs. This system is regarded as the equivalent of the Lower Vindhyans.

The Kurnools have been sub-divided into four series, composed mainly of limestones with subordinate shales and sandstones.

TABLE 22.—THE KURNOOL SYSTEM

Series	Stages
Kundair	Nandyal Shales
	Koilkuntla Limestones
Paniam	Pinnacled Quarztites
	Plateau Quartzites
Jammalamadugu	Auk Shales
	Narji Limestone
Banganapalali	Banganapalli Sandstones

The Banganapallis are rather coarse sandstones of grey and brown colours, sometimes shaly, feldspathic or ferruginous. They contain abundant small pebbles of quartz, jasper, chert and slate, derived mostly from the Cheyair Series. They are the main source of diamonds in the Kurnool formations. Some of the exposures have been extensively worked for diamonds which industry was active in several places in this region till about the year 1840.

The lower beds of the Jammalamadugu Series consist of limestones—the Narji Limestones—of various colours, especially blue-grey, buff and fawn. They are about 120 m thick at Narji where the rock is much quarried for building purposes. They are succeeded by the Auk Shales of buff and purplish colours. The Paniam Series, developed around Paniam, (Panem) and Undutla, comprises sandstones and quartzites. The topmost series, named after the Kundair, since it occupies the valley of that river, has a thickness of 150 to 200 m. The lower third of this is a compact fine-grained limestone (Koilkuntla) wihle the upper part consists of impure limestones and calcareous shales named after the small town of Nandyal.

Outcrops of Kurnool rocks, sometimes called the PALNAD SERIES, are developed in Palnad, in the north-east of the Cuddapah basin, stretching on both sides of the Krishna river. They lie unconformably on the Cuddapahs and contain a diamond-bearing conglomerate at the base. Their thickness is considerable but probably not as much as 6,000 m estimated by Heron. They consist of limestones and shales, the limesstones beinghigh-calcium ones (75-90 per cent calcium carbonate with a little magnesium carbonate) similar to the Narji, Sullavai and Bhima limestones.

JEYPORE — BASTAR — RAIPUR

There are several exposures of sedimentary rocks of Purana aspect in the region between the Godavari and Mahanadi valleys. These are apparently the remnants of a large and continuous spread, isolated by erosion. The sediments appear to have been derivea from the southwest and west. They were originally regarded as mainly the equivalents of the Cuddapahs. They are now correlated with the Kurnools, particularly as lithological similarities are good.

Bastar.—The succession in this area has been named the INDRAVATI SERIES by N.V.B.S. Dutt (1963) who has dervided them as below :—

JAGADALPUR STAGE (= Nandyal) (200-250 m	Upper Shale, with quartzite intercalations Purple shales and limestones Purple, fine grained dolomite Basal purple shales
KANGER STAGE (= Koilkuntla) upto 140 m	Grey laminated shales
TIRATGARH STAGE (= Paniam) 50-110 m	Purple shale with layers of platy quartzite Basal Quartzite, conglomerate, sub-arkose

Chhattisgarh.—The rocks in the large exposure of the Raipur-Drug-Bilaspur area were originally divided into a lower Chanderpur Stage and an Upper Raipur Stage. Dutt (1964) has now divided them further. The units have circular outcrops and the dips are radial towards the north where the youngest stage is exposed.

TABLE 23.—SUCCESSION IN THE RAIPUR SERIES

Stage	Description
RAIPUR (450 m)	Greenish grey, shaly limestone, fine grained in the lower part and purple in the Upper (Seen in the area around Raipur and at shallow depth in wells at Bhilai)
KHAIRAGARH (Variable)	Current bedded sub-arkose with 10-15% feldspar. Outcrop is arcuate
GUNDERDEHI (180 m)	Splintery calcareous shale, with thin sandstone laminae near the top
CHARMURIA (300 m)	Grey, fine-grained, thin-bedded limestones; becomes shaly towards the top. The Mahanadi follows the junction of this with the lower sandstone between Dhamtari and Mohdi.
CHANDARPUR (300 m)	Medium feldspathic sandstone or sub-arkose with conglomerate at the base. Shale layers in the upper part.

A general correlation of the strata in the Cuddapah, Bastar and Raipur basins is given below :

Cuddapah Basin	Bastar Basin	Raipur Basin
UPPER KURNOOL	INDRAVATI SERIES	RAIPUR SERIES
Nandyal (100-300 m)	Jagdalpur (100-150 m)	Raipur Lst. (450 m) Khairagarh Sst. Gunderdehi Sh. (180 m)
Koilkuntla (90 m)	Kanger (140 m)	Charmuria (300 m)
Paniam (60 m)	Tiratgarh (110 m)	Chandarpur (300 m)

In the Bara Pahar and Phuljar hills to the east of the Chhattisgarh area, the Chanderpur Series attains a thickness of 1,500 m and consists of conglomerates, quartzites, sandstones and selicious shales. The upper part is 300 m thick, of pink and buff shales and shaly limestone with a bed of granular quartzite at the base. The diamonds formerly won from the Hirakund in the Mahanadi river above Sambalpur were evidently derived from the rocks of the Barapahar. These rocks have been disturbed and faulted near their margin with the gneisses. It is likely that in this area kimberlite pipes are present from which the diamonds come. It would be easy to locate these pipes by means of geophysical prospecting methods.

BHIMA SERIES

Named after the Bhima river, a tributary of the Krishna, this Series is developed in the Gulbarga and Bijapur Districts. It occupies an area of roughly 4,200 sq. km, lying over the Archaean formations.

The rocks are divided into a lower and an upper series by W. King and R.B. Foote, but recent work by C. Mahadevan shows that a three-fold division is preferable (*Jour. Hyd. Geol. Surv.* V.)

Upper (100 m)	Black, blue, buff and purple shales with local sandstones at the bottom and flaggy limestones at the top.
Middle (165 m)	Creamy, grey, bluish and buff limestones and flaggy limestones.
Lower (110 m)	Sandstones and green and purple shales. The bottom beds are conglomeratic while the topmost beds are often calcareous.

The Lower Bhimas are sandstones and shales, laid down in a gradually deepening sea. The middle division, consisting mainly of limestones, was deposited in deeper waters, probably as chemical sediments. At this period the basin of deposition attained its greatest extent and depth, for some of the beds overlap the earlier Bhimas and lie directly on the gneiss. The upper division points to the contraction and shallowing of the basin, the deposits being mainly shales.

The eastern and southern parts contain only the lower and middle divisions while the upper division is found in the north and west. The deposits are nearly horizontal or low-dipping over large areas but show high dips and evidence of disturbance in the neighbourhood of some faults and at the junctions with the Deccan traps.

The Bhimas are devoid of fossils, though the constituent beds are well suited to the preservation of organic remains. The Kaladgis (which are referable to the Cuddapahs) lie to their west but nowhere in contact with them. The lithology, horizontal disposition and unmetamorphosed nature of the of the Bhimas point to their being the equivalents of the Kurnool formations.

SULLAVAI SERIES

There is a group of rocks called the Sullavai series in the Godavari valley, consisting of slates, quartzites, sandstones and conglomerates. They are well exposed near Sullavai and in the Dewalmari hills, where the quartzites recall the appearance of the Pinnacled Quartzites of the Kurnools. They have a thickness of 360-480 m, and overlie the Pakhals unconformably in the synclinal folds of the latter.

CORRELATION OF THE VINDHYANS

The Vindhyans are developed in two main areas : one comprises the Vindhyans of Rajasthan and Central India which are continued to the south of Bundelkhand and into Bihar. The lower part of the system contains marine deposits, including limestones, while the upper part consists of shallow-water deposits partly formed in a semi-arid climate represented mainly by red sandstones and shales with which gypsum is occasionally associated. The other is the Cuddapah basin in the Andhra State where the Kurnool System is developed and the Raipur-Bastar basin of M.P. The Vindhyans of Northern India have yielded a few primitive fossils.

The Cambrian formations of the Salt Range, especially the PURPLE SANDSTONES, bear a striking resemblance to the Upper Vindhyans of Central India as pointed out by C.S. Fox (*Rec. G.S.I.* LXI, p. 173, 1929) and to some strata in the Cambrian HORMUZ SERIES of Iran.

Amongst the unfossiliferous sediments of the Lesser Himalayas, the Nagthat Series (Jaunsar Series) and a part of the Haimantas are considered to be the equivalents of the Vindhyans. The Nagthats are part of the original Jaunsar Series, named by R.D. Oldham. They overlie the Chandpur Series unconformably in the Chakrata area and in the Krol belt. They comprise conglomerates, pebbly sandstones, arkose, grits, clay-slates, green and purple phyllites. Current bedding and ripple marks are frequently seen in the arenaceous types. The green and purple sandstones exposed on the southwestern flanks of the Mussoorie syncline have a good resemblance to the Vindhyan sandstones of the Peninsula.

The Garhwal and Krol nappes are exposed in the Sub-Himalayas of U.P. The Garhwal nappe contains the crystalline rocks of Askot which include augen-gneisses, para-gneisses and various schists and are intruded by a granite which may be the equivalent of the Almora-Dutatoli granite which cuts through the Chandpur Series. The crystalline rocks of Askot may therefore be of the same age as the Chandpur and Daling Series (Validya, 1962).

KROL NAPPE
Calc-zone of Pithoragarh : (=calc-zone of Tejam) Slates, limestone and magnseite ⎧ Deoban,
⎨ Shali,
⎩ Naldera

Berinag Guartzite : Purple phyllites, and quartzites, sericite and chlorite-schists, ⎧ Nagthat
biotite-schists and amphibolites ⎩ Jaunsar

GARHWAL NAPPE
Crystalline Zone of Askot (Danya formation) :
Chlorite-sericite-biotite schists, para-gneisses and augen-gneisses ⎧ Chandpur
⎩ Daling

The Krol nappe consists of the quartzite zone of Berinag and the calc-zone of Pithoragarh. They are correlated respectively with Nagthat-Jaunsar and with Shali-Deoban-Naldera limestones, the last (Naldera) being a part of the Lower Simla Slates. The Kakarhatti limestone of Simla area and the Deoban Limestone of Chakrata are regarded as of the same age. The Pitharagarh, Shali and Deoban Limestones have yielded stromatolites similar to *Collenia* which is of Upper Proterozoic to Lower Cambrian age.

The Haimantas, which continue up into the Parahio Series of Middle Cambrian age, are of Pre-Cambrian to Lower Cambrian age. They are described under the Cambrian System.

Along the Jadh Ganga in map sheet No. 53M/SW there are fluvo-deltaic rocks overlying the Haimantas and having a resemblance to the Vindhyans and the Nagthats. They include conglomerates and purple and violet flaggy quartzites with current bedding, ripple marks, etc. They may be of Cambrian age.

South of the Katmandu Valley, below the Chandragiri limestone of the part of that name, there occur quartzites underlain by ripple-marked green and purple phyllites and slightly metamorphosed banded shaly and sandy limestones resembling the Semris of the Sone Valley. There are also talcose and chloritic phyllites enclosing boulders and lenses of marble which strongly resemble the Rohtas Limestone. As the Chandragiri Limestones have yielded cystids and other fossils of Ordovician age, the underlying phyllites and limestones which show resemblance to the Smeri Series are most likely to be of Cambrian age.

V.H. Boileau who worked as an officer of the Geological Survey of India in the Simla, Mandi and Dharmasala areas has expressed the opinion that, the Khaira quartzites of Mandi and the Kaimurs are roughly of Middle Cambrian age and the Megal Black Dolomite, the Alsindhi fawn limestones and red shales and the Lower Shali Limestones may be Upper Cambrian or Ordovician. The Lower Maroon Marls, red clays and the Bhagsunath and Guma Limestones, all of which underlie the Khaira quartzites, are equated with the Semris and are regarded as Lower Cambrian to Upper Pre-Cambrian. This correlation seems to be based on the central idea that the Blainis and the formations equated with them belong to a period of glaciation in the Upper Vindhyan (Middle to Upper Cambrian age) and not in the Upper Carboniferous. In northern Kumaon the formations consisting of conglomerates, quartzites and dolomites, found in the Ralam Pass between the Lissar and Gori valleys, and resting on the Martoli phyllites, have been called the Ralam Series by Heim and Gansser. The RALAM SERIES consists of basal conglomerates, quartzites and dolomites. The conglomerates contain quartz pebbles up to the size of a man's head in red or black siliceous ground-mass and attain a thickness of 100 metres. They pass upwards into grey, purple and green massive quartzites 500 to 800 metres thick, overlain by orange coloured massive dolomite 50 to 100 metres thick. The Ralam Series is exposed only in the region between Milam in the west and the Lissar river in the east. It is considered to represent partly the upper-most Pre-Cambrian and partly the Lower Cambrian, thus corresponding roughly to the Vindhyan System.

ECONOMIC MINERAL DEPOSITS

Diamond.—For many centuries past, diamonds have been won from the Vindhyan and Kurnool formations. They are found as pebbles in the Banganapalli group of the Kurnools and in the conglomerates separating the diherent series of the Upper Vindhyans in Panna, Central India, as also just outside the Cuddapah basin in Sambalpur in Orissa. The original source of the diamonds which came to be deposited in the conglomerates is apparently pipe like intrusions of ultrabasic rocks like the one at Majhgawan near Panna.

Several such pipes may be present in the Panna area and in the Cuddapah basin.

The Wajra Karur region in Anantapur is still reputed to yield occasional diamonds to diligent searchers who visit and comb the area after the rains each year. In former centuries there was much activity in this region as evidenced by numerous crushing platforms and enormous amounts of crushed debris strewn around them. Details of old diamond workings will be found in the writings of V. Ball, who has also summarised the information in the volume on economic geology in the Manual of the Geology of India by Medlicott and Blanford. Geophysical work by the G.S.I. located a diamond bearing pipe here in 1962. Several more must be present in this region.

A few years ago V. S. Dubey reported (*Q. J. Geol. Min. Met. Soc. Ind.*, XX, 1-6, 1949) that the Mahjgawan occurrence near Panna was a pipe similar to the diamond-bearing pipes of South Africa. It has since been examined also by the Geological Survey. This pipe occupies a slightly depressed circular area covering several acres. The rock exposed at the surface is a tufaceous vesicular serpentinous and calcareous material. A shallow borehole revealed that at a depth of about 30 m the tufaceous material gives place to a hard, dark greenish grey, compact, brecciated rock consisting mainly of serpentine (with perhaps some unaltered olivine). Chemical analysis shows a composition similar to that of kimberlite. It is undoubtedly the mother rock of diamond. Another pipe has been discovered in the same region and it is very likely that several others may be found. The pipes are of Kaimur or pre-Kaimur age as they have yielded the diamonds found in the conglomerates at the base of the Rewa and Bhander Series (Mathur, 1963).

Pyrite.—The Bijaigarh Shales of the Lower Kaimurs contain a bed, 1 to 1.3 m thick, of sedimentary pyrite of fine grain. This bed has been found to crop out at two places about 5 km apart, on the Kaimur scarp near Amjhor a few miles south of Dehri-Sone, and at Kasisiya Koh about 14 km from Amjhor. The material is of good quality, containing around 45% sulphur, without any arsenic. As the Bijaigarh Shales extend over a large stretch of country, and as the sulphur bed is a sedimentary horizon within the shales, it is thought that this horizon may be found extensively under the Upper Kaimur Sandstone with forms a large plateau here, and may contain many million tons, unless the occurrence proves to be erratic on investigation by drilling. A part of this deposit has been proved to contain several million tons.

Coal.—Streaks of coaly matter were found by Fox in 1928 in the Bijaigarh Shales on the Kaimur scarp near the Japla limestone quarries. Some years later the Kajrahat limestone was found to contain some saucer-shaped carbonaceous discs with coaly matter in the centre. Since then, thin beds of coal and carbonaceous shales have been discovered in other places. One locality is 4 km east of Parsi (24° 25′ : 82° 53′) where a band 1.6 m thick is exposed for 60-70 m along a stream course and another is a borehole in the foundations of the Rihand dam. The better part of the coal gave on analysis 9-13 per cent volatile matter and 24-30 per cent fixed carbon, the rest being ash. The material shows algal spores and cuticles and tracheids of vascular plants. These occurrences are of interest in proving that plant remains exist both in the Semris and Kaimurs, though no workable coal seam may be expected (Mathur, *Indian Minerals, G.S.I.,* **18**, 158-165, 1964).

Limestone.—The limestones of the Vindhyans are amongst the most important sources of raw materials for the lime and cement industry in India — *e.g.*, in the Sone Valley in Bihar and U.P., in Rewa, in Jabalpur, in Guntur and in the Bhima Valley in Hyderabad.

The Narji limestones of the Guntur and Kurnool districts are capable of yielding excellent building stones. The flagstones quarried near Jammalamadugu, Yerraguntla (Cuddapah district), Betamcherla (Kurnool district) and other places, popularly known in Madras as 'Cuddapah slabs' are widely used as paving stones, fence stones, steps and table-tops. These 'slabs are derived from the Jammalamadugu and Kundair formations of the Kurnools. They are easily split into slabs half an inch to four inches in thickness and up to 2.5 by 1.2 m in size. They are fine-grained calcareous slates capable of taking a fairly good polish. Similar slabs are worked at Shahabad in Hyderabad.

Building and decorative stones.—Some of the limestones of the Lower Vindhyan and Lower Bhander stages show spherulitic structures, the concentric shells of which display different colours. Beautiful stones of this kind, found at Sabalgarh near Gwalior, have been used in the inlaid decorations in the buildings of Agra. Some of the buildings of Chitorgarh have been built of Nimbahera limestones. The limestones of the Palnad region (Guntur district), particularly the Narji limestones, are excellent building and ornamental stones, some varieties with deep red, chocolate, green, cream and grey colours yielding very attractive fine-grained marbles. These limestones have been used in the Buddhist sculptures of Amaravati in that district.

The white Vindhyan sandstone of Khatu in Jodhpur yields an excellent flagstone eminently suitable for delicate carving and making the perforated windows and stone screens common in the buildings of Rajasthan. Exquisite carvings have been executed in the sandstones from the Mirzapur district and used in the dargah at Maner near Dinapore and in the architectural monuments at Sasaram and other places in U.P. and Bihar.

The Vindhyan sandstones, especially of the Bhander Series, constitute a great storehouse of excellent stones which, because of their regular bedding, uniform grain, pleasing colours, easy workability and durabiilty, have been very extensively used all over Northern India as building stone. They are worked in many areas in Bundi, Kotah, Dholpur, Bharatpur, Jaipur, Bikaner in Rajasthan and in Mirzapur and other areas in U.P. The stones have cream, light grey and red colours and may show streaks and spots of red or creamy tints. Some are thin bedded, yielding slabs, others thick bedded, suitable for columns and beams. They have been used in the Buddhist Stupas of Sarnath, Barhut and Sanchi ; in the palaces, forts and mosques at Agra, Bharatpur, Delhi, Lahore, etc., and in many buildings in the cities of the Ganges Valley. Fatehpur Sikri has been constructed almost entirely of red sandstone. The modern administrative buildings of the Government of India at New Delhi and even the war-time barracks have used them extensively. They are so adaptable that they have yielded stones for paving, panelling and roofing, for window and door sills, beams, pillars, fenceposts, milestones, telegraph poles, fountains, water troughs and for many other uses.

The Auk shales in the Kurnool district yield good fire-clays and yellow ochers. Near Banganapalle, the shales contain some beds of rather impure, slightly clayey chalk and excellent nodular limestone.

Glass Sand.—Some Vindhyan sandstones near Allahabad, U.P. on disintegrating by weathering, yield good sands which are being used for the manufacture of glass. The deposits cocupy an area of over 100 sq. miles and extend into the neighbouring States. If specially purified, some of these sands can be used also for optical glass manufacture.

SELECTED BIBLIOGRAPHY

(Cuddapah and Vindhyan Systems)

————Symposium on the Cuddapah Basin. *Jour Ind. Geosci. Assoc.* (Hyderabad) **4**, 160 pp., 1964.

Aswathanarayana, U. Isotopic Ages from the Eastern Ghats and Cuddapahs of India. *Jour. Geophys. Res.* **69**, (16), 3479-3486, 1964.

Auden, J.B. Vindhyan sedimentation in the Sone Valley, Mirzapur District, *Mem.* **62**, (2), 1933.

Bajpai, M.P. The Gwalior Traps. *Jour. Geol.* **43**, 61-75, 1935.

Bhanumurthy, Y.R. Geophysical investigations for possible diamondiferous volcanic rocks in Vajrakarur area, Anantapur district, Andhra Pradesh. *Jour. Ind. Geosci. Assoc.* **4**, 131-138, 1964.

Coulson, A.L. Barytes in the Ceded Districts, Madras. *Mem.* **64**(1), 1933.
Coulson, A.L. Asbestos in the Ceded Districts, Madras. *Mem.* **64**(2), 1933.

Dutt, N.V.B.S. Stratigraphy and correlation of the Indravati Series of Bastar. *Jour. Geol. Soc. India*, **4**, 35-48, 1963.
Dutt, N.V.B.S. Suggested succession of the Purana Formations of Chhattisgarh. *Rec. G.S.I.* **93**(2), 143-148, 1964.
Datta, A.K. Life in the Vindhyan Period. *Sci. Cult.* **18**(7), 329, 1953.

Foote, R.B. Geology of Southern Mahratta country. *Mem.* **12**, 1876.

Heron, A.M. Geology of N.E. Rajputana. *Mem.* **45**(1), 1917.
Heron, A.M. Vindhyans of Western Rajputana. *Rec.* **65**, 457-489, 1932.
Heron, A.M. Gwalior and Vindhyan Systems in S.E. Rajputana. *Mem.* **45**(2), 1922.
Heron, A.M. Mineral Resources of Rajputana, *T.M.G.I.I.* **29**, 1933.
Heron, A.M. Geology of Central Rajputana. *Mem.* **79**, 1953.
Heron, A.M. Synopsis of the Purana formations of Hyderabad. *Jour. Hyd. Geol. Surv.* V. (2), 1949.
Howell, B.F. Evidence from fossils of the age of the Vindhyan System. *Jour. Paleont. Soc. India.*, **1**, 108-112, 1956.

King, W. Kadapah and Kurnool formations in the Madras Presidency. *Mem.* **8**, 1872.
Krishnan, M.S. and Swaminath, J. The Great Vindhyan Basin of Northern India. *Jour. Geol. Soc. India*, **1**, 10-30, 1958.

Mahadevan, C. Some aspects of the Puranas and Archaeans of South India. (Pres. Address, Geology Section). *Proc. Indian Science Congress* (Allahabad), 1949.
Mahadevan, C. The Bhima Series in the Gulbarga District, Hyderabad. *Jour. Hyd. Geol. Surv.* V (1), 1946.
Mallet, F.R. Vindhyan Series as exhibited in the North-western and Central Provinces of India. *Mem.* **7**(1), 1869.

Mathur, S.M. Coaly matter in the Vindhyan System. *Indian Minerals (G.S.I.)*, **18** (2), 158-165, 1964.
Mathur, S.M. et. al. Algal structures in Fawn Limestones, Semri Series, Mirzapur District *Rec. G.S.I.* **87**(4), 819-822, 1962.
Mathur, S.M. Geology and sampling of the Majhgawan diamond deposit, Panna District, M.P. *Bull. G.S.I.* **21A**, 59, 1963.
Medlicott, H.B. Vindhyan rocks and their associates in Bundelkhand. *Mem.* **2**(1), 1859.
Middlemiss, C.S. Geology of the Sub-Himalayas of Garhwal and Kumaon, *Mem.* **XXIV** (2), 1890.

Oldham, R.D. Flexible sandstone or itacolumite. *Rec.* **22**, 51-55, 1889.
Oldham, R.D. Geology of the Sone Valley in Rewah State and parts of Jubbulpore and Mirzapur. *Mem.* **31**(1), 1901.

Rao, A.N. The eastern margin of the Cuddapah Basin. *Jour. Ind. Geosci. Assoc.*, **5**, 29-36, 1965.

Sahasrabudhe, Y.S. The Kaladgis of Southern Maharashtra. *Jour. Ind. Geosci. Assoc.* **4**, 13-20, 1964.
Sitholey, R.N., Shrinivastava, P.N. and Verma, C.P. Micro-fossils from the Vindhyans. *Proc. Nat. Inst. Sci. India.*, **19B**, 193-202 1953.

Vaidyanadhan, R. Geomorphology of the Cuddapah Basin. *Jour. Ind. Geosci. Assoc.* **4**, 29-36, 1964.
Valdiya, K.S. Outlines of the stratigraphy and structure of the southern part of Pithograph, Almora District, U.P. *Jour. Geol. Soc. India*, **3**, 27-48, 1962.
Vemban, N.A. A chemical and petrological study of some dykes in the pre-Cambrian Cuddapah rocks. *Proc. Ind. Acad. Sci.* **23-A**, 1946.
Vredenburg, E. Geology of the State of Panna with reference to the diamond-bearing deposits. *Rec.* **33**, 261-314, 1906.
Vredenburg, E. Suggestions for the classification of the Vindhyan System. *Rec.* **33**, 255-260, 1906.

CHAPTER IX

THE PALAEOZOIC GROUP :
CAMBRIAN TO CARBONIFEROUS

THE CAMBRIAN SYSTEM

Named after Cambria (i.e. Wales) in the British Isles, this system is well developed in Northern Europe, in the Mediterranean region, in eastern as well as western North America, in the Himalayas, China and elsewhere. The three major divisions of this system are named after fossils in Europe and after places in North America :

North America			Atlantic region.	Pacific region.
Upper	...	POTSDAMIAN	*Olenus*	*Dikellocephalus*
Middle	...	ACADIAN	*Paradoxides*	*Ptychoparia Olenoides*
Lower	...	GEORGIAN	*Olenellus*	*Olenellus*

Some comparatively thin strata, called TREMADOCIAN and containing the first graptolites (*Dictyonema*), are considered to form the uppermost part of the Cambrian in British Isles but are put at the base of the Ordovician in France.

There is generally a well marked discordance at the base of the Cambrian strata indicating a period of disturbance and mountain building. In contrast with the older formations, the Cambrians contain a rich fauna the dominant elements of which are trilobites and brachiopods. Other groups well represented are foraminifera, sponges, echinoderms (cystids, crinoids), worms, gastropods and pelecypods. Archaeocyathids (which are probably corals) are in several places so abundant as to form limestone reefs, as in Australia and Morocco. The brachiopods are abundant being represented by *obolus*, *Lingula* etc. The trilobites are, however, of great importance for stratigraphical purposes as they can be used for dividing the strata into stages. The important genera are *Olenellus*, *Pardoxides*, *Ptychoparia*, *Conocoryphe*, *Olenoides*, *Redlichia*, *Olenus*, *Dikellocephalus*, *Agnostus*, *Ogygia*, etc., only a few genera continuing into the strata above. The fossil assemblages indicate that three major life-provinces related to depth can be distinguished. Good exposures of the Cambrian, found along the borders of the Baltic shield (Oslo, Leningrad, Esthonia), Poland, Silesia, Bohemia, Wales, etc. have been studied in detail and are used for reference and comparison of similar strata discovered in other parts of the world.

Fossiliferous marine Palaeozoic rocks are absent from the Peninsula except for one or two small patches of Lower Permian age near Umaria in Rewa and possibly also a part of the Upper Vindhyan which may be of Cambrian age. The Cambrian System has been studied in the Salt Range, Kashmir and the Spiti Valley where it is represented by richly fossiliferous beds.

SALT RANGE

Of the three areas mentioned above, the Salt Range is the most easily accessible. Wynne studied the region over 70 years ago and many Geologists have followed him during the years that have elapsed since then. The last to make a comprehensive study was E.R. Gee who mapped the Salt Range and the neighbouring region during the thirties of the present century. Before proceeding to describe the stratigraphy of this region, it would be advantageous to give a brief description of its geographical and structural features.

The Salt Range constitutes the southern edge of the Potwar plateau between East Longitudes 71° and 74°. The northern limits of this plateau are formed by the Kalachitta hills, while the eastern and western limits are delineated by the rivers Jhelum and Indus. The Salt Range forms a series of irregular ridges which are convex towards the south, overlooking the Mianwali plains. These ridges attain an average height of 750 m to 900 m, the highest point being Mount Sakesar (32° 32′ : 71° 56′) 1,525 m above sea-level. The more important of the ridges are named the Chambal, Nili, Rohtas, Pabbi, Sakesar, etc. Though the Salt Range terminates at Mari on the Indus, the formations are continued beyond the Indus where the rocks have an E.-W. trend in the Chichali and Shingar ranges. Farther west the trend veers to the south forming a series of ranges which are convex to the east and south and which are called the Maidan, Marwat, (Nilaroh), Sheikbudin and Bhattani Ranges. The Khasor range lies between the Indus river and the Marwat Range.

The Salt Range is a highly disturbed folded and faulted structure whose southern face is an over-fold thrust towards the Mianwali plains. Its northern side consists of gently dipping trata merging into the Potwar plateau which, for the most part, exposes Siwalik and Murree strata. The southern face of the Salt Range presents a series of escarpments rising abruptly from the plains and exposing Cambrian strata and a fairly continuous succession from the Permo-Carboniferous to the Eocene. There are numerous cross-faults along which block faulting has taken place. Several ravines cut the range in a radial direction and some of these undoubtedly follow zones of faulting.

The arcuate form of the Salt Range is to be attributed to the great Himalayan movements which compressed the strata and made them flow over some distance towards the south, the eastern and western ends having been held back by wedges of ancient rocks which lie underneath and which may be called the Kashmir and Mianwali wedges. Outcrops of ancient rocks belonging to the Delhi System are found near Sarghoda (32° 6′ : 72° 40′) and in the Chiniot and Sangla hills not far from the eastern end of the Salt Range.

The Potwar plateau is occupied partly by the Soan (or Sohan) Syncline filled with Tertiary sediments having a width of some 80 km. The intensity of folding and faulting increases towards the north where the compression was most intense, as shown by Pinfold (*Rec.* 49, pp. 137-159), who distinguishe the following zones from north to south.

1. Kalachitra Anticlinorium
2. Isoclinal Zone
3. Faulted Zone
4. Anticlinal Zone
5. Soan Syncline
6. Salt Range

This region has been studied in some detail, and several boreholes have been put down in it in search of petroleum. The so-called Isoclinal and Faulted Zones reveal the presence of numerous strike faults, giving a false impression of isoclinal structure because of repetition of strata. W.D. Gill (1953) has shown that the Isoclinal Zone is a misnomer and should really be included in the faulted zone. A few anticlines in this area are petroliferous, e.g. Khaur (33° 16' : 72° 27'), Dhulian (33° 1' : 72° 21') and Joya Mair (33° 1' : 72°45'). To the south of these is the Synclinal Zone wihch narrows in a westerly direction, being narrowest between Kalabagh and Kohat where some of the folds close up. The axes of the structures have a general E.-W. trend, but may be N.W.-S.W. in some parts of the area.

The Kohat region to the northwest of the Potwar plateau shows a very complicated structure in which Eocene and Mezosoic rocks are brought up in anticlines. Lower Eocene rocks are present underneath where they are not exposed, for small amounts of oil, apparent.y derived from Lower Eocene limestones, are found in the rock-salt of Kohat. The highly disturbed rocks of this belt continue westwards across the Indus into the Samana Range.

STRATIGRAPHY

The oldest beds exposed in the Salt Range are of Cambrian age as some of them contain trilobites and brachiopods of this age. They are, in several places, directly underlain by saline marl with beds of gypsum and rock salt. Upon them rest marine strata ranging in age from Upper Carboniferous to the Eocene. The succession becomes more complete as one proceeds from east to west. The Upper part of the scarps is composed of either Permian or Eocene limestone. When followed from the top of the scarps in a northerly direction into the Potwar plateau, the Eocene strata are overlain successively by the Murrees and Siwaliks.

At the eastern extremity, near the Jhelum river, the Eocene beds rest directly on the Cambrian. Some distance to the west, the Olive Series of Upper Carboniferous age appears as a thin bed and becomes gradually thicker further west. The glacial boulder beds of Talchir age and the Speckled Sandstones are first seen near Khewra, while in the Nilawan ravine, about 75 km from the eastern end, the Products beds make their first appearance. The places which have given their names to the stages of the Products limestone occur at different distances west of Nilawan (Nila Wahan). The Productus beds attain their full development near Kundghat (40 km west of Nilawan ravine) where Triassic beds first appear. A little farther west, near Amb, Jurassic strata are to be seen. This gradual thickening and the fuller succession

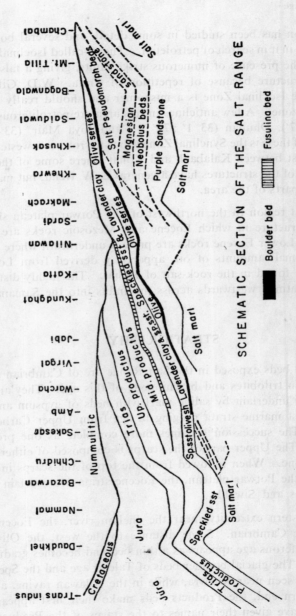

FIG. 1.—Schematic Representation of the Geological Succession in Different Parts of the Salt Range.

TABLE 24.—STRATIGRAPHICAL SUCCESSION IN THE SALT RANGE (AFTER E. R. GEE)

Formation		Description of Strata	Thickness (m)
Pleistocene and Recent		Clays, sandstones, conglomerates, alluvium and loess. Mainly of fresh water and aeolian origin	Up to 300
		—————— Unconformity ——————	
Pliocene and Miocene	Siwaliks	Sandstones, clays and conglomerates, the last in Middle and Upper Siwaliks. Earlier deposits lacustrine and later ones fresh water and alluvial	3,000-4,800
	Murrees		200-600
		—————— Unconformity ——————	
Middle and Lower Eocene	Nummulitic Formation	Foraminiferal limestones and shales with gypsum; coal occasionally in lower part; thin laterite bed at the base	Up to 550
		—————— Unconformity ——————	
Lower Cretaceous, Jurassic and Triassic		Fossiliferous limestones, dolomites, shales, sandstones and carbonaceous beds; mainly marine	Up to 285
Permian—Products Limestone		Fossiliferous limestone, with subordinate marls and shales; marine	60
		—————— Unconformity ——————	
Upper Carboniferous	Lavender Clays	Clays and subordinate sandstones	100-135
	Speckled Sandstones	Sandstones, grits and clays	22
	Conularia beds	Sandstones, shales and carbonaceous beds	30-105
	Talchir boulder-beds	Conglomerates and sandstones, shaly	
		—————— Unconformity ——————	
Cambrian (parly Pre-Cambrian)	Salt Pseudomorph beds	Red shales and flags with salt Pseudomorphs; some gypsum	Up to 100
	Magnesian Sandstones.	Dolomitic sandstones and subordinate shales	Up to 75
	Neobolus beds	Fossiliferous shales and sandstones, somtimes glauconitic and dolomitic	20-50
	Purple Sandstones	Red to maroon sandstones and flags	75-135
		—————— Original sedimentary contact, often sheared ——————	
	Saline Series	Upper gypsum-dolomite beds with oil-shales	Up to 45
		Middle Salt Marl and Rock-Salt Beds	Up to 240
		Lower gypsum-dolomite beds with oil shales	Up to 225

of marine strata in the west indicate that the sea gradually retreated westward during their deposition and that it was deepest in the west. A schematic section longitudinally along the Salt Range is given in figure 1.

There are four important stratigraphic breaks in the succession in the Salt Range — the first between the Cambrian and the Talchir horizon, the second below the Upper Jurassic, the third below the Eocene and the fourth below the Murrees. The general stratigraphic succession is given in Table 24

The Cambrian sediments include Purple Sandstones which are considered as deposits of a semi-arid climate, and also dolomites, some fossiliferous shales sandstones and shales containing salt pseudomorphs. These are all regarded a marine deposits, laid down generally in shallow water. There are no beds to re present the Ordovician, Silurian, Devonian, and Lower Carboniferous ages Glacial boulder beds of Talchir age rest directly over the Cambrian strata and are succeeded by shales and sandstones containing leaf impressions and spores of plants, and by the Eurydesma and Conularia beds. The marine facies continue and Speckled Sandstones and Lavender Clays of Permian age were then deposited. The basin of deposition then became deeper and the Productus Limestones were formed. The Upper Permian is marked by sandy calcareous materials indicating shallow water deposition. The Ceratite Limestone of Triassic age was then deposited and the sea was gradually regressing westwards. The Upper Triassic beds as well as the Kingriali Dolomites are of shallow water origin. Then followed a short period of sub-aerial conditions when laterite was formed on the surface of the exposed beds. This underlies the Variegated Shales of Jurassic age. The Upper Jurassic is represented by the fossiliferous Baroch Limestone. A slight break, marked by broken shell fragments, intervened between these and the succeeding Belemnite beds of Lower Cretaceous age which are shallow water formations containing glauconite. The Belemnite beds are present in the western Salt Range and also in the Surghar Range They are succeeded by massive sandstones with intercalations of carbonaceous shale containing plants and mollusca. The earth movements of the Upper Cretaceous brought about a retreat of the sea to the west and north, but the presence of Maestrichtian fossils in the western Salt Range indicates that such retreat was not complete.

In the early Eocene, estuarine and shallow water conditions prevailed but this was preceded by a short period of sub-aerial weathering as indicated by the presence of laterite. During the early Eocene some coal seams were formed in the western Salt Range. Over the greater part of the area, however the Ranikot period was one of extensive marine conditions, but towards the end of that period there was a regression of the sea in several places. Marine conditions are again indicated during the Laki period, but in the northern Potwar region, the Upper Laki saw the deposition of rock-salt, gypsum and dolomite The Charat beds (Laki to Middle Kirthar) are mostly of fresh water origin containing vertebrate and molluscan fossils. The overlie the Shekhan Limestone of Upper Laki age and are followed by the massive Kohat Limestone of Middle Kirthar age and by the Sirki Shales of Upper Kirthar age. Gypsum beds are here intercalated with Meting Shales (Laki) and with red and purpl sandstones of Lower Chharat beds. Earth movements seem to have taken place at the end of the Ranikot times, when a ridge was formed in Waziristan, for there are no Laki beds there. Laki beds are, however, found on both sides of that ridge but they are somewhat different from each other. The Kohat side became a closed basin during the Laki times and received waters only intermittently in the western and southern parts so that conditions for the deposition of gypsum and salt were present, as in the Bahadur Khel area. Another marine incursion took place in the Upper Kirthar when limestones and shales were deposited in the northern Potwar area.

The strong earth movements which took place at the end of the Eocene uplifted the northern regions. Murree strata were then laid down during the Oligocene and early Miocene times in the brackish waters of the foredeep which was formed in front of the rising mountains. The Murrees are seen overlapping the Laki beds in some places. During the succeeding period, the Siwaliks were laid down in the same basins and the waters in them were gradually becoming fresh. The large thickness of the coarse sediments in the Siwaliks indicates that the basins were depressed to keep pace with the deposition.

It may be noted here that the Saline Series in the Salt Range consist of three stages : a lower Gypsum-Dolomite stage containing beds of gypsum, anhydrite, dolomite, variegated gypseous clays and oil shales ; a middle Salt Marl stage consisting of red marl with thick seams of rocksalt ; and an upper Gypsum-Dolomite stage consisting of massive white or grey gypsum and dolomite with oil-shales. A thin bed of decomposed diabase called the Khewra Trap is found in the upper stage. In the Kohat region, on the other hand, the Saline Series consist of two stages, the junction between the two being considered as tectonic by Murray Stuart (1919). The lower stage consists mainly of rock-salt with at least six seams of salt having a thickness of 230 m or more, while the upper stage contains gypsum, dolomite, impure limestones, green shales and oil-shales, having a thickness of about 30 m. The salt beds in the Kohat region occur fairly regularly along the cores of anticlines and faulted folds in Eocene rocks, often accompanied by Mesozoic rocks. Table 25 shows the Cambrian succession in the Salt Range consisting of five series including the Salt Marl.

TABLE 25.—CAMBRIAN SUCCESSION IN THE SALT RANGE.

Salt Pseudomorph Shales (up to 105 m)	Red to purple and greenish silty shales with casts or Pseudomorphs of salt crystals showing on bedding planes. Contain some gypsum.
Magnesian Sandstones (up to 75 m)	Well bedded cream coloured dolomitic sandstones, sandy dolomites and subordinate shales.
Neobolus Beds (up to 45 m)	Fossiliferous grey shales, sandy shales, sandstones, which may be micaceous, dolomitic and glauconitic Pebble-bed at the base. Characterised by *Neobolus* and other barachiopods as well as some trilobites.
Purple Sanstones (75-140 m)	Fine grained pink, purple and maroon sandstones. Lower part shales and flagstones.
Saline Series (Salt Marls) (up to 450 m)	(c) Upper Gypsum-dolomite with oil-shales (and decomposed diabase, Khewra Trap). (b) Pink, red or purple Salt marl with beds of rocksalt (a) Lower gypsum-dolomite with oil-shales.

SALT MARL OR SALINE SERIES

The Cambrian succession, which is well-exposed in the eastern part of the Salt Range, includes the Salt Marl as the oldest member. It is a mixture of powdery fine-grained marl and fine red sand, which when dry, has the consistency of red brick. The material is soft and homogeneous and does not contain coarse sand or pebbles. The marl, when treated with hydrochloric acid, effervesces strongly, leaving a residue of red mud. The marl forms a practically unstratified mass, conspicuously red to dull purple or maroon in colour, and contains grains of sodium chloride, gypsum and carbonates of calcium and magnesium. Indications of stratification in the marl are given by the presence of layers of salt, gypsum or dolomite. Though no bedded structure is seen in general, sections in the mines sometimes show bedding and contortions of the layers. Occasionally, there are green and grey elongated streaks and patches in the marl. There are also anastomosing and filmy stringers of gypsum in the marl indicating the tendency of the gypsum to segregate. The dolomite in the marl forms honey-combed lumps and it has been pointed out by Middlemiss that there is complete gradation between the lumps and the streaky patches. The inference is that these patches are the result of distintegration of the layers of dolomite and their assimilation by the marl. The dolomite first becomes dotted with punctures which gradually become enlarged to produce a honey-combed or spongy structure, the holes being filled with gypsum. In some places bituminous shale is found in the upper division of the Saline Series, and there are also thin beds of highly altered purplish trap (diabase). The layers of trap may attain a thickness of about 3 m.

The Saline Series is best developed at Khewra in Eastern Salt Range where the lower portion shows beds of pure rock-salt which is colourless to pale pink. The impure earthy bands included in the salt are locally called *kallar*. The upper portion contains numerous *kallar* intercalations and the rock-salt shows the presence of sulphate and chloride of magnesium.

Gypsum.—The gypsum in the Saline Series is generally pure, but it might sometimes show a gradation to limestone and dolomite. It is compact and massive to saccharoidal, white, grey, dark bluish grey or pink and sometime even variegated. Plates of selenite are occasionally found, while in some case the interior of the mass may consist of anhydrite. The beds are often massiv with obscure bedding or vague contortions. Though the bedding is generally parallel to the surface of the underlying salt, it does not follow the structur of foliation of the salt. The beds and lenses of gypsum contain excellent smal doubly-terminated crystals of quartz as at Mari, Kalabagh, Sardi, Khussak Katta, Saiduwali, etc. These crystals sometimes contain inclusions of anhy drite from which it may be inferred that much of the gypsum was originall anhydrite or that the inclusions were converted into anhydrite during the crystallisation of the quartz.

Salt.—The beds of rock-salt are often massive and may sometimes b as much as 30 m thick. The salt is pink to white in colour in the Salt Rang with rare greyish patches, the different beds showing different degrees

THE PALAEOZOIC GROUP—CAMBRIAN TO CARBONIFEROUS

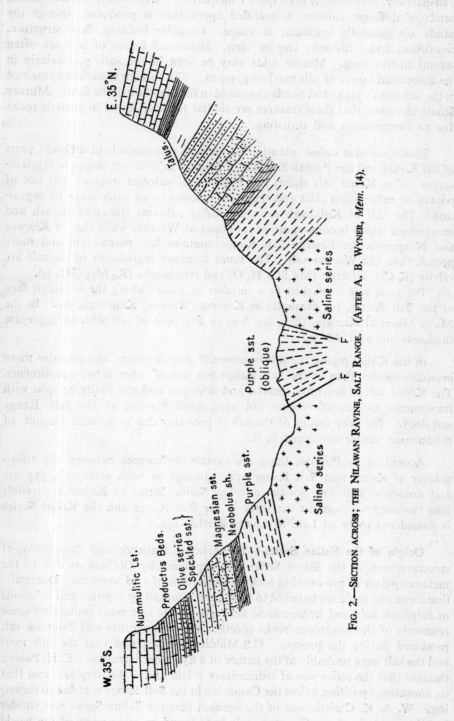

Fig. 2.—Section across; the Nilawan Ravine, Salt Range. (After A. B. Wynne, *Mem.* 14).

transparency, occasionally even quite transparent. Where there are alternating bands of different colours, a stratified appearance is produced, though the bands are generally lenticular in shape. Irregular bedding, flow structure, contortion and schistosity can be seen. Individual lenses of salt are often several metres long. Minute folds may be seen in the salt, particularly in the associated layers of salt marl or gypsum. The foreign materials contained in the salt form streaks and bands elongated in the direction of the flow. Murray Stuart considers that these features are similar to those found in gneissic rocks due to compression and thrusting.

Stuart has also called attention to the fact that the salt in different parts of the Kohat and the Punjab Salt Range belongs to different stages of crystallisation. The Kohat salt shows the presence of calcium sulphate but not of potash or magnesium salts and therefore belongs to an early stage of deposition. The salt of Kalabagh near the Indus contains traces of potash and magnesium which become more pronounced at Warcha, while that of Khewra and Nurpur in the Eastern Salt Range contains less magnesium and more potash than the Warcha salt. The more common ingredients in the salt are sylvite ($K Cl$), kieserite ($Mg SO_4 H_2 O$) and langbeinite ($K_2 Mg_2 (SO_4)_3$).

The rock salt is worked in a number of places along the southern face of the Salt Range, for example, at Khewra, Warcha, Kalabagh, etc. In the Mayo Mines at Khewra, there are four or five beds of salt with an aggregate thickness of over 60 m.

In the Kohat region the salt is generally grey in colour and contains more insoluble matter than in the Salt Range but less of other saline constituents. The Kohat salt is generally sheared and schistose and can easily be split with implements and therefore does not need much blasting as the Salt Range salt does. The grey colour of the salt is probably due to a small amount of bituminous matter contained in it.

According to Pinfold, there are certain differences between the stratigraphy of Kohat and Salt Range areas, though in both areas salt, gypsum and dolomite beds are associated. The Saline Series of Kohat is divisible into two stages as against the three in the Salt Range and the Kohat Series is considered to be of Laki to Lower Kirthar age.

Origin of the Saline Series.—The great disturbance and complexity of structure seen in the Saline Series was explained by Oldham as due to the metamorphism of pre-existing rock by the action of acid vapours. Dolomite, limestone and shale are believed to have been attacked by vapours and solutions of sulphuric acid and hydrochloric acid, giving rise to a marl containing some remnants of the calcareous rocks together with the gypsum and common salt produced during the process. C.S.Middlemiss suggested that the salt marl and the salt were probably of the nature of a hypogene intrusive. E. H. Pascoe thought that the series was of sedimentary origin and of Tertiary age, and that its anomalous position below the Cambrian in the Salt Range was due to thrusting. W. A. K. Christie was of the opinion that the Saline Series was similar in all respects to the sedimentary salt beds found in other parts of the world and that the marl belonged to the last phase of desiccation of an inland sea basin. He also held that the well known plasticity of salt under pressure was responsible for the obliteration of the bedded character and for its acquiring lenticularity and flow structure. It may therefore be concluded that the Saline Series is of sedimentary origin and that the peculiar structures exhibited by the beds are the result of intense tectonic disturbances.

Fossils.—The salt marl and the salt have not yielded any megascopic fossils and it is only the limestones that have occasionally shown *Nummulites* and other foraminifera of Eocene age. Gee reported that Eocene foraminifera were discovered in the Salt Marls (*Rec. 66*, pp. 32, 66, 117). In the Jaba nala area of Daud Khel, Nummulitic limestone with fossils is seen to pass along the strike into massive gypsum (*Rec. 69*, p. 63). Beds of gypsum have also been recorded as intercalated with foraminiferal limestones near Bahadur Khel (*Rec. 65*, pp. 113-114). Davies and Wadia (1929) found foraminifera within the gypsified nummulitic limestone in the gypsum-dolomite stage at Bahadur Khel.

Carbonised remains of plants have been found in the Salt Marl of the upper part of the Saline Series in the Dandot and Khewra scarps (*Rec. 66*, p. 117). Similar plant fragments were also found in the shale band intercalated with gypsum beds in the Nilawan and in the scarp north of Dhak.

B. Sahni and his associates have carefully examined the various members of the Saline Series for their micro-fossils. They state that they obtained the samples of materials to be tested from mine workings and borehole cores and took every care to assure that there was no contamination from extraneous sources. All the samples of marl, rock-salt, gypsum, dolomite and even bituminous shale were found to contain microfossils which consisted of shreds of angiospermous wood, gymnosperm tracheids, grass-like cuticles and chitinous parts of insects. In their opinion these micro-fossils could not be assigned an age as early as Cambrian. Fox has advocated the view that these plant fragments could have been carried down into the strata of the Saline Series by percolating waters and could have been incorporated from exposed materials during the movement of salt along fault planes. It is, however, difficult to imagine such solutions penetrating into beds lying at a depth of several hundred feet, now encountered in fresh mine openings or drill-hole cores.

Age of the Saline Series.—The age of the Saline Series of the Salt Range has been a matter of controversy among Indian Geologists for many years. The work of Wynne (1878) led him to the conclusion that in the esatern part of the Salt Range the Saline Series was of Cambrian age, as in several sections here the Saline Series underlies the Purple Sandstone. Wynne also expressed the opinion that the salt in the Kohat region was of Eocene age. Many years later, Koken and Noetling expressed the opinion that the Saline Series of Salt Range was also of Eocene age and its position under the Cambrian succession was due to thrusting. This hypothesis was supported by Zuber and Holland 1914) and later by Pascoe (1920). Several other geologists have aslo made contributions to this problem and amongst them may be mentioned Middlemiss (1891), Murray Stuart (1919), Christie (1914), Pinfold (1918), Davies and Wadia (1929), Fox (1928), Anderson (1926), Cotter (1933). This was the the subject of two symposia held under the auspices of the National Academy of Sciences and the Indian Academy of Sciences in 1944 and 1945 (Published by the former in 1944 and 1946) in which several geologist took part. The Salt Range and Kohat areas were mapped by E.R. Gee of the Geological Survey of India between 1930 and 1940 and the results of his work appeared in several communications.

In the Cis-Indus region the Saline Series appears at various stratigraphical horizons but mostly below the Cambrian sequence, as for instance, near Khewra and Khussak. Farther west, in the Sakesar and Tredian hills, it underlies the Talchir boulder-bed or the Speckled Sandstone, and the Talchirs transgress on the various members of the Cambrian sequence. Middlemiss has observed that the junction of the Saline Series with the Purple Sandstone had almost always a brecciated appearance, the top of the marl being often full of fragments of the overlying sandstone. He stated that there is no interbedding between marl and Purple Sandstone and that the junction zone does not show a conformable passage of original deposition. The Salt Marl occupies the cores of folds and flexures in Amb, Dandot and other places. In the Amb glen, the marl is found to underlie the Talchir boulder bed and below it are Permian strata in an inverted condition. At Vasnal, the inliers of the Saline Series occur below Nummulitic strata. It transgresses on the Nummulitic strata between Dandot and the Makrach valley, while near Daud Khel, the Nummulitic limestone appears to pass laterally into massive gypsum in one place. In other sections red and grey marls with gypsum are overlain by Talchir conglomerates. Gee states, however, that the gypsum deposits in this ergion belong to two different ages, namely, Eocene gypsum which is massive, pure and light coloured with intercalation of greenish clay shales of Laki age, and Cambrian gypsum which is pink coloured and contains quartz crystals and intercalations of red and blue marl and is generally found below the Talchirs. The basal Talchirs are reported to be pink (this colour being attributed by Gee to derivation from the marl) and to contain boulders and pebbles of gypsum derived from the denudation of the Cambrian gypsum beds.

In the Chittidil-Sakesar-Amb area, the Cambrian sequence with the Saline Series is repeated thrice because of faulting. The Saline Series is overain by the Purple Sandstone or by Talchir conglomerates. In a section 2 km north of Chittidil Rest House, the junction between the Saline Series and the Talchirs is an undisturbed sedimentary contact. In other sections close by, the Talchirs transgress gradually on the various stages of the Cambrian Section. Gee considers the sections in this area to represent undisturbed sedimentary sequence without any evidence of thrusting. He has also stated that the upper surface of the dolomite (of the Saline Series) gives the appearance of an eroded surface, the boulders and pebbles derived from which have been incorporated in the succeeding Talchir boulder-beds.

THE PALAEOZOIC GROUP—CAMBRIAN TO CARBONIFEROUS

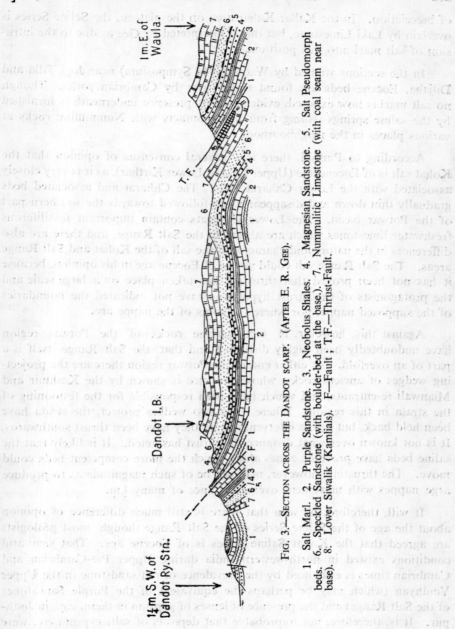

FIG. 3.—SECTION ACROSS THE DANDOT SCARP. (AFTER E. R. GEE).
1. Salt Marl. 2. Purple Sandstone. 3. Neobolus Shales. 4. Magnesian Sandstone. 5. Salt Pseudomorph beds. 6. Speckled Sandstone (with boulder-bed at the base.). 7. Nummulitic Limestone (with coal seam near base). 8. Lower Siwalik (Kamlials). F—Fault; T.F.—Thrust-Fault.

In the Warcha area there are abundant evidences of thrusting, though there is a difference of opinion about the magnitude of the thrust. In the tunnel at the southern end of the Warcha salt mine, the rock-salt, gypsum and dolomite occur intimately associated with Recent gravels. In several places in the Khewra area the junction of the Saline Series with the overlying Purple Sandstone is clearly a disturbed one. At the eastern end, at Jalalpur, the Saline Series is thrust over steeply dipping Middle and Upper Siwalik beds which have been reversed, and the Siwaliks below the thrust show evidences of brecciation. In the Kallar Kahar area on the plateau, the Saline Series is overlain by Laki Limestone, but this is interpreted by Gee as due to the intrusion of salt marl into this position.

In the sections studied by Wadia (1944 Symposium) near Jogi Tilla and Diljaba, Eocene beds are found thrust over by Cambrian rocks. Though no salt marl is now exposed, evidence of its presence underneath is furnished by the saline springs issuing from fault contacts with Nummulitic rocks at various places in the neighbourhood.

According to Pinfold, there is a general consensus of opinion that the Kohat salt is of Eocene age (Upper Laki to Lower Kirthar), as it is very closely associated with the Lower Chharat beds. The Chharat and associated beds gradually thin down and disappear when followed towards the southern part of the Potwar basin. The Lower Chharats contain important fossiliferous freshwater limestones which are absent in the Salt Range, and there are also differences in the nature and character of the salt of the Kohat and Salt Range areas. The Salt Range salt could not be of Eocene age in his opinion, because it has not been proved that thrusting has taken place on a large scale and the protagonists of the thrust hypothesis have not indicated the boundaries of the supposed nappe, nor where the roots of the nappe are.

Against this, however, is fact that the rocks of the Potwar region have undoubtedly been highly disturbed and that the Salt Range itself is a part of an overfold. At either end of the Potwar region there are the projecting wedges of ancient rocks, whose presence is shown by the Kashmir and Mianwali re-entrants, and which have been responsible for the festooning of the strata in this region. Where these two wedges project, the strata have been held back, but in the intervening area they have been thrust southwards. It is not known over what distance the thrust has acted. It is likely that the saline beds have provided zones along which the more competent beds could move. The thrusting, however, may not be of such magnitude as to produce arge nappes with movement over a distance of many km.

It will, therefore, be seen that there is still much difference of opinion about the age of the Saline Series of the Salt Range though most geologists are agreed that the Kohat Saline Series is of Eocene age. That semi-arid conditions existed in north-western India during Upper Pre-Cambrian and Cambrian times is evidenced by the prevalence of red sandstone in the Upper Vindhyan (which may be perhaps the equivalent of the Purple Sandstones of the Salt Range) and the presence of lenses of gypsum in them, *e.g.*, in Jodhpur. It is, therefore, not improbable that deposits of salt, gypsum etc. were formed in a desiccating basin during the Cambrian. The careful examination of the limestones for micro-fossils and the application of new techniques should help in solving this problem.

PURPLE SANDSTONE

The Salt Marl is generally overlain in many places in the Eastern Salt Range by the Purple Sandstones. These are massive fine textured sandstones having buff, dull red and maroon colours. Sometimes they are flaggy and can be split into rather thick slabs, particularly near the basal portion of the formation. The sandstones are up to 140 m thick in the east and 60 m thick in the west. They show current-bedding, ripple-marks and other evidences of deposition in shallow waters in a rather arid climate. The lower beds are shaly and occasionally flaggy. They are entirely unfossiliferous and can be traced as far west as Chidru where they are overlapped by younger beds. They have a resemblance to the Bhander or Rewa sandstones of the Vindhyan System with which they may be homotaxial. There is no gradual passage of the Salt Marl into the Purple Sandstones and there is no interbedding of the two.

The junction zone shows a mixture of the two types of rocks, but as discussed by Middlemiss (*Rec.* **24**, 1891), the material of this zone is of the nature of a breccia as it contains fragments of the sandstone in the marl. The sandstone generally appears to be rather shattered; first the layers adjoining the marl continue to keep their position parallel to the bedding, but are partly detached; farther away the fragments of the sandstone become smaller and more turned in all directions. The fragments are angular and are surrounded by a matrix of the saline marl.

In the eastern part of the Salt Range, up to about Musakhel, the Salt Marl is directly overlain by Purple Sandstone. Further west, in the middle part of the Range, as also in the outcrops inside the range and in the Tredian hills, the Purple Sandstones are absent and the Saline Series is directly overlain by Talchir conglomerates. In many places there is a general discordance between Saline Series and the Purple Sandstones.

NEOBOLUS BEDS

The Purple Sandstones are overlain by dark greenish and purplish shales containing intercalations of dolomite called the NEOBOLUS BEDS from the fact of their containing the primitive brachiopod Neobolus (*N. warthi, N. wynnei*, etc.). They are 6 to 60 m thick and are particularly well exposed in the Khusak hill not far from Khewra. Other fossils in these shales are :—

TRILOBITES : *Ptychoparia richteri, P. Sakesarensis, Redlichia noetlingi, Chittidillia plana, Conocephalus warthi*
BRACHIOPODS : *Lingula warthi, Lingulella wanniecki, Mobergia granulata, Discinolepis granulata, Orthis warthi.*
PTEROPOD : *Hyolithes wynnei.*

The fossil assemblage indicates a Middle Cambrian age, mainly the lower part thereof.

MAGNESIAN SANDSTONES

The succeeding Magnesian Sandstones are prominently displayed in the scarps of the Eastern Salt Range. They are mainly cream coloured massive dolomitic sandstones or arenaceous dolomites and flags, sometimes showing fine lamination, but include also thin shale bands of green to dark colour. They are 75 m thick in the east, diminishing to 24 m in the west. They show fucoid and annelid markings and contain the Cambrian gastropod named *Stenotheca*.

SALT PSEUDOMORPH SHALES

Succeeding the Magnesian Sandstones conformably, there are bright red to variegated shales with laminated sandstone layers. The thickness is up to 105 m. The shales contain cubic pseudomorphs or casts which represent replacement of salt crystals by clay on the shores of an enclosed marine basin which was drying up. The crystalline form of the salt crystals is shared by both the upper and lower surfaces of each bedding plane. Excellent sections of this zone may be seen near Pidh, Dandot, Nilawan ravine, etc.

Parts of the Cambrian succession are seen also in the Mari-Indus and Kalabagh areas where the Salt Marl occurs with gypsum. The southern part of the Khasor Range, especially a little to the northwest of Saiduwali in Dera Ismail Khan district, shows a thick sequence (120 m) consisting of Purple Sandstone, flaggy dolomite, bituminous shale, greenish grey shale and massive white to pink gypsum. It is not known how much of this is Cambrian and how much post-Cambrian.

KASHMIR

Palaeozoic rocks are exposed on the northern flanks of the anticlines trending in a N.W.-S.E. direction from beyond Hundawar in the northwest of the Kashmir valley to the Lidar valley. Good sections are seen in the Lidar valley, in the Basmai anticline in the Sind valley, in the Vihi district and in the Shamsh Abari syncline on the border of the Kashmir valley.

The Cambrian succession is best seen in the Hundawar area where the Dogra Slates are conformably succeeded by clay slates, greywackes and quartzites containing annelid tracks and some badly preserved organic remains which may be of Lower Cambrian age. They pass upwards into thick-bedded blue clays, arenaceous clays and thin-bedded fossiliferous limestones. The fauna is particularly rich in trilobites :

TRILOBITES : *Agnostus* sp., *Conocoryphe frangtengensis*, *Tonkinella kashmirica*, *Microdiscus* sp., *Anomocare hundwarensis*, *Solenopleura lydekkeri*.

BRACHIOPODS : *Acrothele* aff. *Subsidus*, *Obolus kashmiricus*, *lingulella* sp., *Botsfordia* cf. *coelata*.

Also the Pteropod *Hyolithes* and the cystid *Eucystites*.

Cowper Reed who has described the fauna (*Pal. Ind.* N.S. XXI, Mem. 2, 1934) states that it is largely endemic. It belongs to deep sea facies while the Salt Range fauna is of shelf facies and the spiti fauna of intermediate character.

In the Sind Valley, Vihi and Banihal areas, the Cambrian is either absent or is represented by disturbed and foliated shales and slates without identifiable fossils.

THE PALAEOZOIC GROUP—CAMBRIAN TO CARBONIFEROUS

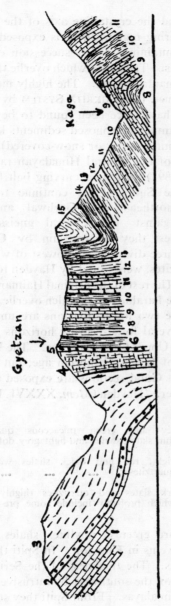

FIG. 4.—SECTION ON THE PARAHIO RIVER, SPITI.
(AFTER H. H. HAYDEN, *Mem.* 36, Pt. 1.)

1. Cambrian dolomite. 2. L. Silurian conglomerate. 3. Red (L. Silurian) quartzite. 4. Silurian limestone. 5. Muth quartzite. 6. Products shales. 7. L. Trias and Muschelkak. 8. Daonella Shales. 9. Daonella limestone. 10 Grey shales. 11. Dolomite with Tropites beds. 12. Juvavites beds. 13. Coral limestone. 14. Monotsi shales. 15. Quartzite series.

TETHYS HIMALAYAN ZONE
SPITI

Haimanta System.—Beyond the crystalline axis of the Himalayas, in the Spiti Valley and the neighbouring region, there is exposed a great synclinal basin with N.W.-.S.E. axis containing a full succession of strata from the Pre-Cambrian to the Cretaceous. The strata which overlie the Vaikrita System here have been called the Haimanta System. The highly metamorphosed and granitised groups of rocks named the VAIKRITA SYSTEM by Griesbach, which were thought to be Archaean have since been found to be mainly Cambrian as they pass laterally into the unmetamorphosed sediments known as the HAIMANTA SYSTEM (*Haimanta* meaning snowy or snow-covered). The Haimantas occur on the northern slopes of the Central Himalayan range between Spiti and Kulu (*Mem.* XXIII, 49, 1891) forming a curving belt from the Chandra River to the SSE across the Sutlej. They continue to the north-west into Lahaul and to the northeast into Garhwal and Kumaon. On the southwest they abut against the Central gneiss and crystalline schists, while on the northwest they are overlain by Ordovician strata. Griesbach divided them into three divisions the lowest of which, consisting of thick conglomerates and quartzites, were found by Hayden to be of Ordovician age, and had to be discarded. The rest of the original Haimantas now constitute the two lower divisions while the Parahio Series which overlie them conformably forms the upper division. The two lower divisions are unfossiliferous while the upper division contains several fossil-bearing horizons indicating Middle Cambrian (and perhaps part of Upper Cambrian) age. The Middle and Lower divisions should therefore be of Lower Cambrian age, but may be partly of uppermost Pre-Cambrian age. Good sections are exposed along the Parahio river, a tributary of the Spiti river. Hayden (*Mem.* XXXVI, 1904) has classified them as below.

		Metres
Haimanta System	Upper —Grey and green micaceous quartzites thin slates, shales and light grey dolomites...	360
	Middle —Bright, red and balck shales with some quartzites	300
	Lower— Dark slates and quartzites (highly folded) which probably include some pre-Cambrian	600-900

The LOWER HAIMANTAS are greenish phyllites, shales and thick bedded quartzites traversed by quartz veins in Kumaon. In Spiti they are clay slates interbedded with quartzite beds. The thickness of the Series is of the order of 900 m. They are bordered on the southwest by intrusive granites forming the root zone of the Central Himalayas. From Spiti they stretch into Lahaul to the northwest and into Garhwal and Kumaon to the southeast. They have been metamorphosed to mesograde rocks near the margin of the granite intrusives.

The MIDDLE HAIMANTAS form a very conspicuous horizon because of their bright red and black colour visible from a distance. They consist of

THE PALAEOZOIC GROUP—CAMBRIAN TO CARBONIFEROUS

ed and pink quartzites and shales, the colour being due partly to the limonite and hematite formed from the pyrite contained in the shales. Orange coloured patches and streaks in the red bands add to their brightness. These bands stand out amidst black carbonaceous slates which form the adjacent beds. They pass upwards into greenish grey shales and quartzites with pink shale partings. The Middle Haimantas are about 300 m thick and are well developed in Spiti and Bashahr where they form a useful stratigraphic horizon.

The Middle Haimantas pass upwards into the UPPER HAIMANTAS (PARAHIO SERIES) which are mainly grey and green quartzites, slates shales with thin beds of grey dolomite. They contain several fossiliferous horizons. Though, in general, the fossils are rather crushed, they are identifiable and fairly rich in trilobites. The full section of the Parahio Series described by Hayden is given below :

TABLE 26.—PARAHIO RIVER SECTION OF UPPER HAIMANTAS

Lithological units	Thickness in feet
19. Conglomerate	—
18. Quarzites and siliceous shales	50
17. Grey dolomite weathering brownish red	20
16. Flaggy sandstone, quartzite and siliceous slate	40
15. Grey dolomite weathering brownish red	30
14. Siliceous slates with grey quartzite bands and thin beds of pink dolomite	250
13. Dark siliceous slates with fragmentary fossils	10
12. Siliceous slates and flaggy quartzites	30
11. Siliceous and argillaceous slates with trilobites	6
10. Grey slaty quartzite capped by thin dolomite	50
*9. Slates, siliceous above and argillaceous below, with trilobites	30
8. Dark grey quartzite	60
7. Pink shaly dolomitic limestone, with trilobites	12
*6. Calcareous quartzite with *Lingulella* and trilobites, underlain by a narrow band of limestone and slates with trilobites	10
5. Grey micaceous quartzite with thin bands of mica-schists	150
*4. Slates alternating with narrow bands of limestone, with *Lingulella* and trilobites	10
3. Slate, chiefly siliceous, and quartzite	150
*2. Dark slate with trilobites	30
1. Red and green slaty quartzite with fossils in the uppremost beds	250
	1,188

Fossils are found only in the zones marked with an asterisk. The fauna rich in trilobites of which *Oryctocephalus* and *Ptychoparia* are particularly well represented. Most of the trilobite species are not known elsewhere. The ssiliferous beds range in age from Middle to Upper Cambrian and the fossils clude :

Trilobites ... ... *Agnostus spitiensis, Microdiscus griesbachi, Redlichia noetlingi, Oryctocephalus salteri, Ptychoparia spitiensis, P. Stracheyi, P. consocialis, Conocephalites memor, Anomocare conjunctiva, Olenus haimantensis*

Brachiopods ... *Nisusia depsaensis, Lingulella haimantensis, L. spitiensis, Acrotreta parahioensis, Obolella cf. crassa, Acrothele praestans*

Echinoderms ... *Eocystitis* sp.   Pteropod : *Hyolithes*

KUMAON

Garbyang Series.—The Haimantas are represented in the region north o Kulu and in Lahaul and also in the Kumaon Himalayas. In Kumaon thei equivalents are the Garbyang Series, named by Heim and Gansser after th village of that name in the Kali valley. They are exposed over a long distanc from east of the region of the Nampa peaks to Nanda Devi and beyond. The consist generally of slaty to phyllitic, fine grained, calcareous sandstone an argillaceous dolomite, the latter containing green chloritic bands weathe ing to a brown colour. The chloritic bands are probably partly metamorphose basic tuff. The Garbyang Series is found to be thickest in the Kali sectio where it attains nearly 5,000 m but is reduced to barely 1,500 m in the Go valley above Milam. That it is of Cambrian age is proved by the occurrenc of badly preserved and flattened gastropod shells in the sericitic calcareo phyllites and by the presence of crinoidal limestone in the middle part in tl Dhauli valley. The first calcareous sandstone bed above the Garbyang Seri has furnished undoubted Ordovician fossils. Hence the whole of the Garbyar Series is taken to be of Cambrian age.

BURMA

No fossiliferous Cambrian rocks have so far been discovered in Buim but part of the Chaung Magyis of the Shan States, the Mergui Series of Merg and also some volcanic rocks may probably belong to this age.

THE BAWDWIN VOLCANICS

In the neighbourhood of Bawdwin ($23° 6' : 97° 18'$) in the Tawngpe State near the China border, the Chaung Magyis are overlain by the Pangy beds, consisting of quartzites, grits, sandstones and shales and some rhyoli grits. These grade perfectly into the Bawdwin volcanic series and have general N.W.-S.E. strike. They are composed dominantly of tuffs and su ordinate rhyolites containing clear grains of quartz and having a brown chocolate colour. They are seen to have suffered much crushing and at tl surface are soft and light grey owing to decomposition. The rhyolites a tuffs are probably the effusive phase of the Tawngpeng granite which is expos at a distance of 8 km west of the volcanics.

The Bawdwin lead-zinc-silver ore-body occupies a prominent fault a shear zone (the Bawdwin fault) which is over 2,400 m long and 150 m wi in which the tuffs and rhyolites have been replaced by Pb-Zn-Ag ore. $ gradations can be found from solid ore through partly replaced tuffs to u replaced volcanic rocks. The ore is mainly a fine grained mixture of argen ferous galena and sphalerite with some chalcopyrite, often showing signs considerable crushing. The ore-body is lens-shaped and about 900 m lo and of varying width. It is cut up by two faults, the Yunnan fault in the nor and the Hsenwi fault in the south, into three sections ; the northern sectio called the *Shan lode* has an average width of 6 m ; the central section, t *Chinaman lode* has an average width of 15 m but is in places 42 m ; the southe

PLATE I
CAMBRIAN FOSSILS

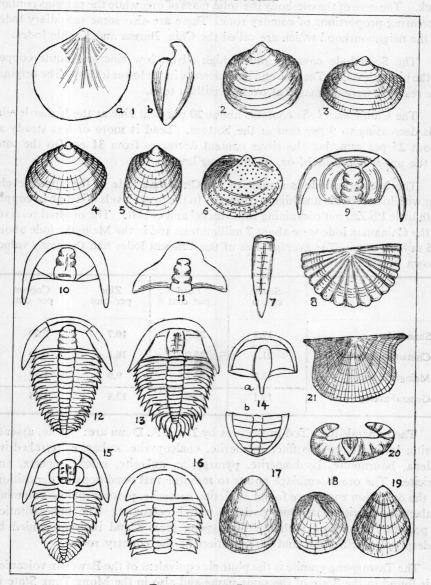

EXPLANATION OF PLATE I

1. *Orthis warthi* (6). 2. *Neobolus warthi*. 3. *Neobolus wynnei*. 4. *Schizopholis rugosa*. 5. *Lingula warthi*. 6. *Mobergia granulata*. 7. *Hyolithes wynnei* (1.3). 8. *Orthis marhaumensis* (2.5). 9. *Hoefria noetlingi*. 10. *Ptychopariu spitiensls* (0.7). 11. *Conocephalus warthi*. 12. *Ptychoparia salteri* (3). 13. *Oryctocephalus salteri* (2.5). 14. *Microdiscus griesbachi* (6). 15. *Anomocare hundwarense* (3). 16. *Conocoryphe sejuncta*. 17. *Lingulella haimantensis* (3). 18. *Acrothele praestans* (3). 19. *Botsfordia coelata* (2). 20. *Pseudotheca waageni* (1.5). 21. *Nisusia depsaensis* (1.5).

section, called the *Meingtha lode* is 6 m wide. The ore-body as a whole strikes N.N.W. and dips towards the west, with a northerly pitch. The hanging wall is well defined while the foot-wall is often indefinite and grades into the country rock. The core of the ore-body is a solid mass of ore, while the margins contain increasing proportions of country rock. There are also some subsidiary lodes in the neighbourhood which are called the Chin, Burma and Kachin lodes.

The Shan lode contains rather high silver, low zinc and some copper in the higher levels. The zinc and lead increase in the lower levels. The original ore reserves were estimated at over 20 million tons.

The Chinnaman lode contains about 20 per cent zinc at the higher levels, this decreasing to 9 per cent at the bottom. Lead is more or less steady at about 25 per cent, but the silver content decreases from 31 oz. (to the ton) in the upper levels to 14 oz. in the lower levels.

The Meingtha lode is similar to the Chinaman lode in the upper levels, but with lower Pb-Zn and higher copper. In the lower levels it is high in copper with little Pb-Zn, but containing some nickel and cobalt. The original reserves in the Chinaman lode were about 7 million tons and in the Meingtha lode about 1.6 million tons. The average ores of the different lodes had the assay values shown below :

	Silver oz/ton	Lead per cent	Zinc per cent	Copper per cent
Shan lode	17.8	21.5	10.7	2.09
Chinaman lode ...	21.1	25.0	16.1	0.40
Meingtha lode ...	13.0	15.2	9.0	1.97
General average ...	19.1	22.7	13.8	1.05

The minerals identified in the ores by Dr. J. A. Dunn are: pyrite, arseno-pyrite, lollingite, gersdorffite, sphalerite, chalcopyrite, cubanite, tetrahedrite galena, bournonite, boulangerite, pyrargyrite, ankerite, calcite, quartz, and sericite. The ore assemblage points to mesothermal conditions of deposition. In the oxidation zone were found anglesite, cerussite, pyromorphite, calamine malachite, azurite, massicot, goslarite and brochantite. The mineralisation is probably connected with the Tawngpeng granite and is accompanied by widespread silicification and sericitisation of the country rock.

The Tawngpeng granite is the plutonic equivalent of the Bawdwin volcanics It is found in the State of the same name and also in the Mong Tung State in South Hsenwi. It is a biotite-granite containing no tourmaline. Of about the

J. C. Brown : Geology and ore deposits of Bawdwin mines. *Rec.* 48, 121-178, 1917.
A. B. Colquhoun : *Trans. Amer. Inst. Min. Eng.*, 69, 211, 1923.
M. H. Lowman : *Op. cit.*, 56, 181, 1917.
A. B. Colquhoun : *Mining Mag.* 44, 329-333; 45, 23-26, June and July, 1931.
J. A. Dunn : A microscopic study of the Bawdwin ores, *Rec.* 72, 333-359, 1937.
E. L. G. Clegg : A note on the Bawdwin mines, Burma. *Rec.* 75, paper 13, 1941.

same age are the siliceous tuffs containing angular quartz fragments in a fine-grained chloritic groundmass, which occur near the Lagwe Pass on the Burma-China frontier.

Many of the islands of the Mergui Archipelago show granite, porphyry, rhyolite and agglomerate. The products of this volcanism have been deposited with the sediments of the Mergui Series. There are also some felsites and amphibolised basic rocks which are intrusive into the Merguis and may be semi-contemporaneous with them.

THE ORDOVICIAN AND SILURIAN SYSTEMS

The strata which are now referred to the two systems Ordovician and Silurian were formerly described under the Silurian which embraced both. In eastern Europe the Silurian is called Gothlandian. As already mentioned, the DICTYONEMA horizon (Tremadocian) is sometimes included in the Ordovician. If not, the Ordovician may begin with the stage containing *Niobe* and *Euloma* (Trilobites) which lie about the Dictyonema beds. The name Silurian was proposed in 1835 by the British geologist Roderick Murchison, after the name of an ancient Welsh tribe.

The faunas of these eras include many corals — *Cyathophyllum, Zaphrentis, Heliolites, Favosites, Halysites* etc., some of these forming limestone reefs. The graptolites however are the most characteristic fossils, some 30 graptolite zones being used for classifying these strata. The remains of graptolites, which were free-swimming pelagic organisms, are found in fine shales. The crinoids are very abundant while in some places Cystids attain importance (*Pleurocystites* and *Echinosphaerites*). The Trilobites continue to be quite important, the common genera being *Asaphus, Calymene, Illaenus, Phacops, Dalmanites,* etc. Brachiopods begin to come in fairly abundantly (*Orthis, Chonetes, Spirifer, Atrypa, Pentamerus, etc.*) in rather shallow water environment. Several genera of Cephalopods make their first appearance — *Orthoceras, Cyrotoceras, Trochoceras, Lituites*; and also the ammonoids *Agoniatites, Tornoceras, Anarcestes,* especially in the Mediterranean geosyncline. Lamellibranchs and gastropods are not important but the peculiar form *Conularia* whose systematic position is not clear, is interesting. These two systems are well developed along the border of the Baltic shield, and along the north European geosyncline, with its northern branch along Scandinavian and Caledonian mountains and the southern branch along Poland, Thuringia, Harz etc. There is also the primitive mediterranean geosyncline whose deposits are noticed in Spain, the Pyrenees, Morocco, Carnic Alps etc.

Three or four distinct facies of these strata are noted : Sandy shales and fine sandstone of the *flysch facies* (*flysch* being a Swiss term for beds liable to slip) formed in shallow water, of large thickness and poor in fossils ; a *Calcareous facies* formed in shallow marine basins rich in fossils, especially Corals, trilobites and brachiopods as in Gothland and Esthonia ; a *black shale facies*, of shallow to deep water origin, usually associated with graptolites, as in Scotland, and the Lake district (England). In some places volcanic flows and pyroclastics are also found.

The end of the Silurian was marked by the *Caledonian orogeny* in Europe and America. In many areas there is a great discordance while in others the deposits acquire shallow water and lagoonal characteristics, heralding the continental sedimentation of the Old Red Sandstone.

The standard divisions of the Ordovician and Silurian are the following.—

Silurian	Downtonian Ludlowian Wenlockian Llandoverian	Salopian (Valentian)
Ordovician	Ashgillian Caradocian Llanvirnian Llandeilian Skiddavian	Bala Llandeilo Arenig

SPITI

As already remarked, the Spiti area shows a full succession of Palaeozoic rocks. The Cambrian formations described already are overlain by a thick series of shallow water deposits consisting of conglomerates, quartzites and grits and these in turn by shales and limestones, this whole succession being referable to the Ordovician and Silurian. The lower, arenaceous part is about 500 m thick while the upper calcareous and argillaceous strata have a thickness of 150-200 m.

TABLE 27.—LOWER PALAEOZOIC SUCCESSION IN SPITI

		Metres
Upper Silurian	Muth Quartzite	
Silurian	8. Grey siliceous limestone weathering red.	24
	7. Grey limestones, weathering brown, with brown and red marls	21
	6. Grey coral limestone	15
	5. Shaly limestone with brachiopods, gastropods and corals	9
	4. Hard grey dolomitic limestone... ...	12
	3. Dark grey limestone with cystids ...	12
———?———	2. Dark foetid limestone with trilobites and brachiopods	60
Ordovician	1. Shaly and flaggy sandstone with plants and *Orthis*	45
	0. Flaggy quartzites and siliceous shales passing down into red quartzites with conglomerates at base (unfossiliferous)	450
Upper (and partly Middle) Cambrian	Shales slates, quartzites, etc.	360
Lower Cambrian	Red and black slates and quartzites ...	300
	Dark slates and quartzites	600-900

Practically all the beds in the above succession contain fossils. Bed No. 2 is particularly rich in trilobites and brachiopods and is referable to Caradocian age. The beds above it contain cystids and brachiopods which indicate a transition from Ordovician to Silurian. The siliceous limestone (Bed No. 8) below the unfossiliferous Muth Quartzite shows *Favosites, Pentamerus* and other fossils on weathered surfaces and is of early Wenlock age, the Upper Silurian being part of the succeeding Muth Quartzite. The fossils in the two systems are mentioned below :—

Trilobites	...	...	*Asaphus emodi* var. *milamensis, Illaenus brachioniscus, I. punctulosus, Calymene nivalis, Cheirurus mitis*
Brachiopods	...	...	*Orthis (Dinorthis) thakil, Dalmanella testudinaria, Leptaena rhomboidalis, L. trachealis, Strophomena chamerops, Rafinesquina umbrella, R. aranea, R. muthensis, Plectambonites himalensis, Christiania nux*
Lamellibranchs and Gastropods	...		*Pterinea thanamensis, Lophospira himalensis, L. pagoda, Bellerophon ganesa, Conradella* aff. *obliqua.*
Cephalopods	...	...	*Orthoceras kemas, Cyrtoceras centrifugum, Gonioceras* cf. *anceps*
Bryozoa	...	...	*Ptilopora, Phylloporina, Ptilodactya*
Actinozoa	...	...	*Streptelasma* aff. *corniculum, Heliolites depauparata*
Pteropods	...	...	*Tentaculites*

SILURIAN

Trilobites	...	...	*Encrinurus* aff. *punctatus, Calymene* sp.
Brachiopods	...	...	*Orthis (Plectorthis) spitiensis, Dalmanella basalis, Orthis calligramma, Leptaena rhomboidalis, Orthotetes* aff. *pecten, Pentamerus oblongus*
Lamellibranchs	...	...	*Palaeoneilo* cf. *victoriae*
Gastropods	...	...	*Euomphalus* cf. *triquetrus*
Cephalopods	...	...	*Orthoceras* cf. *annulatum*
Actinozoa	...	...	*Lindstroemia* sp., *Zaphrentis* sp., *Propora himalaica, Favosites spitiensis, Halysites wallichi, H. catenularia* var. *kanaurensis*

In the Ordovician, the brachiopods are abundant and trilobites comparatively much less common and mollusca rare. The Silurian fauna is rich in corals with remarkable North American affinities. Though the faunas of the two systems have local characters, they show much more affinity to those of North America than Northern Europe.

NORTHERN KUMAON

Ordovician.—In Northern Kumaon the Ordovician is represented by the SHIALA SERIES which occurs between the Garbyang series and the characteristic red shales of the Silurian. In the Shiala Pass near Kuti, it consists of 400 to 500 m of shales intercalated with grey and greenish marly limestone sometimes containing breccia of crinoidal fragments. From near the top of the division, Heim and Ganseer collected well preserved and characteristic Ordovician fossils which include *Calymene* cf. *Douvillei, Orthis Thakil, Orthotetes Pecten, O. Orbignyi, Rafinesquina* aff. *Subdeltoidea, Leptaena Sphaerica,* (very abundant) *Sowerbyella Umbrella* etc.

Silurian.—This System is represented by the VARIGATED SERIES which is the same as the Red Crinoidal Limestone of Griesbach. This consists of a repetition of limestones and marls and siliceous shales of varied colours from top to bottom, red colour being dominant and conspicuous from a distance. In the Shiala Pass this consists of 200 to 300 m of red, grey and green shales with layers of white limestones, capped by brown-weathering dolomite and 500 to 600 m of shaly lenticular limestone with fragments of crinoids. Further east at Gunji, this system is reduced to less than 50 m of violet coloured sandy shales which overlie the Garbyang Series with a disconformity. East of the Kali valley, in Nepal, the thickness is only 100 m or so. It is also known to be found in the Zanskar Range to the northwest and is apparently repeated by folding and thrusting further to the north. Because of its shaly nature, it has been intensely folded in the sections exposed in the Zanskar Range. The Varigated Series is considered to be mostly of Silurian age.

KASHMIR

Ordovician.—The Ordovician is present in parts of Kashmir though the exposures have not yielded good fossils. In the Lidar valley it may underlie the fossiliferous Silurian exposed along the flanks of an anticline in which a complete Silurian to Triassic succession is found. A similar anticline is found also in the Basmai area in the Sind valley. At Trehgam and its neighbourhood, on the northern limit of the Shamsh Abari syncline, there are greywackes, slates, and limestones which contain some crushed and fragmentary fossils, including *Orthis* cf. *calligramma* and some crinoid stems.

Silurian.—The Lidar valley fold near Eishmakam clearly exposes Silurian rocks, composed of arenaceous shales and impure limestones. The fossils include orthids, strophomenids, corals and fragmentary crinoids, which indicate a Silurian age. The Shamsh Abari syncline also contains fossiliferous Silurian slates and greywackes but the fossils are mostly crushed and obliterated. Elsewhere in Kashmir, the Cambrian strata are overlain by the Muth Quartzites of Upper Silurian to Devonian age, or by the Agglomeratic Slates or Panjal Trap of Middle Carboniferous or later age. *Didymograptus* and *Monograptus* have been reported in horizons 9 m apart, from an undisclosed locality. They indicate Ordovician and Silurian age respectively.

SALT RANGE, HAZARA AND BALUCHISTAN

A reef limestone of Upper Silurian to Middle Devonian age has been found recently near Nowshera (about 40 km east of Peshawar (34° N: 72° E). It forms a low ridge 30-70 m high, with strata dipping at 45° north. The lower part is micaceous phyllites with thin layers of crinoidal limestone while the upper part is a massive pink limestone. The lower 30 m of this is dolomite containing some brachiopods, *Camarotoechia* and *Atrypa* being dominant. Above it is the reef limestone which is full of spherical stromatoporoids and tabulate corals including *Heliolites, Thamnopora, Favosites, Alveolites* etc. which can be recognised on weathered surfaces. In the topmost layers brachiopods and some cephalopods are also noticed. Above the pink limestone comes a grey medium grained dolomitic quartzite. The limestone-dolomite reef has been previously described as a slightly metamorphosed marble without close examination. Similar limestone has also been found near Swabi in the Mardan district in the same region (Teichert and Stauffer, Science No. 1701, p. 1287, 1965). Farther south, in the Baluchistan Arc, there is no record of strata of any age between the Pre-Cambrian and the Permo-Carboniferous.

There is a stratigraphic gap in these regions from the Cambrian to the Upper Carboniferous, the latter being represented by glacial boulder beds.

BURMA

The Shan States of Burma contain well developed Ordovician and Silurian Systems. The Tawngpeng System (Pre-Cambrian to early Cambrian) is overlain by sandstones, shales and limestones which are subdivided as shown below :—

NYAUNGBAW LIMESTONE
HWE-MAUNG PURPLE SHALES
UPPER NAUNGKANGYI STAGE
LOWER NAUNGKANGYI STAGE
NGWETAUNG SANDSTONE

Of these the middle three divisions are well developed, the other two being local.

The Lower Naungkangyi Stage.—This is seen in a series of exposures on the Shan Plateau and consists of marls and limestones. Good sections are to be observed in the Gokteik gorge and in the valley of the Nam Pangyum. Amongst the fossils of this stage are :—

Brachiopods	*Orthis irravadica, O. subcrateroides, Leptaena ledetensis, Rafinesquina imbrex, R. subdeltoidea, Schizotreta, Plectambonites quinquecostata*
Trilobites	*Calymene birmanica, Cheirurus dravidicus, Asaphus, Phacops*
Cystids and Crinoids	*Aristocystis dagon,* and several species of *Heliocrinus* and *Caryocrinus*
Bryozoa	*Diplotrypa sedavensis, Phylloporina orientalis*

The fossil assemblage shows affinities with the Lower Ordovician of the Baltic region of Europe, and very little with that of Spiti in which the important elements are trilobites and cystids.

The Upper Naungkangyi Stage.—This has a wide distribution in the Shan States and shows two facies — a shaly facies west of Lashio and a calcareous shaly one east of the same place. The western outcrops are about 300 m thick and show evidences of crushing and compression. The chief fossil organisms in them are :—

Brachiopods	*Lingula* cf. *attenuata, Dalmanella testudinaria* var. *shanensis, Orthis calligramma, Strophomena* sp., *Rafinesquina subdeltoidea, Porambonites sinuatus, Plectambonites sericea*
Trilobites	*Agnostus* cf. *glabratus, Calymene birmanica, Illaenus lilunesis, Cheirurus submitis, Phacops dagon, Pliomera insangensis*
Crinoids	*Heliocrinus, Caryocrinus*
Bryozoa	*Diplotrypa palinensis, Phylloporina, Ceramopora*

PLATE II
ORDOVICIAN AND SILURIAN FOSSILS

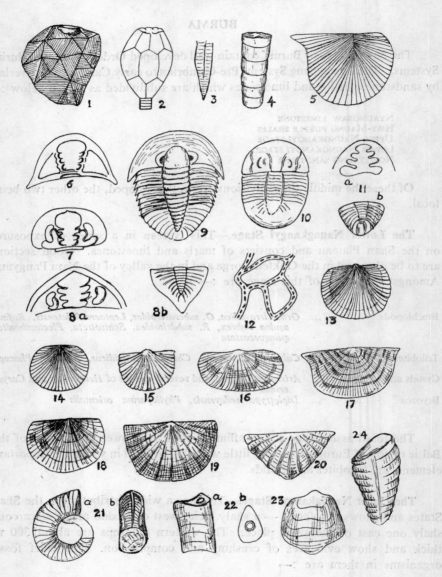

Explanation of Plate II

1. *Heliocrinus qualus* (0.3). 2. *Caryocrinus avellana* (0.7). 3. *Tentaculites elegans* (3). 4. *Orthoceras commutatum* (0.7). 5. *Pterinea konghsaensis* (0.7). 6. *Encrinurus konghsaensis* (0.7). 7. *Phacops (Pterigometopus) dagon* (0.7). 8. *Phacops (Dalmanites) longicaudatus* (1.5). 9. *Oxygites birmanicus* (0.7). 10. *Illaenus liluensis* (2). 11. *Calymene nivalis* (1.5). 12. *Halysites wallichi* (1.5). 13. *Orthis (Dalmane'lla) basalis*. 14. *Orthis (Dalmanella) mansuyi* (1.5). 15. *Schuchertella planissima* (3). 16. *Plectambonites sericea* (2). 17. *Leptaena trachealis* (6). 18. *Strophomena chamerops*. 19. *Rafinesquina umbrella*. 20. *Plectambonites himalensis*. 21. *Bellerophon ganesa*. 22. *Gonioceras anceps* (0.5). 23. *Triplecia uncata* (0.7). 24. *Cyrtoceras centrifugum* (0.7).

The purple calcareous shales in the eastern area are also roughly of the same age as the Upper Naungkangyis as evidenced by their fossil content.

Brachiopods ... ... *Dalmanella* sp., *Plectambonites sericea, Orthis testudinaria, O. subcrateroides.*

Nyaungbaw Limestone.—These limestones overlie the Naungkangyis and contain *Camarocrinns Asiaticus* and other fossils which indicate that they are also Ordovician in age.

SOUTHERN SHAN STATES

The Ordovician is developed also in the Southern Shan States where it is represented by the Mawson Series, Orthoceras Beds and Pindaya Beds.

Mawson Series.—In the eastern part of the Mawson highland and farther south, there are calcareous shales and limestones containing *Orthoceras, Actinoceras, Oxygites, Pliomera, Orthis, Cyrtolites, Helicotoma* etc. These beds are of Ordovician age and contain the lead deposits of Mawson described by Dr. J. Coggin Brown (*Rec. G.S.I.* LXV, 394-433, 1930).

Orthoceras Beds and Pindaya Beds.—On the western limb of the Mawson anticline, there occur purple argillaceous limestones and shales containing crinoid stems and species of *Orthoceras, Diplograptus and Monograptus,* The Orthoceras Beds are definitely of Middle Ordovician age and their equivalents are known in Yunnan, South Manchuria and in the Baltic region.

Bands of calcareous shales, slates and thin limestones occur in the Pindaya Range, bordered by Permo-Carboniferous limestones. The enclosed fauna has close relationships with that of the Naungkangyi Beds, and includes cystids, trilobites, brachiopods, etc., among which may be mentioned :

Orthis cf. *irrevadica, Ptychoglyptus shanensis, Yeosinella consignata, Christiania* cf. *tenuicincta, Rhinidictya* cf. *nitidula ; Favosites ; Caryocrinus ; Illaenus ; Dictyonema.*

Several of the fossils are in a bad state of preservation. The Silurian of Burma has the following sub-divisions :

Zebingyi Beds
Namshim Sandstones and Marls
Pangsha-pye Graptolite Beds

The GRAPTOLITE BEDS are found near Hsipaw and other places in the Namtu valley and in the Loi-lem range east of Lashio, overlying the Upper Naungkangyis. Just beneath the Graptolite-bearing bands there is a bed containing trilobites. The Gaptolite Beds are white shales of about 15 m thickness, containing abundant fossils indicating a Llandovery age :

Graptolites ... ... *Diplograptus modestus, D. vesiculous, Climacograptus medius, C. tornquisti, C. rectangularis, Monograptus concinnus, M. gregarius, M. tenuis, Rastrites peregrinus, Cyrtograptus* sp.
Brachiopods ... ... *Dalmanella elegantula, Dalmanella mansuyi, Stropheodonta mcmahoni, S. feddeni*
Trilobites ... ... *Phacops (Dalmanites) hastingsi, Acidaspis shanensis*
Ostracods ... ... *Beyrichia* sp.

The NAMSHIM (NAMHSIM BEDS, of the same age as the Wenlock beds of England, contain two divisions, the lower consisting of sandstones of varying degrees of coarseness and containing several trilobites and brachiopods. The former include *Calymene Blumenbachi, Encrinurus Konghsaensis, Cheirurus* cf. *Bimucronatus, Phacops (Dalmanites) Longicaudatus*! etc. The upper Namshim stage is marly in composition and sometimes rests directly on the Naungkangyis. These marls contain a rich fauna :

Bryozoa	... ...	*Fenestella* sp.
Trilobites	... ...	*Encrinurus konghsaensis, Calymene blumenbachi (Cheirurus bimucronatus)*
Brachiopods	... ...	*Lingula lewisi, Orthis rustica, Orthis biloba, Dalmanella elegantula, Leptaena rhomboidalis, Strophomena corrugata, Pentamerus* cf. *oblongus, Atrypa reticularis*

The ZEBINGYI STAGE comprises limestones and shales containing numerous fossils which show affinities to Upper Silurian and Lower Devonian of Europe, especially of the Mediterranean region. On the whole, the Devonian affinities are strong.

Graptolites	... ...	*Monograptus dubius* and other species.
Trilobites	... ...	*Phacops (Dalmanites) swinhoei, P. shanensis.*
Brachiopods	... ...	*Atrypa marginalis, A. subglobularis, Stropheodonta comitans, Meristina* sp.
Cephalopods	... ...	*Orthoceras* aff. *commutatum* and other species.

Also abundant *Tentaculites* (pteropod) and several lamellibranchs.

The Burmese Silurian fossils resemble the Silurian ones of Northern Europe and England and are unlike the Himalayan fossils of the same age. The Burmese fauna is rich in graptolites, while the corals which are common in the latter are scarce in the former.

THE DEVONIAN SYSTEM

The Devonian system was laid down after the Caledonian mountain building movements so that there is often a marked stratigraphical break at its base. Three distinct and contrasted facies are seen in Europe : the shallow marine facies of S.W. England, the Franco-Belgian Ardennes, Schiefengebirge, Bohemia, etc. characterised by sandstones and shales of shallow deposition : the muddy (shaly) deep water facies seen in Montagne-Noire, Pyrenees, Morocco, Carnic Alps, etc. ; and the continental Old Red Sandstone facies of Scandinavia, the Baltic region, Scotland, and Ireland showing fiuviatile and lacustrine sediments.

The O.R.S. facies contains some plants, Crustacea (*Eurypterus, Pterygotus*), Ostracods, and several genera of fishes belonging to the Selacians, Ganoids, Placoderms and Dipnoans. In the marine facies the graptolites have practically disappeared. Several genera of corals such as tetracorals, stromatopora and tabulate corals are of importance as reef-builders. The pelecypods are represented by Aviculids and Megalodontids etc. ; the gastropods by Pleurotomarids, capulids and Bellerophontids. Brachiopods enter ir fair abundance, *Spirifer, Orthis, Athyris, Atrypa, Stringocephalus* etc

THE PALAEOZOIC GROUP—CAMBRIAN TO CARBONIFEROUS

being common. Cephalopods also become abundant *Gomphoceras, Cyrtoceras, Phragmoceras, Anarcestes, Agoniatites, Parodiceras, Gephyroceras, Mantioceras* etc.), the species of some of these being useful for stratigraphical zoning.

The type marine succession is that found in the Franco-Belgian Ardennes where the strata have been divided as shown below:

Divisions		Fossils
Upper	Famennian	*Spirifer Verneuili* ; *Cheiloceras* ; *Clymenia* ; *Entomis* (Ostracod).
	Frasnian	*Sp. verneuili, Athyris concentrica, Orthis striatula, Rhynchonella cuboides* ; *Gephyroceras, Mantioceras.*
Middle	Givetian	*Cyathophyllum, Favosites* ; *Megalodus* ; *Atrypa reticularis, Stringocephalus burtini.*
	Eifelian	*Spirifer cultrijugatus* ; *Calceola Sandalina.*
Lower	Coblenzian	*Phacops, Homalonotus* ; *Avicula* ; *Bellerophon* ; *Spirifer, Athyris.*
	Gedinnian	*Spirifer sulcatus.*

In the Indian region, Devonian strata are found at several places in the Himalayan belt and in Burma. Shore deposits are represented by the Muth Quartzites of Spiti, Kumaon and Kashmir. Limestones with brachiopods and corals are exposed in Spiti, Chitral and Burma while a lagoonal facies is also seen in Burma (Wetwin Shales.)

SPITI—KUMAON

A group of hard, white unfossiliferous quartzites conformably overlies the Silurian rocks containing *Pentamerus Oblongus* and other fossils of early Wenlock age and is overlaid by fossiliferous Lower Carboniferous rocks. The quartzites are 150 m thick and are known as the *Muth Quartzites.* They are mainly Devonian in age but the lower part is Upper Silurian.

In northern Kumaon the Devonian and the Upper Silurian are represented by the Muth Series which has the same characters as in Spiti. In the Kuti region this series is 800 to 1,000 m thick, composed of quartzite with dolomite layers which weather to a brown colour, overlain by massive white quartzites. The top quartzite is sometimes replaced by dolomite and varies considerably in thickness up to 200 m. Though the overlying Kuling Shales of Permian age are conformable to them, there is almost everywhere a considerable stratigraphical gap representing a period of erosion and non-deposition. The Devonian is represented by a crinoidal limestone in the Lipu-Lek area. The full section of the Muth Quartzite in northern Kumaon is considered to represent the Upper Silurian and a part of the Devonian.

In Upper Spiti and Kanaur, there are dark fissile limestones containing abundant *Orthotetes* (*O. crenistria*) and other brachiopods such as *Atrypa aspera, Strophalosia* etc. These are Middle to Upper Devonian in age.

In Byans, near the Nepal border, Devonian fossils have been found in some dark limestones near Tera Gadh camp. These include : *Atrypa reticularis, A. aspera, Pentamerus* cf. *sublinguifer, Camarophoria* cf. *phillipsi, Rhynchonella* (*Wilsonia*) cf. *omega, orthis* aff. *bistriata* ; and also the cephalopod *Orthoceras* and the coral *Favosites*.

KASHMIR

Devonian strata are found overlying the Silurian in the Lidar valley and in the Shamsh Abari syncline. They are hard white quartzites having a thickness of up to 600 m. They are lithologically similar to the Muth Quartzites and have yieldedl *Sptrifer, Athyris, Phacops* and some mollusca.

CHITRAL

In the State of Chitral on the Afghan frontier, Devonian rocks are well developed and comprise thick limestones which are underlain by quartzites, sandstones and conglomerates. These latter are unfossiliferous and apparently represent older Palaeozoic rocks. The Devonian rocks, which are generally brought into juxtaposition with conglomerates of much younger age by a great fault, comprise beds of limestone with corals and brachiopods. The fossils include :

Trilobites	... ...	*Proetus chitralensis*
Brachiopods	... ...	*Spirifer* aff. *primaevus, S.* cf. *robustus, Athyris* cf. *subconcentrica, Pentamerus sieberi, Orthis* cf. *praecursor* var. *sulcata*, species of *Productella* and *Dalmanella, Stropheodonta phillipsi, Orthotetes hipponyx, Chonetes* aff. *embryo, Chonetes McMahoni*
Gastropods	... ...	*Loxonema, Euomphalus, Pleurotomaria*
Actinozoa	... ...	Several species of *Cyathophyllum*

Also crinoid stems and *Fenestella*.

BURMA

Plateau Limestone.—In the Federated Shan States the Silurian rocks are succeeded by the Plateau Limestones which occupy a large area. They are mainly of dolomitic composition, but argillaceous and arenaceous intercalations are known in places. The typical Plateau Limestone is a hard, light grey, fine-grained and granular rock which has been crushed and is traversed by thin veins of calcite. It contains a few traces of fossils including corals and foraminifera.

The Plateau Limestone extends in age from Devonian to Lower Permian. The brecciated nature and paucity of fossils do not permit of the separation of beds belonging to different ages. The total thickness is about 900 m. It was probably a limestone, later dolomitised to a large extent. The lower part contains two fossiliferous patches, *viz.*, Padaukpin Limestone and Wetwin Shales which have yielded Devonian fossils.

PLATE III
DEVONIAN FOSSILS

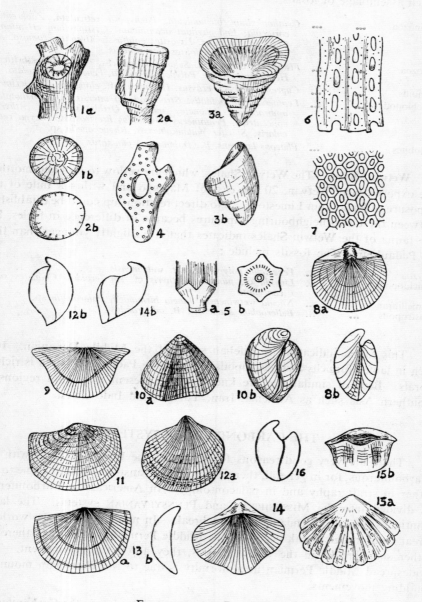

EXPLANATION OF PLATE III

1. *Cyathophyllum caespitosum* (1.5). 2. *Amplexus hercynicus* (1.5). 3. *Calceola sandalina*
  Pachypora reticulata. 5. *Hexacrinus pyriformis* (1.5). 6. *Fenestella polyporata*
  7. *Polypora populata* (3). 8. *Orthis (Schizophora) striatula* (2). 9. *Stropheodonta
  terstrialis* (0.5). 10. *Athyris chitralensis* (0.5). 11. *Leptaena rhomboidalis*. 12. *Atrypa
  inosa chitralensis*. 13. *Orthothetis umbraculum*. 14. *Spirifer murchisonianus* (0.5). 15. *Pen-
  merus (Gypidula) brevirostris* (0.5). 16. *Spirifer verneuili*.

Padaukpin Limestone.—This is exposed in a small area near Padaukpin situated at a distance of 2 km from Wetwin railway station, and contains a rich assemblage of fossils :

Actinozoa	... ...	*Cyathophyllum birmanicum, Pachypora reticulata, Zaphrentis cornicula, Endophyllum acanthicum, Cystiphyllum cristatum, Calceola sandalina, Favosites goldfussi, Alveolites ramosa, A. suborbicularis, Heliolites interstinctus.*
Bryozoa	... ...	*Fistulipora tempestiva, Selenopora coelebs, Fenestella arthritica, Hemitrypa inversa, Polypora populata, Fenestropora isolata.*
Crinoids	... ...	*Cupressocrinus* cf. *crassus, Hexarinus* aff. *elongatus, Taxocrinus.*
Brachiopods	... ...	*Leptaena rhomboidalis, Stropheodonta interstrialis, Orthotetes umbraculum, Chonetes minuta, Orthis striatula, Athyris concentrica, Pentamerus (Gypidula) brevirostris, Atrypa reticularis, Spirifer padaukpinensis, Rhynchonella* sp.
Trilobites	... ...	*Phacops latifrons, P. (Dalmanites) punctatus*

Wetwin Shales.—The Wetwin Shales, which are yellow to buff and mottled are exposed near Wetwin, 20 km east of Maymyo and within a mile of the exposure of Padaukpin Limestone. No direct relationship could be established between these two neighbouring exposures because of difference in facies, but the fauna of the Wetwin Shales indicates that it is slightly younger than that of Padaukpin. The fossils include :

Bryzoa	... ...	*Fenestella polyporata* var. *wetwinensis*
Brachiopods	... ...	*Lingula* cf. *punctata, Athyris* cf. *spiriferoides, Chonetes subcancellata*
Lamellibranchs	... ...	*Nucula wetwinensis, Janeia birmanica, Palaeoneilo* sp.
Gastropods	... ...	*Bellerophon shanensis, B. admirandus*

This fauna indicates the Eifelian stage of the Middle Devonian. It is rich in lamellibranchs and gastropods whereas the Padaukpin fauna is rich in corals. Beds of similar age are known to be present in several regions in Southern Asia such as Armenia, Iran, Yunnan and Indo-China.

THE CARBONIFEROUS SYSTEM

There are very good reasons for grouping the Lower Permian with the Carboniferous, for in general, there is a perfect transition from the one to the other in stratigraphy and in palaeontology. In America, the Carboniferous is divided into the MISSISSIPPIAN and PENNSYLVANIAN systems. The latter continues into the Permian without a break. In many areas in the world, a great stratigraphic break occurs in the Middle Permian, and though there are other such breaks in the Carboniferous, they are not so prominent. The widespread Middle Permian unconformity is due to the Hercynian mountain building movements.

The Hercynian orogeny took place in a series of pulses in the Carboniferous and Permian. The first important phase is that just below or just above the Westphalian. In Western Europe the Hercynian orogeny affected Southwest England, France and Germany, where the Armorican chains have a N.W.-S.E. trend. In Central Europe the mountains constitute the Variscan or Hercynian chain (after the Harz Mountains) with N.E.-S.W. trend.

THE PALAEOZOIC GROUP—CAMBRIAN TO CARBONIFEROUS

At the end of the Devonian period there were two regions of contrasted facies of deposition in Europe, viz., a northern one with a tropical climate constituting the Scandinavian, Baltic and British regions in which the OLD RED SANDSTONE is developed. The other was the Mediterranean geosyncline region which extends from southwest England through France and Central Europe into the Mediterranean area, and northwest Africa with a branch extending into Russia and the Urals. The Old Red Sandstone region became a land area covered with extensive lakes and shallow seas in which dolomitic limestones were deposited. These strata constitute the DINANTIAN, in which corals and brachipods are abundant. In the S.W. of England as well as in France there is a sandstone-shale facies of the FLYSCH type to which the name CULM FACIES has been given. This shows shales and sandstones of shallow water deposition, containing lacustrine fauna as well as carbonaceous matter. In the Mediterranean geosynclinal region the deposits are mainly shales containing a fairly rich ammonoid fauna.

At the end of the Dinantian period most of the European region became a vast continental area in which swamp and brackish water deposits with shales and coal seams were laid down, occasionally with marine intercalations. The strata are divided into NAMURIAN, WESTPHALIAN and STEPHANIAN. These contain the main coal deposits of Europe and also of Northern Asia and North America.

Fauna and Flora.—The flora of the Carboniferous has been studied in great detail, especially in Europe. The European flora is characterised by *Lepidodendron, Sigillaria* (Lycopodiales), *Calamites, Walchia* (Conifer *Cordaites* and several pteridosperms *Pecopteris, Neuropteris, Alethopteris, Odontopteris,* etc.). It extends westward into North America and eastward in the Urals and beyond. The plants have enabled palaeobotanists to divide the Westphalian and Stephanian into numerous zones. The appearance of *Callipteris* has been taken as indicating the commencement of the Permian. The continental fauna include crustacians (*Estheria, Leia,* etc.), insects, fresh water pelecypods (*Anthracomya, Carbonicola, Naiadites*), amphibians (*Branchiosaurus, Actinodon, Archegosaurus*) etc. and some reptiles (*Mesosaurus, Proterosaurus* etc.). Most of the invertebrate phyla are well represented in the marine fauna. The Fusulinids are characteristic of the Upper Carboniferous and Lower Permian ; *Fusulinella* appear in the Dinantian ; *Fusulina* in the Moscovian ; *Schwagerina* in the Uralian, and *Neo-Schwagerina* and *Sumatrina* in the Lower Permian. The first apperance of *Pseudo-Schwagerina* is now taken as marking the beginning of the Permian era. The corals include *Chaetates, Lithostrotion, Zaphrentis, Amplexus,* etc. which form massive reef limestones in several places. In fact, the corals have been used in England for the division of the Dianantian strata into zones. The crinoids and blastoids are locally abundant. Bryozoa (*Fenestella, Protoretepora*) occur sometimes in large numbers. The trilobites, which are now a declining group, are represented by *Phillipsia*. A few pelecypods (*Aviculopecten, Pecten, Schizodus, Posidonomya*), and gastropods (*Bellerophon, Murchisonia, Euomphalus*) are also important. The brachiopods are very abundant and are represented by a number of genera amongst which may be mentioned *Productus, Chonetes Spirifer, Syringothyris, Athyris,* etc. The Cephalopods (Goniatites) are important as zone fossils in the deep marine facies, and amongst them may be mentioned *Pericyclus, Beyrichoceras, Nomismoceras, Glyphioceras, Homoceras, Reticuloceras, Gastrioceras* and *Schistoceras.* Several of the Carboniferous genera disappeared before the end of the era and were replaced by others in the Permian.

In Russia, the Carboniferous is developed in the Moscow basin, the Ural geosyncline and in the Donetz basin. In the Moscow basin there is an unbroken succession from the Devonian into the Lower Carboniferous which are calcareous and contain *Spirifer medius,* some species of *S. tornacensis* group, *Pericyclus* (goniatite) and *Zaphrentis.* These are overlain by plant bearing beds which also show Dinantian plants. Limestones follow, with *Productus giganteus* and other fossils which show a Namurian age. The Middle Carboniferous is also calcareous, containing some fusulines and *Spirifer mosquensis* (Westphalian). A similar succession is seen in the Urals. The Upper Carboniferous of the Moscow basin, known as the GJEL or GSHEL STAGE, is dolomitic and characterised by S. *Supramosquensis,* followed by strata containing *Schwagerina princeps* and *Macrodon.* In the Urals the Upper Carboniferous is a fine-grained pelagic limestone. Its lower part is the OMPHALOTROCHUS BED containing *Spirifer marcoui,* followed by beds with *Productus cora* and other species. The upper beds contain Schwagerines and a rich Cephalopod fauna including *Agathiceras, Medlicottia, Pronorites* and *Gastrioceras.*

The Carboniferous is very thick in the Donetz basin. The Dinantian is partly marine and partly plant-bearing, containing *Spirifer medius, Productus giganteus* and *Dibunophyllum.* Then come thick shales and sandstones (containing plants) and a few marine intercalations containing *Reticuloceras* and *Gastrioceras,* indicating Visean and Namurian ages. The Westphalian is also similar, characterised by coal beds, plant fossils as well as Fusulines and *Spirifer mosquensis.* The Upper Carboniferous has a lower horizon with *S. mosquensis* and upper beds with *Schwagerina* and Stephanian flora. The top beds contain flora with Permian elements. The Donetz basin is thus a transition area between the marine Carboniferous of the Moscow-Urals region and the coal bearing basins of Europe.

The chief divisions of the Carboniferous and their characteristic fossils are as follows :

DIVISIONS	ZONES	BRACHIOPODS etc.	AMMONOIDS
Stephanian (=Uralian)	St. Etienne Rive de Gier	*Schwagerina* *Sp. Supramosquensis*	
Westphalian (=Moscovian)	D. La Houve C. Bruay B. Anzin A. Vicoigne	*Spirifer mosquensis*	*Anthracoceras* *Gastrioceras*

PLATE IV

DEVONIAN AND CARBONIFEROUS FOSSILS

Explanation of Plate IV

Atrypa reticularis. 2. Rhynchonella subsignata (1.5). 3. Productella baitalensis. 4. rtina heterocycla. 5. Conocardium rhenanum. 6. Phacops latifrons. 7. Euomphalus diatus (1.5). 8. Janeia birmanica (0.7) CARBONIFEROUS (9). Pleurotomaria Worthenia tralensis. 10. Fusulina (Schellwienella) kraffti (4), section. 11. Michelinia mansuyi. 5)., sections. 12. Fenestella eichwaldi (1.5). 13. Lonsdaleia indica (1.5) section. 14. hwagerina fusiformis (4). 15. Spirifer fasciger. 16. Spiriferella rajah. 17. Protoretepora pla enlarged.)

THE PALAEOZOIC GROUP—CAMBRIAN TO CARBONIFEROUS

Namurian	...	Sp. mosquensis (rare) Prod. giganteus	Reticuloceras Eumorphoceras
Visean	... ...	Prod. giganteus	Glyphioceras Beyrichoceras
Tournalsian		Spirifer tornacensis	Pericyclus Gattendorfia
Etroeungt		Sperifer medius	Clymenia Wocklumeria

In India the Carboniferous System is developed in a few places in the Himalayan region and in the Shan States of Burma as well as in Lower Burma. The end of the era was marked by mountain building movements which brought about great changes and initiated marine sedimentation in the Himalayan area and continental sedimentation in the lands of the Southern Hemisphere.

SPITI

Kanawar System.—The Muth Quartzites of Devonian age constitute a conspicuous horizon in the Himalayan area. They are overlain in Spiti by a series of limestones, shales and quartzites, called the KANAWAR SYSTEM, which is subdivided as follows :—

Po Series (600 m)	...	FENESTELLA SHALES THABO STAGE : quartzites and shales with plants
Lipak Series (600 m)	...	Limestones and shales with *Syringothyris cuspidata,Productus,*etc.

Lipak Series.—Named after the Lipak river, and well exposed near the junction of that river with the Sutlej, this series consists mainly of limestones and quartzites with subordinate shales. The limestones are generally hard, dark grey and contain thin siliceous bands. The lower portion contains some corals and brachiopods which have proved difficult to extract from the rocks. In the upper part there are thin shales and limestones which have yielded good fossils among which are :

Brachiopods	... ...	*Procdutus cora, P. semireticulatus, Chonetes hardrensis, Syringothyris cuspidata, Orthotetes* sp. *Derbyia* cf. *senilis, Spirifer kashmiriensis, Strophomena analoga, Reticularia lineata, Athyris roysii, A. subtilita.*
Trilobite	... ...	*Phillipsia* cf. *cliffordi.*
Lamellibranchs	... ...	*Conocardium, Aviculopecten* etc.
Pteropod	... ...	*Conularia quadrisulcata.*

The fossils show that they are of Lower Carboniferous age.

Po Series.—The Po Series, which overlies the Lipak Series, has two subdivisions. The lower, called the THABO STAGE, contains the plant fossils *Rhacopteris ovata* and *Sphenopteridium furcillatum* which are regarded as Lower Carboniferous. The upper, known as FENESTELLA SHALES, because of the richness in some horizons of the bryozoa called *Fenestella*, is of Upper Carboniferous age and contains the following fossils :—

Brachiopods	... ...	*Productus scabriculus, P. undatus, Dielasma* sp., *Spirigera* cf. *gerardi, Spirifer traingularis, Reticularia lineata*

G—30

The Po Series is succeeded by a conglomerate which contains pebbles and boulders of the underlying formations and represents a short stratigraphical break.

KASHMIR

Syringothyris Limestone.—In the Palaeozoic anticline of the Lidar valley in Kashmir, the Muth Quartzites are overlain conformably by thinbedded limestones called the SYRINGOTHYRIS LIMESTONES. They are well exposed at Eishmakum and Kotsu, southeast of Srinagar, and apparently extended further out but are covered by the younger Panjal trap and alluvium. Beds of the same age are found also near Banihal in the Pir Panjal where they attain a thickness of 900 m. The strata are to be correlated with the Lipak Series of Spiti.

Fenestella Shales.—The Syringothyris Limestones are followed by a thickness of 600 m of quartzites and shales the latter being often calcareous. They are exposed in the Lidar valley and near Banihal.

The lowest beds of these are unfossiliferous but above them come shales full of *Fenestella* and also brachipods, corals, etc. The upper beds are mainly quartzites with shale intercalations. The more important fossils in the Fenestella Shales are :

Polyzoa	... ...	*Fenestella* and *Protoretepora ampla*.
Trilobite	... ...	*Phillipsia* sp.
Brachiopods	... ...	*Productus cora, P. scabriculus, P. undatus, P. spitiensis, Spirifer lydekkeri, S. triangularis, S. varuna, Strophalosia, Aulosteges*, etc.
Lamellibranchs	... ...	*Aviculopecten, Modiola, Pecten*, etc.

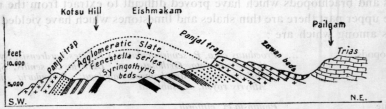

FIG. 5.—SECTION ACROSS THE LIDAR VALLEY ANTICLINE, KASHMIR.
(AFTER C. S. MIDDLEMISS, *Rec.* 40).

The Fenestella Shales are in several places conformably associated with volcanic agglomerates which are semi-contemporaneous with them. Their exact age is somewhat uncertain because of the special characters of its fauna, but it must be somewhere in the Middle or Upper Carboniferous.

CHITRAL

The Devonian rocks are followed conformably by marine Carboniferous strata in Chitral, comprising the Chitral Slates and Sarikol Shales. Amongst them are beds containing *Fusulina* and *Bellerophon*.

BURMA

Northern Shan States

The upper part of the Plateau Limestone is of Carboniferous to Permian age, but as mentioned already, it has not been found possible to subdivide the formation in a satisfactory manner. The Lower Plateau Limestone, which is dolomitic, passes up by perfect gradation into the finely crystalline, grey or blue-grey, calcitic Upper Plateau Limestone. This formation occurs in a number of hills and ridges and contains *Fusulina elongata* and some usually ill-preserved corals and brachiopods in places. Amongst the fossils which have been identified are :

Foraminifera	... ...	*Fusulina elongata*
Anthozoa	... ...	*Lonsdaleia indica, Syringopora* sp. *Zaphrentis* sp.
Bryozoa	... ...	*Fenestella* cf. *perelegans, Hexagonella ramosa, Polypora* cf. *ornata.*
Trilobite	... ...	*Phillipsia* sp.
Brachiopods	... ...	*Spirifer striatus, S. fasciger, Martinia dispar, Reticularia lineata, Spiriferina cristata, Spriigera roysii, Spirigerella derbyi, Schizophoria indica, Oldhamina* sp., *Productus cora, P. cylindricus, P. gratiosus, Chonetes grandicosta, Dielasma biplex, Camarophoria* sp. *Notothyris simplex, Marginifera* sp.
Cephalopoda	... ...	*Xenapsis carbonaria*
Lamellibranchs	... ...	*Pseudomonotis, Pecten, Schizodus*
Gastropods	... ...	*Pleurotomaria, Murchisonia, Holopella, Neritopsis*

The fauna is Permo-Carboniferous and shows some affinity to the Lower and Middle Productus Limestone of the Salt Range.

Southern Shan States

The Upper Plateau Limestone is found also in the Southern Shan States near Taunggyi where it has the same characters as in the north. A fossiliferous locality on the Taunggyi-Loilem road yielded abundant bryozoa but only a few brachiopods. The fauna found includes *Fusulina*, corals, *Productus* and *Lyttonia* which indicate Carboniferous to Permian age.

Tenasserim

The Moulmein limestone beds with associated sands and clays are about 80 m thick and are are found overlying the Mergui Series near Moulmein. They are exposed in the Thampra hill and the hills to its south, and also in the islands of the Mergui Archipelago. They have yielded fossils from north-west of Tharabwin which indicate an Upper Carboniferous age.

Foraminifera	... ...	*Schwagerina oldhami*
Anthozoa	... ...	*Lonsdaleia salinaria, Lithostrotion* sp.
Brachiopods	... ...	*Spirifer* sp., *Productus sumatrensis, Athyris* sp.
Gastropods	... ...	*Bellerophon* sp., *Murchisonia* sp.

TABLE 28.—Rough Correlation of Palaeozoic Strata

PERMIAN	...	Permian of Spiti, Kashmir ; Speckled Sandstone, Olive Series, Productus Limestone ; Zewan Beds, Panjal Trap ; Damuda System, Upper Plateau Limestone ; Krol Series (?)
UPPER CARBONIFEROUS		Talchir, Tanakki, Blaini and other Tillites
CARBONIFEROUS	...	Po and Lipak Series ; Lower Plateau Limestone
DEVONIAN	...	Deovnian of Chitral and Tethys Himalayas ; Muth Series ; Padaukpin Limestone and Wetwin Shales (Burma)
SILURIAN	...	Silurian of Burma, Kashmir and Spiriti (Variegated Series)
ORDOVICIAN	...	Ordovician of the Tethys Himalayas ; Shiala Series ; Strata in Chandragiri Pass (Nepal) ; Bawdwin Volcanics and Tawng Peng Granite (?) ; Ordovician of Shan States
CAMBRIAN TO UPPER-PRE-CAMBRIAN		Cambrian of Tethys Himalays ; Haimanta System ; Garbyang, Ralam and Martoti Series and Budhi Schists ; Cambrian and (?) Saline Series of the Salt Range ; Nagthat, Chandpur and Chail Series ; Jaunsar Series ; Deoban, Shali and Krol (?) Series ; Simla, Dogra, Attock slates ; Daling-Darjeeling, Baxa Sinchula-Jainti (Bhutan) ; ? Mergui Series. Vindhyan and Kurnool Systems.

UNFOSSILIFEROUS PALAEOZOIC STRATA

KASHMIR-HAZARA

Tanawal Series.—A formation of considerable thickness, composed of phyllites, quartzites, quartz-schists and conglomerates, occurs in a number of places in western Kashmir and Hazara. The rocks are more or less metamorphosed and folded up with the Dogra Slate Series. The quartzites, however, appear to be silicified Infra-Trias limestones or perhaps the equivalents of the Muth Quartzites. They are entirely unfossiliferous. In the Pir Panjal, members of this series have been observed to pass laterally into the Agglomeratic Slates. Hence, though their age is not known with certainty, they may represent a part of the gap between the Cambrian and Upper Carboniferous. The Tanawal Series is overlain by the Tanakki Conglomerate which is now regarded as the equivalent of the Talchir Boulder Bed and the Blaini Boulder Bed.

SIMLA-GARHWAL

Jaunsar Series.—In the Simla-Garhwal region the Palaeozoic is probably represented by the Jaunsar Series, which is an unfossiliferous assemblage of slates, sandstones and quartzites, resting on the Simla Slate Series. The Jaunsars of Garhwal are divided as below :

Jaunsar Series	NAGTHAT—Purple and green phyllites, quartzites, sandstones and conglomerates
	CHANDPUR — Phyllites, schists, banded quartzites, tuffs and lavas
	MANDHALI — Limestones, slates, phyllites, grits and boulder bed

THE PALAEOZOIC GROUP—CAMBRIAN TO CARBONIFEROUS

The Nagthats bear resemblance to the Tanawals of Hazara, and the Chandpurs to the Chails and part of the Dalings. Farther east in Garhwal, the Chandpurs rest on the Barahat Series, consisting of quartzites, limestones and some lavas, which may be the equivalents of the Nagthats, since the line of separation is a thrust fault.

The Jaunsars have been affected in some places by tectonic disturbances of pre-Krol (*i.e.*, Pre-Permian) age, as a result of which they have acquired the Aravalli strike direction. It is apparent that the rocks of Rajasthan and of Garhwal have been affected by the same post-Vindhyan tectonic movement.

It would be convenient at this stage to review the salient points in the Palaeozoic (pre-Upper Carboniferous) stratigraphy of the Indian region. During the Vindhyan times (presumably partly uppermost Pre-Cambrian and partly Cambrian) there were three major marine areas in the Peninsular part, the first in the Cuddapah basin, the second in the Chhattisgarh basin and the third forming the great Vindhyan basin. A part of the Vindhyan basin may underlie the younger strata in the Ganges Valley of Bihar. Its northern part has become involved in the Sub-Himalayan region where the strata have yet to be untangled from the confused thrust sheets in which rocks of other ages are also found. The representatives of the Vindhyans may be the Chandpurs and Nagthats. We have no information at all whether any part of the Palaeozoic is represented here, for the formations contain no fossils so far as known, and the succeeding Infra-Krol and Krol rocks are tentatively regarded as Permian or Mesozoic. The great Vindhyan basin was quite distinct from the northern open ocean which then occupied the region of the Central Himalayas and it probably dried up during the Cambrian. The Salt Range area may represent the northwestern edge of this basin. Lower Palaeozoic rocks are not present in Hazara while the earliest rocks exposed in the Baluchistan area are of Permo-Carboniferous age.

In the Sub-Himalayan and Central Himalayan regions of Kumaon, between Almora and Rakshas Lake, Heim and Gansser have distinguished the following formations. Of these, the unfossiliferous or poorly fossiliferous rocks occur below the Central Himalayan thrust sheet and the fossiliferous rocks of the Tethyan Facies above the thrust :—

MUTH SERIES	...	Brown quartzites with dolomite layers, capped by white quartzites. A special facies of this is the crinoidal limestone of Lipu-Lekh — *Silurian to Devonian.*
VARIEGATED SERIES	...	Repetition of marls and shales of red, green, grey colours with layers of limestone — *Silurian.*
SHIALA SERIES ...	...	Calcareous sandstones and thin layers of lenticular limestone with crinoid fragments — *Ordovician*
GARBYANG SERIES	...	Slaty phyllites, Calcareous sandstones, sandy and argillaceous dolomites — *Cambrian*
RALAM SERIES ...	...	Basal conglomerates, greywackes, orange coloured dolomites — *Lower Cambrian* and possibly *partly Pre-Cambrian*
MARTOLI SERIES	...	Phyllites, Calcareous phyllites, quartzites ; probably equivalent to the *Algonkian*

The Central Himalayan zone was occupied by an open ocean in which fossiliferous Lower Palaeozoic strata were deposited such as are known in

Northern Kumaon, Spiti, Kashmir and Chitral, extending into the Hindukush and Northern Iran. The Shan area of Burma did not belong to the above geological province but to another which included Yunnan, Southern and Southwestern China and Indo-China. There does not seem to have been any intermingling of fauna between these two basins, at least until the Devonian. Much more detailed work will have to be done in the Himalayas before a clearer picture can be obtained of the distribution of the strata of various ages and of the palaeogeography of the earlier geological periods.

SELECTED BIBLIOGRAPHY

Christie, W. A. K. Salt deposits of Cis-Indus Salt Range. *Rec.* **44**, 241-262, 1914.
Cotter, G. de P. Presidential Address to the Geology Section, *Proc.* 18th *Indian Science Congress*, 1931.
Cotter, G. de. P. Geology of part of the Attock district west of Long. 73° 45′ *Mem.* **55**(2), 1933.
Davies, L. M. Notes on the geology of Kohat. *Jour. A.S.B.* **20**, 207, 1926.
Fox, C.S. Contributions to the geology of the Punjab Salt Range. *Rec.* **61**, 147-179, 1928.
Gee, E. R. Saline Series of N.W. India, *Curr. Sci.* II, 460-463, 1934.
Griesbach, C. L. Geology of the Central Himalayas. *Mem.* **23**, 1891.
Hayden, H. H. Geology of Spiti. *Mem.* **36**(1), 1904.
Heim, A. and Gansser, A. Central Himalaya. *Denkschr. Naturf, Gesell*, Zurich, **73**(1), 1936.
Holland, T.H. General report of the G.S.I. for 1902-03, p. 26, 1903.
La Touche, T. D. Geology of Northern Shan States. *Mem.* **39**, (2), 1913.
Middlemiss, C.S. Geology of the Salt Range with a reconsidered theory of origin and age of the Salt Marl. *Rec.* **24**, 19-41, 1891.
Pascoe, E. H. Petroluem in Punjab and N.W.F.P. *Mem.* **40**(3), 1920.
Pinfold, E.S. Notes on the stratigraphy and structure of N.W. Punjab. *Rec.* **49**, 137-160, 1918.
Stuart, M. Potash Salts of the Salt Range and Kohat. *Rec.* **50**, 28-56, 1919.
West, W. D. Some recent advances in Indian Geology. *Curr. Sci.* III, 412-416, 1934.
Wynne, A. B. Geology of the Salt Range in the Punjab. *Mem.* **14**, 1878.
Wynne, A. B. Trans-Indus Salt region. *Mem.* **11**(2), 1875 ; *Mem.* **17**(2), 1880.
───── Symposium on the Saline Series. *Proc. Nat. Acad. Sci.* XIV (6), 1944; XVI (2-4), 1946.

FOSSILS

Redlich, K. Cambrian fauna of the Salt Range. *Pal. Ind.* N.S. 1 (I), 1899.
Reed F. R. C. Cambrian fossils of Spiti. *Pal. Ind.* Ser. XV, Vol. VII (1), 1910.
Reed, F. R. C. Ordovician and Silurian fossils of the Central Himalayas. *Pal. Ind.* Ser. XV, Vol. VII(2), 1912 ; N.S. VI (1), 1915.
Reed, F. R. C. Cambrian and Ordovidian fossils from Kashmir. *Pal. Ind.* N.S. XXI (2), 1934.
Reed, F. R. C. Lower Palaeozoic fossils of N. Shan States. *Pal. Ind.* N.S. II (3), 1906.
Reed, F.R.C. Ordovician and Silurian fossils of N. Shan States. *Pal. Ind.* N.S. VI (1), 1919.
Reed, F. R. C. Lower Palaeozoic fauna of S. Shan States, *Pal. Ind.* N.S.XXII (3), 1936.
Reed, F.R.C. Devonian faunas of N. Shan States, *Pal. Ind.* N.S. II (5), 1906.
Reed, F. R. C. Devonian fossils from Chitral and the Pamirs. *Pal. Ind.* N.S. VI (2), 1922.

CHAPTER X

THE GONDWANA GROUP

Introduction

After the deposition of the Vindhyan rocks and their uplift, there was a great hiatus in the stratigraphical history of the Peninsula. At the end of the Palaeozoic Era, *i.e.*, towards the Upper Carboniferous, a series of changes took place which brought about a redistribution of the land and sea and which was also responsible for the mountain-building movements called the *Hercynian* or *Variscan*. At this time there existed a great Southern Continent or a series of land masses which were connected closely enough to permit the free distribution of terrestrial fauna and flora. This continent, which included India, Australia, South America, Antarctica, South Africa and Madagascar, has been called *Gondwanaland*. It shows evidence of the prevalence of the same climatic conditions and distribution of the same type of deposits from the Upper Carboniferous to the Jurassic. The era with which we are now concerned began with a glacial climate for we find the deposits commencing with a glacial boulder-bed which has been recognised in all the above-mentioned lands. The bulk of the strata which followed the glacial conditions was laid down as a thick series of fluviatile or lacustrine deposits with intercalated plant remains which ultimately formed coal seams. The basins of deposition must have been shallow, and slowly oscillating for we find each cycle of deposition starting with coarse sandstones and proceeding through shales to coal seams. The plant remains embedded in these sediments have remarkably close affinities in all the lands mentioned, and comprise *Glossopteris, Gangamopteris, Neuropteridium, Gondwanidium,* etc. This floral assemblage, called the 'Glossopteris flora,' is very characteristic of the deposits of the lower part of this group. The amphibian and reptilian faunas of this era are strikingly similar and point to unrestricted inter-migration. For instance, according to Prof. Von Huene, there is an extraordinary resemblance between the Dinosaurs of Madhya Pradesh in India and those of Madagascar, Brazil, Uruguay and Argentina. This Gondwana continent seems to have persisted through the greater part of the Mesozoic era and to have been broken up during the Cretaceous, either by the sinking of marginal strips or by the drifting apart of the component parts. The close faunal and floral affinities of the Gondwana strata in India and the Southern continents will become apparent when we deal with their equivalents in Africa, South America and Australia.

NOMENCLATURE AND EXTENT

The name Gondwana was introduced by H.B. Medlicott in 1872 in a manuscript report, but appeared for the first time in print in a paper by O. Feistmantel published in 1876 (*Rec.* IX, Pt. 2, p. 28). It is derived from the kingdom of the Gonds, a great and ancient tribe who still inhabit the Central Provinces (Madhya Pradesh) where these formations were studied by Medlicott.

The name has also been extended to the large continent which existed in the uppermost Palaeozoic and the Mezosoic times in the Southern Hemisphere.

The rocks forming this Gondwana group are of fluviatile or lacustrine nature and were deposited in a series of large river or lake basins which later sank along trough-faults amidst the ancient rocks. It is to this faulting that we owe the preservation of the Gondwana strata with their rich coal seams in this country.

Distribution.—The Gondwana rocks are mainly developed along two sides of a great triangular area, the third side of which is formed by the northern part of the east coast of the Peninsula, *i.e.*, from the Godavari valley to the Rajmahal hills. The northern side of this corresponds roughly to the Damodar, Sone and Narmada valleys, trending nearly E.-W. while the southwestern side runs along the Godavari valley with a N.W.-S.E. trend. In the interior of this triangle is a subsidiary belt along the Mahanadi valley. These long and narrow tracts are really a series of faulted troughs. Other groups of exposures are found along the Himalayan foot-hills of Nepal, Bhutan and Assam, and also in Kashmir. Further, Upper Gondwana rocks are seen in a series of detached outcrops along the east coast of India, between Cuttack and Cape Comorin, in the Rajmahal hills, Madhya Pradesh, Rewa, Saurashtra, Kutch, and also in Ceylon.

Division of the Gondwanas.—Though the Gondwanas are generally referred to as a System, their extent and magnitude in space and time entitle them to be considered as a major Group. They span the time from the Upper Carboniferous to the Jurassic or Middle Cretaceous and comprise strata whose total thickness is from 6,000 to 7,000 m. At least one of their major divisions has the status of a geological "System," which can be separated into series and stages. It is therefore appropriate to refer to the Gondwanas as a 'Group' as in the case of the standard Palaeozoic or Mesozoic Groups.

The Gondwana group is divided into two major divisions based mainly on palaeontological evidence. This two-fold classification has been strongly supported by C. S. Fox in his monograph on the coalfields of India. The line of divisions is taken as above the Panchet Series, the lower portion being characterised by the Glossopteries flora and the upper by the Ptilophyllum (Rajmahal) flora. There is also some evidence that the faulting and folding of the Gondwana basins took place mainly in post-Permian and Mahadeva times.

A tripartite classification was suggested by Feistmantel and followed by E. Vredenburg in his 'Summary of the Geology of India' (1910, p. 50), where Lower, Middle and Upper Gondwanas are shown as equivalent to the Permian, Triassic and Jurassic systems of Europe. In support of this is the intervention of arid continental deposits containing Triassic reptiles and amphibians in the middle formations of the Gondwanas, the beds above and below them indicating more normal conditions. The base of the Panchet Series is thought to correspond roughly to the base of the Trias. Though these factors make the tripartite division plausible, the evidence of the flora is entirely in favour of a two-fold division. Disconformities and stratigraphic gaps separate the Panchets and Mahadevas from each other and from the beds above and below them, but the magnitude of the gaps is not known. Arkell has suggested that the whole of the Jurassic may be missing. Though the details of the stratigraphy of the different Gondwana areas in India differ to some extent, it is very clear that the Glossopteris flora practically died out during the Panchet and was replaced completely by the more advanced *Thinnfeldia-Ptilophyllum* flora which characterises the Upper Gondwana times.

Sub-divisions of the Gondwanas.—The Lower as well as the Upper Gondwanas have each been divided into three or more series of formations. In the ascending order they are the Talchir, Damuda (Damodar) and Panchet in the lower division, and the Mahadeva, Rajmahal, Jabalpur and Umia series in the upper division. Each series consists of stages which have received different names in different areas. Table 29 shows at a glance the correlation of these stages and their position in standard stratigraphical scale.

TALCHIR SERIES

Tillite.—The lowest series of this group is named after the coalfield and the former State of Talcher (Talchir) in Orissa, where it was first studied. Its lowest member is a tillite or boulder-bed, which is succeeded by shales and sandstones. This boulder-bed forms a conspicuous and characteristic datum line in the geology of the Peninsula, and is in general 15 to 60 m thick. Its presumed equivalents are found in the Salt Range, in Kashmir-Hazara (Tanakki boulder-bed), in Garhwal (Mandhali beds) and in Simla (Blaini boulder-bed). The Boulder-bed consists of an unsorted mixture of boulders, pebbles, rock fragments and clay, the boulders often showing facets and striae of glacial origin. In most of the peninsular occurrences there is evidence, according to C.S. Fox, of this bed being to some extent a resorted glacial deposit, whereas in the Salt Range the ice carrying the moraines seems to have deposited the materials directly in a shallow sea. The Salt Range occurrence contains boulders and pebbles of granite, Malani rhyolite, etc., which have apparently been derived from the region of the Kirana hills about 80 km to the south and also from some parts of Rajasthan. At Pokharan (27° 0' : 71° 55') near the western border of Jodhpur and at Bap (27° 22' : 72° 24') about 65 km farther to the north-east, there are certain dark sandstones overlain by boulder-beds containing ice-scratched boulders. The Bap beds contain boulders of Vindhyan limestone and therefore assigned to the age of Talchir Tillite. The Pokharan boulder-bed may probably be much earlier *i.e.*, Vindhyan or Pre-Cambrian. Boulder-beds of the same age have been recorded also to occur in the Kosi valley near Barahakshetra, in the Lachi Series of N. Sikkim (Lachi—28° 1' : 88° 45') and in the Sub-Himalayan region as far east as E. longitude 96°. In these regions the Gondwanas are generally overthrust by the Siwalik rocks.

Shales and Sandstones.—The Talchir boulder-bed is overlain by shales and these in turn by sandstones, the total thickness of these being 150 to 200 m. The shales are greenish in colour and usually break up into thin pencil-like or prismatic fragments, for which reason they are often called 'needle-shales'.

TABLE 29—GONDWANA SUCCESSION IN INDIA

Cretaceous	L	UMIA JABALPUR CHUGAN	Satyavedu-Tirupati-Pavalur Sriperumbudur-Raghavapuram-Vemavaram Golapalli-Budavada
	U	RAJMAHAL (? Chikiala-Kota)	
Jurassic	M L	HIATUS (?)	
Triassic	Rh U	MAHADEVA	Yerrapalli-Maleri, Bagra-Denwa, Tiki, Parsora, Dubrajpur
	M		Pachmarhi, Yerrapalli Panchet
Permian	U	RANIGANJ	Kamthi ; Chintalpudi ; Himgir ; Motur ; Bijori ; Daigaon ; Pali
	M	BARREN MEASURES	Iron stone shales ; ? Motur
	L	BARAKAR KARHARBARI	
Carboniferous		TALCHIR	
	U	TALCHIR-TILLITE	

The shales are arenaceous, micaceous or calcareous, and sometimes grade into sandstones which are also generally greenish or greenish-brown in colour. The sandstones generally contain grains of undecomposed feldspar which furnish evidence of very cold conditions of deposition. The shales and sandstones may be intercalated with each other but the latter are more common in the upper part, indicating a general coarsening of the sediments as deposition went on. In the later part of the Talchir times, however, the glacial conditions seem to have given place to milder climate as evidenced by the presence of fossil plants.

Distribution.—The Talchir beds are found in most of the Lower Gondwana areas of the Peninsula in the faulted troughs, and also sometimes as outliers on the gneisses of the neighbouring regions. It is thought that the deposition of a series of moraines in the early Talchir age was responsible for the formation of a number of more or less connected lakes and swamps which received the sediments of the succeeding (Damuda) age.

Talchir fossils.—The plant fossils of the Talchirs occur only in the upper part and show an assemblage which is fairly distinct from that of the Damuda (Damodar) stage, but closely allied to that of the immediately succeeding Karharbari beds which are considered to form the lower portion of the Barakars and which are well represented in the Giridih coalfield. The important localities of Talchir fossils are Karaon (Deoghur coalfield), Rikba (Karanpura), Latihar (Auranga), Nawadih (Hutar), Behia-Baragaon, 30 km northwest of Anukpur (Sohagpur), Goraia (south of Pali in Rewa) and Kuppa (Sonada).

At Rikba and other places the following fossils have been found :—

Pteridospermae	...	*Glossopteris indica, G. communis, G.* sp., *Gangamopteris cyclopteroides* var. *major, G. cyclopteroides, G. angustifolia, G. buriadica, Vertebraria indica* (which is now believed to be the rhizome of *Glossopteris*)
Cordaitales	... ...	*Noeggerathiopsis hislopi*
Incertae	... ...	*Samaropsis* sp. (seeds)

In the Punjab Salt Range the Talchir is represented by the boulder-bed and shales which lie above the Salt Pseudomorph zone. Near Kathwai, from the shales immediately overlying the boulder-bed, several types of spores and cuticles have been discovered. The following fossils were also obtained from the same locality, about 8 m above the boulder-bed.

Equisetales	... ...	*Schizoneura* sp.
Filicales ?	... ...	*Alethopteris* sp.
Pteridospermae	...	*Glossopteris communis, G. indica, Gangamopteris buriadica, Vertebraria indica, Noeggerathiopsis (Cordaites)* sp.
Inertae	...	*Ottokaria Kathwaiensis, Samaropsis emarginata, Cordaicarpus sahnii.*

The Blaini boulder-bed of the Simla Hills, the Tanakki boulder-bed of Kashmir-Hazara and the Mandhali beds of Tehri-Garhwal, the boulder-beds of Lachi and at a few places in the Himalayan foot-hills of Assam are also considered to be their equivalents. The location of these may be taken as indicating the proximity of the northern shore-line of the Gondwanaland, but allowance must be made for their overthrust which should have carried them some distance south of their original location.

MARINE PERMO-CARBONIFEROUS

Umaria.—An interesting discovery made by K. P. Sinor some years ago is the occurrence of a 3 - m thick band of highly fossiliferous marine sandstones and clays lying on the boulder-bed in a railway cutting west of Umaria. More recently, Ghosh (Science and Culture, XIX, p. 620, 1954) found a simlar marine bed at Anukpur and Manendragarh some distance east of Umaria but it is intercalated with boulder-beds. The Umaria marine bed overlies the Talehir boulder-bed but passes upwards witaout any visible breack into the overlying Barakar rocks. It contains four thin horizons packed with fossils, comprising only a few genera — *Productus, Spiriferina, Reticularia* and a few small gastropods, the first of these being the most common. The fossils are stunted and probably represent the remnants of a marine fauna whose habitat was gradually becoming fresh water by inundation from the rivers flowing into this arm of the sea from the surrounding land. The fauna shows, according to Cowper Reed, an admixture of characters of Carboniferous and Permian age but the species are all new, with highly individual characters. From the nature of the fauna it is thought that this area was connected with the sea of the Salt Range and that the age of the beds more or less corresponds to the Karharbari stage of Giridih, to the Eurydesma and Conularia beds of the Salt Range and to the Gangamopteris beds of Kashmir. The more important fossils in these are :—

Brachiopods ... ... *Productus umariensis* (abundant), *P. rewahensis, Spirifer narsarhensis, Reticularia barakarensis, Athyris* aff.; *protea*
Gastropods ... ... *Pleurotomaria umariensis*
Also fish remains and crinoid stems

Manendragarh.—Near Manendragarh in the Hasdo valley, about 150 km E.S.E. of Umaria, there is another exposure of marine rocks containing the following fossils — *Spirifer hasdoensis, Protoretepora ampla, Eurydesma hobartens, E. playfordi* (ana other spp.), *Aviculopecten squamulifer, Euomphalus* sp. etc. The fossils from this locality and Umaria are closely allied to those of the Lyons group in the N.W. Basin of Australia which are assigned a basal Permian age (Sahni and Dutt, 1962). The Australian and Indian areas of deposition may have been connected by shallow water, probably through the area of the present Mahanadi or Godavari trough.

Daltonganj.—A find of *Fenestella* of Permo-Carboniferous age has been reported from a 3-metre-thick band in a stream exposure of Talchir beds near Rajhara Colliery (22° 10′ : 84° 4′) in the Daltonganj coalfield, Bihar. This is in the Damodar-Sone valley and the marine incursion may have come from the Himalayan area.

Rajasthan.—Palaeozoic rocks are exposed in the Jodhpur-Bikaner area in the north and in Jaisalmer in the south. In the Jodhpur area, the PHALODI SANDSTONES, which are the equivalents of the Upper Vindhyan, overlie the basement. They are red current-bedded sandstones with minor clay and gypsum bands. They are overlain by the grey, siliceous dolomites with chert bands and nodules called the Phalodi Limestones which are identical with and lie west of the BILARA FORMATION, well exposed near Sojat and Odania. It is believed to be Lower Palaeozoic. In the Jaisalmer area the Phalodi Sandstones are apparently not represented while the equivalent of the Bilara is the BIRMANIA FORMATION, called after Birmania (26° 10′ : 70° 58′) which is 60 km. south of Jaisalmer. The BADHAURA FORMATION follows over a large unconformity. It consists of marine calcareous sandstones with brachiopods in the lower part and of yellow grits and brown clays with molluscs and cephalopods in the upper part and is Permo-Carboniverous in age. At Pugal, 20 km northeast of Badhaura, a borehole went through 100 m of shales which are unfossiliferous but contain pollen indicating Permian age. Another unconformity separates the Badhaura from the overlying MAYAKOR FORMATION, which is identical with the Lathi Formation of Jurassic age. It consists of 10 m of yellow-brown to black sandstones containing fossil wood. The succession in Rajasthan is shown below.

MAYAKOR FORMATION (LATHI) : 10 m of yellow-brown and black sandstones with fossil wood —Jurassic

BADHAURA FORMATION : 20 m Yellow sandstones and brown clays with mollusca and cephalopods in the upper part ; calcareous sandstones with *Spirifer, Productus, Derbyia, Straparollus*, etc. in the lower part —Permo-Carboniferous

BILARA FORMATION (BIRMANIA) : 90 m Grey dolomite with chert layers and nodules — ? Palaeozoic

PHALODI SANDSTONE (JODHPUR) : 50 m Red current-bedded sandstones and clay bands with gypsum —? Vindhyan

DAMUDA SYSTEM

These formations attain a considerable thickness and are of great economic importance. They comprise four series *viz.*, Karharbari, Barakar, Barren Measures and Raniganj. This System takes its name from the Damodar river (a tributary of the Hooghly river which flows through the Bokaro, Jharia and Raniganj fields) and is the most extensive and best developed division of the Gondwanas.

Karharbari Series.—Above the Talchir Series there is a distinct unconformity which is succeeded by the Karharbari beds in the Giridih coalfield, the name being derived from a village in this area. Here it forms the lower portion of the coal-bearing Barakar rocks and is separated from the typical Barakars by a thickness of barren sandstone. Though forming a distinct stage as decided on palaeobotanical evidence, this is stratigraphically more allied to the overlying Barakars than to the Talchirs.

This series consists of pebbly grits and sandstones which attain a thickness of 60 to 120 m and contain intercalated coal seams two of which are important and are being worked. It has been recognised in the Karanpura, Hutar, Daltonganj, Umaria, Mohpani and Shahpur fields.

The fossils found in this series are :—

Equisetales	*Schizoneura gondwanensis, S. wardi*
Pteridospermae	*Glossopteris indica, G. decipiens, G. longicaulis, Gangamopteris cyclopteroides,* var. *major, G. angustifolia, G. buriadica, Vertebraria indica, Gondwanidium (Neuropteridium) validum*
Cordaitales	*Noeggerathiopsis hislopi, N. stoliczkana, N. whittiana*
Coniferales	*Buriadia sewardi, Moranocladus oldhami*
Incertae	*Callipteridium* sp., *Ottokaria bengalensis, Arberia indica, Samaropsis milleri, S. raniganjensis, Cordaicarpus indicus*

The Gangamopteris beds at Khunmu and Nagmarg in Kashmir contain the Amphibians *Archegosaurus ornatus, Lysipterygium deterrai, Actinodon resinensis* and the ganoid fishes *Amblypterus Kashmirensis* and *A. symmetricus* and also the following plants :—

Pteridospermae	*Glossopteris indica, Gangamopteris kashmirensis, Vertebraria indica*
Cordaitales	*Noeggerathiopsis hislopi*
Ginkgoales ?	*Psygmophyllum haydeni*

These beds are intercalated with pyroclastics and are overlain by the Panjal traps. The Gangamopteris beds of Golabgarh Pass and Marahom are overlain by the marine Zewan beds of Middle Permian age.

BARAKAR SERIES

The name is derived from the Barakar river which cuts across this stage in the Raniganj coalfield. It consists of a thickness of 750 m of white to fawn coloured sandstones and grits with occasional conglomerates and beds of shale in the Jharia coalfield. The sandstones often contain more or less decomposed feldspars. Because of their uneven hardness, the sandstones weather with a rough surface and produce potholes in stream-beds. This stage contains much carbonaceous matter in the form of streaks, lenticles and seams of coal. In the Jharia coalfield, the Barakars include at least 24 seams of coal, each more than 1.2 m in thickness, and it has been calculated that over 75 m of coal are present in the toal thickness of some 600 m of strata.

This is the chief coal-bearing stage in practically all the Lower Gondwana areas of India, including **Darjeeling, Buxa Duars** and **Abor Hills**. In the last mentioned place the base of the Barakars shows intercalations of marine beds containing anthracolithic fauna. The Barakars consist of sandstones with false-bedding, shales and coal-seams which appear in this order and are repeated over and over again. The sandstones sometimes contain trunks of trees but generally they lie flat. The Barakar seams are best developed in the Jharia coalfield, where the ratio of the thickness of coal to that of the strata is as high as 1 : 8. Occasionally very thick seams occur, such as the **Kargali** seam of Bokaro and the **Korba** seam of Hasdo valley each of which is about 100 feet thick. In several cases the coal seams are associated with beds of fire-clay.

The Barakars seem to have been laid down in a series of large shallow lakes probably connected by streams. The coal appears to be due to the accumulation of large amounts of debris of terrestrial plants accumulated under quiescent and stagnant conditions. Though coal is so abundant in the Barakar strata, plant fossils are found only in some localities and animal fossils seem to be rare.

Among the more important fossil plants in the Barakars are :

Equisetales	...	*Schizoneura gondwanensis, S. wardi, Phyllotheca griesbachi*
Sphenophyllales	...	*Sphenophyllum speciosum*
Pteridospermae	...	*Glossopteris indica, Glossopteris communis, G. ampla, G. Retifera, Gangamopteris cyclopteroides* (only in the lower beds), *Sphenopteris polymorpha*
Cycadophyta	...	*Taeniopteris danaeoides, T. feddeni, Pseudoctenis balli.*
Cordaitales	...	*Noeggerathiopsis hislopi, N. whittiana, Dadoxylon indicum*
Ginkgoales ?	...	*Rhipidopsis ginkgoides*
Incertae	...	*Barakaria dichotoma, Dictyopteridum sporiferum, Cordaicarpus* cf. *cordai*

The Barakars are found at several places along the Sub-Himalayan zone in Nepal, Sikkim, Bhutan and the Assam Himalaya. In the Darjeeling area they are found, with occasional coal seams, at Pankabari and other places. A glacial boulder bed has been noted at Tindharia at the base of the Gondwanas.

In the Rangit valley of Sikkim, plant-bearing Lower Gondwanas with pebble beds below and coal seams above have been discovered at Naya Bazar, Khemgaon and other places. The plants include *Glossopteris, Vertebraria* and *Schizoneura*. In the same area poorly preserved Spiriferids, *Fenestella*, etc. have been found in some exposures between the Rangit and Tista rivers (A.M.N. Ghosh—Preliminary Notes on Rangit Valley Coalfield, Sikkim. *Ind. Minerals*, 6, No. 3, 1953). Upper Carboniferous and Permian marine fossils have been collected by Wager in the Lachi ridge. The former include *Productus, Athyris, Spirigerella*, etc., while the latter show *Waagenoconcha purdoni, Camarotoechia* sp., *Syringothyris lydekkeri, Fenestrellina*, etc. Between the two fossiliferous beds are 200 m of pebble beds consisting of pebbles and coarse angular grains of quartz and feldspar in a fine siliceous silt. These are thought to represent the Talchir boulderbeds. Maclaren found bluish limestone boulders containing Permo-Carboniferous marine fossils at the mouth of the Subansiri gorge. These yielded *Productus, Spirifer, Dielasma, Reticularia, Chonetes*, etc., and are the equivalents of the Kuling Shales of Spiti. The source of this limestone has now been ascertained by B. Laskar to be in the Ranganadi basin about 32 km S.W. of the above locality just north of the Tertiary belt. The Gondwana rocks are overthrust by the Miri quartzites (Purana ?) and by rocks resembling the Buxa Series.

In the Sireng river valley in the Abor Hills also similar limestones have been found. These may be continuous with the exposures in the Dikrang valley to the southwest. Lower Gondwanas with the characteristic plant fossils, sometimes with crushed coal, occur in the Sela Agency where they lie over the Tipams and are overridden by Daling schists ; in the Kala Pani and the Bor Nadi in Bhutan and in the Aka Hills north of Tezpur and in the Bharati valley.

The Barakars in the Himalayan foot-hills are generally thrust over the Siwaliks or other Upper Tertiary (*e.g.*, Tipam) sediments, and are in turn overridden by more ancient rocks such as the Buxa and Daling Series.

BARREN MEASURES (IRONSTONE SHALES)

The Barren Measures, which intervene between the Barakar and Raniganj series in the Jharia coalfield, are about 600 m thick, being entirely barren of coal seams, but containing streaks of carbonaceous matter. They consist mostly of sandstones, which are somewhat less coarse than the Barakar type. They are represented in the Raniganj coalfield by the Ironstone Shales whose thickness is about 420 m. Their representatives are thinner still in the Karanpura fields and farther west. They consist generally of carbonaceous shales with clay-ironstone nodules which are sideritic at depth, but when oxidised at and near the surface become limonitic. These are, in places, rich enough to form workable iron-ore which was used in the blast furnaces of the Bengal Iron Co. (since amalgamated with the Indian Iron & Steel Co.) situated at Kulti. The ironstone contains about 35-40 per cent iron. The Barren Measures are seen in the Jharia and Karanpura fields but when followed in the coal-fields farther west, they merge into the overlying Raniganj Series which are also barren of coal seams.

The fossils plants found in the Barren Measures are :

Lycopodiales	...	*Bothrodendron* sp.
Pteridospermae		*Glossopteris indica, G. ampla, Gangamopteris cyclopteroides*
Cordaitales	... ...	*Noeggerathiopsis hislopi*

Motur Stage.—In the Satpura area the Barren Measures seem to be represented by the Motur Stage which is also devoid of coal. This consists of white sandstones with intercalated layers of red, yellow and carbonaceous shales as in the Pench valley. In the Tawa valley of the Betul district, the Moturs do not contain red clays but show brownish and greenish sandstones and buff to greenish buff clays which are often calcareous. In South Rewa, the beds occurring between the Barakars and the Pali and Daigon beds (Raniganj age) are apparently to be referred to the Barren Measures.

RANIGANJ SERIES

This is typically developed in the Raniganj coalfield where it attains a thickens of over 1000 m. It is of about the same thickness in the Satpura area where it is known as the Bijori Stage, but is thinner (570 m) in the Jharia coalfield. In the type area it consists of sandstones, shales and coal-seams, the sandstones being definitely finer grained than those of the Barakar Series. Valuable coal-seams occur in these strata only in the Raniganj coalfield. The coal is higher in volatiles and moisture than the Barakar coal, and there are certain seams, like the Dishergarh, Poniati and Sanctoria seams, which yield excellent, long flame, steam coals.

Fossil wood (*Dadoxylon*) has been found in the upper part of this Series both in the Raniganj and Jharia fields and in the Motur beds of the Pench and Tawa valleys.

Typical fossils of the Raniganj Series are :

Equisetales ...	... *Schizoneura gondwanensis, Phyllotheca indica*
Sphenophyllales	... *Sphenophyllum speciosum*
Filicales ? ...	... *Alethopteris roylei*
Pteridospermae	... *Glossopteris indica, G. communis, G. browniana, G. retifera, G. angustifolia, G. stricta, G. tortuosa, G. formosa, G. divergens, G. conspicua, Gangamopteris whittiana, Vertebraria indica, Sphenopteris hughesi, S. plymorpha, Pecopteris phegopteroids*
Cycadophyta ...	... *Taeniopteris danaeoides, T. feddeni*
Cordaitales ...	... *Noeggerathiopsis hislopi*
Ginkgoales ? ...	... *Rhipidopsis densinervis*
Coniferales ...	... *Buriadia* (*Voltzia*) *sewardi*
Incertae ...	... *Palaeovittaria kurzi, Belemnopteris woodmasoniana, Dictyopteridium sporiferum,* (*Actinopteris*) *bengalensis, Samaropsis raniganjensis*

The Raniganj Series is represented by the Bijori Stage in the Satpuras ; by the Kamthi beds of Nagpur and the Wardha valley in Chanda, the Pali beds in South Rewa ; the Himgir beds in the Mahanadi and Brahmani valleys ; the Almond beds occurring just south of the Pachmarhi scarp ; and the Chintalpudi sandstones of the Godavari valley.

The Kamthi Beds (named after Kamptee near Nagpur) comprise red and grey argillaceous sandstones and conglomerates with interstratified shales. The beds contain patches and nodules of ferruginous material. They extend down into the Wardha-Godavari valley where it is difficult to separate them from the lithologically similar Upper Gondwanas, and where this facies of rocks may include both Raniganj and Panchet Series. Their lithology has led to the confusion of correlating them with the Pachmarhi sandstones and Supra-Panchets. They contain only impressions of plants and practically no carbonaceous matter. The fossils found in the Kamthis are :— Equisetales :

PLATE V

GONDWANA PLANTS

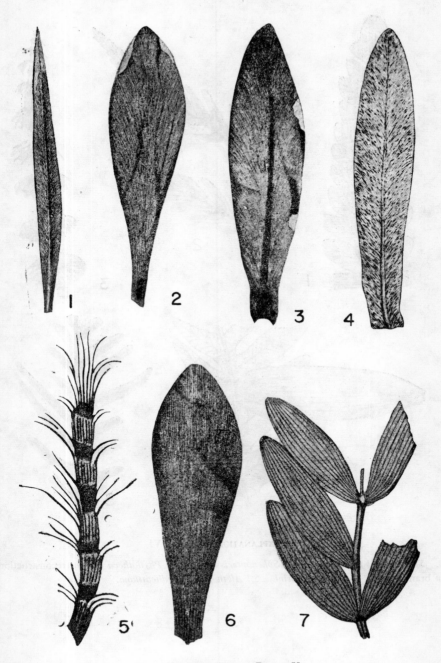

EXPLANATION OF PLATE V

1. *Glossopteris angustifolia.* 2. *Gangamopteris major.* 3. *Gangamopteris cyclopteroides.* 4. *Glossopteris decipiens.* 5. *Phyllotheca indica.* 6. *Noeggerathiopsis hislopi.* 7. *Schizoneura gondwanensis.*

PLATE VI

GONDWANA PLANTS

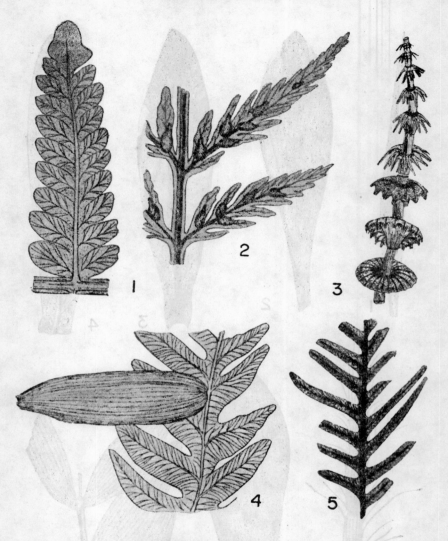

EXPLANATION OF PLATE VI

1. *Sphenopteris polymorpha.* 2. *Sphenopteris tenuis.* 3. *Phyllotheca Sahnii* (reconstruction of branch). 4. *Alethopteris norini.* 5. *Alethopteris medlicottiana.*

PLATE VII

GONDWANA PLANTS

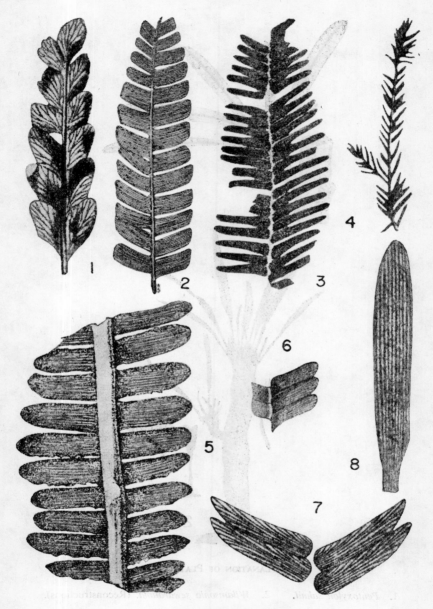

Explanation of Plate VII

1. *Thinnfeldia*. 2. *Nilssonia*. 3. *Otozamites hislopi*. 4. *Palissya jabalpurensis*. 5. *Pterophyllum rajmahalense*. 6. *Ptilophyllum cutchense*. 7. *Otozamites* (enlarged). 8. *Podozamites lanceolatus*.

PLATE VIII

GONDWANA PLANTS

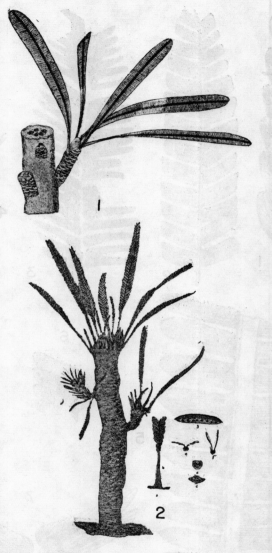

EXPLANATION OF PLATE VIII

1. *Pentoxylon sahnii.*   2. *Wllamsonia sewardiana.* (Reconstructions).

Phyllotheca indica ; Pteridospermae : *Glossopteris indica, G. communis, G. browniana, G. stricta, G. Ampla, Gangamopteris hughesi, Vertebraria indica* ; Cycadophyta : *Taeniopteris* cf. *macclellandi, T. dunaeoides, T. feddeni* ; Cordaitales : *Noeggerathiopsis hislopi.*

The Pali Beds (named after Pali near Birsinghpur railway station on the Katni-Bilaspur line) consist of coarse feldspathic sandstones associated with ferruginous and argillaceous bands, the latter yielding plant fossils of Raniganj age such as, *Noeggerathiopies hislopi, Danaeopsis hughesi, Thinnfeldia odontopteroides.*

The Himgir Beds of the Raigarh-Himgir coalfield are composed of red sandstones and shales of Kamthi facies which overlie the Barakars unconformably. They contain a Raniganj flora which includes : Equisetales: *Schizoneura gondwanensis* ; Pteridospermae : *Glossopteris indica* ; *G. browniana, G. angustifolia, G. communis, Vertebraria indica, Pecopteris lindleyana*, and *Sphenopteris polymorpha.*

The Bijori beds in the Chhindwara district, comprising sandstones, micaceous flags and shales which are sometimes carbonaceous have been found to contain remains of the labyrinthodont *Gondwanosaurus bijoriensis* and the following plant fossils : Equisetales : *Schizoneura gondwanensis* ; Sphenophyllales : *Sphenophyllum speciosum* ; Pteriodospermae : *Glossopteris communis, G. damudica, G. angustifolia, G. retifera, Gangamopteris,* sp., *Vertebraria indica* ; Incertae : *Samaropsis* cf. *parvula.*

The Almod beds were originally referred to the Panchet by H.B. Medlicott, but are now classified with the Bijoris by H. Crookshank.

THE PANCHET SERIES

The Panchet Series succeeds the Raniganj Series with a slight unconformity and sometimes overlaps on to the Barakars. The rocks of this Series, which have a total thickness of 450 to 600 m, rest upon the Raniganj Stage of the Raniganj coalfield and constitute the Panchet hill which is a prominent landmark. They comprise greenish, buff, and brownish sandstones and shales in the lower part, and greyish micaceous and feldspathic sandstones and shales in the upper part. The lower sandstones are often false-bedded and contain no coalseams or carbonaceous matter. They have yielded the primitive amphibian *Indobrachyops panchetensis* and skulls and bones of *Lystrosaurus* (reptile). The fossils definitely indicate a Lower Triassic age for the Panchet Series. They are not known in the Jharia coalfield, but their equivalents are found in Bokaro and Auranga. *The Almod beds* in the Pachmarhi area may probably be of Panchet age in part. The *Mangli beds* of the Wardha valley, composed of fine red and yellow sandstones and grits which are sometimes used as ornamental building stones and are more or less similar in appearance to the beds of Kamthi, are also referable to the Panchets on fossil evidence, since they contain *Brachyops laticeps* (a labyrinthodont) and *Estheria* (crustacean). The lower part of the Panchet beds are found near Maitur, northwest of Asansol, where plant fossils (*Glossopteris, Schizoneura*, etc.) are found, with close

Damuda affinities, and also *Pecopteris concinna* and *Cyclopteris pachyrachis*. A slightly higher horizon near Deoli, called the *Deoli beds* (also in the Raniganj field on the Damodar river) has yielded the following fossils :—

Labyrinthodonts	...	*Gonioglyptus longirostris, G. hyxleyi, Glyptognathus fragilis, Pachygonia incurvata, Pachygnathus orientalis.*
Reptiles	... ...	*Dicynodon orientalis, Epicampodon indicus.*
Crustacea	... ...	*Estheria mangliensis*

The Parsora Stage.—This stage is named after the deserted village of Parsora north of Pali in South Rewa. These beds consist of medium grained sandstone with micaceous and ferruginous bands. They overlie the Pali beds (of Raniganj) age with the intervention of several hundred m of unfossiliferous strata. Some of the ferruginous beds contain fossil plants which have a distinctly younger aspect than the Damuda flora. *Glossopteris* appears to be absent but *Noeggerathiopsis (Cordaites) hislopi* and a few other forms persist. *Danaeopsis (Thinnfeldia) hughesi* and *Thinnfeldia odontopteroids* are abundant, which impart a Triassic aspect. The beds of Chicharia (this village being 10 km north of Parsora), which may be either of the same horizon or somewhat younger, contain *Thinnfeldia sahnii* which is referred to the Lower Triassic by Seward (*Rec.* 66, pp. 235-243). The age of the Parsora and Chicharia beds is still a moot question, but it may be the same as that of Maleri *i.e.*, Middle to Upper Triassic.

MAHADEVA SERIES

Pachmarhi Stage

This series is named after the Mahadeva hills on which is situated the celebrated Mahadeva temple near Pachmarhi, which latter place has given the name to the stage which forms its lower part. The upper portion includes both the Denwa and Bagra Stages.

The Pachmarhi Stage forms the magnificent scarp above which the town of Pachmarhi is situated, and attains a thickness of about 750 m. It is of the nature of a huge lenticular mass of sandstone between the Denwa and Bijori, beds, and consists of red and buff sandstones with some red clays near the base and top. There are layers of haematitic clay and platy veins of hard dense ferruginous matter which, on weathering, resemble broken pieces of pottery. The Pachmarhi and other stages of the Mahadevas are entirely devoid of carboneceous matter, though the clayey layers sometimes show leaf impressions. The sandstones are generally somewhat coarse grained and tinted in various shades of red because of the disseminated ferric iron. Some of the Pachmarhi sandstones are of good quality for building and have been used at Pachmarhi, Warora and other places.

In the Damodar valley the Mahadeva Series is known by the vague term Supra-Panchet. In the Rajmahal hills it is represented by the *Dubrajpur Sandstone*.

Maleri (Marweli) Stage

Continental formations of Mesozoic age are exposed in the Pranhita-Godavari valley near and below Sironcha, along a N.W.-S.E. belt. The MALERI STAGE near Marweli which overlies the Kamthi strata, consists of red clays and subordinate calcareous sandstones. The red clay beds have yielded remains of Dipnoan, Subholostean and Pleuracanth fishes, Rhynchosaurian (*Paradapedon*), Phytosaurian and other reptiles, Metaposaurid Amphibians and coprolites, and also unionids and fossil tree trunks. The reptiles indicate a Carnic to Middle Noric age (Jain, Robinson and Roy Chowdhury, 1964).

South-southeast of the Maleri outcrop along the same strike, but separated by a fault, is a restricted exposure of red clays near Yerrapalli, which belongs stratigraphically to a lower horizon, now named the YERRAPALLI FORMATION. The fossils in these clays are distinct from those of Maleri and comprise Dicynodont and Erythrosuchoid reptiles and a large amphibian (? Capitosaur). These are referable to the *Cynognathus Zone* of the Beaufort Series of the Karroo System of South Africa, referable to the late Lower Triassic. The Maleri Stage is overlain by the Kota and Chikiala Beds.

Denwa and Bagra Stages

These two correspond roughly to the Maleri Stage. The Denwa Stage conformably overlies the Pachmarhis and consists of about 360 m of pale brownish or greenish yellow, bright, mottled red clays with subordinate bands of white and yellow sandstones which are often calcareous. When followed southwards, the Denwa clays and the overlying Jabalpur Stage pass under the Deccan trap and then emerge as thinner beds in the Pench valley. The Denwas contain remains of *Mastodonsaurus indicus* (allied to *Capitosaurus* and *Metapias*) and some obscure plant remains. The labyrinthodont remains point to Upper Triassic or Rhaetic age.

Bagra Stage

This stage consists of pebble-beds and conglomerates with red jasper in a matrix of red sandy clay, and lies unconformably on the Archaeans or on Lower Gondwanas. Followed southwards, the Bagras pass laterally into the Denwas so that they may be considered as shore-line deposits partly equivalent to, and partly younger than, the Denwas. The sandstones of the Tamia scarp lying above the Denwas, though lithologically similar to the Pachmarhis, are considered to be part of the Upper Denwa beds.

Kota Stage

In the Pranhita-Godavari valley, the Kota stage, named after Kota which is 8 km north of Sironcha, is about 600 m thick and occurs above the Maleri beds probably with an unconformity. The constituent strata are mostly sandstones and grits of light to brown colour, with red clay bands and a few limestone beds. They sometimes contain carbonaceous clays and thin coal seams and also occasionally gypsum.

Plant fossils have been found near Gangapur and Anaram :—

Cycadophyta	... ...	*Ptilophyllum acutifolium, Taeniopteris spatulata.*
Coniferales	... ...	*Elatocladus (Palissya) jabalpurensis, E. conferta, E. (Taxites) tenerrima, Retinosporites indica, Araucarites cutchensis.*

The Crustacean *Estheria* and also ganoid fish remains are found in a yellow limestone exposure near Kota :— *Lepidotus deccanensis, L. breviceps, L. longiceps, Tetragonolepis oldhami, Depedius egertoni.*

About 6 m below the limestone strata containing the fish remains, there is a bed containing dinosaurian bones. All these fossils indicates a Liassic age for the Kota Beds.

RAJMAHAL SERIES

Rajmahal Stage

The type area is the Rajmahal hills at the head of the Ganges delta near the border of Bihar and Bengal. This series consists of 450 to 600 m of basaltic lava flows with intercalated carbonaceous shales and clays, some of these being silicified and porcellanoid. Two of the flows near Taljhari are of pitchstone. The total thickness of these intercalated sedimentary beds is only 30 m, each bed being 1.5 to 6 m thick. The intertrappean sediments between the lower four or five flows contain plant remains, fossil wood and unionids. The more important plant fossils found in the chert beds near Nipania (24° 36′ : 87° 33′), Amajhola, Kalajhor, etc. are :—

Equisetales	... ...	*Equisetites rajmahalensis*
Lycopodiales	... ...	*Lycopodites gracilis, Lycoxylon,* sp.
Filicales	... ...	*Marattiopsis macrocarpa, Gleichenites gleichenoides, Cladophlebis denticulata, Coniopteris hymenophylloides, Sphenopteris hislopi, Pecopteris lobata, Protocyathea rajmahalense*
Pteridospermae ?	... ...	*Danaeopsis rajmahalensis, Thinnfeldia,* sp.
Cycadophyta	... ...	*Ptilophyllum acutifolum, P. cutchense, Otozamits bengalensis, Dictyozamites falcatus, D. indicus, Taeniopteris lata, T. spatulata, T. musaefolia, T. morrissii, T. ovata, T. crassinervis, Nilssonia (Pterophyllum) princeps, N. rajmahalensis, N. morrisiana, N. medlicottiana, N. bindrabunensis, N. fissa*
Coniferales	... ...	*Elatocladus (Palissya) conferta, Retinosporites (Palissya) indica, Pagiophyllum peregrinum, Brachyphyllum expansum*
Caytoniales	... ...	*Sagenopteris bambhanii*
Gymnospermous stems and cones	...	*Nipanioxylon guptai, Pentoxylon sahnii, Nipaniostrobus sahnii, Ontheodendron florini, N. Masculostrobus rajmahalensis, Sakristrobus sahnii, Carnoconites* sp.
Incertae	... ...	*Rajmahalia paradoxa, Podozamites lanceolatus*

The Rajmahal Traps bear a great resemblance to the Deccan Traps in composition and vary from dolerites to basalts, depending on their texture. Phenocrysts or aggregates of feldspar and pigeonite are irregularly distributed in a ground mass of labradorite, pigeonite, augite, magnetite, glass and also palagonite and chlorophaeite. Some flows are vesicular, the cavities being filled with calcite, chalcedony and analcite. The topmost flow is 75 m thick but the others are thinner, the average being about 23 m. Pumice, variolite and pitchstone are sometimes found. The traps are lateritised in some places and the laterite was at one time used as ore for iron smelting. The traps extend to the east and southeast but have been faulted down and covered by Cretaceous and Tertiary strata in the Ganges delta. The thickness of the sedimentary cover increases eastwards to over 3,600 m, probably as a result of step faulting.

Sahni expressed the opinion that none of the Rajmahal plants indicate a younger age than Upper Jurassic, Spath considered them as of Upper Neocomian age basing his opinion on the imperfectly preserved ammonites associated with the East coast Gondwanas which are their equivalents. They may be Upper Jurassic to early Cretaceous.

Chikiala Stage

This is named after a village some 16 km north of Kota and consists of brown and buff sandstone, generally ferruginous, and some conglomerates, the thickness being 150 m. It does not contain shale beds and the basal conglomerate seems to indicate an unconformity, though the junction between Kota and Chikiala is always covered by alluvium. Some unimportant coal beds are occasionally present in the Chikiala sandstones.

JABALPUR SERIES

This series is divided into two stages, the lower called CHAUGAN STAGE and the upper JABALPUR STAGE. They consist of white and light coloured clays and massive soft sandstones. Some of the shales are carboncaeous, and such of the coal seams as occur (for example in the Hard and Morand rivers in the Satpura region) are not of economic importance. Plant fossils occur in both the stages, the Chaugan Stage fossils resembling those of Kota. The two stages contain the following fossils :—

Pterdospermae ?	...	*Thinnfeldia* sp.
Cycadophyta	...	*Dictyozamites indicus, Taeniopteris spatulata, T. crassinervis, Anomozamites* cf. *nilssonia, Nilssonia princeps, N. orientalis, N. rajmahalensis*
Coniferales	...	*Pagiophyllum divaricatum*

Fossils of the Jabalpur stage :—

Filicales	...	*Gleichenites gleichenoides, Cladophelebis medlicottiana, Eboracea lobifolia*
Cycadophyta	...	*Ptilophyllum acutifolium, Otozamites hislopi, Williamsonia indica.*
Coniferales	...	*Elatocladus jabalpurensis, E. plana, Brachyphyllum expansum, Pagiophyllum* cf. *peregrinum, Retinosporites indica, Araucarites cutchensis, A. macropterus, A. latifolius.*
Ginkgoales ?	...	*Ginkgoites lobata, Phoenicopsis* sp.
Incertae	...	*Podozamites lanceolatus*

Kathiawar.—In Dhrangadhra and Wadhwan in the western part of Gujarat there is a large outcrop of nearly horizontal sandstones occupying an area of 2,500 sq. km. They are, for the most part, older than the Umia beds of Kutch and homotaxial with the Jabalpur beds. These beds are 300 m thick, the lower half being of Jabalpur age and the upper half of Umia age. The lower beds comprise soft yellow sandstones with white specks of kaolonised feldspar, ferruginous concretions and intercalations of carbonaceous shales and thin coal seams. There are ferruginous shales locally enriched to masses of haematite. The shales contain plant fossils of the Jabalpur facies and also several common to Jabalpur and Umia stages :—

| Filicales | ... | ... | *Cladophlebis whitbyensis* |
| Cycadophyta | ... | ... | *Ptilophyllum cutchense* |

The upper beds consist of gritty harsh sandstones of purple or dark colour with layers of conglomerate. This upper division is probably equivalent to the Umia beds of Kutch.

These Gondwana beds are overlain by the *Wadhwan Sandstones* which are probably the equivalents of the Bagh beds. In the Idar State there are some sandstones called HIMMATNAGAR (AHMEDNAGAR) SANDSTONES from which the zerophytic ferns — *Matonidium indica* and *Weichselia reticulata* of Wealden age, have been obtained. These are probably to be correlated with the Umia plant beds of Kutch and the Barmer Sandstone of Rajasthan, the latter containing some angiosperms.

Kutch.—The uppermost division of the Upper Gondwanas, which are developed in Kutch interstratified with marine beds are called the UMIA SERIES after Umia village 50 miles N.W. of Bhuj. The plant beds are probably of Lower Cretaceous age, and below the Aptian beds, according to the earlier workers.

Rajnath (1933) has divided the Umia Series into three stages as below :

Bhuj Stage	...	...	...	Beds with Palmoxylon Beds with Ptilophyllum Zamia beds	...	Cretaceous
Ukra Stage	...	...	...	Calcareous shales with marine fossils		Aptian
Umia Stage	...	...	...	Barren Sandstones and shales Trigonia beds ... ... Barren Sandstones ... Green oolites and shales Barren Sandstones		Tithonian
Katrol Series	...	...	...	...Tithonian to Kimmeridgian		

The lower (UMIA) STAGE consists of Barren Sandstones and some marine beds containing *Trigonia* belonging to Tithonian and early Cretaceous age. The overlying UKRA BEDS are of marine origin and contain ammonites of Aptian age.

The Upper beds (BHUJ STAGE) have yielded Bennettitalean and coniferous plant remains amongst which are *Ptilophyllum* (*Zamia*). The flora is allied to that of the Jabalpur beds. Silicified palms (*Palmoxylon mathuri*) also occur.

The chief plant fossils found in the Umia Series are :—

Filicales	...	...	*Cladophlebis whitbyensis*
Cycadophyta	...	...	*Ptilophyllum acutifolium, P. cutchense, Taeniopteris vittata, Williamsonia blanfordi*
Coniferales	...	...	*Brachyphyllum expansum, Elatocladus plana, Retinosporites indica, Araucarites cutchensis, A. macropterus*
Incentae	...	...	*Actinopteris* sp., *Pachypteris specifica*

These beds also contain the remains of a saurian, *Plesiosaurus indica*.

Orissa.—At the head of the Mahanadi delta near Cuttack, there occur Upper Gondwana rocks called the ATHGARH BEDS, the exposures being to some extent obscured by laterite and alluvium. They comprise sandstones, grits, conglomerates and some white or reddish clays, the last of which have yielded some fossils :—

Filicales	... ...	*Marattiopsis macrocarpa, Gleichenites gleichenoides, Cladophlebis, indica, C. whitbyensis, Rhizomopteris balli* (probably the rhizome of a fern).
Coniferales	... ...	*Retinosporites indica.*

The first four of these are Rajmahal species and therefore these beds are referred to that age.

About 8 km west of Cuttack there are carbonaceous shales traversed by basalt dykes, the shales resembling the Rajmahal inter-trappean beds.

The Athgarh sandstones have been used to some extent in the celebrated temples of Puri (Jagannath — mutilated in English into Juggernaut !) and Bhuvaneshwar. The Sun temple at Konarak is built partly of these while the Jain caves with sculptures in the hillocks called Kandagiri and Udayagiri, near Bhuvaneshwar, have been hewn out of the same sandstones.

Godavari district.—The Upper Gondwanas are found between Rajahmundry and Vijayawada, resting unconformably upon the Kamthi (Chintalpudi) sandstones. They comprise three divisions, the GOLAPILLI SANDSTONES below, RAGHAVAPURAM SHALES in the middle and the TIRUPATI (TRIPETTY) SANDSTONES above. The lower division comprises about 350 feet of orange to brown sandstones and grits, enclosing a flora allied to the Rajmahal. The Raghavapuram Shales which succeed them consist of 150 feet of white and buff shales, sometimes variegated, and purplish arenaceous shales. They contain plants as well as marine fauna like Cephalopods and Molluscs. Since 1960 they have yielded foraminifera which are said to indicate Lower Cretaceous age. The Tirupati Sandstones (about 50 m thick) overlie either the Golapalli Sandstone or the Raghavapuram Shales. They are red to brown sandstones and conglomerates, unfossiliferous on the Tirupati hill 36 km northeast of Ellore) but some outlaying exposures of these in the neighbourhood contain *Trigonia* and other fossils. The fossils found in these stages are :—

Fossils of the Golapilli Stage :—

Filicales	... ...	*Marattiopsis macrocarpa, Cladophlebis (Alethopteris) indica*
Cycadophyta	... ...	*Ptilophyllum acutifolium, P. cutchense, Taeniopteris ensis, Dictyozamites falcata, Nilssonia (Pterophyllum) morrisiana, Williamsonia* sp.
Coniferales	... ...	*Elatocladus (Palissya) conferta, Retinosporites (Palissya) indica, Araucarites macropterus*

Fossils of the Raghavapuram Stage :—

Filicales	... ...	*Cladophlebis (Pecopteris) reversa*
Cycadophyta	... ...	*Ptilophyllum acutifolum,* ? *Pterophyllum* sp., *Otozamites abbreviatus, Taeniopteris spatulata, T. macclellandi*
Coniferales	... ...	*Elatocladus (Taxites) tenerrima*
Ginkgoales ?	... ...	*Ginkgo crassipes*
Incertae	... ...	*Pachypteris ellorensis, Podozamites lanceolatus*

Amongst the animal fossils found in these beds are fish scales and several mollusca including — *Leda, Mytilus, Trigonia interlaevigata, Solen, Tellina, Pecten,* etc. The foraminifera include *Haplophragmoides* and *Ammobaculites.*

Fossils of the Tirupati Stage :—

Several fossils have been found in this stage, mostly mollusca and molluscoidea, amongst which are some ammonites and belemnites and also *Trigonia ventricosa, T. smeei, Inoceramus, Pseudomonotis, Lima* and *Pecten.*

Ongole Area.—In this region there are four patches in which Upper Gondwanas are found. They are near Kandukur, Ongole, Vemavaram and Guntur. The best Upper Gondwana exposures are found near Budavada 38 km N. by E. of Ongole. The lower beds, *i.e.*, BUDAVADA SANDSTONES are buff coloured, and are the marine equivalents of the Golapilli Stage. They are succeeded by thin-bedded, fissile, purplish, variegated shales, called the VEMAVARAM SHALES containing an abundant flora closely related to the Raghavapuram Shales. Overlying these are the PAVALUR SANDSTONES which are brown and red sandstones apparently unfossiliferous.

In the vicinity of Pavalur are said to occur large blocks of sandstones containing marine fossils belonging to the genera *Belemnites, Cerithium, Ostrea, Rhynchonella,* etc., but the rock is dissimilar to the Pavalur Sandstone and may possibly be younger.

The Budavada Sandstones have yielded the following plant fossils (all Bennettitalean) :— *Ptilophyllum acutifolium, Taeniopteris (Angiopteridium) sptatulata, Otozamites* sp. and *Dictyozamites indicus.*

The Vemavaram Shales have yielded a fairly rich assemblage of fossils amongst which are :—

Filicales	... ...	*Cladophlebis (Alethopteris) indica, Sphenopteris* sp., *Dicksonia* sp.
Cycadophyta	... ...	*Taeniopteris spatulata, T. ovata, Ptilophyllum aacutifolium, P. cutchense, Pterophyllum distans, Dictyozamites indicas, Zamites proximus.*
Coniferales	... ...	*Retinosporites indica, Brachyphyllum rajmahalensis, Araucarites* sp.

L. F. Spath examined, a few years ago, the collections of ammonites made from near Budavada by Foote, from Raghavapuram by King and from Raghavapuram, Vemanavaram and Budavada by L.A.N. Iyer (*Pal. Ind., N.S.* IX, No. 2, pp. 827-829, 1933). From the first two collections he has described *Pascoeites budavadensis* and *Gymnoplites simplex* respectively, both being new fossils (genera as well as species) referred to Lower Cretaceous from their evolutionary characteristics. Iyer's collections contained poorly preserved fossils in variegated shales having a non-marine aspect. The collection consisted of mostly new genera and species and not particularly well suited for satisfactory determination : *Holcodiscus* cf. *perezianus, H.* cf. *caillaudianus, Hoplites,* cf. *borowae, H.* cf. *beskidensis, H. codazzianus, Lytoceras* sp. cf. *vogdti, Pascoeites budavadensis, P. crassus,* and *Gymnoplites simplex,* etc.

According to L. F. Spath, the ammonites indicate an Upper Neocomian age, and consequently also for their equivalents the Rajmahal and Sriperumbudur beds. The Jabalpur and Tripetty (Tirupati) beds become Aptian. There is thus a strong difference of opinion on the age of Rajmahal beds. Arkell (Jurassic Geology of the World, London 1956, p. 384) has expressed the opinion that there is a big stratigraphical gap in the Upper Gondwanas from the Upper Triassic to the Neocomian, the Jurassic being unrepresented. The Stanvac Oil Company who conducted drilling in the Ganges delta also believe that the strata associated with the Rajmahal Traps (encountered at depth in the boreholes) may be of the uppermost Jurassic or early Cretaceous age.

Madras-Chingleput.—There are two occurrences near Madras which are referable to the Upper Gondwanas. The lower stage, named after *Sriperumbudur (Sripermatur)* 40 km W.S.W. of Madras, contains marine animals and plant remains, especially in the eastern part of the basin. The invertebrate fossils found here are *Leda, yoldia, Tellina, Lima, Pecten*, etc. One or two ammonites similar to those in the Raghavapuram Shale are also said to have been found by Bruce Foote from loose boulders here, but a search in 1940 did not reveal any animal fossils except mollusca, nor the boulders referred to. The plant fossils found are :—

Filicales	... ...	*Cladophlebis whitbyensis, C. indica, C. reversa*
Pteridospermae ?	...	*Thinnfeldia* sp.
Cycadophyta	... ...	*Taeniopteris spatulata, T. macclellandi, Pitlophyllum acutifolium, P. cutchense, Dictyozamites indicus, Otozamites abbreviatus, O. bunburyanus, Pseudoctenis (Pterophyllum) footeanum*
Coniferales	... ...	*Pagiophyllum (Pachyphyllum) peregrinum, Brachyphyllum rajmahalensis, B. rhombicus, Elatoctadus conferta, E. plana, Araucarites cutchensis, A. macropterus*
Ginkgoites ?	... ...	*Ginkgo crassipes*

The area was studied by M. S. Venkataram in 1939-40. The beds occur as patches spread over nearly 2,000 sq. km in Chingleput and North Arcot districts, the largest patch occurring at and around Sriperumbudur. The strata are composed mainly of white to pink clays, shales and feldspathic sandstones. They have apparently been laid down in shallow basins on an irregular floor with an easterly slope and are mainly lacustrine in character. The age deduced from the plant fossils is the same as that of the Tabbowa beds of Ceylon.

The Upper stage forms the SATYAVEDU (SATTIVEDU) BEDS about 55 km N.W. of Madras, consisting of purple mottled ferruginous sandstones and conglomerates which contain fragmentary plant fossils. They are underlain by the Sriperumbudur beds and therefore thought to be the equivalents of the Tirupati Stage.

Trichinopoly.—Small areas of Upper Gondwana beds are also found near Uttattur village in Trichinopoly (Tiruchirapalli) district where they consist of micaceous shales, grey sandstones and grits containing calcareous concretions.

They rest on Archaean gneisses and are overlain by marine Cenomanian beds, and contain :—

Filicales	... ...	*Cladophlebis indica,*? *Actinopteris*
Cycadophyta	... ...	*Ptilophyllum acutifolium, P. cutchense, Otozamites* sp., *O. rarinervis, O. abbreviatus, Anamozamites* sp., *Dictyozamites indicus, Taeniopteris spatulata*
Coniferales	... ...	*Elatocladus conferta, E. plana, Retinosporites indica, Araucarites cutchensis*

These indicate the same horizon as the Vemavaram and Sriperumbudur beds.

Ramnad.—Bruce Foote found exposures of yellow and buff shales resembling the Upper Gondwana shales of Uttattur near Sivaganga but was unable to examine them in detail.

In 1951, V. Gopal mapped the area and collected plant remains from these beds. The sediments consist of boulder-beds and conglomerates at the base, followed by micaceous sandstones and by alternating grits and shales in the upper part. The plant fossils have been described by Gopal and Jacob (*Rec.* G.S.I. 84 (4), 1955) who state that they are of Kota age. The most important of them are : —*Cladophlebis lobata, C. reversa, Taeniopteris spatulata, T. maclellandi, T. densinervis, Ginkgoites crassipes, Brachyphyllum expansum, Elatocladus plana, Podozamites lanceolatus,* etc. These strata have been called the SIVAGANGA BEDS.

CEYLON

Upper Gondwana strata, called the TABBOWA SERIES[1], are found to occupy an area of about 2.5 sq. km north and east of the Tabbowa tank, 13 km N.E. of Puttalam. They consist of sandstones, conglomerates, shales and nodular limestones. The strata are apparently faulted into the gneiss, but are generally covered by alluvium so that their actual extent is not known. The same rocks have recently been found near Andigama[2] some 25 km south of Tabbowa and are believed to continue further south.

The sandstones and shales contain plant impressions, while thin coal seams have been noted near Andigama. The plant fossils were examined and described by Seward and Holttum[3] in 1922 and a further collection has been studied by K. Jacob who states that the following have been identified :—

Filicales	... ...	*Sphenopteris* sp., *Coniopteris* sp., *Cladophlebis* sp., *C.* cf. *browniniana, C. reversa*
Cycadophyta	... ...	*Taeniopteris spatulata, Nilssonia* cf. *schaumburgensis.*
Coniferales	... ...	*Araucarites cutchensis, Brachyphyllum mamillare, Elatocladus plana, E.* sp. *Desmiophyllum,* (? *Podozamites*) sp.

Several of the species agree with those described by Feistmantel from the East Coast Gondwanas of Madras, which are referred to the Kota Stage.

---

1. E. J. Wayland : The Jurassic rocks of Tabbowa. *Ceylon Jour, Sci.* XIII, B. Pt. 2, 1924.
2. P. Deraniyagala : *Proc. Ind. Sci. Congress*, 1940, Pt. 4, p. 77.
3. A.C. Seward and R. E. Holttum : Jurassic plants from Ceylon. *Q.J.G.S.*, LXXVIII, Pt. 3, 1922.

Seward and Holttum agree with this, but Jacob thinks that the age may be slightly younger, particularly as the presence of *Cladophlebis* cf. *browniana* and *Nilssonia schaumburgensis* give the assemblage a newer aspect than Kota. A final decision on the age of the Upper Gondwanas of Ceylon and Madras must however await the results of further detailed work.

IGNEOUS ROCKS IN THE GONDWANAS

Dolerite and Basalt.—Most of the Gondwana coalfields are traversed by hypabyssal intrusives of basic rocks — dolerites and basalts, which may be sometimes olivine-bearing — as dykes and sills. They are common in the Satpura, Sone-Damodar, Assam and other fields north of the Satpura axis and are comparatively rare in the Godavari and Mahanadi valleys. Some of the dykes are affected by faults while others pass through them without interruption, so that in general they may be regarded as later in age than the faults. Thick sills are occasionally found, especially in Rewa and Satpura regions, at the junction of dissimilar formations. Dykes are also known to pass into sills in the Satpura area. They generally follow straight courses and appear to be controlled by fractures in the basement rocks than in the Gondwana strata.

The intrusives of the Satpura and Rewa areas are undoubtedly of Deccan trap age, while those of the Damodar valley and Assam are thought to be of Rajmahal age, though evidence on this point is not clear. In the Rajmahal area the traps are dolerites, basalts and andesites, very similar to the Deccan traps in their chemical and mineralogical characters and their age may be Lower to Middle Jurassic.

Mica-lamprophyre.—Another type of igneous rock is also found as dykes and sills in the coalfields of the Damodar valley, Giridih and Darjeeling foothills. This is the 'mica-peridotite,' a lamprophyric or mica-rich ultrabasic rock containing altered olivine, calcite (or dolomite), bronze coloured mica and much apatite. Fresh and unaltered rock is rather rare even at depth, being dark grey, hard and tough. At the surface and in mine workings, as commonly seen, it is buff coloured and soft, containing nests of mica. Mica-peridotite dykes are generally of small thickness (1 to 1.8 m) and have a tendency to form anastomosing veins, lens-like masses and thin flat sheets at the junction of coal seams and sandstones or in the coal seams themselves. They seem to have come up through faults and to have particularly preferred the coal horizons and their junctions. They give evidence of high fluidity and high temperature, as they have destroyed, coked, or otherwise rendered the coal useless at and near their contact. Alongside some of the sills the coal has been devolatilised and converted into a kind of coke (*jhama*) for a distance of as much as 2 m from the contact. The high fluidity of the intrusive is evidenced by the intricate ramifications of the veins and sheets, which may often be less than an inch thick, traversing the cracks amidst the *jhama* which has developed columnar structure. The rock seems also to have spread out into sheets more in the lower seams than in the upper, in the same area. In contrast with these, the dolerite dykes are scarcely harmful to the coal and they are not found to penetrate the coal seams as thin veins.

The mica-peridotites are seen only in the Damodar valley and in the Lower Gondwanas of the eastern Himalayan region. They seem to be nearly of the same age as the dolerites.

GONDWANAS IN OTHER CONTINENTS

From the descriptions which follow, it would be noticed that strata of Gondwana age are extensively developed in all the southern continents *viz.*, Australia, South Africa, South America and even in Antarctica. Our knowledge of the last named region is now rapidly accumulating through international cooperation. In all these continents the equivalents of the Gondwana group begin with glacial deposits in the Upper Carboniferous and Permo-Carboniferous Thereafter there is an amelioration of climate and the strata laid down during the Permian contain workable coal seams formed under cola temperate conditions. The Triassic was generally a period of drought and of continental conditions; the strata include brightly coloured sandstones and clays indicating a high degree of oxidation and enclose remains of amphibians, reptiles, fishes and silicified fossil wood. The Jurassic period shows milder climate with the development of a new flora, but though carbonaceous materials occur, coal seams are not common. The Gondwana era generally closes with extensive development of basic volcanic flows which often attain great thicknesses. In India, however, the Gondwana era is considered to continue well into the Cretaceous because of the fact that the younger flora developed during the Jurassic is found to continue into some deposits whose age is referable to the Middle Cretaceous (Aptian or Cenomanian).

AUSTRALIA

The equivalents of the Gondwana group are found in several areas in the eastern part of Australia, including Queensland, New South Wales and Tasmania. The best development is that in the coalfields of New South Wales and the neighbouring areas. In this region it is possible to distinguish a marine as well as a terrestrial facies in the Upper Carboniferous as shown in Table 30.

The lowest beds, known as Lower Burindi, comprise conglomerates mudstones, tuffs, etc. and often contain marine fossils. The age is roughly Tournaisian as determined by such fossils as the Corals *Lithostrotion, Zaphrentis, Syringopora,* etc. These are overlain by the Lower Kuttung, composed of volcanic conglomerates, tuffs, lavas, etc. the last attaining a thickness of nearly 880 m. Their marine equivalents are the Upper Burindi beds consisting of sandstones, mudstones and oolitic and crinoidal limestones. The terrestrial Lower Kuttung facies contains *Lepidodendron veltheimianum, Stigmaria, Archaeocalamites,* etc. which indicate a Visean age. These are followed by the Upper Kuttung beds which enclose two important glacial horizons separated by non-glacial beds. The glacial beds contain a *Rhacopteris* flora with *Calamites* and *Cardiopteris*. Their marine equivalents are the Emu Creek beds containing *Spirifer, Strophalosia, Reticularia, Stutchburia, Phillipsia* and Fenestellids.

THE GONDWANA GROUP

TABLE 30.—COMPOSITE SECTIONS OF THE CARBONIFEOURS (N.S.W.)

Terrestrial	Marine	Age
Upper Kuttung (4,700) ft.) Main glacial Paterson toscanite Lower glacial	Emu Creek Mudstones and tuffs with *Spirifer, Reticularia,* and *Syringothyris, Stutchburia*	Moscovian
Lower Kuttung (5,000 ft.) Volcanics, Lavas, tuffs and conglomerates	Upper Burindi Limestones with intercalated sandstones and clays (*Lithostrotion* etc.)	Visean
Lower Burindi (5,000 ft.) Tuff, mudstones, oolite, conglomerates, with marine fossils	Lower Burindi (7,000 ft.) Fossiliferous limestones, mudstones, tuffs and conglomerates	Tournaisian

Permian.—In the Hunter Valley area (N.S.W.) the basal beds which contain some glacial boulder-beds possibly equivalent to the Upper Kuttung, are represented by the Lochinvar formation. They are overlain by the Lower Marine formation consisting of conglomerates, grits, sandy shales, tuffs and lava flows and containing *Eurydesma, Dielasma, Linoproductus, Chonetes* and Corals. Above these come the Lower or Greta Coal Measures followed by the Upper Marine group of crinoidal limestones, mudstones and glacial beds. These are succeeded by the Upper Coal Measures (Tomago and Newcastle Coal Measures). The Hunter Valley Permian section is 5,300 m thick and contains the most important coal formations of Australia.

Both the Lower and Upper Coal Measures contain *Glossopteris, Gangamopteris, Phyllotheca, Sphenopteris, Cordaites,* etc.

In the Bowen basin of Queensland the Permo-Carboniferous and Permian section (Some 4,500 m thick) consists of basal volcanics and glacials overlying unconformably the conglomerate beds of Carboniferous age. The volcanics are associated with shales containing Rhacopteris flora. These are followed by Middle Bowen Coal Measures and marine beds. The coal measures contain *Glossopteris* and *Gangamopteris* while the marine beds have yielded *Strophalosia, Terrakea* (productid), *Streptorhynchus, Martiniopsis* and *Stenopora*. The upper Bowen beds are tuffceous sandstones, and shales with coal seams. They have yielded several species of *Glossopteris, Gangamopteris, Phyllotheca, Sphenopteris,* etc. The Marine facies developed in the eastern portion contains *Martiniopsis, Strophalosia, Terrakea, Taeniotherus, Anidanthus,* etc. A similar fauna is developed also in the southern end of the western part of the basin. In this latter, *Eurydesma* is found in the Lower Bowen (Dilly beds).

The Triassic is well developed in the Cumberland basin near Sydney. It consists of the Narrabean group of Lower Triassic age (over 600 m) thick which contains *Estheria* as well as *Glossopteris* and *Schizoneura*. The fish remains in the upper part of the Narrabean are considered to be not older than the lower Triassic. The succeeding formations are the Hawkesbury Series formed of sandstones, shales and carbonaceous shales about 240 m thick. These are succeeded by the *Wianamatta* group which is also 240 m thick and probably of Upper Triassic age.

The Hunter-Bowen Orogeny extensively folded the Upper Palaeozoic strata and was accompanied by granitic intrusions. This orogeny is responsible for the marked unconformity which is almost always seen at the base of the Triassic.

TASMANIA

The succession of formations in Tasmania is shown in the accompanying Table. The whole of the Permian section totals about 750 m thick and consists of a quartzite-limestone suite. The environment varies from lacustrine or fluviatile to shallow marine. The Hunter-Bowen orogeny whose locus was in the mainland, influenced Tasmanian deposits and raised up the basin of deposition inducing erosion in some places, so that there is always a marked disconformity below the Triassic. The *Granton Limestone, Grange Mudstone* and *Woodbridge Glacials* are correlated with the different parts of the Marine Series of N.S.W. ,on the evidence of fossils. These fossils include *Fenestella, Stenopora* and other bryozoa. In the Hobart area *Eurydesma* occurs. Other marine fossils include *Taeniotherus, Strophalosia, Polypora,* and *Protoretepora ampla*.

In the Woodbridge Glacials abundant erratics are found as well as a number of fossil trees. The Ferntree mudstone follows on the Risdon and is characterised by regular varved sediments formed in a shallow sea. The Cygnet Coal Measures contain thin coal seams and several species of the Glossopteris flora.

TABLE 31—THE GONDWANA GROUP IN TASMANIA

Age	Formations	Equivalents
Jurassic ...	Dolerite flows and sills	Ipswich Series
Up. & Mid. Trias	*Feldspathic Sandstone* (800 ft.) with *Thinnfeldia, Phyllotheca, Cladophlebis, Johnstonia, Phoenicopsis.*	
L. Trias ...	*Knocklofty sandstone* and shale (700 ft.) with *Thinnfeldia, Phyllotheca.*	
	—Disconformity—	
Tartarian ...	*Cygnet Coal Measures* (200 ft.) with *Glossopteris* etc.	Upper Coal Measures
Kazanian ... Kungurian ...	*Ferntree mudstone* (300 ft.)	......
Artinskian (Risdon formations)	*Woodbridge glacials* (400 ft.) with *Stenopora*	Upper Marine Series
	Grange Mudstone (300 ft.) with *Wyndhamia*	
	Porter's Hill mudstone (50 ft.)	Part of Lower Marine Series
	Granton limestone and marl with *Eurydesma* and *Stenopora.*	
Sakmarian ...	Basal Glacial beds (400 ft.)	
	—Unconformity—	

LOWER PALAEOZOIC ROCKS

THE GONDWANA GROUP

The Triassic formations contain fish and Labyrinthodont remains as well as *Phyllotheca, Thinnfeldia, Cladophlebis*, etc. The formations are lacustrine in nature. The feldspathic sandstone at the top of this system is associated with shales which are carbonaceous and contain several coal seams, some of which have been spoiled by intrusions of dolerite. The plant fossils found in this formation are of typical late Triassic aspect including *Thinnfeldia, Sphenopteris, Ginkgo, Sagenopteris*, etc. This flora is allied to that of Wianamatta Series of N.S.W. and the Ipswich Series of Queensland.

The Jurassic in Tasmania continues on from the Triassic and is terminated by thick doleritic intrusions and flows. The dolerite is of tholeiitic composition consisting mainly of labradorite, pigeonite and subordinate augite.

SOUTH AFRICA
The Karroo System

The Gondwana group in South Africa, known as the Karroo System consists of three major series, *viz.* the Ecca, Beaufort and Stormberg. The Karroo System is underlain by the Witteberg Series of white sandstones and shales of Devonian age. The " Lower Shales " underlying the Dwyka Tillites were originally included in the Dwykas but are now relegated to the Witteberg. The succession is given in the following table. The Karroo System shows very considerable differences in thickness in the vast areas which it occupies, viz., The Union of South Africa, South West Africa, Angola, Bechuanaland, Rhodesia and East Africa, for the conditions of deposition in the different parts of this great basin must have differed very much from one area to another.

Dwyka Series.—The Karroo System commences with the tillites and fluvo-glacial beds which are called the Dwyka Series. They are generally 100 to 250 m thick but might rarely be 420 m thick. They are absent in Rhodesia and Kaokoveld. They are continental in character in the Karroo and Transvaal regions but marine in Namaland where they contain *Eurydesma* and other fossils. In Bechuanaland they contain glacial varvites and thin coal seams.

The lower portion of the Dwyka Series is of Upper Carboniferous and Permo-Carboniferous age. In the eastern part of the basin the glaciers seem to have come from the region of the Indian Ocean and in the western part from the north and northwest. In some areas *Gangamopteris cyclopteroides* has been found at the base of the tillites. The Eurydesma beds have yielded *Eurydesma, Conularia, Productus, Acrolepis, Palaeoniscus*, etc.

The Lower Dwykas pass upwards into the " White Band, " also called the Mesosaurus beds. They are about 200 m in thickness and are characterised by aquatic reptiles of the genera *Mesosaurus* and *Noteosaurus*. Locally the shales carry *Glossopteris, Lepidodendron* and *Dadoxylon*.

The Dwykas correspond to the Tubarao Series of Brazil. The great ice sheet of the Namaland region is supposed to have extended into South

Brazil, which was a contiguous area in the Carboniferous, Permian and Triassic. The "White Band" and the succeeding Upper Shales are of Lower Permian age and considered to be part of the Dwyka Series.

Ecca Series.—This series is extremely variable in thickness in different areas, being only 100 m thick in Namaland but 1,500 m thick in the Cape Province. The Lower Eccas are dark unfossiliferous shales with dolomite lenses. The Middle Eccas are dark shales and contain the major coal measures of South Africa with several coal seams — e.g., the Witbank coalfields region, where the seams are up to 12 m in thickness. There are also bituminous oil shales. The Upper Eccas are mainly bluish shales. The plant fossils found in Middle Ecca Series include *Lycopodiopsis derbyi*, *Lepidodendron*, *Glossopteris*

TABLE 32—THE KARROO SYSTEM IN SOUTH AFRICA (After Maack)

(Thickness shown in metres)

Age		Cape Province	Namaland	Kaokoveld
Lias	STORMBERG	Stormberg Volcanics (1,400) (Drakensberg)	Mandelstein decke (30) of Kub-Hoch-anas.	Kaoko Volcanics (650)
Up. Trias.	STORMBERG	Cave Sst. (250) Red Beds (500)	...	Main Sst (50)
Noric Carnic	STORMBERG	Molteno Beds (600)	HIATUS	
Ladinic Anisic (Mid. Trias.)	BEAUFORT	Up Beaufort (400) (Burghersdorp)		
Seythic (L. Trias.)	BEAUFORT	Mid. Beaufort (200)		
Up. Permian	BEAUFORT	L. Beaufort (600)		
L. Permian	ECCA	Up. Ecca Mid. Ecca (1,500) (Coal) L. Ecca	Karroo Sst. and Shales (100)	Shales and Sst. (100)
Up. Carbon.	DWYKA	Up. Shales white Band (200) (*Mesosaurus*)	Grey Kuppe Shales (*Mesosaurus*)	Shales, Sst. (50) and limestones (*Mesosaurus*)
	DWYKA	Tillites and glacials (460)	Black Shales (Coal) Varvites	Shales with thin coals
	DWYKA		Eurydesma beds Tillites and fluvo-glacials (500)	Sst. and coloured shales Conglomerate (non-glacial) (80)
L. Carbon. ?		L. Shales		
Devonian		Witteberg Sst.		

Gangamopteris, Sphenopteris, Neuropteridium, Sphenophyllum, Cordaites, Schizoneura, etc., which are typical components of a Permian flora.

Beaufort Series.—The Eccas pass conformably upwards into the Beaufort Series whose maximum thickness is about 2,000 m. This series extends from Upper Permian to the Lower Trias in age. The Lower Beaufort, which is up to 1,200 m thick, consists of sandtones and shales of different colours and shows three palaeontological zones. This also contains reptiles (*Endothiodon*) and Glossopteris flora. The Permian-Trias boundary occurs within or above the top of the Lower Beaufort. The Middle Beaufort consists of red green shales with a maximum thickness of 300 m. It contains the *Lystrosaurus* zone with several reptiles and amphibia of Lower Triassic age. The Upper Beaufort consists of sandstones, arkose and coloured shales with a maximum thickness of 500 m. The *Procolophon* and *Cynognathus* zones, rich in reptile remains, occur within these. The Glossopteris flora found in these strata are Triassic in age and includes *Schizoneura, Thinnfeldia, Pterophyllum, Ginkgoites*, etc. The Beaufort Series is missing in the northern and northwestern areas.

Stormberg Series.—There is generally a stratigraphical break between the Beaufort and Stormberg Series covering the Middle Triassic. The lower part of the Stormberg Series are the Molteno beds, the lower portion of which contains some coal seams with *Thinnfeldia, Taeniopteris*, etc., while the upper portion consists of red beds with reptiles (*Thecodontosaurus, Massospondylus*, etc.). They are followed by Cave Sandstones which are fine-grained, massive, bright coloured sandstones of eolian origin indicating a desert climate. The Cave Sandstones contain some reptiles and fish remains as well as silicified wood, their age being Rhaetic. The upper part of the series consists of volcanic lows called Stormberg or Drakensberg volcanics which attain a thickness of 1,400 m in places. They comprise basaltic lavas as well as melaphyres and porphyrites. In some places in the Cape Province they contain inter-trappean sandstones. The lava flows indicate a period of tension when Gondawnaland broke up. They are of Lower Jurassic age and similar lava flows are also found in Brazil and Argentina. The Cape Province also contains intrusives of Late Cretaceous or early Tertiary age comprising granites, nepheline syenites, alkali basalts, kimberlite etc., some of which traverse the Karroo formations. It is not known whether these are to be considered a later manifestation of the Stormberg igneous activity.

EAST AFRICA

In NORTHERN RHODESIA the rocks of the Karroo System are well developed. They begin with glacial beds which are succeeded by the Wankie Series which are partly Lower Permian and partly Middle Permian and contain coal seams. The Upper Permian is represented by Madumbisa Shales. After an unconformity, the Triassic beds follow — Escarpment Grits-Mudstone Group, Pebbly Arkose and Forest Sandstone. They are overlain by 600 m of lava flows of Stormberg (Jurassic) age.

In SOUTHERN RHODESIA the Wankie Series is represented by the Lower Metabola beds and Busse Series while the Upper Metabola are of Upper Permian age. The Upper Karroos are represented by Escarpment Grits.

The Karroos are seen in MOZAMBIQUE in the western part and across the northern area, largely along the Zambesi basin. TANGANYIKA shows an excellent development of the Karroos whose eight sub-divisions have been named K 1 to K 8. The first five divisions together with the conglomerates at the base are of Ecca age and contain coal seams and Glossopteris flora. The Lower Beaufort (K 6) contains *Endothiodon, Dicynodontia, Pareisauria* and *Dadoxylon* wood. The Middle Beaufort is marked by a hiatus while the Upper Beaufort (K 7) consists of coarse sandstones. The Stormberg Series is partly feldspathic sandstones and marls with fossils, and partly lava flows. These formations continue into NYASALAND. In BECHUANALAND the Karroo rocks are partly hidden under desert sands but exposures are seen in the southwest, east and northeast. The rocks generally thicken in a westerly direction. Dwyka tillites, Ecca Coal Measures with coal seams and Stormberg lavas are seen. The lavas have been encountered in several boreholes underneath the desert sands. In UGANDA only a few small occurrences of Karroo rocks are known. A borehole at Entebbe went through 300 m of Karroos without reaching their base. They gave remains of the *Glossopteris* flora and only thin coal layers and carbonaceous shale. In KENYA the Karroos occur along a strip in the east, being the continuation of the Tanga beds of Tanganyika. The bottom beds (Taru Grits) are of Upper Permian age, unconformably overlying basement schists. They are overlain by thick sandstones and grits of Triassic age, the Permian — Trias sequence being called the Duruma Sandstone Series. The subdivisions include the Mijiva Chumvi Beds (L. Trias) and the Mariakani and Mazeras Sandstones (U. Trias). The Triassic beds have yielded the crustacean *Estheria,* and the plant remains *Dadoxylon* and *Equisitites.* There is probably an unconformity between the Triassic and the Jurassic, the latter being marine and mainly of Bajocian to Kimmeridgian age. These have yielded corals, brachiopods and ammonites. The Jurassic sea was connected with the Tethys through Eritrea and Arabia on the one hand and through Kutch and Rajasthan on the other. A land mass which was a southerly continuation of Arabia has since disappeared by faulting. There is a striking similarity between the Jurassics of Madagascar and Kutch and East Africa, and particularly between the first two.

The Karroo System in ANGOLA is divided into the three series, as shown in the accompanying table (Table 33).

The tillites, which are 10-30 m thick, are overlain by black shales containing *Noeggerathiopsis* and *Gondwanidium.* The Cassanje Series which comprise the greater part of the Triassic, are bright coloured and contain several species of fishes in the lower part, fragmentary plant remains in the middle and *Estheria* in the upper part. The Lunda Series shows unfossiliferous silicified sandstones in the lower part, while the upper part consists of dolerites which are mainly intrusive.

TABLE 33. THE KARROO SYSTEM IN ANGOLA

Angola		Belgain Congo	South Africa
Lunda Series.	Upper — Dolerites Lower — Sandstones, Shales	Up. Kwango ...	Stormberg
Cassanje Series	Upper-Sandstones with *Estheria*. Middle-Sandstones with plants. Lower—Shales with fishes	L. Kwango ... Lualaba Series	Stormberg Up. Beaufort Mid. Beaufort
Lutoe Series.	Lunda-black shales ... Tillites	Lukuga Series.	Ecca, Dwyka

In MADAGASCAR the Karroo formation is exposed along a narrow linear tract in the western part, parallel to the western coast. The Permo-Carboniferous and Permian are represented by the Sakoa Series which show tillites, black shales and Lower Gondwana plants. A marine horizon intercalated in this series contains *Productus, Spirifer*, etc. This is followed unconformably by the Sakamena Series which is of Upper Permian and Lower Triassic age. This shows a large number of Gondwana plants as well as several reptiles, the latter in the Rhinesuchus beds and Tangasaurus beds corresponding to the Lower Beaufort. In the extreme north there are intercalated marine horizons which contain *Xenaspis, Cyclolobus* (Upper Permian) and other Ammonites belonging to the *Otoceras, Gyronites* and *Flemingites* zones (Trias.)

The Sakamena Series is overlain by the Isalo group which has been divided into four units extending from the upper part of the Lower Trias through the Upper Trias into the Bathonian. These contain both continental and marine deposits. The former show dinosaur remains as well as fossil wood belonging to *Dadoxylon, Cedroxylon* and other genera in the continental beds ; and *Macrocephalites, Anabacia*, etc., in the marine beds. In the Jurassic there was definite marine connection between Kenya and Madagascar on the one hand and Kutch on the other as attested by common fossils which include Brachiopods, Cephalopods, Molluscs, etc. It is thought that during the Jurassic, East Africa was closely connected with the Mediterranean region and with India mainly through Eritrea and Arabia through which a shallow arm of the Tethys extended.

SOUTH AMERICA

The Gondwana history of South America indicates that, though South Africa and South America are now separated by the deep and wide South Atlantic ocean, they had common and extraordinarily similar geological history over a long period of time extending from the Carboniferous to the Cretaceous. In South America the great Parana basin occupies a large part of central, western and southern Brazil (parts of Minas Gerais, Mato Grosso, Goias, Santa Catarina, Rio Grande do Sul, Parana and Sao Paulo) and parts of the adjoining regions of Argentina, Uruguay and Paraguay. The total area of the Gondwana basin exceeds 1.5 million sq. km. and the total thickness of sediments exceeds 2,000 m. The sediments are overlain by 1,000 m of basaltic lavas, the beginning of the igneous activity heralding the termination of the Gondwana history. Sedimentation was mainly continental (glacial, fluviatile, lacustrine, eolian) and was extraordinarily uniform over large areas. Though this basin is characterised by rather localised marine invertebrate fauna which makes it difficult to make precise correlations with South Africa, yet the Glossopteris flora and the reptilian fauna have characteristic Gondwana affinities and show that the two regions had very similar geological history over the vast stretches of time encompassed by the Gondwana era.

BRAZIL

The Gondwana group in Brazil is called the Santa Catarina System divided into three major divisions.

3. Sao Bento Series.
2. Passa Dois Series.
1. Tubarao Series.

The Tubarao Series rests on pre-Devonian granite and on the Parana Series containing Devonian marine fossils. The lower portion of the series is called the Itarare group composed of thick glacial and intercalated sediments which contain a Rhacopteris flora mixed with elements of the Glossopteris flora and fish remains. There are several characteristic species of Gondwana plants, e.g., *Gangamopteris, Noeggerathiopsis, Gondwanidium, Sphenopteris, Lepidodendron, Lycopodiopsis derbyi*, etc. This is followed by the Rio Bonito formation containing local coal seams as well as glacial beds. The Palermo silts succeed the Rio Bonito formation and consist mainly of silts and glacial beds. The whole of the Tubarao Series is assigned to the Upper Carboniferous.

The marine horizons in this series occur in the south and contain a fauna related to the *Spirifer supramosquensis* zone of the European Uralian (Permo-Carboniferous). In the western and northwestern part of the basin, the facies representing the Tubarao Series is rather different and is called the Acuidauana formation. This consists generally of red sandstones and tillites the colour being due to the oxidation of the sediments. It contains only rare fossils and some silicified wood.

The succeeding group of beds, which is of Permian age, is called the *Passa Dois Series* whose thickness is about 300 m. This comprises the Irati Shale formation at the base and the Estrada Nova and Rio do Rosto groups above. The Irati Shale, which consists of black bituminous shales with *Mesosaurus* and other reptiles is a marker horizon corresponding to the White Band of South Africa. At places it contains *Conularia, Schizodus* and other fossils which indicate a marine origin. The other fossils include Crustacea, fish and fossil wood (*Dadoxylon*, etc.).

The succeeding Estrada Nova group consists of red and green shales and silts. This is developed mainly in Matto Grosso and South Goias. Its upper portion, called the Terezona Member, consists of fine sandstones, silicified limestones and grey-blue marls. This contains fish remains, silicified trees and some lamellibranchs. The upper stage of the Passa Dois Series is the Rio do Rasto group which contains reptilian fossils as well as several lamellibranchs which may be estuarine. The flora found in this includes *Taeniopteris, Phyllotheca, Glossopteris, Gangamopteris*, etc. indicating an Upper Permian age. The Estrada Nova beds are considered to be the equivalent of the Ecca formation of South Africa and of the Lower Patquia of Argentina.

The Sao Bento Series are constituted by eolian sands associated with basaltic flows. The lowermost horizon, which is probably of Upper Keuper age, is called the Santa Maria red beds which are probably the equivalents of the Molteno formation of South Africa. These consist mainly of current-bedded sandstones and red shales containing *Dadoxylon, Cedroxylon* and *Zuberia* (*Thinnfeldia*). There are also reptile bones belonging to several genera (*Rhynchocephalia, Dicynodontia, Cynodontia, Pseudosuchia*, etc.). These are regarded as somewhat later than the Beaufort reptiles, the Beaufort period being represented in Brazil by the well marked hiatus between the Rio do Rasto and Santa Maria beds.

The Santa Maria beds are succeeded by the Botucatu Sandstones which are red coloured and uniform in grain with stratification as in sand dunes. They are intercalated with some basalt flows. They are undoubtedly composed of desert sands, being the most extensive formation in the Parana basin, covering 1.2 million sq. km. and progressively overlapping lower formations in a northerly direction. There are local intercalations of lacustrine beds.

The Botucatu Sandstones are overlain by the Serra Geral formation which are largely composed of basaltic flows with intercalations of desert sands. They form plateau-like spreads in South Brazil and have varying thickness, the maximum being 900 m. They continue into the North Argentine pampas and cover over 1 million sq. km. The basalts are of uniform composition and consist of labradorite, augite, some pigeonite, olivine and accessory magnetite and apatite and a small amount of glassy matter. Ophitic texture is common in the denser parts. There are also amygdaloidal flows with zeolites.

Some nephelene-bearing syenitic intrusive and effusive rocks related to these are found to have cut through the Sao Bento and Tubarro Series in some places, though they are also largely to be found traversing Pre-Cambrian rocks. Some phonolites cut through the Botucatu Sandstones and these are mainly of Jurassic age. The Botucatu Sandstones are considered to be of Rhaetic age while the Serra Geral volcanics are Lower Jurassic.

In the centre of the Parana basin the lavas are covered by reptile-bearing sandstones called the *Bauru Sandstones* which are of Upper Cretaceous (Seno-

TABLE 34.—THE SANTA CATARINA SYSTEM (Brazil)

Age	Group	Formation	Member/Details
Cretaceous		Bauru sandstone (= Caiua formation)	
		Disconformity	
Jurassic	SAO BENTO	Serra Geral Volcanics and desert sandstones	
Up. Trias	SAO BENTO	Botucatu Sandstone (also Piramboia facies)	
Up. Trias	SAO BENTO	Santa Maria formation (= Molteno)	
		Disconformity	
Permian	PASSA DOIS	Rio do Rasto	Morro Pelado member
Permian	PASSA DOIS	Rio do Rasto	Serrinha formation (*Solenomorpha* and *Glassopteris* flora)
Permian	PASSA DOIS	Estrada Nova	Terezina member (varvites — *Lycopodiopsis derbyi*)
Permian	PASSA DOIS	Estrada Nova	Serra Alta formation (*Lycopodiopsis, Lepidodendron, Dadoxylon*, etc.)
Permian	PASSA DOIS	Irati ...	Irati Shale (= White Band) with *Mesosaurus*
Up. Carbon.	TUBARAO	Guata ...	Palermo Silts (partly glacial)
Up. Carbon.	TUBARAO	Guata ...	Rio Bonito (Coal seams, local glacials)
Up. Carbon.	TUBARAO	Itarare ...	Itarare group (Impure Glossopteris & Rhacopteris floras ; 3 marine horizons) (= Aquidauana facies)
		Disconformity	
Devonlan		Parana Series	Ponta Grossa formation Furnas Sandstone

nian age). There is a distinct unconformity at the base of the Bauru Sandstones.

ARGENTINA

The Gondwana rocks are found in the Sub-Andean belt as well as in the pre-Cordilleran region. In the former the Gondwanas are overthrust by Palaeozoic rocks and complexly folded and faulted. They overlie the Devonian unconformably. In the pre-Cordilleran region the Gondwanas were originally called the Paganzo System but have since been renamed the *Patquia Formation*. They are 900 m thick. The basal portion of the Patquia is largely glacial in origin. The underlying Tupe Series of Lower Carboniferous age is also largely glacial but there is an unconformity between the two. Marine intercalations in the glacials contain a fauna closely related to the *Spirifer supramosquensis* and *Euomphalus subcircularis* fauna indicating more frigid conditions than their equivalents in the Amazon valley. They are referable to Uralian age. There are also plant fossils which contain a mixed European Carboniferous and Glossopteris flora similar to that of Brazil and referable to the Sakmarian. The upper part of the Patquia formation consists of glacial tillites, fluvo-glacial and varved shales. These contain a pure Glossopteris flora which extends up to the top of the formation and are therefore the equivalents of the Passa Dois Series.

Overlying the Patquia formation are the so-called 'Argentine Rhaetic' but this has now been shown to contain four floral groups extending over the whole range of the Triassic. These flora have the greatest affinity to similar Australian flora but also show fairly close resemblance to Indian and South African flora. These Triassic beds begin after an erosional unconformity. In the upper part there are reptile zones while the uppermost portion is composed of thick basaltic flows. Table 35 shows the succession of Upper Gondwana strata in the pre-Cordilleran region of Argentina.

TABLE 35.—UPPER GONDWANAS IN PRE-CORDILLERAN ARGENTINA

Rhaetic	... ...	Gualo-Rio Blanco formations (400 to 1,200 m.)
Upper Keuper	... ...	Ischigualasto-Cacheuta formations (200 to 600 m.) with reptile horizons
Lower Keuper	... ...	Rastros — Potrerillos (400 to 700 m.)
		Hiatus
Middle Trias	... ...	Ischichuca — Cerro de las Cabras (600 m.)
Permo-Trias	... ...	Famatina volcanics, etc. (700 m.)

FALKLAND ISLES

The Falkland Isles contain strata similar to those of South Africa and Argentina both in Gondwana and pre-Gondwana times. Table 36 gives the rough stratigraphic equivalents. The Gondwana System here begins with the Lafonian tillite (equivalent of the Dwyka tillite) which overlies the Bluff Cove beds of Upper Carboniferous age. The tillite contains both semi-angular and rounded pebbles probably deposited in a marine environment. The general direction of ice movement appears to have been northwards.

TABLE 36.—THE GONDWANAS IN FALKLAND ISLANDS

	Falkland Isles	S. Africa
U. Triassic	West Lafonian beds	Molteno / U. Beaufort
L. Triassic	Bay of Harbour beds	M. & L. Beaufort
U. Permian	Choiseul Sound and Brenton Loch beds	L. Beaufort / Upper M. Ecca
L. Permian	Lafonian Sandstone	Lower M. Ecca
Basal Permian	Black Rock Shales	L. Ecca / U. Dwyka Shales
Permo-Carb.	Lafonian Tillite	Dwyka Tillite
U. Carb.	Bluff Cove beds	Lower Shales
L. Carb.	Port Stanley beds	Witteberg Series
M. Devonian	Port Philomel and Fox Bay beds	Bokkeveld Series
L. Devonian	Port Stephen beds	Table Mountain Series

The overlying Black Rock Slates are of basal Permian age equivalent to the Upper Dwyka and the Lower Ecca. This series contains a cherty horizon which is thought to be the equivalent of the white weathering cherty zone called the " White Band " in South Africa, but this horizon has so for not yeilded any organic remains.

Above the last formation comes the Lower Permian Lafonian Sandstone, which is a thin bedded friable brown feldspathic sandstone whose upper portion shows alternation of sandstone and shaly bands. These are also unfossiliferous. The Choiseul Sound and Brenton Loch beds which consist of finely bedded silt-stones containing *Glossopteris, Dadoxylon, Phyllotheca*, etc., are referable to Upper Permian age and may be the equivalents of the Upper Ecca and the Lower Beaufort beds.

The succeeding Bay of Harbour beds and West Lafonian beds are thick (2,400 m), coarse sandstones with interbedded mudstones, also containing several species of the Glossopteris flora. They range in age from Permo-Trias to Upper Trias. Lithologically the two series are similar but they have been separated only on palaeobotanical grounds.

Basic dykes similar to those found in the Karroo System cut the Devonian and the Lafonian sediments. It is known that these are later than the Lafonian sediments and therefore probably of Jurassic age. They must represent the dyke facies of the Stormberg lavas but there are no lava flows in the Falkland Isles.

ANTARCTICA

The Antarctic continent situated at the South Pole is entirely surrounded by oceans except that the Palmer Peninsula is connected through the Scotia Arc (Island Arc) to the Argentine Andes. The whole continent is under an ice sheet whose thickness exceeds 3,000 m in places. Two large indentations, the Ross Sea and the Weddell Sea which are marked by great ice shelves separate western from eastern Antarctica, their eastern margins really forming the western flanks of the great Trans-Antarctic mountain system. Rock exposures occur only at some places along the coasts and on a few peaks of mountains which rise above the ice cover. Information on Antarctica was meagre until about 1956 when the programme of the International Geophysical Year brought scientific teams to this continent for systematic studies aided by all modern facilities in transport and equipment provided by many Governments.

West Antarctica is composed of Mesozoic and Cenozoic rocks. Most of it is a continuation of the Andes Mountains through the Scotia arc, with a core of Mesosoic rocks and showing volcanic rocks of Upper Tertiary and Recent ages on the seaward side.

East Antarctica is a shield area having much in common with South Africa, India and West Australia. Precambrian rocks occupy most of its area except a zone close to the Trans-Antarctic mountains. The Precambrians consist of gneisses, schists, granites and charnockites, the last being abundant in the sector between meridians 10° and 100° in which are Enderby Land and part of Wilkes Land.

The Trans-Antarctic Mountains have a core of Precambrian rocks intruded by Precambrian and Lower Palaeozoic granites. The following general succession has been established in the region of the Beardmore glacier near the Ross Sea.

TABLE 37.—THE GONDWANAS IN ANTARCTICA

	FERRAR GROUP—Includes a number of thick sills of dolerite and flows of basalt (3000 m.) —Jurassic,	
	FALLA FORMATION—Repeated sequence of conglomerate. sandstones and siltstones, with 120 m basal arenites overlain by carbonaceous shales (1000 m.) —Trias to U.Permian	
BEACON GROUP ...	BUCKLEY COAL MEASURES—Arkosic sandstones silts and shales with coal seams yielding Glossopteris flora (800 m) —L. Permian	
	PAGODA TILLITE — Tillite overlain by alternating greenish siltstone and fine sandstone with some shale bands (150 m) —Basal Permian	
	LOWER BEACON — Mainly quartz-arenites with some siltstones etc. contains Terebratulids, fishes etc .of Deronian age (300 m.) —U. Carboniferous to Devonian	
HOPE GRANITE	... Intrusive granite, monzonite and aplite —L. Palaeozoic.	
BYRD GROUP	... Mainly limestones and dolomites containing Archaeocyathinae (600 m.) —Mid. to L. Cambrian	
BEARDMORE GROUP	... Greywackes, arkose, and low grade schistose rocks (1500 m.) —L. Cambrian to U. Pre-Cambrian	

The late Pre-Cambrian formations consist of greywackes, arkose and mildly metamorphosed schistose rocks. They are succeeded by Lower to Middle Cambrian limestones called the BYRD GROUP which occur as transported blocks in the Bearmore glacier and have now been traced to their source in the mountains. They contain several species of Archaeocyathinae, some species being identical with those in similar strata in the Adelaide Geosyncline in Australia and in early Cambrian strata in the Cape Province of South Africa. These rocks have been intruded by the HOPE GRANITE which includes granites and quartz-monzonites which are probably Ordovician in age.

The Cambrian strata are overlain by the great BEACON GROUP ranging in age from the Devonian to the Jurassic and having a total thickness of about 5,000 m. The lowest member of this is the LOWER BEACON SANDSTONE consisting of sandstones and siltstones containing marine fossils of Devonian to Carboniferous age. This is overlain by the PAGODA TILLITE which is early Permian and corresponds to the Talchir Tillite. The beds above the tillite are greenish siltstones, fine sandstones, arkose and some shale bands. Further up are the BUCKLEY COAL MEASURES, 800m thick, consisting of cyclically deposited strata with coal seams. These have yielded many plant remains of the

Glassoptesis flora which point to a Lower Permian age corresponding to the Barakars. The FALLA FORMATION succeeding the coal-bearing strata consist of a lower series of arenites, a middle portion of shales and an upper portion showing a repetition of sandstone, shales and siltstones. This formation is considered to be of Permian to Triassic age. The FARRAR GROUP which comes at the top of the Beacon Group consists entirely of a succession of volcanic flows and numerous thick sills of dolerites (individual sills being 15 to 450 m thick) of theoleiitic composition. The thicker sills sometimes show lower layers composed of coarse hypersthene- and olivine-dolerites. The Volcanic series is of Lower Jurassic age. In the Hope Bay area and along the coast of Graham Land are fissile shales containing remains of *Elatocladus, Brachyphyllum, Nilssonia, Otozamites* and several other genera of Jurassic age perhaps extending into the Neocomian. It will be seen that there is a great similarity, lithological and paleontological, between the Beacon Group and the Gondwana Group of India and the Karroo System of Africa which are all of the same age. The Beacon Group is succeeded by Cretaceous and Tertiary rocks which occur to the west of the Trans-Antarctic mountains.

A general comparison of the Gondwana stratigraphy of all the above regions shows that there was a period of severe glaciation in the Upper Carboniferous and Permo-Carboniferous. The glacial beds consist of tillites and fluvo-glacials and are followed by fine grained sediments or varvites reminiscent of slow glacial deposition. The Permian is the period of main coal formation in Gondwanaland and is characterised by the very distinctive Glossopteris flora with its numerous genera and species, which developed from the hardy survivors of the more cosmopolitan Upper Palaeozoic flora much of which had been wiped out during the intense cold of the Upper Carboniferous glaciation. The Triassic is characterised by continental sediments formed under arid desert conditions. They are generally brightly coloured sandstones and shales showing a high state of oxidation. They contain the remains of amphibia, reptiles and fishes which must have lived in and around the gradually contracting lakes and river valleys of the period. The Triassic arid climate brought about a marked change in the flora, the characteristic members of the Glossopteris flora being replaced by the Thinnfeldia — Ptilophyllum flora. The Jurassic period gives evidence, at least in some areas, of the return to milder climate, for the strata contain carbonaceous material, though scarcely any coal seams. In large parts of South America and South Africa, however, the deposits of Jurassic age are desert sandstones and thick volcanics. Marine beds are generally absent except in the marginal regions, as in East Africa, Madagascar and Western India (Kutch). The volcanics are comparatively of less importance in India (Rajmahal volcanics).

There is a considerable mass of data which lead to the conclusion that South America and South Africa were contiguous or were part of one land mass, as has been pointed out by Du Toit and more recently by Caster (See paper in *Bull. Amer. Mus. Nat. Hist.* **39**, part 3, 1952). The remarkable similarity and even identity of numerous species in the flora points to the unrestricted migration over lands which were close to each other and were not

Table 38
CORRELATION OF GONDWANA STRATA

	S. AFRICA		AUSTRALIA		BRAZIL		ARGENTINA		INDIA	
Jurassic	Drakensberg Volc.	STORMBERG	Volcanics etc.		Serra Geral Volc.	SÃO BENTO	Volcanics		Jabalpur Kota Rajmahal Volc.	
Up. Trias Rhaetic	Cave sst.								Mahadeva (Pachmarhi)	MAHADEVA
Noric	Up. Molteno Red Beds		Wianamatta (Up. Ipswich)	HAWKESBURY	Botucatu sst. & Volcanics		Gualo-Rio Banco			
Carnic	Molteno		Arenito Hawkesbury (L. Ipswich)		Santa Maria Red Beds		Ischigualasto-Cacheuta			
Mid. Trias Ladinic	U.Beaufort Cynognathus Procolophon	BEAUFORT	Narrabeen				Rastros-Potrenillos		Panchet	PANCHET
Anisic	M.Beaufort Lystrosaurus		Hunter-Bowen Orogeny		FalseSanta Maria		Ischichuca-Cerro-de-las-Cabras			
L. Trias (Scythic)	Lower Beaufort Cysticephalus Endothiodon Dinocephalus		Newcastle Coal (Dempsey-Tomago)	KAMILAROI	Rio-do-Rasto	PASSA DOIS	Famatina Volcanics			
Up. Permian	Ecca	ECCA	Up. Marine		Estrada Nova		Patquia		Raniganj	DAMUDA
M. Permian	White Band		Greta Coal measures		Inati Shale				M. Damuda	
L. Permian	Dwyka	DWYKA	L. Marine (Bacchus Marsh) (Lochinvar)		Tubarao	TUBARAO	Patquia		Barakar Karharbari	
Up. Carb.	Lower Shale		Up. Kuttung Kanimblan Orogeny	KUTTUNG			Up. Tupe BajoVelis Asturian orogeny		Talchir Hercynian orogeny	TALCHIR
M. Carb.	Witteberg		L. Kuttung (Burindi)		Parana		Tupe			
L. Carb.										
Devonian										

/// Hiatus   ××× Tillites   ≡≡≡ Coal

separated by any large water barriers. The reptile and fish faunas also show much similarity.

The great continent of Gondwanaland apparently began splitting up in the Jurassic and early Cretaceous. The vast outpourings of lavas in the Jurassic were a manifestation of the tension to which the crust was subjected, resulting later in the separation and drift of the continents. The marine Cretaceous occurring along the coasts of India, Western Australia, East Africa and Argentine show the marine transgressions which took place over the separating continents during the Cretaceous. There are also evidences of marginal foundering along the eastern costs of Brazil, the western coasts of the southern parts of Africa, as well as the western coast of India, during the Tertiary era.

STRUCTURE OF THE GONDWANA BASINS

The Gondwana rocks occupy tectonic troughs with faulted boundaries arranged along linear zones, the magnitude of the faults on the two major sides being very unequal. This has the effect of producing a dip of the strata towards the faulted side with the greater throw. Thus, in the Gondwana belt of the Damodar valley, the faults run E.-.W. and the strata dip generally towards the more faulted southern boundary.

In the Pranhita-Godavari Valley the direction of the faults is roughly N.W.-.S.E. and the dip of the strata is towards the northeast. In the Chhattisgarh-Mahanadi basin the trend of the faults is N.W.-.S.E., the Gondwana strata lie mainly to the northeast of the major faults and dip to the southwest, while those lying to the south of the faults dip towards the northeast. In all cases the major faults conform to the direction of strike of the gneissic country rocks. In addition to the major trough faults which form the boundaries of the various coalfields, there are also numerous other faults which cut across the strata. The age of the trough faults has not been determined but they are supposed to be largely post-Lower Gondwana. It is possible that the troughs along the Godavari and Mahanadi valleys were in existence in Upper Carboniferous to early Permian times, because a shallow water connection is indicated between the Permian marine basin of central India and that of Western Australia to explain the close relationship of the fauna of the Umaria marine beds to that of the Lyons Group in the Carnarvon basin. In almost all the coalfields one side of the trough is much shallower than the other and may even be unaffected by faulting ; for instance, in parts of the Raniganj coalfield, the northern boundary shows the strata in an undisturbed condition of original deposition without any faulting, while the southern side has a throw estimated at 2,750 m. In the Jharia coalfield the southern boundary fault has a throw of some 1,500 m. In some of the basins a certain amount of unconformity and overlap may be seen in the strata towards the margins. The present coalfields are generally the remains of much larger basins and owe their preservation from denudation to the trough faults. The strata generally dip at a low angle but they may show higher inclination near faults and intrusions. Most of the faults within the basins are of normal type, but some are tear, sag or hinge faults. They are referable to two major groups, one set trending W.N.W.-E.S.E. and the other W.S.W.-E.S.E., the first set being generally the more prominent. The chief direction of faulting, especially in the eastern coalfields, is E.N.E.-W.S.W. The age of the faulting in the Damodar valley coalfields is generally post-Panchet and pre-Mahadeva and in some cases post-Mahadeva. There are also post-Deccan trap faults in the Satpura area.

The dolerite dykes intrusive into the Lower Gondwanas in the Satpura coalfields are probably related to the Deccan Trap. In the eastern coalfileds the dolerite intrusions are considered to be related to the Rajmahal traps. It is also thought that the Deccan traps originally extended to as far east as Lohardaga in the Ranchi district where the outliers have been completely lateritised. If the Deccan traps did extend so far out, it is not unreasonable to argue that dykes connected with them may also have extended into the Damodar valey coalfields. As the Rajmahal and Deccan traps have practically identical mineralogical and chemical characteristics, so far as known, the only way to distinguish them would be with the help of trace elements. The lamprophyre dykes and sills are also of much the same age, but as they are present only in the Damodar valley and Himalayan coalfields their origin should be sought for in eastern India. They appear to be more or less contemporaneous with the Rajmahal traps.

Most of the Gondwana basins in the Peninsula are free from folding disturbances though occasionally such are to be found near the more pronounced faulted margins of the troughs. In the Himalayas, however, the Gondwanas have been severely affected by the Tertiary mountain building movements. The Gondwanas in Sikkim, Bhutan and other Himalayan areas are thrust over the Siwalik rocks and are in turn overthrust by Palaeozoic rocks and crystalline schists. In consequence, the coal seams in these have suffered crushing and devolatilisation and have been converted into semianthracitic material.

CLIMATE AND SEDIMENTATION

The Gondwana era was initiated by a glacial climate during which a vast continental ice sheet covered a large part of Gondwanaland. So far as India is concerned the ice sheet may have covered Rajasthan and Madhya Pradesh as well as the Eastern Ghats area. Glaciers appear to have flowed out from Rajasthan towards Salt Range where the Talchir Boulder-bed overlying the Cambrian succession or the Speckled Sandstone, contains pebbles and boulders derived at least in part from Rajasthan. The boulder-beds of roughly the same age, which have been found in the Kashmir-Hazara region, in the Simla hills, in the Sub-Himalayas of Sikkim, Bhutan and further east, apparently mark the northern limits of Gondwanaland at that time though some of these may have travelled conisderable distances towards the south from their original position because of overthrusting. It is likely that most of the boulder-beds in the above-mentioned areas were laid down by glaciers in the sea.

The Talchir boulder-beds in the Damodar valley contain large quantities of quartzites and gneissic rocks, the former having a great resemblance to the Vindhyan quartzites in Sone valley to the N.W. of the coalfields. These materials may have travelled towards east or southeast from the Vindhyan highlands. The boulder-beds in the Godavari valley coalfields are supposed to contain rocks derived from the Eastern Ghats region. Much further work is necessary to determine the exact nature and composition of the boulders and pebbles and to trace their source.

The climate gradually warmed up during the Talchir period when fluvio-glacial sediments and varved clays were laid down. The prevailing greenish tints of the rocks indicates that the climate was still quite cold. Though there are plant impressions in the beds immediately overlying the boulder-beds, the flora was apprently still scanty during the Talchir period.

In the Damuda period the climate was definitely warmer and more humid and permitted the growth of luxuriant vegetation. An enormous amount of vegetation was carried by streams into the swamps and lake basins to form the coal beds. This period marked the zenith of development of the Glossopteris flora, for we find the Raniganj beds in Bengal have yielded the largest number of genera and species of the Glossopteris flora.

The Damuda Series attains a total thickness of over 2,000 m in the eastern coalfields of the Damodar valley but the original thickness may have been appreciably larger. This series becomes thinner when followed westwards into the Karanpura field and farther west. The strata consist of sandstones containing kaolinised feldspars followed by shales and then by coal. The succession is repeated many times and during the Barakar period there may have been as many as 50 or 60 cycles of sedimentation. The Barakar strata in the Jharia coalfields contain at least 25 coal seams each being a part of a sedimentary cycle. This would have been possible only if there were repeated sinking of the basin of sedimentation. The nature of the flora as well as the presence of undecomposed or partly decomposed feldspars in the sediments indicate that the climate was cold temperate. In the eastern coalfields which have been studied in some detail, it is seen that the coal seams as well as the associated strata generally thicken in a westerly direction. The seams are more numerous and less pure in the west than in the east, and also split up into thinner seams westward. These indicate that the source of the sediments was somewhere in the west and that the basin of sedimentation was deeper and quieter towards the east. The drainage seems to have flowed from the west to east in the Damodar valley. It probably found an outlet into what is now the head of the Bay of Bengal or in the northeast where there are several small coalfields in West Bengal and Santhal Parganas. Indeed, Blanford has suggested that the coal-bearing rocks might extend under the Gangetic alluvium into the Himalayan region of Sikkim which must have been the limit of land in the Damuda times. In the case of the Godavari valley, it would appear that the drainage was in a northwesterly direction towards the tract which formed an arm of the sea extending from the Cambay region along the Narmada valley into Madhya Pradesh. It is thought likely that a reversal of drainage towards the east and southeast took place along the Godavari and Mahanadi valleys in the Jurassic. However, detailed studies have yet to be made for determining the source of sediments and the trend of the drainage systems during the Gondwana era.

In all the Gondwana coalfields of India the available evidence points to the fact that the vegetation had travelled some distance before being deposited ultimately to form coal. In no case has any upright tree stem been found in the coal seams nor roots extending into the under-clay. It is true that stems are found but they generally lie more or less flat on the top of the coal seams. Most of them are silicified except the cortical portion which has been carbonised. Owing to pressure the stems have generally been crushed to an elliptical section and they indicate that they have drifted from the place where they originally grew. The nature of the coal seams is indicative of their ' drift ' origin. All Gondwana coals contain high ash and even the best seams contains not less than 5 or 6 per cent ash. The ash is inherent in the coal, being more or less uniformly distributed in the coal matter and therefore very difficult to eliminate by ordinary washing processes. Regarding the proportion of coal to the strata in which they occur, it may be said that it is generally high. In the Barakar stage of the Damodar valley coalfields, the proportion of coal to strata is roughly 1 to 8 or 10. In the Raniganj Series the proportion is less ranging from 1 : 20 to 1 : 35. Some of the seams in the Barakars are very thick, as for instance the Kargali and Bermo seams in the Bokaro coalfields, which have thickness of 15 -30 m, and the Korba seam in the Korba coalfield which is over 30 m thick. In the southeastern part of the Jharia coalfield the coalescence of several seams has produced one seam which is 26 m thick.

The coal-bearing rocks of the Godavari valley are apparently continued underneath the Deccan traps into the Badnur and Chhindwara areas. There is little doubt that coalbearing rocks are present underneath the Deccan traps, in an area of many square kms, but no serious attempt has so far been made to locate them.

Even during the upper part of the Damuda period, the areas outside the Damodar valley seem to have experienced a somewhat drier climate for we find the Kamthis, which are the equivalents of the Raniganj Series, are composed of reddish ferruginous sandstones. They are generally barren of coal seams and only rarely show some streaks of carbonaceous matter.

The succeeding Panchet Stage definitely marks a change of climate. There is a slight unconformity between the Raniganj and Panchet beds in the Raniganj coalfield and probably in other areas. The Panchets enclose practically no carbonaceous matter and appear to have been laid down in flood plains and shallow lakes. The presence of labyrinthodonts in them indicates a period of gradual drying up of the sedimentary basins.

The Mahadeva Series, represented by the Supra-Panchet and the Pachmarhi beds which are of Triassic age, are definitely sediments of an arid climate. They consist of ferruginous sandstones often with thin layers of hematite and are entirely devoid of carbonaceous matter. There is a hiatus between the Panchets and the Mahadevas though its magnitude is at present a matter of conjecture. After this, in the Jurassic, there is an indication of a return to more favourable and moist conditions. The Glossopteris flora practically died out during the dry period of the Triassic and a new flora (Thinnfeldia — Ptilophyllum flora) gradually established itself in the Rajmahal times, and continued well into the Cretaceous. During the Rajmahal times, however, the earth's crust seems to have experienced tension in Gondwanaland resulting in the outpouring of vast quantities of lavas and volcanic materials. Lavas of this period are found in the Ganges delta but their full extent is not known.

PERMO-CARBONIFEROUS FLORAS

Prior to the Middle Carboniferous, a considerable degree of uniformity prevailed in the flora of the different parts of the world. These floras began to differentiate into local groups. With regard to Australia, Walkom [1] has identified a succession of three distinct floras in the Carboniferous and the Permian, consisting of :

(1) The *Lepidodendron veltheimianum flora* of lower Carboniferous age.

(2) The *Rhacopteris flora* or Lower to Middle Carboniferous age characterised by *Rhacopteris, Cardiopteris, Noeggerathia, Adiantites*, etc. This is already differentiated from the Eur-American flora of Carboniferous age which contains *Neuropteris, Pecopteris, Alethopetris*, etc.

(3) The *Glossopteris flora* commencing from the Permo-Carboniferous and attaining its maximum in the Permian. This is entirely different from the Rhacopteris flora.

The Permian and Permo-Triassic contain four distinct floras each with its own region of distribution. They are as follows:—

(1) **The Eur-American Folra** which is the best known of the four, occupied the eastern United States and the whole of Europe as far as the Ural mountains, Turkestan and Iran. Its areas is now separated by the North Atlantic.

(2) **The Angara (Kuznetz) Flora** which occupied Asia to the east of the Urals, down to the Pacific coast and to the north of Korea and Mongolia.

(3) **The Cathaysian (Gigantopteris) Flora.** This extended from Korea and northern China southwards into Indo-China, Thailand and Sumatra in Asia, and into western North America down to Okhlahama and Texas. Its western border was in Kansu in western China. Its area is now separated by the North Pacific.

(4) **The Glossopteris Flora** which prevailed over all the present southern continents and also India. The area of its distribution is now widely separated by the Atlantic and Indian oceans.

As already stated, the Eur-American flora has been well studied and is the best known, the characteristic members being *Lepidodendron, Pecopteris, Neuropteris*, etc. It met the Angara flora in the Urals region and the Cathaysian flora in western China and Chinese Turkestan. Its western limit was in southeastern United States.

---

[1] Walkom, A. B. The succession of Carobniferous and Permian floras in Australia. *Proc Roy. Soc. N.S.W.*, 78(1/2), pp. 4-13, 1944.

The Angara flora characterised the Asiatic region known as Angaraland. In the early stages it occupied only part of Siberia but extended gradually down to the Pacific coast and as far south as the Nanshan mountains in China where it displaced the Cathaysian flora. Its characteristic members are *Psygmophyllum, Callipteris* and *Czekanowskia,* but it contains some elements of the other floras including even the Glossopteris flora which latter is supposed to have travelled through a land bridge over Kashmir and the Pamir region.

The Cathaysian flora, typically developed in Shensi, is found in Korea, the greater part of China, Indo-China, Thailand, Malaya and Sumatra. In the earlier stages it had much affinity with the Eur-American flora but gradually developed its own characteristic elements in the later stages — *Lobatannularia, Protoblechnum, Gigantopteris, Saportaea, Chiropteris, Tingia* etc. The best known species of this flora, viz., *Gigantopteris nicotinaefolia,* developed in the Middle Permian. This flora extended along the Pacific coast of Asia and over the Bering Straits into western North America. It met the Eur-American flora in Chinese Turkestan and the Glossopteris flora in western China as well as in New Guinea. It was gradually swamped in the later stages by the Angara flora which spread down to the Pacific coast of Siberia and into northwestern China.

The Glossopteris flora is the purest of the four, for it contains the least admixture with the other floras. It must have been derived from the few elements which survived the Permo-Carboniferous glaciation of the southern hemisphere. Spores of *Glossopteris* and other genera are found in the shales associated with tillites. It probably originated in Antarctica which apparently occupied the central region of Gondwanaland. The earliest members of this flora to appear in India are *Gangomopteris cyclopteroides, Glossopteris indica* and *Noeggerathipsis,* in the Talchirs. In the succeeding Karharbaris these continued, while *Gondwanidium* and *Buriadia* appear, as well as *Schizoneura, Ottokaria, Cardaicarpus,* etc. The flora attained its best development in the Raniganj period which contains the largest number of species. The majority of the plants are Pteridosperms, with practically no Filicales and only a few Conifers. The flora began to die out during the Panchet and most of the species disappeared during the succeeding Mahadeva. *Thinnfeldia (Dicroidium)* appeared in the Panchet while *Ptilophyllum, Williamsonia, Cladophlebis,* etc., appeared in the Rajmahal times. Only a few species of *Glossopteris Phyllotheca, Dadoxylon* and *Schizoneura* continued into the Upper Gondwanas. The Cycadophyta attained their full development in the Upper Gondwana, while Filicales and Conifers also became important. In the later stages appeared *Elatocladus, Brachyphyllum, Araucarites,* etc. which continued well into the Cretaceous in the Umia Series of western India.

Most of the Mesozoic Gymmosperms and Cycadophyta have since disappeared. The Angiosperms began to appear and gained importance in the Cretaceous, being now the dominant plant group.

The Tethys separated India from the regions of western and southern China, Burma and Malaya which at present are its close neighbours. The only land connection between India on the one hand and China and Angaraland on the other in Lower Gondwana times was through the volcanic islands (of Panjal Trap) which appear to have been present in the Kashmir region. These lands came close together only in the Cretaceous and Tertiary when the Tethys was obliterated and raised into a mountain belt. This explains the absence of any inter-mingling of the Permian floras across northeastern India and the very different geological history of S.W. China, Burma etc., during the Mesozoic.

ECONOMIC MINERALS IN THE GONDWANAS

Clays.—The main importance of the Gondwana System centres around coal, but there are also deposits of various types of clays. The coal seams are often associated, especially in the Barakar Stage, with important beds of fire-clay which are worked for the manufacture of refractory bricks. Other types of clay available from the Gondwana strata are useful for making bricks, pottery, terra cotta and china-ware. Factories utilising clay of various descriptions are situated in the Raniganj, Jharia and Jabalpur areas, Fuller's earth is obtained from the Jabalpur and Katni areas while the Upper Gondwanas near Madras contain bentonitic clays with bleaching properties. White clay and moulding sand are obtained from Mangalhat in the Rajmahal hills.

Sandstone.—The Barakar, Raniganj, Kamthi and Pachmarhi sandstones are used locally as building material, though they are not, in general, comparable in quality to Vindhyan sandstones. The Ahmednagar Sandstone of Idar State has been used in the delicate tracery that adorns the mosques of Ahmedabad. A sandstone of satisfactory durability is the Athgarh Sandstone which has been employed to some extent in the magnificent temples of Puri, Bhuvaneshwar and Konarak. It is fine grained and is suitable also for carving. It is in the Athgarh Sandstones that the caves at Khandagiri, near Bhuvaneshwar, have been carved out. The Tirupati sandstones and the Satyavedu sandstones have also been used for building purposes.

The Barakar sandstones are also, in places, suitable for making millstones and abrasive stones. They may occasionally also serve as sources of good quartz sand if the impurities are small or could be easily separated.

Iron-Ore.—The beds of sideritic iron-ore and their oxidised outcrops occurring in the Ironstone Shales of the Raniganj coalfield have been worked as iron-ore for the blast furnaces of the Barakar Iron Works and its successor the Bengal Iron Co. Similar ironstones occur also in Auranga and Hutar coalfields. Pockets of limonitic iron-ore and of ochre are also found in the red sandstones of Kamthi and Mahadeva age. The reserves of ironstones in the Raniganj coalfield may be estimated at about two thousand million tons. They contain about 40 to 45 per cent iron and 16 to 19 per cent silica and are of a low grade compared to the rich hematite ores in the Iron-ore Series of Singhbhum-Keonjhar-Bonai.

Coal.—Most of the coal in the Gondwanas is found in the Damuda (Damodar) System, *i.e.*, both in the Barakar and Raniganj Series, the former being the more important one. Coal seams occur also in the Kota, Chikiala, Jabalpur and Umia Stages but they are of small extent and thickness and generally of inferior quality.

Coal seams are developed in practically all the areas where the Barakar Series occurs. Raniganj Series coal is important only in the Raniganj coalfield though found also in Jharia, Bokaro and a few neighbouring fields. Coal of Jabalpur age is found as unimportant seams in the valley of the Hard river, and of Kota and Chikiala age in the Godavari drainage area. Thin seams occurs also in the Umia beds of Cutch.

Coal consists of carbonised remains of vegetation accumulated either *in situ* or transported by water and deposited. Practically all Indian coal sems to be of the latter type. Chemically it consists essentially of carbon and hydrogen with subordinate amounts of oxygen and nitrogen. These are combined in very complex ways. Coal is usually banded, the bands being dull and bright. The dull bands are composed of *durain* which is organic matter mixed with extraneous mineral matter which latter constitutes the 'ash' of coal. The bright bands known as *vitrain* are much purer and are high in volatiles. There are two other constituents also, one being *clarain*, a silky or satiny looking material, and the other *fusain* which looks like soft friable charcoal and soils the hands when handled.

For ordinary commercial purposes it is necessary to have a knowledge of the proximate composition of coal, determined according to certain standardized empirical procedure. In this the moisture, volatile matter, fixed carbon and ash are determined. The calorific value, either in British Thermal Units (per lb.) or in Calories (per kilogram or lb.) is also generally found, as well as the caking tendency. The colour of ash gives an idea of the amount of the iron present and indirectly its tendency to clinker on grates when burnt. It is necessary also to know the percentage of phosphorus and sulphur present, and the form (pyritic, sulphate or organic) in which the sulphur occurs.

The ultimate analysis determines the percentage of elements present (O, H, C, N) and is useful in evaluating the suitability of the coal for certain purposes, particularly in the chemical industry.

The general characters of coal of the Barakar and Raniganj Series are shown below :—

Barakar	Raniganj
Low moisture (1 to 3 per cent)	High moisture (3 to 8 per cent or more).
Low volatile (20 to 30 per cent)	High volatile (30 to 36 per cent).
High fixed carbon (56 to 65 per cent)	Medium fixed carbon (50 to 60 per cent).
Excellent steam coal and often excellent coking coal.	Generally poorly caking, though some are moderately so ; good gas coal and long flame steam coal.

The classification adopted by the Coal Grading Board in India is used for trade purposes. The scheme in use is shown in Table 39.

Most of the Gondwana coals are good steam and gas coals, none being anthracitic. In the Himalayas, however, the coal has generally been crushed by earth movements and is in consequence low in moisture and volatiles, but very friable. The best coking coals are practically confined to Jharia, Giridih and Bokaro fields while some from the Raniganj and Karanpura fields are semi-coking and can with advantage be blended with Jharia coal to yield good coke. A part of the coal in Giridih is exceptionally good, being very low in phosphorus and useful for the manufacture of ferroalloys.

In the Bengal and Bihar fields, where a large number of seams lie one over the other, it is found that the lower seams are generally higher in fixed carbon and lower in moisture and volatiles than the ones above, which may be explained as due to the effect of pressure and compaction. When a region is subjected to folding, as in the Mohpani field of the Madhya Pradesh, these effects are still more marked in comparison with a neighbouring field.

TABLE 39.—COAL CLASSIFICATION (INDIAN COAL GRADING BOARD)

Grade	Low volatile	High volatile
Selected	Upto to 13 per cent ash Over 7,000 cal.	Up to 11 per cent ash Over 6,800 cal.
Grade I	13 to 15 per cent ash Over 6,500 cal.	11 to 13 per cent ash Over 6,300 cal. Under 9 per cent moisture
Grade II	15 to 18 per cent ash Over 6,000 cal.	13 to 16 per cent ash Over 6,000 cal. Under 10 per cent moisture
Grade III	Inferior to the above	Inferior to the above

GONDWANA COALFIELDS OF INDIA

As already mentioned, the coalfields are found along certain linear tracts and can be divided into groups. The names of the fields given below are those shown on the map accompanying Fox's 'Lower Gondwana Coalfields of India' (*Mem.* 59).

Himalayan area.—Abor, Miri, Daphla and Aka, Bhutan hills, Buxa Duars and Darjeeling. These fields are of little economic importance as they are inaccessible and the seams are folded, faulted and crushed.

North Bengal.—Hura, Gilhuria, Chuparbhita, Pachwara and Brahmani. These are all small and unimportant.

Damodar valley.—The northern group includes Kundit Kuraia, Sahajuri, Jainti, Giridih (Karharbari), Chope, Itkhuri and Daltonganj. This is much less important than the southern group which includes Raniganj, Jharia, Bokaro, Ramgarh, North and South Karanpura, Auranga and Hutar. These are at present the most important producing fields and contain almost all the coking coal reserves of the country.

Madhya Pradesh (Eastern group).—In this area the middle series of the Gondwana System are well developed and the coalfields appear in the strata beneath them. They include Tatapani, Ramkola, Singrauli, Korar, Umaria, Johilla, Sohagpur, Sanhat and Jhilmili.

Between the extensive Sohagpur field and the northern border of the Chhattisgarh area there is a large spread of Talchir rocks so that some of the following appear as outliers on the Talchirs : Jhagrakhand, Kurasia, Koreagarh, Bisrampur (Sirguja), Bansar, Lakhanpur, Panchbhaini, Damhamunda and Sendurgarh.

Madhya Pradesh (Western or Satpura).— Mohpani, Sonada, Shahpur, Dulhara, Patakhera, Kanhan valley and Pench valley. Of these, Mohpani is an isolated field lying north of the Satpuras, the others lying along their southern border. Pench valley is being actively exploited at the present time.

Mahanadi valley.—Hasdo-Rampur (Sirguja), Korba, Mand river, Raigarh (N. and S.), Himgir, Ib river (Rampur) and Talchir. Some of these are potentially important.

Wardha-Godavari valley.—Bandar, Warora, Wun, Ghughus-Telwasa, Chanda, Ballarpur, Wamanpalli, Sasti-Rajura, Antargaon, Tandur, Sandrapalli, Kamaram, Bandala-Allapalli, Lingala, Singareni, Kottagudem, Damarcherla, Ashwaraopeta and Bedadanuru.

In this area there is an extensive development of Kamthi and younger rocks, the coal measures appearing as isolated patches separated by and from underneath the younger rocks. There is every probability that several of these coalfields are much more extensive than they appear, and extend underneath the younger rocks. The extent of the Godavari valley fields is therefore likely to be considerable.

THE RANIGANJ COALFIELD

This is the easternmost field in the Damodar valley and is situated around Asansol, about 210 km northwest of Calcutta. It covers about 1,550 sq. km of proved coal-bearing area. It is surrounded on three sides by Archaean rocks but on the east it passes beneath alluvium and laterite where its extension is a matter of speculation to be proved by drilling. Table 40 shows the succession of the formations exposed in the field.

The Raniganj coalfield is faulted down on the south and west, the southern boundary being a series of faults, running *en echelon*, indicating a throw of 2,700 m near the Panchet hill. Over the greater part of the northern side, the Gondwana boundary is one of original deposition, modified of course by later erosion. The oodest beds are found in the north, and are overlapped by younger beds in a southward direction, the general dip being also southward. Besides the boundary faults, there are also oblique and cross faults in the field. The main dislocation probably took place in the Jurassic. The field is traversed by many dolerite and micaperioditite dykes, the latter having produced much damage to coal. The intrusives are later than the faults and may be of Rajmahal or Deccan trap age.

TABLE 40—GONDWANA SUCCESSION IN THE RANIGANJ COALFIELD

Formation	Description	Maximum thickness
		Metres
Supra Panchets.	Red and grey sandstones and shales	300
Panchet	Micaceous yellow and grey sandstones, red and greenish shales	600
Raniganj	Grey and greenish soft feldspathic sandstones, shales and coal seams	1,050
Ironstone shales	Dark carbonaceous shales with ironstone bands	360
Barakar	Coarse and medium grey and white sandstones, shales and coal seams	630
Talchir with boulder-bed at the base.	Coarse sandstones above and greenish shales and sandy shales below	300

Coal.—Coal seams, most of which have two or more local names, occur both in the Barakar and Raniganj stages. The seams of the Barakar Series are named as follows from below upwards :— Pusai ; Damagaria-Salanpur A ; Bindabanpur-Salanpur B ; Gopinathpur-Salanpur C-Kasta ; Laikdih-Shampur 5-Ramnagar ; Shampur 4 ; Chanch-Begunia-Shampur 1. The seams in the Raniganj stage are, from below upwards :—Taltor ; Sanctoria-Poniati ; Hatnal-Koithi ; Dishergarh-Samla ; Bara Dhemo-Raghunathbati-Manoharbahal-Rana-Poriarpur-Satgram-Jotejanaki-Dobrana-Sonpur ; Sripur-Toposi-Kenda Chora-Purushottampur ; Lower Dhadka-Naramkuri-Banbra-Sonachora-Bonbahal ; Bara Chak-Nega-Jameri Raniganj-Lower Kajora-Jambad-Bowlah- Bankola ; Gopalpur-Upper Dhadka-Satpukhuria-Ghusik-Searsol-Upper Kajora ; Hirakhun-Narsamuda.

Of the above, several of the Barakar seams and in particular the Ramnagar, Laikdih, and Begunia seams are caking. The Sanctoria-Poniati and Dishergarh seams belonging to the Raniganj stage are excellent steam coals which can also be blended with good Jharia coals for making coke. Estimates of the coal reserves by the Geological Survey of India in 1953-54 in this field, including that already exploited, are :

Kinds of coal	Million tons	
	300 m depth	600 m depth
Coking (grade I and better)	288	548
Non-coking (grade I and better)	2,759	4,838
Non-coking (other)	4,949	8,432
Total	7,996	13,818
Coal still in ground	7,456	13,807

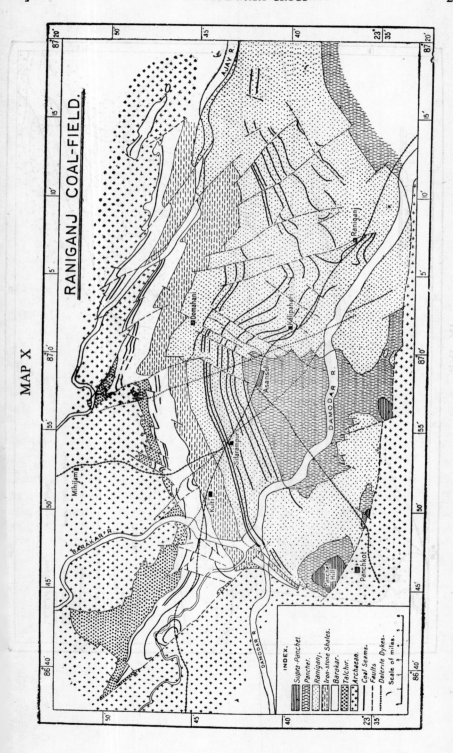

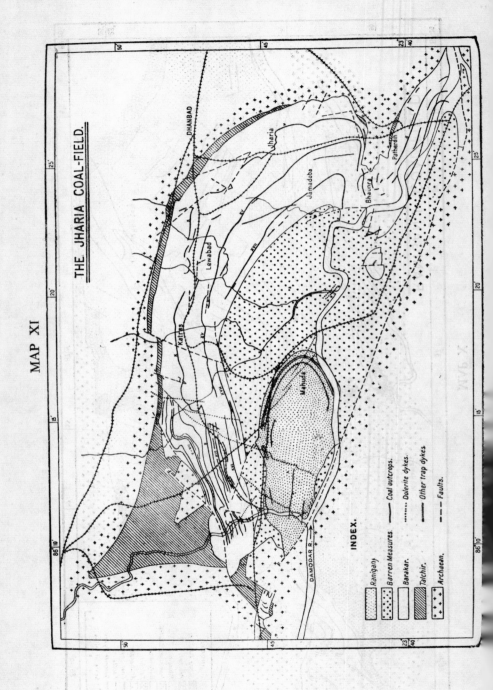

This coalfield has been worked since about 1800, the total amount raised to the end of 1950 being around 350 million tons, from available statistics. The difference between this and the figure obtainable from the Table represents coal lost in working, by fire, flooding, etc.

JHARIA COALFIELD

This is the most important coalfield of India, being responsible for something like 40 per cent of the total Indian production, besides being the most important storehouse of the best coking coal. It is situated about 275 km west of Calcutta, the town of Dhanbad lying on its northeastern corner. The field is roughly sickle-shaped, being about 20 km N.-S., and 38 km E.-.W. The total area of Gondwana rocks exposed is 710 sq. km of which Barakars occupy 220 sq. km, and Raniganj Series 55 sq. km. The other strata exposed are Talchirs and Barren Measures. The Talchirs are found along the northern and western margins with a maximum thickness of 240 m, the boulder-bed being 15 m thick. The Barakar Series and the Barren Measures have each a thickness of 600 m while the Ranigunj Series is 560 m thick, the former exposed in the north and east of the field. The Raniganj measures form an ellipitcal outcrop in the southwestern part of the field.

The Jharia coalfield is faulted against the Archaeans along the southern margin, where the strata dip inward towards the centre of the field. Two small horsts of gneiss are found, one in the northeastern and the other in the northwestern part. There are also faults on the other sides as well as cross-faults within the field, most of them being of the ' sag-fault ' type. A large number of dykes of dolerite and mica-periodite traverse the field.

Though coal seams occur in both the Barakar and Raniganj stages, those in the Barakars are by far the most important. This stage contains more than 25 seams, the more important of which are numbered I to XVIII from below upwards, seams X and above being of good quality. Amongst the best seams in this field (indeed in India) are XI, XIV, XIV-A, XV, XVII and XVIII.

The total available reserves and those which were originally present in the Jharia field, as per estimates made in 1953 by the Geological Survey of India are as follows. The available data are confined to a depth of only 600 m from the surface.

The field has been exploited since about 1895, the amount raised up to 1960 being about 700 million tons.

	Million tons	
	0 to 300 m	0 to 600 m
Grade I and better	2,005	2,682
Other grades	6,112	9,672
Original reserves, total	8,117	12,354
Reserves now in ground	7,442	11,679

THE COAL RESOURCES OF THE GONDWANAS

According to recent estimates the reserves of all Gondwana coal in seams, 0-25 m or more in thickness, are of the order of 90,000 million tons, but if the calculation is restricted to seams of 1.2 m thickness and over, the reserves would be of the order of 35,000 million tons. Considering coals of comparatively good quality, viz., those containing 20 per cent or less ash, the latter figure will be reduced perhaps to about 10,000 million tons. Good quality coking coals are confined practically to the eastern coalfields of the Damodar Valley — Raniganj, Jharia, Bokaro and Giridih, but the last is nearing exhaustion. The coals of the Karanpura and Kanhan valley fields are coking to semi-coking. All the other fields are so far known to contain only non-coking coals, but some of them can be blended with good Jharia coking coal to yield coke of suitable strength for large blast furnaces. An estimate made in 1950 by the Metallurgical Coal Committee puts the reserves of coking coal in India at 2,000 million tons in the ground, to recover all of which it will be necessary to adopt srtingent conservation measures.

It is only the Raniganj and Jharia fields which have been mapped satisfactorily. There are now some data for reserves at depths of more than 600 m. Our knowledge of the other coalfields is still unsatisfactory but they are being mapped in detail and investigated by drilling to estimate the correct order of reserves. It is however certain that the present estimates of reserves will be appreciably increased as a result of detailed investigations.

SELECTED BIBLIOGRAPHY

———— Indian Coals, their nature and origin. *Fuel Res. Inst.* India. 1949
Adie, R. J. (Ed.) Antarctic Geology. North Holland Publ. Co., 1965, 770 pp.
Ahmad, F. Glaciation and Gondwanaland. *Rec. G.S.I.* 86, p. 637-674, 1960.
Ahmad, F. Palaeogeography of the Gondwana period with special reference to India. *Mem. G.S.I.*, 90, 1961. 142 pp.
Colbert, E.H. Relationship of the Triassic Maleri fauna. *Jour. Palaeont. Soc.* India, 3, 68-81, 1958.
Cotter, G. de. P. A revised classification of the Gondwana System *Rec.* 48, 23-3, 1917.
Fox, C.S. The Jharia Coalfield. *Mem.* 56, 1930.
Fox, C.S. The Natural History of Indian Coal, *Mem.* 57, 1931.
Fox, C. S. The Gondwana System and related formations, *Mem.* 58, 1931.
Fox, C.S. The Lower Gondwana Coalfields of India. *Mem.* 59, 1934.
Gee, E. R. Geology and Coal resources of the Raniganj Coalfield *Mem.* 61, 1932.
Gee, E. R. Coal. *Rec.* 76, Paper 16, 1945.
Ghose, C. A petrochemical study of lamprophyres and associated intrusive rocks of Jharia coalfield. *Q.J.G.M.M.S.I.* 21(4), 133-147, 1959.
Gopal, V. and Jacob, K. Geology of the Upper Gondwanas in the Ramnad District, Madras, *Rec.* 84(4), 1955.
Grindley, G.W. Geology of Queen Alexandra Range, Beardmore Glacier, Ross Dependency, Antarctica. *N.Z. Jour. Geol. Geophys.* 6, 308-347, 1963.
Hadley, J. B. (Ed.) Geology and Palaeontology of the Antarctic. *Amer. Geophys. Union, Antarctic Rec. Mono.* No. 6, 1965 280 pp.
Jain, J. L., Robinson, P.L. and Roy Chowdhury, T.K. A new vertebrate fauna from the Triassic, Deccan, India, *Q.J.G.S.* (London), 120, 115-124, 1964.
King, W. Coastal region of the Godavari District. *Mem.* 16(3), 1880.
Mayr, E. (Editor). The problem of land connections across the South Atlantic, etc. *Bull. Amer. Mus. Nat. Hist.* 99(3), 1952.
Mehta, D.R.S. and Narayanamurthy, B.R. Revision of mapping of the Raniganj and Jharia Coalfields. *Mem.* 84.
Mukherjee, K.K. Petrology of the lamprophyres of the Bokaro coalfield. *Q.J.G.M.M.S.I.* 23, 69-87 1961.

Raja Rao, C.S. and Purushottam, A. Pitchstone flows in the Rajmahal Hills. Santhal Parganas, Bihar, *Rec. G.S.I.*, 91(2), 341-349, 1956.

Teichert, C. (Editor). Symposium on the Gondwana System, 19*th Int. Geol. Congr.* (Algiers), 1952.

FOSSILS

Huxley, T. H. *et al.* Vertebrate fossils from the Panchet and Kota-Maleri rocks; Labyrinthodonts of the Bijori group; Reptilia and amphibia of the Maleri-Denwa groups. *Pal. Ind. Ser.* IV, Vol. I, Nos. 1-5, 1865-1885.

Feistmantel, O. Fossil flora of the Gondwana System *Pal. Ind. Ser.* II, XI, XII, Vol. I-IV, 1863-1886.

Seward, A. C. Lower Gondwana plants from Golabgarh Pass, Kashmir, *Pal. Ind. N.S.* IV, (3), 1912.

Seward, A. C. and Sahni, B. Revision of Indian Gondwana plants. *Pal. Ind. Ser N.S.* VII (1), 1920.

Sahni, B. Revision of Indian fossil plants — Coniferales. *Pal. Ind. N.S.* XI 1928-31.

Reed, F. R. C. Permo-Carboniferous Marine fauna from the Umaria Coalfield. *Rec.* 60, 367-398, 1928.

CHAPTER XI
UPPER CARBONIFEROUS AND PERMIAN SYSTEMS
GENERAL

The term PERMIAN was introduced in 1841 by Sir Roderick Murchison to designate the strata overlying the Carboniferous formations in the province of Perm in Russia, north of the Caspian Sea and southwest of the Urals. These beds are marine in character in contrast with the NEW RED SANDSTONES of England and the ROTHELIEGENDE of Germany, which are their equivalents but of continental facies. The major divisions of the Permian are shown below:—

	W. Europe	Mediterranean	Russia
Upper Permian	Thuringian (Zechstein)	Bellerophon Limestone	(Tatarian) Kazanian
	Saxonian (U. Rothliegende)	Groden Beds (with volcanics)	Kungurian
Lower Permian	Autunian (L. Rothilegende)	Sosio Beds (Sicily) Trogkofel Beds	Artinskian (Sakmarian)
Upper Carbon.	Stephanian	Schwagerina Beds Auernig Begds	Uralian

The period from the Middle Carboniferous to the end of the Permian was one of great crustal disturbance known as the Hercynian (Variscan or Armorican) revolution. The different major phases of this, as identified in Europe are :

Pfalzic — Between Rothliengende and Zechstein
Saalic — Autunian
Asturic — Upper Carboniferous
Sudetic — Between Visean and Namurian
Bretonic — L. Carboniferous

Of these, the Sudetic, Asturic and Saalic have produced major disturbances in different areas. The Appalachians of North America, the Armorican chains of Britain and France, the Vosges, Harz mountains, the Urals, Tien Shan, Kun Lun, Karakorum and certain other ranges in Central and Eastern Asia were all formed during this revolution, though some of them like the Kun Lun and the Karakorum were again uplifted during the Himalayan revolution. Structural and stratigraphical data seem to indicate that the Appalachians lay not far to the west of the British Isles and in structural continuity with the Armorican chains during this period.

The Eurasian continent (*Laurasia*) at the time probably included part of North America and Greenland but it was separated from *Angaraland* (Siberian Shield) by the Ural geosyncline. This geosyncline was compressed and uplifted during the Permian times. North-eastern Asia was apparently connected with western North America as indicated by the distribution of Permo-Carboniferous floras. The Mediterranean region, which was already a marine basin, became more clearly a geosynclinal basin of deposition in the Permian and continued as such until folded into mountain chains in the Himalayan revolution of the Tertiary era. This marine basin, which has been given the name of *Tethys* by Eduard Suess, stretched from Spain and Northwest Africa in the west, through the regions now occupied by the Alps, Carpathians, Caucasus, Elburz, Zagros and Himalayas into the great island arcs of Indonesian archipelago and New Guinea. To the south of this great ocean lay another great land mass, called *Gondwanaland*, which apparently comprised India, Australia, South and East Africa, Antarctica and South America. The members of this southern continent contain fluviatile and continental deposits characterised by very closely related flora and continental fauna (fishes, amphibians and reptiles) and known as the Gondwana or Karroo formations.

FLORA AND FAUNA

The continental facies of the Permian, being deposits laid down under arid conditions, are not rich in fossils. A few places containing marine deposits are very richly fossiliferous, as in Texas, Sicily, Eastern Alps, Armenia, Perm, the Punjab Salt Range and Timor Island. But, whether continental or marine, the Lower Permian strata are continuous with those of the Upper Carboniferous. For this reason, the Permian is now taken as beginning with those strata in which *Pseudo-schwagerina* (fusulinid) or *Callipterts conferta* (plant) make their first appearance. Fusulinids and Cephalopods serve as zone fossils, while Brachiopods, though very abundant are not of the same importance.

The flora, though not rich, contains quite a number of genera and species belonging to Pteridosperms, Conifers, Cycads and other groups. During this period there were four major foral groups in different parts of the world, the Glossopteris flora being the most specialised amongst them (*see* Chapter on the Gondwana group).

Amongst the fauna, all the major phyla except the mammalia are represented. The corals include the important genera *Waagenophyllum, Favosites, Michelinia, Pachypora* etc. Echinoderms are abundant in some areas (e.g. Timor) and include *Archaeocidaris, Eocidaris, Cyathocrinus, Ceriocrinus, Timorocrinus, Schizoblastus*, etc. The marine faunas contain numerous brachiopods including *Productus* (*P. artiensis, P. cancrini, P. horridus*) *Marginifera, Spirifer, Chonetes, Strophalosia* (*S. horrescens, S. goldfussi*), *Athyris*, etc., some species being restricted to particular beds. The aberrant genera, which take the peculiar form assumed by the Rudistids in the Cretaceous, are characteristic — *Richthofenia, Geyerella, Scachinella*, and also *Lyttonia* and *Oldhamina*. The more important pelecypods and gastropods are *Lima, Gervillea, Eurydesma, Myophoria; Bellerophon, Stachella, Euphemus, Pleurotomaria*, etc. The Cephalopods, though not as abundant as the brachiopods, are stratigraphically important. The characteristic genera are *Xenaspis, Medlicottia, Pronorites, Popanoceras, Agathiceras, Gastrioceras, Propinacoceras*. The Arthropoda are represented by *Phillipsia, Cheiropyge* (Trilobites) *Eurypterus* and several Ostracods. Fish remains (*Palaeoniscus, Amblypterus, Acanthodes*) are often found in lacustrine strata. Amphibians are quite important, being represented by *Archegosaurus Actinodon* etc. Reptiles, which had already made their appearance, became the dominant land animals in the Permian ; they include *Proterosaurus, Meso saurus, Pareiasaurus, Haplodus, Dicynodon, Endothiodon, Palaeohatteria* etc.

Continental Facies : Western Europe. Over a large part of Germany France and Britain, the Stephanian strata pass upward into the Permian The latter comprise lagoonal, lacustrine and aeolian deposits in the lower strata (Rothliegende) and shallow epicontinental marine deposits in the Upper (Zechstein). The Rothliegende are brightly coloured conglomerates, sandstone and shales. A stratigraphic and faunal break occurs between the Lower and Upper Rothliegende, and this should really be considered as the division between the Palaeozoic and Mesozoic, in the opinion of many stratigraphers. The Zechstein is a limestone-dolomite formation containing an abundant fauna rich in individuals but poor in genera and species, and generally improverished owing to the stringent living conditions in enclosed marine basins. The Zech stein comprises the famous Kupferschiefer (copper bearing shales) near the base and thick beds of salt-gypsum-anhydrite (such as those of Stassfurt) in the upper part.

The Lower Rothliegende in the Autun area in France, called the AUTUNIAN contains carbonaceous shales and sandstones with thin coal layers, indicating somewhat more humid conditions of deposition than in Germany. The Upper Rothliegende, known as the SAXONIAN, is similar to that of German where it occupies large areas, often covered by younger formations. The Permian in the British Isles is similar to the formations in Germany, but the lower part (Autunian) is missing, while the upper part becomes more conti nental when followed from east to west and contains salt, gypsum and anhy drite. The upper strata pass imperceptibly into the similar looking Bunte (Lower Triassic).

Russian Marine Facies. Permian strata are typically exposed in the Perm Artinsk-Kazan region north of the Caspian Sea and also in the Moscow basin and in the Urals. The Urals were folded and uplifted at the beginning of the Lower Permian and again in the Middle Permian. In the Perm region there is a discordance above the Uralian, and the ARTINSKIAN strata contain brachio pods (*Productus artiensis, Chonetes*) and Cephalopods (*Pronorites, Agathicera Propinacoceras* etc.). In the region of Nizhni-Novgorod (Gorky), the Permian overlies the Uralian without a break and contains fusulines, corals and brachio pods but no cephalopods. The Artinskian is overlain by the KUNGURIAN without a break, but the latter contains dolomite and gpysum in the upper part, indicating the arrival of continental conditions. The fossil in the Kungu rian are pelecypods and gastropods but no cephalopods. In the Pechor valley there are marine beds (with *Productus, Pseudomonotis,* etc.) with inter calated plant beds containing *Pecopteris, Noeggerathiopsis,* and *Gangamopter* which indicate land connection with Gondwanaland, perhaps through the

Kashmir region. The KANANIAN, which succeeds the Kungurian, shows two facies — an eastern lacustrine one with Kupferschiefer, showing plants, vertebrates and pelecypods, and a western one with marine fossils such as *Productus cancrini, Strophalosia, Athyris, Dielasma, Schizodus*, etc. The two facies meet in the Vaitka area west of Perm. The Kazanian fauna does not bear much resemblance to the Zechstein as the two marine areas were not directly connected. The topmost beds, the TATARIAN (or Tartarian) are lagoonal and transitional, i.e. Permo-Triassic.

Mediterranean Marine Facies. Excellent development of the marine facies is seen in a few places. At Sosio in Sicily a few exposures containing a rich fauna of lamellibranchs, gastropods, cephalopods, and brachiopods indicate the existence of strata of Artinskian to Kungurian age. A complete and richly fossiliferous marine sequence is exposed in the gorge of the Araxes near Djulfa in Armenia, ranging unbroken from the Upper Devonian to the Triassic. Other occurrences have been studied in the Carnic Alps, Dalmatia and Greece. The Timor Island in Indonesia has shown the existence of Permian strata exceptionally rich in corals, echinoderms, cephalopods etc. but the strata are highly disturbed. There is also a fine development of the marine Permian in Texas which has been exhaustively studied, its main divisions being Wolfcamp, Leonard-Hess, Guadalupe and Ochoa from below upwards.

The Upper Carboniferous revolution in the Indian region brought about a renewal of sedimentation in parts of peninsular India and this continued up to the Lower Cretaceous. In the Himalayas, the sea which had regressed during the Devonian times encroached on land after the great revolution. In Kashmir, however, land conditions prevailed for some time during the Permian and part of the Trias, accompanied by volcanic activity. The Kashmir Sea perhaps acted as a land bridge between Gondwanaland and Angaraland during the Permo-Triassic, for we have evidences of a certain amount of intermingling of the characteristic floras of these two regions.

In the region of the Lesser Himalayas, relics of glacial deposits of Upper Carboniferous times are found in a few places. The deposits laid down in this region during the Mesozoic, are however, mostly unfossiliferous so that it is difficult to ascertain their age. It is possible that they were deposited in a isolated basin in which the conditions may have been unfavourable to life. These basins presumably dried up during the Upper Mesozoic.

The Upper Palaeozoic unconformity in the Himalayan area is of varying magnitude in different places. It is of very short duration in the Spiti Valley, for the Po Series of Lower Middle Carboniferous age is followed by Permo-Carboniferous strata including the Productus Shales. In Kumaon the Carboniferous strata are missing, the eroded surface of the Muth Quartzite being overlain directly by Permian rocks. In Kashmir there are the Syringothyris Limestone or Fenestella Shales. In other places the unfossiliferous Tanawal Series is seen below the Permo-Carboniferous strata. In Hazara the Pre-Cambrian Dogra Slates are often overlain, with marked unconformity, by glacial boulder beds called the Tanakki Conglomerates which are considered to be the equivalents of the Talchir boulder-beds. In Chitral also there is an unconformity at the top of the Carboniferous strata, the latter being followed by the Infra-Trias and the Trias. In the Salt Range there is a hiatus between the Salt Pseudomorph shales and the Talchir boulder-bed, the latter being followed by the Speckled Sandstones and by a fine development of marine Permian.

SPITI—THE KULING SYSTEM

The Upper Carboniferous of Spiti, as already noted, is marked by an unconformity conglomerate. This conglomerate rests on the Po Series or earlier rocks and is followed by Permian strata constituting the Kuling system.

Calcareous sandstone.—The lowest members of the Kuling system are grits and quartzites, overlain by calcareous sandstones having a thickness of about 30 m, and containing fossils which have affinities with those of the Middle Productus Limestones :

Productus sp., *Spirifer fasciger, S. nitiensis, S. marcoui, Dielasma latouchei Aulosteges gigas, Spirigera gerardi.*

Productus Shales.—The calcareous sandstones are succeeded by a group of brown or black carbonaceous and siliceous shales called the Productus Shales which have a thickness of 30-60 m and form a fairly persistent horizon in Kashmir, Spiti, Kumaon and Nepal. They enclose, especially in the lower part, a rich and characteristic Permian fauna of the same age as the Middle and Upper Productus beds of the Salt Range, but yet of a different facies. About 9 m below the top, there is a horizon having a thickness of barely 0.3 m, from which concretions containing cephalopods have been obtained, this being in fact the only cephalopod yielding horizon in the shales. Amongst the fossils of the Productus Shales are :

Brachiopods	... ...	*Productus purdoni, P. abichi, P. gangeticus, Spirifer rajah, S. fasciger, Spirigera gerardi, Marginifera himalayensis, Chonetes lissarensis.*
Cephalopods	... ...	*Xenaspis carbonaria, Cyclolobus oldhami, C. krafffi, C. haydeni*

KUMAON

In northern Kumaon the Muth Quartzite formation is generally conformably overlain by the black *Kuling Shales* of Permian age. The line of junction is always very sharp and represents a considerable lapse of time which may extend from some part of the Devonian to the lower part of the Permian. Basal conglomerates which are said to underlie the Kuling Shales in Spiti are not found in Kumaon. This tsratigraphical interval is partly filled, in some places, by sediments like the Po and Lipak Series as in Spiti, or by the Panjal Traps as in Kashmir.

The Kuling Shales or (Productus Shales) are usually from 30 to 50 m thick, but may be less as at Tinkar Lipu, or as much as 100 m thick as at the Lebong Pass west of Kuti. They appear to represent a deep marine deposit and are generally poor in fossils. The fossils got from them include *Cyclolobus oldhami* and *C. walkeri* both of which indicate Upper Permian age. A lower

PLATE IX
PERMO-CARBONIFEROUS FOSSILS

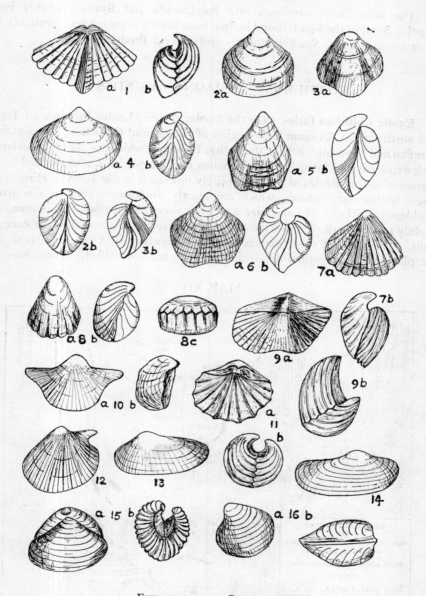

EXPLANATIONS OF PLATE IX

1. *Spiriferina cristata*.   2. *Spirigerella grandis*.   3. *Spirigerella derbyi*.   4. *Athyris royssi*.   5. *Dielasma biplex*.   6. *Reticularia lineata*.   7. *Camarophoria purdoni*,   8. *Hemiptychina himalayensis*.   9. *Syringothyris cuspidata*.   10. *Marginifera himalayensis*.   11. *Enteletes tschernyscheffi*.   12. *Pseudomonotis middlemissi* (0.3).   13. *Parallelodon brennensis* (0.3).   14. *Sanguinolites kashmiricus* (0.3).   15. *Palaeocorbula difficilis* (0.5).   16. *Astartila ovalis* (0.5).

horizon of the Permian is found in this region only near **Kalapani** where the shales have yielded *Productus himalayanus*, *P. abichi*, *P. semireticulatus*, and *Spirifer tibetanus*.

The same facies continues into **Painkhanda** and **Byans** probably into Nepal. Some of the fossils found in Spiti have been recognised here, particularly *Chonetes lissarensis*, *Spirifer nitiensis* and several Productids.

NORTH-WEST KUMAON — HUNDES

Exotic Chitichun facies.—On the border of the Hundes Province of Tibet and northwestern Kumaon is the region of Kiogar and Chitichun. It contains Permian and Mesozoic rocks of the Tethys Himalaya which are closely associated with extrusive and intrusive basic igneous rocks in which are embedded huge masses of sedimentaries of a facies entirely unknown in the Tethyan Himalaya. These sedimentary masses, which are mostly limestones, vary in size from boulders to hillocks of some size and even sheet-like masses. This area lies slightly to the north of the strike continuation (to the southeast) of the zone of Spiti. The roots of these masses may lie somewhere in Tibet, though so far the place of their original deposition has not been identified. These masses,

MAP XII

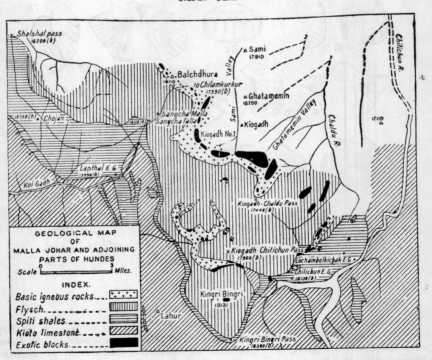

known as the "exotic blocks" have been described by A. von Krafft and C. Diener. Their presence here has been explained as due to the shattering of the rock formations in the Tibetan region by great volcanic explosions and transport over long distances by lava flows to their present position. An alternative hypothesis advanced by Diener and later supported by Suess (" The Face of the Earth ", English Edition Vol. IV, pp. 561-567) is that they represent the remnants of thrust sheets or *nappes* from the north. This view is strongly supported by Heim and Gansser who surveyed the area during their traverses in the Himalayas and Tibet in 1936. They state that the limestone masses are of the nature of *Klippen*, lying among much younger, mainly Cretaceous, sediments though completely surrounded and often intimately penetrated by igneous rocks. They have found evidence that the intrusives must have penetrated these rocks mainly before thrusting, though it is likely that some of them were injected along the thrust sheets while the thrust movements were actually in progress. Though these blocks apparently lie pell-mell, bedding planes can often be followed between neighbouring blocks. They show contact metamorphism and serpentinisation. They generally show a northward dip, indicating that the thrust movements were directed to the south.

The exotic blocks comprise pink, red and white limestones of all ages from Permian to the Cretaceous, this type being called the ' Chitichun facies ' by Heim and Gansser. Those of Permian age contain several characteristic brachiopods. The Upper Triassic limestones are extraordinarily like the red and white Hallstatt marble of the Alps both in lithology and fossil content. It is remarkable that this Chitichun facies should be closely related to the Alpine type and not to the Tethys Himalayan facies lying to its south. This indicates that the conditions of life and sedimentation in the marine zone which stretched from Tibet to the Eastern Alps were extraordinarily uniform but that they were markedly different in the adjoining area of the Tethys Himalayan zone.

The Permian limestones forming the peaks Chitichun I, Malla Sangcha, etc. have yielded a rich fauna comprising the following :—

Trilobites	... ...	*Phillipsis middlemissi, Cheiropyge himalayensis.*
Brachiopods	... ...	*Productus semireticulatus, P. chitichunensis, P. gratiosus, P. abichi, Marginifera typica, Lyttonia nobilis, Spirifer fasciger, S. wynnei, S. tibetanus, Martinia* cf. *glabra, M. elegans, Reticularia lineata, Spirigeralla grandis, S. derbyi, Athyris royssii, A. subexpansa, Enteletes waageni, Camarophoria purdoni, Hemiptychina himalayensis, Dielasma elongatum, Notothyris triplicata, Richthofenia* sp.
Cephalopods	... ...	*Nautilus hunicus, Xenaspis carbonaria, Cyclolobus walkeri.*
Anthozoa	... ...	*Amplexus coralloides, Zaphrentis beyrichi, Clisiophyllum* sp., *Lonsdaleia indica.*

MOUNT EVEREST REGION

The several Everest expeditions have gathered a considerable amount of information on the geology of North Sikkim and the neighbourhood of Mt. Everest. The top of Mount Everest is composed of massive, arenaceous limestone dipping gently towards the north and continuing laterally into North Sikkim. This is called the *Mount Everest Limestone*, whose thickness must be 300 to 600 m. It is well bedded and contains feldspathic sandy bands. In Mount Everest it has been metamorphosed to crystalline limestones calc-schists and banded hornfels. Its age is Carboniferous in the main and possibly Permo-Carboniferous in part.

The Mount Everest Limestone is underlain by the *Everest Pelitic Series* of 1,200 m thickness, consisting mainly of shaly and slaty rocks with calcareous and sandy bands. It is injected profusely by granitic rocks. Its age is probably Carboniferous or earlier.

The Everest Limestone is overlain conformably by the *Lachi Series* which covers a large area north of Everest and in North Sikkim. It has an aggregate thickness of 600 m and contains glacial boulder-beds, a lower fossiliferous imestone bed (15 m thick) and an upper fossiliferous calcareous sandstone (90 m thick) containing Upper Permian fossils. It is overlain by the *Tso Lhamo Series* of Triassic age.

The fossils in the Lachi Series include :—

LOWER HORIZON	... Several corals including *Straparollus lachiensis*.

UPPER HORIZON (CALCAREOUS SANDSTONE) :—

Bryozoa ...	... *Fenestrinella internata, Goniocladia* sp.
Brachiopods ...	... *Spirifer (Spiriferalla) rajah* (numerous), *Spirifer (Neospirifer) musakheylensis, Marginifera himalayensis, Productus (Waagenoconcha) purdoni, Uncinunellina jabiensis, Syringothyris lydekkeri.*
Lamellibranchs	... *Parallelodon* aff. *tenuistriatum, Pleurophorus* sp. *Aviculopecten hyemalis.*
Gastropod ...	... *Pleurotomaria* aff. *orientalis.*

The succession in the Everest region is shown below :—

Tso Lhamo Ser es ...	⎧ Dark limestones and shales (Triassic) ⎨ Quartzites and shales (120 m) ⎩ Calcareous sandstones (90 m) with Upper Permian fossils
Lachi Series ...	⎧ Pebble beds (180 m) ⎨ Limestones and shales (15 m) ⎩ Quartzites, silts and shales (180 m)
Mt. Everest Limestone ...	Massive arenaceous limestones (300 - 600 m). Carboniferous to Permo-Carboniferous.
Mt. Everest Pelitic Series	Slaty rocks with bands of limestone and sandstone, injected by granite (1,200 m) Mainly Caronbiferous
Lower calcareous Series ...	Limestones ,shales, etc., extensively injected by granite.

In the Rangit Valley of Sikkim, coal-bearing Barakar rocks have been found in association with calcareous rocks containing some Productids, *Neospirifer, Fenestella,* etc.

ASSAM HIMALAYA

Our knowledge of the geology of the Assam Himalaya was, till recently, confined to the material collected during military expeditions into the northeastern frontier tracts. Since 1950 some geological traverses have been made in these areas. The fossiliferous limestone boulders of Permian age found by MacLaren in the Subansiri river gorge have now been traced to their source n the Ranganadi basin about 32 km southwest of the gorge. The fossils include *Productus, Spirifer, Chonetes, Reticularia,* etc. indicating the same age as the Kuling System. Similar limestones have also been found in the Sireng valley in the Abor Hills. Since these rocks are thrust over the Tertiaries it is evident that they must have travelled some distance southwards from the place of their original deposition.

KASHMIR — PANJAL VOLCANIC SERIES

The great revolution of the Upper Carboniferous age converted a large part of Kashmir into a land area with volcanic conditions reigning. The volcanic activity was at first of the explosive type and contributed fragmentary products which were deposited as agglomerates and pyroclastics. The later activity was mainly in the form of lava flows. It reached its climax in the early Permian ; then waning gradually, it died off finally in the Upper Triassic. While the igneous action was prevalent in certain areas, the sea was encroaching in others, so that we find marine sediments of Permian and Triassic ages side by side with products of volcanism.

Distribution.—The volcanic series is extensively developed in the Pir Panjal range whence its name is derived; to the west of the Zanskar range up to Hazara ; in Ladakh and Baltistan ; also in many areas around the Jhelum valley.

Varying age.—The earliest manifestations of volcanism seem to be of Middle to Upper Carbonifreous age, *e.g.*, as early as the latter part of the Moscovian in the Lidar valley and Upper Carboniferous near Nagmarg. The trap is seen to overlap beds of various ages and its upper limits are different in different areas. In some places it underlies the Gangamopteris or Zewan beds. Near Sonamarg it is higher up in the Permian, while north and west of the Wular lake it extends to Lower, Middle and Upper Triassic and is intercalated with beds of the respective ages.

Agglomeratic Slates.—The volcanic agglomerates are gritty or greywacke-like and grade frequently into slates. They contain angular fragments of quartz, feldspar, quartz-porphyry, granite, limestone, devitrified glass, etc. They were once regarded as of glacial origin. But the absence in them of ice-scratched pebbles and boulders and the presence of devitrified glass fragments and volcanic material, have tended to support the view that they are pyroclastic and derived from volcanic explosions and rearranged by subaerial agencies. Some of these may be welded tuffs.

The Agglomeratic slates are generally unfossiliferous but well-preserved fossils have been obtained from them in about half a dozen localities, the most important of which are : (1) Near Kismar hamlet at Nagmarg where they overlie the Silurian and are succeeded by the Panjal traps, the upper portion being fossiliferous ; (2) in the anticline of the Marbal valley overlying the Fenestella Shales ; (3) in the Kolahoi-Basmai anticline where they overlie the Syringothyris Limestone, the higher beds being interbedded with the trap ; (4) in the Golabgarh pass in Pir Panjal overlying the Fenestella Shales.

The fossils found in these areas are :—

Brachiopods	... ...	*Spirifer nitiensis, S. Kismari, S. fasciger ; Productus cora* var. *lineatus, P. scabriculus, P. undatus, Dielasma* sp., *Chonetes* sp., *Syringothyris cuspidata, S. nagmargensis, Derbyia regularis Camarophoria dowhatensis.*
Bryozoa	... ...	*Protoretepora ampla, Fenestella* sp.
Lamellibranchs	... ...	*Lima* sp., *Pinna* sp.,

Panjal Trap.—The Panjal Trap consists of bedded flows of green, purple and dark colours. The lavas are sometimes amygdaloidal or porphyritic though compact fine-grained varieties are the most common. Interbedded with them are pyroclastics and occasonally sedimentary strata. The flows vary in thickness up to 6 to 9 m and are generally lenticular. Locally they attain an enormous thickness, the maixmum being estimated at 2,100 m in the Uri district.

The lavas are andesitic to basaltic in composition, but acid and ultra-basic varieties have also been found. The acid varieties found in the area around the Kashmir valley include rhyolites, dacites and trachytes. The ferromagnesian minerals and feldspars are generally seen to have been chloritised and epidotised. Amygdales of chlorite, epidote, delessite, zoisite and quartz are found in some flows. Interstitial glass is fairly common, but generally devitrified.

Gangamopteris Beds.—Beds containing Lower Gondwana plants occur intercalated with pyroclastics in several places and are overlain by the Panjal Traps. The following localities are well known as yielding typical flora ; Golabgarh pass (Pir Panjal), Gulmarg, Khunmu and Risin, Nagmarg, Bren and Marahom. Near Nagmarg on the Wular lake, the soft arenaceous beds underlying these plant beds contain *Productus cora, P. Srabriculus, Spirifer nitiensis, Derbyia, Syringothyris cuspidata* and *Fenestella* which give an unmistakable indication of Uralian to Lower Permian age of the fauna. The slaty plant beds contain *Gangamopteris kashmirensis, Glossopteris indica, Psygmophyllum haydeni, Vertebraria*; also the Amphibian *Archegosaurus ornatus* and the fishes *Amblypterus kashmirensis* and *A. symmetricus*. These beds are therefore of Lower Permian (Karharbari) age.

The Apharwat ridge near Gulmarg has yielded *Gangamopteris, Glossopteris, Alethopteris, Cordaites* and *Psygmophyllum* from beds underlying the lava flows and overlying the Tanawal Seris. At Bren, near Srinagar, a *Eurydesma*-bearing horizon is just below the Gangamopteris bed. At Risin and Zewan in the Vihi district, the Gangamopteris bed underlies fossil-bearing Permian limestone and contains the amphibian *Actinodon risinensis* and the fish *Lysipterigium deterrai*.

The Gangamopteris beds have different positions with reference to the volcanics — above the volcanics at Khunmu and Golabgarh pass, below them at Nagmarg and Bren, or intercalated with them in one or two places — and their fossil content points to the same age as that of the Talchir and Karharbari beds. Their age is Upper Carboniferous to Lower Permian. The Gondwana Sandstones become calcareous near the top and pass gradually into a marine limestone called the Zewan Beds.

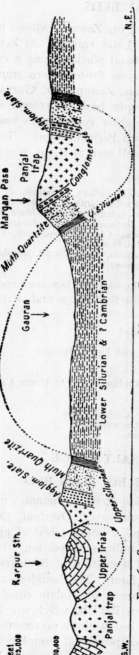

Fig. 6.—Section through the Naubug Valley and Margan Pass. (After C. S. Middlemiss, Rec. 40).

ZEWAN BEDS

The Zewan Beds take their name from Zewan, a village in the Vihi district. They extend also into the Sind and Lidar valleys. At Zewan they lie on a silicified limestone (novaculite), the basal portion being a crinoidal limestone overlain by beds containing the bryozoa *Protoretepora ampla*. This horizon is well exposed in the Golabgarh pass, Zewan spur, Guryal ravine, etc. and contains numerous colonies of *P. ampla*, besides *Lyttonia nobilis, Spiriferina, zewanensis* and *Derbyia* sp. Above these come shales and thin limestones about 120 to 150 m thick, with a rich Permian fauna. The Upper Permian beds are missing at Zewan but are well seen at Barus, a short distance away.

The fauna includes :

Brachiopods	... ...	*Productus cora, P. indicus, P. spiralis, P. gratiosus, P. abichi, Spirifer rajah* (abundant), *S. nitiensis S. fasciger, Marginifera himalayensis* (abundant) *M. vihiana, Spirigera gerardi, S. subexpansa, Dielasma latouchei, Dielasma hastatum, Camarophoria purdoni, Chonetes lissarensis, C. laevis, Spiriferina cristata, Lyttonia nobilis.*
Bryozoa	... ...	*Protoretepora ampla, Fenestella* aff. *fossula*
Corals	... ...	*Amplexus, Zaphrentis.*

The fauna of the Zewan beds shows that they correspond to the Middle and Upper Productus Limestone of the Salt Range and to the Chitichun Limestone.

The succession in Kashmir is summarised below :

Zewan beds with *Protoretepora ampla* bed near the base.	Middle and Upper Permian
Gangamopteris beds	
Agglomeratic Slates with Panjal Trap	Permo-Carboniverous to Upper Carboniferous
Fenestella shales	Middle Carboniferous
Syringothyris Limestone ...	Lower Carboniferous
Muth Quartzite	Devonian

THE SALT RANGE

GLACIAL BOULDER BED

The Lower Palaeozoic of the Salt Range culminates in the Salt Pseudomorph Shales of presumably Cambrian age. Overlying this, with the intervention of an unconformity, is found a boulder bed of glacial origin which lies on one of the stages of the Cambrian succession. It is heterogeneous in composition, being composed of a mixture of boulders and pebbles in a matrix of fairly comminuted rock flour. The boulders are of various size upto 60 cm or more across and consist of Malani rhyolite and crystalline rocks from Rajasthan and Southern Punjab, which are often striated and faceted by ice action. The boulder bed attains a maximum thickness of 60 m. Both in constitution and stratigraphical position it resembles the Talchir boulder bed of the Peninsula. The presence in it of rocks from Rajasthan is proof that the boulders were derived from that region and transported by glaciers. The age of the boulder bed is indicated by the presence of spores, leaf impressions and other remains of Lower Gondwana plants in the beds immediately overlying it.

THE OLIVE SERIES

EURYDESMA HORIZON.—The Upper Palaeozoic in the eastern end of the Salt Range is represented by the Olive Series, consisting of dark greenish, yellow and white, or spotted sandstones and having a thickness of 50 m. At its base is the boulder bed which, in certain cases, may represent resorted and re-deposited glacial material. Near Pidh (in the Eastern Salt Range) the sandstones immediately overlying the boulder bed contain numerous casts of marine bivalves belonging to the genus *Eurydesma* and other fossils, viz : *Eurydesma cordatum, E. hobartense, E. punjabicum, E. subovatum, Pterinea* cf. *lata, Nucula pidhensis, Aviculopecten* sp., *Astartila* cf. *ovalis, Cardiomorpha penguis* ; (Brachiopod) *Dielasma dadanense* ; (Bryozoa) *Fenestella fossula*.

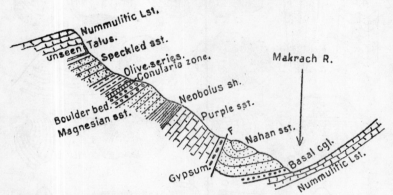

FIG. 7.—SECTION ACROSS THE MARKACH VALLEY, SALT RANGE. (AFTER C. S. MIDDLEMISS, *Rec.* 24).

A similar Eurydesma-bearing zone is associated with the Upper Carboniferous or lowermost Permian in Kashmir. The fossil contents of these two horizons show a great deal of similarity as will be seen from the following :

Salt Range	Kashmir
Fenestella fossula	F. fossula
Dielasma dadanense	D. lidarense
Aviculopecten cunctatus	A. cunctatus
Cardiomorpha penguis	C. sp.
Astartila cf. ovalis	A. ovalis
Eurydesma cordatum var. mytiloides ...	E. cordatum var. mytiloides
E. subobliqua	E. cordatum var. Subquadrata
E. punjabicum	E. globosum

In New South Wales, Australia, two *Eurydesma* horizons are known ; the lower one with *E. hobartense* occurs just below the Gangamopteris bed and is Upper Carboniferous ; the upper one with *E. cordatum* is assigned a Permian age. Similar beds also occur associated with the Dwyka conglomerate of South Africa. The Kashmir *Eurydesma* horizon is closely allied to the Lower Permian of the Kolyma Province in Siberia, the brachiopod fauna of the two being related.

PLATE X

PERMIAN FOSSILS

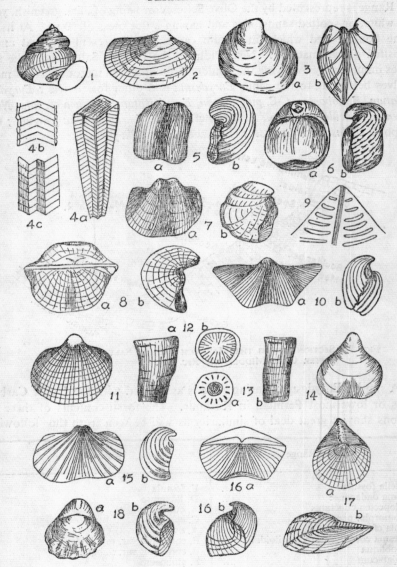

EXPLANATION OF PLATE X

1. *Pleurotomaria nuda* (1.5). 2. *Sanguinolites tenisoni* (0.7). 3. *Eurydesma ellipticum* (0.4). 4. *Conularia warthi* (b and c enlarged). 5. *Productus cora* (0.3). 6. *Productus purdoni* (0.3). 7. *Productus gangeticus* (0.3). 8. *Productus indicus* (0.3). 9. *Lyttonia nobilis* (0.3). part of ventral valve. 10. *Spirifer musakheylensis* (0.3). 11. *Reticularia indica* (0.3). 12. *Zaphrentis beyrichi* (0.7). 13. *Amplexus coralloides* (0.7). 14. *Martinia elegans*. 15. *Strophalosia costata* (0.5). 16. *Spirifer marcoui* (0.3). 17. *Streptorhynchus pelargonatus* (0.3). 18. *Notothyris warthi* (0.7).

Conularia Zone.—The boulder bed passes upwards, in certain places, into calcareous sandstone and black shales containing concretions enclosing well preserved fossils among which *Conularia* is the most abundant. The *Conularia* horizon is about 30 cm thick and a few feet above the boulder bed and thus also slightly younger than the Eurydesma horizon. It is well seen near Choya Saidan Shah, Mt. Chel. Ratuchha, etc. The fossils in this horizon include :

Gastropods	... ...	*Conularia laevigata, C. warthi, C. punjabica, C. salaria, Pleurotomaria nuda, Bucania warthi.*
Lamellibranchs	... ...	*Pseudomonotis subradialis, Sanguinolites tenisoni*
Vermes	... ...	*Serpulites warthi*
Brachiopods	... ...	*Spirifer vespertilio, Martiniopsis darwini, Chonetes cracowensis, Discinisca warthi*

Near Kathwai, the Conularia beds yielded *Glossopteris* and *Gangamopteris* impressions and the following bivalves : *Palaeomutela oblonga, Palaeonodonta salaria, P. subquadrata, P. singularis.* These indicate, according to Cowper Reed, a Lower Permian horizon. The Conularia fauna is also represented in the marine beds intercalated with Upper Palaeozoic glacial beds in Australia.

THE SPECKLED SANDSTONE SERIES

Boulder-bed.—West of the Nilawan ravine the position of the Olive Series is taken by the Speckled Sandstones. In their typical development they include a basal portion comprising the *Boulder Bed*, a middle portion with *Speckled Sandstones*, and an upper portion with *Lavender Clays*. The Boulder Bed has the same stratigraphical position as that under the Olive Series, and fine exposures are seen near Makrach and to its west. It is seen also near the western end of the Salt Range.

Speckled Sandstones (*sensu stricto*).—These are 60 to 120 m thick and consist of reddish to brownish sandstone with green and purple patches. Numerous small concretionary masses are seen on the weathered surface. They contain thin intercalations of shales of purple, lavender and grey colours and also gypseous bands. The shale bands become prominent and thick in the upper part which is spoken of as Lavender Clays.

Lavender Clays.—These lavender coloured shaly rocks and associated sandstones display current bedding, ripple marks and other evidences of shallow deposition. They are well seen in many places west of Nilawan and up to the Indus, beyond which they become mainly calcareous.

The Olive Series and the Speckled Sandstones are to be regarded as two facies of the same strata. They seem to be the equivalents of the Schwagerina beds of Russia and may be referred to the Upper Uralian or the lowermost Permian. The Eurydesma and Conularia beds are the equivalents of the Nagmarg beds of Kashmir and the Rikba plant beds of the Karanpura coalfield in Bihar which are all correlated with the Sakmarian stage of Russia. There is some difference of opinion amongst geologists whether to include the Sakmarian in the Upper Carboniferous or Lower Permian, but C. S. Fox favoured its inclusion in the Carboniferous.

THE PRODUCTUS LIMESTONE SERIES

One of the best developed normal marine Permian areas in the world is in the Salt Range so that the Middle Permian is often referred to by the name Panjabian, after the Province in which the Salt Range lies. The Permian succession met with in this area is shown in Table 41.

TABLE 41.—PERMIAN OF THE SALT RANGE

	Division	Stage	Age
Productus Limestone Series PERMIAN	UPPER 30-50 m	CHIDRU (Sandstones and marls) JABBI (Sandstones and marls with Cephalopods) KUNDGHAT (Sandstones containing *Bellreophon*, etc.)	Thuringian (Zechstein)
	MIDDLE 60-120 m	KALABAGH (Limestones and marls with crinoids, etc.) VIRGAL (Siliceous limestones)	Panjabian or Saxonian (Rothliegendes)
	LOWER 60 m	KATTA (Calcareous sandstones and arenaceous limestones) AMB (Calcareous sandstone and limestones with *Fusulinae*)	Artinskian
SPECKLED SANDSTONES 100-150 m		LAVENDER CLAY (15-30 m) SPECKLED SANDSTONE (30-90 m) *Eurydesma* and *Conularia* beds BOULDER BED (30-60 m)	Lower Permian to Uralian

Amb.—The lowest beds of the Permian succession constitute the Amb stage which is made up of coarse yellowish and greenish calcareous sandstones with a few layers of coaly shales, well exposed in the Nilawan ravine. Continued westwards, they become gradually more calcareous.

Katta.—The Katta stage forms a transition between the Amb beds and the Middle Productus Limestones. It contains arenaceous limestones, Fusulinids being abundant in this as well as in the upper part of the preceding Amb stage.

Middle Productus Beds.—The most important part of the succession is the Productus Limestone proper, which forms the crags of the outer escarpment in Salt Range. It contains dolomite but the dolomitisation has been responsible for a good deal of obliteration of the enclosed fossils. The limestones are bluish grey to grey while the dolomites are cream coloured. The limestones, which include crinoidal and coralline ones, show fossils on the weathered surfaces, while the intercalated marls contain abundant and easily extractable fossils.

The Productus Limestone proper is divided into two stages, the lower one named after Virgal and the upper after Kalabagh. The equivalents of

PLATE XI
PERMIAN FOSSILS

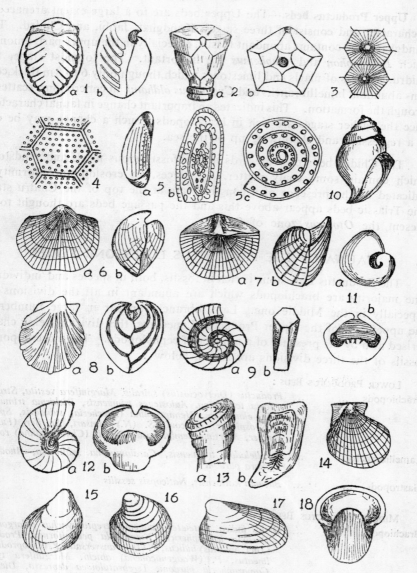

Explanation of Plate XI

1. *Oldhamina decipiens* (0.7). 2. *Cyathocrinus indicus* (0.7). 3. *Lonsdaleia salinaria* (2). 4. *Hexagonella ramosa* (4). 5. *Fusulina kattaensis*: a- specimen (4); b-longitudinal section (7); c-transverse section (10). 6. *Orthis jurasanensis* (0.7). 7. *Derbyia hemisphaerica* (0.3). 8. *Blanfordinia lunwalensis* (0.5). 9. *Euomphalus parvus* (1.5). 10. *Macrocheilus avellanoides* (0.7). 11. *Euphemus indicus* (0.7). 12. *Bellerophon jonesianus* (0.3). 13. *Richthofenia lawrenciana* (0.7). 14. *Lima footei* (0.3). 15. *Schizodus pinguis* (0.7). 16. *Lucina progenetrix* (0.7). 17. *Pleurophorus complanatus* (2). 18. *Stachella semiaurata* (0.7).

these beds are found in the Chitichun limestone in Northern Himalaya, in the brachiopod beds of the Vladimir region in Russia, and the Productus Limestone of the Timor island in Indonesia.

Upper Productus beds.—The Upper beds are to a large extent arenaceous in character and consist of three stages, Kundghat, Jabbi, and Chidru. The Kundghat stage contains abundant fossils, especially in the upper part, amongst which *Bellerophon* and *Euphemus* are important. The topmost stage, the Chidru, consists of marls and limestones which, though only 6 to 8 m thick, contains abundant lamellibranchs and *Cyclolobus oldhami* in concretions scattered through the formation. This indicates an important change in faunal characters, since the earlier stages are rich in brachiopods ; such a change may be due to a regression and shallowing up of the sea.

The Chidru beds pass upwards into unfossiliferous marls and sandstones which contain some coaly matter. In places, an erosion unconformity is indicated by the presence of a conglomerate at the top of the Chidru stage. The Triassic beds appear above this and the passage beds are thought to represent the *Otoceras* zone of Spiti.

FAUNA OF THE PRODUCTUS LIMESTONE SERIES

The Productus Limestone is rich in fossils, both in species and individuals. The majority are brachiopods which are abundant in all the divisions and especially in the Middle one. Lamellibranchs appear in large numbers in the upper part of the Upper Productus Limestone, this division being characterised also by the presence of a number of cephalopods. The more important fossils of the three divisions are shown below :

LOWER PRODUCTUS BEDS :

Brachiopods	*Productus (Dictyoclostus) spiralis, Marginifera vestita, Strophalosia tenuispina, Aulosteges trimuensis, Dielasma trimuense, D. elongatum, Hemiptychina (Beecheria) sublaevis, Spirifer (Neospirifer) marcoui, S. (N.) kimsari, Spirifer (Fusella) niger, Martinia semiglobosa, Athyris (Cleiothyridina) roissyi*
Lamellibranchs	*Parallelodon trimuensis, Cardiomorpha sublimosa, Anodontophora purdoni*
Gastropods	*Bucania kattaensis, Naticopsis sessilis*

MIDDLE PRODUCTUS BEDS :

Brachiopods	*Orthis indica, Enteletes waageni, Streptorhynchus pelargonatus, Derbyia hemispherica, Kiangsiella pectiniformis, Productus (Dictyoclostus) indicus, P. (D.) transversalis, P. (Linoproductus) lineatus, P. (Waagenoconcha) abichi, Marginifera typica, Camarophoria purdoni, Terebratuloidea depressa, Dielasma truncatum, Oldhamina decipiens, Richthofenia lawrenciana, Hemiptychina inflata, Spirifer (Anelasma) wynnei, Athyris (Cleiothyridina), capillata, Martiniopsis punjabica, Spiriferina (Spiriferellina) cristata, Spirigerella grandis, S. derbyi.*
Lamellibranchs	*Allorisma waageni, Schizodus dubiformis, Pseudomonotis waageni, Aviculopecten regularis*
Gastropods	*Pleurotomaria (Mourlonia) punjabica*
Cephalopods	*Metacoceras goliathus, Solenocheilus peregrinus, Xenaspis carbonaria*

UPPER PRODUCTUS BEDS :

Brachiopods	...	*Schizophoria juresanensis, Enteletes sulcatus, Productus (Dictyoclostus) indicas, P. (D.) spiralis. Tschernychewia typica, Productus (Dictyoclostus) aratus, Productus (Linoproductus) cora, Strophalosia horrescens, Chonetes squamulifer, Camarophoria superstes, Lyttonia nobilis, Hemiptychina himalayensis, Spirifer (Neospieifer) warchensis, Spirigerella grandis, S. praelonga*
Lamellibranchs	...	*Schizodus emarginatus, Solemya biarmica, Aviculopecten morahensis*
Gastropods	...	*Pleurotomaria (Worthenia) durga, Bellerophon equivocalis, B. jonesianus, Naticopsis sessilis*
Cephalopods	...	*Foordiceras oldhami, Planetoceras postremum, Episageceras wynnei, Waagenoceras* cf. *oldhami, Xenodiscus plicatus, Cyclolobus oldhami.*

Dr. F. R. Cowper Reed, after examining the extensive collections from the Salt Range, stated that there are no grounds based on faunal characters, for sub-dividing the three major divisions of the Productus Limestones into stages as was done by Waagen. Even the separation of Middle and Upper Productus Limestones is difficult, for the brachiopod faunas are for the most part common to the two. However, the Upper Productus beds are more arenaceous and contain a number of lamellibranchs and cephalopods not present in the Middle division.

The Lower Productus beds are characterised by the presence of *Productus spiralis, Spirifer marcoui* and *S. niger*. The subgenus *Taeniotherus* of *Productus*, and the genus *Aulosteges* appear also to be confined to them. Many of the species found in the Lower Productus beds do not seem to extend into the higher division. The Middle Productus beds are very rich in brachiopods. The sub-genera *Haydenella* and *Cancrinella* of *Productus*, and also *Richthofenia* are confined to them. The Upper division is characterised by *Tschernyschewia* and *Cryptacanthia* and several cephalopods.

Correlation.—The Speckled Sandstones with *Eurydesma* are correlated with the Rikba beds and the Nagmarg beds. They are considered to be the equivalents of the Dwyka Shales of South Africa, the Lower Marine Series of New South Wales and the Sakmarian of Russia. The age of these in the standard scale is Upper Carboniferous to lowermost Permian. The top of the Zewan beds may be correlated with the base of the Upper Productus Limestone. The Lower Productus beds may approximately be equivalent to the Agglomeratic Slates of Kashmir. The Kuling (Productus) Shales of Spiti may be correlated with the Upper Productus beds, while the sandstones underlying the former may be of Middle Productus age. The Upper Productus beds contain ammonoids (*Medlicottia, Cyclolobus, Popanoceras*) which are also found in the Artinskian, but Spath (Catalogue of Cephalopods in the British Museum, 1934, p. 24) expressed the opinion that they are Upper Permian. Grabau correlated the Kuling Shales with the Maping Limestones of China. The Middle and Upper Productus beds are considered the equivalents of the Lopingian of China as some of the brachiopods are common to both groups of formations.

PLATE XII
PERMIAN AND LOWER TRIASSIC FOSSILS

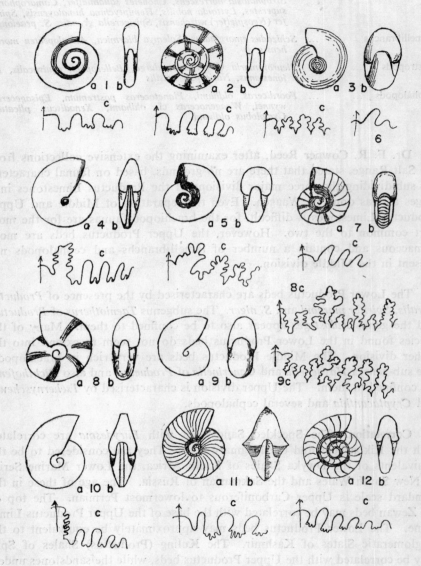

EXPLANATION OF PLATE XII

1. *Xenaspis carbonarius* (0.7). 2. *Xenodiscus carbonarius* (0.7). 3. *Arcestes priscus* (0.4). 4. *Sageceras (Medlicottia) wynnei* (0.7). 5. *Hyattoceras* aff. *cumminsi* (0.4). 6. *Gastrioceras* (suture). 7. *Adrianites (Hoffmannia)* sp. (0.7). 8. *Cyclolobus oldhami* (0.2). 9. *Cyclolobus kraffti* (0.2). LOWER TRIASSIC FOSSILS: 10. *Medlicottia dalailamae* (0.3). 11. *Otoceras woodwardi* (0.3). 12. *Ophiceras tibeticum* (0.3).

KASHMIR — HAZARA
THE INFRA-TRIAS

Great Limestone.—A hill range consisting mainly of limestone runs roughly eastwast for nearly 100 km amidst the red sandstones and clays of the Murree formation, some 25 km south of the Main Boundary Fault. It is a thick-bedded, compact, white to light grey limestone, 600 m thick. The lower part is dolomitic and the upper part siliceous, with bands and nodules of chert. It overlies the Agglomeratic Slate and shows a few intercalated layers of the same formation near the base. It is overlain by Eocene Nummulitic Limestones and shales. It is moderately folded and is marked on its southern margin by a steep fault. The Murrees which surround this formation are also separated by steep thrust faults from the Siwaliks on their south.

The Great Limestone is unfossiliferous and of Sub-Himalayan aspect. As it overlies the Agglomeratic Slates, it is most probably Permian in age and equivalent to the Krol limestone. Its position to the east of the Kashmir Wedge would place it with the Sub-Himalayan formations and not with the Productus and Zewan limestones which are fossiliferous and Tethyan in their relationships. The Infra-Trias Limestone of Hazara, which is also unfossiliferous and overlies the Tanawal Series with Sub-Himalayan and Peninsular affinities, seems to be its close equivalent. The intervention of the Productus Limestone Series in the Salt Range between the Jammu and Hazara areas is due to a south-directed thrust of the strata deposited in a more northern zone.

In western Kashmir and Hazara the Tanawal series is overlain by a boulder-conglomerate containing ice-striated boulders. This formation is probably the equivalent of the Talchir boulder-bed. Over this are found purple sandstones and shales which are in turn followed by 600 m of unfossiliferous dolomitic limestones. This group of rocks is known as the Infra-Trias series as it underlies Triassic formations. The dolomitic rocks are, in places (*e.g.*, in Kaghan) intercalated with flows of the Panjal trap. Their age is therefore, at least in part, Upper Carboniferous to Permian.

SIMLA — GARHWAL
INFRA-KROL AND KROL SERIES

The Upper Palaeozoic unconformity in the Simla region is probably indicated by the presence of the Blaini boulder-bed which may be correlated with the Tanakki conglomerate of Hazara. The Blainis contain also some pink limestones and slates, the latter resembling the slates of the Infra-Krols. The Infra-Krols consist of dark slaty shales with thin bands of quartzite. The Infra-Trias is presumably represented here by the Infra-Krol and Krol series, but it must be pointed out that there is little lithological similarity between the rocks of the two areas.

In the Krol belt of Sirmur area the Krol Series is well developed. Its basal beds are Krol Sandstones, orange to brown in colour, generally soft, and containing fragments of shale. They are followed by limestones which show five sub-divisions with an aggregate thickness of 600 m.

Krol E.—Massive cream coloured limestone, calcareous sandstone and brown shales.
  ,,   D.—Cherty limestone, dark limestone, bleached shales and quartzites.
  ,,   C.—Massive cyrstalline limestone, often sulphurous.
  ,,   B.—Red and green shales with dolomitic limestone.
  ,,   A.—Thin bedded blue limestone, shaly limestone, calcareous and carbonaceous shales.

In Garhwal, the Lower Krols (A) are slaty, the Middle Krols (B) being represented by purple shales and the Upper Krols (C, D, E) by massive limestone as seen around Mussoorie and farther east. The Krols of the two areas are compared below :

	Simla	Garhwal
Krol E	} Upper Krol : Limestone	Massive Limestone
,, D		
,, C		
,, B	Middle Krol : Red Shales	Purple Shales
,, A	Lower Krol : Limestone	Slates

There are two great overthrusts in the Krol belt, the Giri thrust which brings the Blainis and Jaunsars over the Krol series, and the Krol thrust which brings the Infra-Krols over the Kasauli (Tertiary) beds. In the Chakrata area the Krol thrust brings the Krol and older rocks on to the Subathu beds. The Krol Series is seen to be highly folded in the area between these two thrusts.

The Krols show signs of rather shallow water deposition and only rarely contain fragmentary and undeterminable fossils. The foetid limestones occurring in them may be taken as indicating that the basin of deposition was far from favourable for the habitation of organisms.

In the Simla region there are also the *Shali Limestone*, dolomites and slates whose age is unknown, but from whose resemblance to the Krols, the two were correlated. The Shali Limestones are overlain by the *Madhan Slates* and these in turn by the Subathu and Dagshai beds. It is now believed that the Shali limestones are late Pre-Cambrian as they have yielded stromatolites. They are equated with the Deoban limestones.

The Blainis may be roughly correlated with the Permo-Carboniferous, and the Infra-Krol and Krol beds with the Permian. Plant spores found in the Krols have been assigned to the Permian or early Mesozoic.

EASTERN HIMALAYA

In the Lesser Himalayas of Darjeeling, Buxa Duars, Bhutan and some other places farther east, the Upper Carboniferous and Permian are represented by the Gondwanas in which typical Barakar rocks with carbonaceous shales and poor coal seams have been recognised. The Gondwanas are generally thrust over the Siwaliks to their south. It has already been mentioned that in the Rangit valley in northern Sikkim and in the Subansiri valley and in the Abor hills, the Lower Gondwanas are associated with marine Permian rocks.

BURMA : SOUTHERN SHAN STATES

The Upper Plateau Limestone has yielded several Permian corals — *Sinophyllum, Iranophyllum splendens, Wentzella* cf. *timorica* etc. which have clear affinities with the Permian fossils from Southern China and Persia (*Pal. Ind.* N.S. 30 (2), 1941).

SELECTED BIBLIOGRAPHY

PERMO-CARBONIFEROUS AND PERMIAN

Ahmad, F. Glaciation and Gondwanaland, *Rec. G.S.I.*, 637-674, 1960.
Ahmad, F. Palaegeography of the Gondwana period with special reference to India. *Mem. G.S.I.*, **90**, 1961.
Auden, J. B. Geology of the Krol belt. *Rec.* **67**, 357-454, 1934.

Bion, H. S. and Middlemiss, C. S. Fauna of the Agglomeratic Slates of Kashmir. *Pal. Ind.* N.S. XII, 1928.

Dickins, J. M. and Thomas, G. A. The marine fauna of Lyons Group of the Carrandibly Formation of the Carnarvon Basin, West Australia. *Rep. Bur. Min. Res.* (Australia), **38**, 65-96, 1959.
Diener, C. Geological structure of the Chitichun region. *Mem.* **28**, Pt. I, 1898.
Diener, C. Anthracolithic fossils of Kashmir and Spiti. *Pal. Ind. Ser. XV*, Vol. I (2), 1899.
Diener, C. Permo-carboniferous fauna of Chitichun I. *Op. cit.* Pt. 3. 1897.
Diener, C. Fossils of Productus shales of Kumaon and Garhwal. *Op. cit.* Pt. 4, 1897.
Diener, C. Permian fossils of the Central Himalayas. *Op. cit.* Pt. 5, 1903.
Diener, C. Anthracolithic fossils of Shan States. *Pal. Ind. N.S. III*, 4, 1911.
Diener, C. Anthracolithic faunae of Kashmir, Kanaur and Spiti. *Pal. Ind. N.S.* 2, 1915.
Dunbar, C.O. Fusulinids of Lower Productus limestones. *Rec.* **66**, 405-413, 1933.

Griesbach, C.L. Notes on the Central Himalayas, *Rec.* **26**, 19-25, 1893.

Misra, J.S. Shrivastava, B.P. and Jain, S.K. Discovery of marine Permo-Carboniferous in western Rajasthan. *Current Sci.* 30 (7), p. 262-263 1961.

Sahni, M.R. and Dutt, A.K. Argentine and Australian affinities in a Lower Permian fauna from Manendragarh, Central India. *Rec. G.S.I.* **87**(4), 655-670, 1962.

Reed, F.R.C. Permo-Carboniferous marine faunas from the Umaria coalfield. *Rec. G.S.I.*, **60**, 367-398, 1926.
Reed, F. R. C. Upper Carboniferous fossils from Chitral and the Pamirs. *Pal. Ind. N.S.* VI, 4, 1925.
Reed, F.R.C. Fauna of the Agglomeratic slates of Kashmir. *Op. cit.* XX, 1, 1932.
Reed, F. R. C. Fossils from the Productus limestone of the Salt Range. *Op. cit.* XVII,1931.
Reed, F. R. C. Fossils from the Eurydesma and Conularia beds of the Salt Range. *Op. cit.* XXIII, I, 1936.
Reed, F. R. C. Brachiopoda and Mullusca from the Productus Limestones of the Salt Range, *Op. cit.*, XXIII, 2, 1944.
Reed, F. R. C. Anthracolithic faunas of S. Shan States. *Rec.* **67**, 357-454, 1934.

Seward, A. C. and Smith Woodward, A. Premo-carboniferous plants and vertebrates from Kashmir. *Pal. Ind. N.S. II*, 2, 1905.
Sherlock, R.L. The Permo-Triassic Formations. London, 1947.

Waagen, W. Salt Range Fossils (Productus Limestone). *Pa.l Ind. Ser.* XIII, Vol. I, 1-7, 1879-1887.
Waagen, W. Salt Range Fossils — Geological results. *Pal. Ind. Ser.* XIII, Vol. IV, 1889-1891.
Wager, L. R. The Lachi series of N. Sikkim. *Rec.* **74**, 171-188, 1939.
Wynne, A.B. Geology of the Salt Range in the Punjab. *Mem.* **14**, 1878.

CHAPTER XII

THE TRIASSIC SYSTEM

The Triassic System is the earliest of the three Systems which make up the Mesozoic or Secondary era. The name is derived form the three-fold division adopted for it originally in Germany, where it was first studied. Though it is supposed that the Mesozoic group commences at the base of this System, critical study in many areas has shown that there is really no marked stratigraphic or palaeontologic break at the base, but that such a break occurs in the middle of the Rothliegende (Middle Permian). In his exhaustive review of the Permo-Triassic formations, R. L. Sherlock (The Permo-Triassic formations. Hutchinson, London, 1947) has shown that the Upper Permian should really be deemed to commence the Mesozoic. He has actually proposed the name EPIRIC to the new System which would include the Upper Rothliegende and Zechstein and the Triassic. Whether this new monenclature is accepted or not, the important point to note is that the Upper Permian grades in most cases into the Triassic.

Two major facies of the Triassic are developed in Europe. The Germanic facies, occurring in Germany and the British Isles is partly continental and partly epicontinental marine, while the Mediterranean facies is geosynclinal marine and is found in the Alps, Pyrenees and North Africa. The relationship between the two facies is shown below.

Alpine	*Germanic*
Rhaetic	
Noric	
Carnic	} Keuper
Ladinic	
Anisic (= Virgloric)	} Muschelkalk
Scythic (= Werfenic)	Buntsandstein

Germanic facies.—The BUNTSANDSTEIN (or Bunter in England) consists of sandstones, conglomerates and shales of lagoonal facies, with red to violet colours. The main part is really a variegated conglomerate, while the upper beds contain shell banks full of *Lingula, Myophoria* and *Hoernesia*. The middle division, the MUSCHELKALK, is a shell limestone, 300 to 400 m thick, containing crinoids (*Encrinus* etc.), brachiopod (*Spiriferina, Coenothyris*) and pelecypods (*Myophoria, Hoernesia, Lima*, etc.) and also ammonoids. The lower part of this division is the WELLENKALK, so called because of the wavy and thin-bedded character of the limestone. The upper division, the KEUPER, is generally marly with carbonaceous shale intercalations. *Myophoria goldfussi* and *M. kefersteini* characterise the Lower and Upper Keuper respectively. There are also gypsum and anhydrite beds in places. There is practically no Muschelkalk in Britain, the Bunter together with the Upper Permian forming the NEW RED SANDSTONE formation. The Keuper is similar to that of Germany. In Poland and Upper Silesia there is a transition from the Germanic to the Mediterranean facies, the stratigraphy being very useful for correlating the two facies.

Alpine facies.—The Mediterranean or Alpine facies consists almost entirely of limestones deposited in open seas. Four sub-facies are seen in the four major nappes — Bavarian, Hallstatt, Dachstein and Dolomite. The lower Triassic is often a shore facies while the Middle division shows coral or cephalopod limestones or dolomite. The upper Trias forms the Hauptdolomit of the high Alps or the famous Hallstatt limestone and Dachsteinkalk. The Hallstatt limestone is a deep water facies containing numerous ammonites and brachiopods while the Dachsteinkalk is associated with coral reefs and shell limestones.

Life during the Triassic

Flora.—Plant remains are found in certain continental strata, in the Lower Trias. These include the major groups found in the Permo-Carboniferous — equisetales, conifers, cycads, Ginkgos, etc. Calcareous algae like *Gyrosporella, Diplopora, Phyllosporella*, etc. form reefs in the marine calcareous facies.

Fauna.—There is a rich and varied fauna in the Triassic, particularly numerous *Ceratites* and related ammonoids, brachiopods and pelecypods. Calcareous sponges are found in some littoral deposits while corals are well developed, especially Hexacorals and Astracidae such as *Thecosmilia, Montlivaltia, Thamnastrea*. Crinoids and echinoids are abundant, the fossil *Encrinus liliformis* being abundant in the Muschelkalk. Several genera of the brachiopods such as *Spirigera, Spiriferina, Athyris,* Terebratulids and Rhynchonellids, play an important part in the stratigraphy. There is a rich variety of bivalves — *Myophoria, Gervillea, Avicula, Pecten, Mytilus, Modiola, Ostrea, Lima, Megalodon,* and univalves — *Pleurotomaria, Worthenia, Bellerophon,* etc. Cephalopods and especially the Ceratite group, are of importance in the zoning of the formations.

Amongst the vertebrates, fishes are quite abundant. But the pride of place goes to the reptiles, and particularly the Theromorphs and Dinosaurs, which were masters of the land and had developed into huge animals which could walk, run, swim or fly. Some of them are important enough to be used as zone fossils in South Africa — *Endothiodon, Cysticephalus, Lystrosaurus, Procolophon, Cynognathus* etc. Only a few Labyrinthodonts (Amphibia) are present, such as *Brachyops* and *Mastodonsaurus*. The oldest Mammal also appears in the Triassic.

In the fauna of the Himalayan Trias, the ammonoids including the Orthoceras, Goniatite and Ceratite groups constitute the most important and useful fossils. The oldest Belemnite — *Atractites* — occurs here. Ceratites are characteristic of the Trias, the earlier strata containing the more primitive forms and the later strata the more highly developed and ornamented members.

The Lower Triassic ammonoids appear simple and monotonous in appearance in comparison with the variety seen in the Upper Permian. The simple-sutured but robust Lower Triassic types were able to withstand and survive the changes that took place at the end of the Permian, whereas the specialised Permian types died out. The primitive Ceratites seem to have first appeared in Middle to Upper Permian times represented by such forms as *Xenodiscus carbonarius*. *Otoceras* which characterises the earliest Trias in India has a raised ear-like rim around the umbilicus and a ceratitic suture of the type rendered familiar by the common European Muschelkalk fossil *Ceratites nodosus* ; the saddles have a smooth rounded outline while the lobes are slightly denticulated. *Hendenstroemia* and *Pseudosageceras* still retain the ceratitic outline of the suture, but the lobes and saddles become numerous, thus foreshadowing the complex sutures of the *Pinacoecratidae* of Middle and Upper Trias. Another genus, *Episageceras*, which occurs in the Otoceras beds of the Himalaya and the Stachella beds of Salt Range, is a relic of the *Goniatite* group, very closely allied to and derived from the Permian *Medlicottia*. Further up in the Lower Trias the beds are chaarcterised by the large and spirally grooved Ceratite of the genus *Flemingites*.

The Muschelkalk is the age of the typical *Ceratites*, which begin to develop here along two lines ; in one the shell gains strength by corrugations while the sutures remain simple ; in the other shell remains smooth but gains sterngth by the ramifications of the sutural lines. The first group constitutes the typical *Ceratites*, of which large numbers are present both in the Himalaya and Kashmir, being somewhat evolute forms with thick ribs and blunt knobs. In *Sibirites* the sutures remain simple but the shells are highly sculptured. The smooth-shelled forms with sutures tending to complexity are represented by *Gymnites* and *Ptychites*, the former a compressed form and the latter rather globose. The genus *Sturia* resembles *Gymnites* but has a spirally grooved shell.

In the Upper Triassic beds, the *Ceratites* reach their zenith of development, and are represented by a number of beautiful forms in the Himalayan strata. The two lines of development continue here, the contrast becoming stronger than in the Muschelkalk. The genus *Arpadites*, some species of which are of fairly large dimensions, resembles the typical *Ceratites* but has a deep smooth furrow on the outer margin ; in *Ceratites* proper the periphery is either rounded or raised into a keel. The genus *Distichites* is similar to *Arpadites* but has somewhat simpler suture line. In the genera *Tibetites* and *Sirenties*, the shell is sculptured and involute and shows a tendency to become clypeiform in shape. In *Tropites* the shell tends to become globose but the whorls do not completely overlap the umbilical channel. The involution is much more pronounced in *Halorites* and *Juvavites* in which the umbilicus is nearly or almost completely obliterated. Smooth-shelled forms are less prominent than the highly sculptured ones in the Upper Trias, often tending to become globose. The genera *Lobites* and *Didymites* combine a subdued sculpturing with only slightly serrated sutures which however have a large number of inflections. Amongst the smooth clypeiform ammonoids is *Placites* (abundantly represented in the Tropites limestone of Byans) which is distinguished by the multiplicity of auxiliary inflections. In *Carnites* and *Pinacoceras* the inflections are greatly multiplied by the splitting up of the external saddle. The well known forms *Pinacoceras parma* and *P. metternichi* are amongst the ammonoids which exhibit the most complicated sutural lines. They are accompanied by *Bambanagites* which has a sculptured shell but has the same plan of suture through less complicated. A true member of the Ammonite group is represented by *Discophyllites* which is a precursor of the *Phylloceratidae* which attains prominence in the Jurassic.

The vast majority of the specialised Ceratites died out at the end of the Trias as was the case with the earlier ammonoids at the end of the Permian. Only a few ammonoids like the Phylloceratids survived the changes at the end of the Trias.

The marine Triassic is well developed in the northern Himalayan zone in the classic area of Spiti and its sub-divisions are easy of correlation with those of the Mediterranean region of Europe. This belt extends into Kumaon, there being some differences in the lithology near the Nepal border. To the west, in Kashmir, the strata are thicker, but they have not been studied in as much detail as in Spiti, though several important zones have been identified. The Triassic of the Cis-Indus Salt Range is incompletely developed, only the lower division and the lowest part of the middle division being seen. The upper division is, however, seen in the Trans-Indus portion.

The Upper Trias is seen as a thick series of argillaceous rocks in the Zhob-Pishin region of Baluchistan on the west and in the Arakan region on the east. The Triassic rocks assume the red sandstone facies in parts of Yunnan and Szechuan, bearing some resemblance to the Gondwana strata of the same age. A marine facies is however to be seen in the areas bordering on the ' Red basin ' of China and in Tonkin in Indo-China. The Tethyan facies is developed in the Indonesian Archipelago, particularly in Timor Island from which excellent fossil collections have been got. To the west of the Himalayas, the Triassic rocks have been recognised in several places between them and the Alps. Indeed there is a remarkable similarity between the calcareous Tibetan facies and the Hallstatt marble facies in the Eastern Alps not only in fauna but also in lithology ; but this facies is exposed only in the exotic blocks which represent a zone hidden under thrust sheets.

SPITI : THE LILANG SYSTEM

The most complete section of the Trias is exposed in the Spiti-Kumaon belt of the Himalaya north of its main axis where it forms immense escarpments rising often to a height of 3,000 m from the level of the adjoining valleys. Deriving its name from the type section of Lilang in Spiti, the system is seen to comprise black limestones with shale intercalations and to attain a thickness of nearly 1,250 m. It is entirely marine in character and is fossiliferous except in the topmost division which grades imperceptibly into the Lower Jurassic without any change in lithology.

As in Europe, it is divisible into three series, but the divisions are of very unequal thickness. The Lower Trias is only 12 m and the Middle 120 m, while the Upper Trias attains a thickness of 1,100 m.

320 GEOLOGY OF INDIA AND BURMA [CHAP.

The Himalayan Trias has been studied in Spiti, Johar, Painkhanda and Byans. Superb sections are seen in the neighbourhood of Lilang and other places in Spiti and in the Bambanag and Shalshal cliffs further east, though, unfortunately, these are difficult of access from the plains. Table 42 gives details of the general section at Lilang.

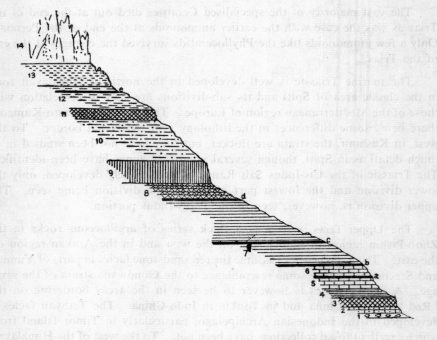

Fig. 8.—GENERALISED SECTION NEAR LILANG, SPITI.

(AFTER A. VON KRAFFT. AS IN DIENER, *Mem.* 36, Pt. 3.).

1. Productus shales. 2. Lower Trias. 3. Muschelkalk. 4. Daonella shales. 5. Daonella limestone. 6. Halobia limestone. 7. Grey beds. 8. Tropites shales. 9. Dolomite with *Lima austriaca*. 10. Juvavites shales. 11. Coral Limestone. 12. Monotis beds. 13. Quartzite series. 14. Megalodon limestone. *a.* Zone of *Joannites thanamensis*. *b.* Zone of *Joannites cymbiformis*. *c.* Bracihopod layer (Grey beds). *d.* Zone of *Tropites sub-bullatus*. *e.* Zone of *Monotis salinaria*.

LOWER TRIAS

The Lower Trias follows conformably on the Productus shales and consists of limestones and intercalated shales. The basal 3 m contain rich fossiliferous zones showing at least three distinct faunal assemblages, *viz., Otoceras Ophiceras, Meekoceras* and *Hedenstroemia-Flemingites* faunas. The Chief fossils in these zones are :

Otoceras zone ... *Otoceras woodwardi, O.* cf. *undatum, Ophiceras sakuntala, Pseudosageceras dalailamae.*

Ophiceras zone ... *Ophiceras sakuntala, O. tibeticum, Xenodiscus radians, Pseudomonotis griesbachi.*

Meekoceras zone ... *Meekoceras varaha, M. markhami, M. lilangense, M. jolinkense Aspidites spitiensis, Koninckites haydeni, Xenodiscus radians*

TABLE 42.—TRIAS OF SPITI

(After Hayden, *Mem.* XXXIV. Pt. 1, and Diener, *Ibid.*, Pt. 3)

Division			Name of beds	Description of beds	Thickness Meters
LIAS			Kioto (= Megalodon)	Massive limestone and dolomite	500
UPPER TRIAS	Noric		Limestone	Massive limestone and dolomite (with *Spirigera noetlingi*, *Megalodon ladakhensis* and *Diceracardium himalayense* 100 to 135 m. above base).	250
			Quartzite series	White and brown quartzites with grey limestones and and black shales (with *Spirigera maniensis*).	100
			Monotis shales	Sandy and shaly limestone with brown-weathering shales and sandstones (with *Monotis salinaria* and *Spiriferina griesbachi*).	100
			Coral limestone	Limestones (with *Spiriferina griesbachi*, and coral and crinoid remains).	30
			Juvavites beds	Brown-weathering shales, limestones and sandstones (with *Juvavites angulatus*).	150
	Carnic		Tropites beds	Dolomitic limestones (with *Dielasma julicum*), Shales and dark limestones with ammonite bed (*Tropites subbullatus*) 130 m above base.	100 200
			Grey beds	Grey shales and shaly limestones with a bivalve bed 100 m above base, and an ammonite bed 16 m above base. (*Spiriferina shalshalensis* and *Joannites cymbiformis*).	150
			Halobia beds	Dark splintery limestone (with *Halobia* cf. *comata*). Bed with *Joannites thanamensis* near base.	45
MIDDLE TRIAS	Ladinic		Daonella Limestone.	Hard dark limestone (with *Daonella indica*).	45
			Daonella shales	Black limestone, shaly limestone and shale (with *Daonella lammeli* and *Ptychites gerardi*).	50
		Muschelkalk	Upper Muschelkalk	Concretionary limestone with shale bands (*Ptychites rugifer*).	6
			Lower Muschelkalk	Dark shales and grey limestones (with *Keyserlingites dieneri*, *Sibirites prahlada*, *Spiriferina stracheyi*).	2
			Nodular limestone	Hard nodular limestone with few fossils.	20
			Basal Muschelkalk	Shaly limestone (with *Rhynchonella griesbachi*).	1

Division		Name of beds	Description of beds	Thickness Meters
Lower Trias	Bunter	Hedenstroemia beds	Limestone with *Pseudomonotis himaica*. Shaly limestones and shales alternating (unfossiliferous). Thin-bedded limestones and shales (with *Hedenstroemia majsisovicsi, Flemingites rohilla* and *Xenodiscus nivalis*).	1 8 2
		Meekoceras zone	Thin-bedded limestones and shales with *Meekoceras varaha* and *Meekoceras lilangense*.	1
		Ophiceras zone	Grey limestone with *Ophiceras sakuntala* and *Pseudomonotis griesbachi*.	.3
		Otoceras zone	Brown limestone with *Otoceras woodwardi*.	.6
Permian		Productus shales	Dark shales with Permian fossils.	?

The three lowest zones are close together and separated from the *Hedenstroemia beds* by 1.5 m of unfossiliferous rock. The Hedenstroemia beds contain some 30 species of cephalopods including *Hedenstroemia mojsisovicsi, Sibirites spitiensis, Pseudosageceras multilobatum, Xenodiscus nivalis, Flemingites rohilla, Aspidites muthianus, Meekoceras* cf. *joharense, Koninckites giganteus.*

These are followed by similar but unfossiliferous beds having a thickness of about 8 m, above which a fossiliferous 1-m bed occurs, containing *Pseudomonotis himaica* and *P. decidense.*

MIDDLE TRIAS

The Muschelkalk.—Within one metre of the last mentioned bed, there is another horizon which yielded *Rhynchonella griesbachi* and *Retzia himaica* which is referable to the basal Muschelkalk. It is followed by hard, nodular limestone, about 20 m thick, similar in lithological characters and stratigraphical position to the Niti limestone of the Niti Pass in Kumaon, and containing only a few fossils. The over-lying beds, consisting of dark shales and limestones, constitute a typical Lower Muschelkalk horizon, with an ammonite-rich lower part and a brachiopod-rich upper part. They contain :

Cephalopods ... ... *Keyserlingites (Durgaites) dieneri, Danubites kansa, Dalmatites ropini, Monophyllites hara, M. confucii, Sibirites prahlada.*
Brachiopods ... ... *Spiriferina stracheyi, Rhynchonella dieneri.*

The Upper Muschelkalk.—This consists of a concretionary limestone and contains a large number of cephalopods and a few brachiopods and mollusca. Amongst the fossils are :

Brachiopods ... ... *Coenothyris* cf. *vulgaris, Mentzelia Koeveskalliensis.*
Cephalopods ... ... *Ptychites rugifer, P. gerardi, Ceratites thuilleri, C. trinodosus, C. (Hollandites) ravana, C. (H.) voiti, Beyrichites khanikoffi, Sturia sansovinii, Buddhaites rama, Gymnites jollyanus, Orthoceras spitiensis, Joannites* cf, *proavus, Proarcestes* aff. *bramantei.*

PLATE XIII
TRIASSIC FOSSILS

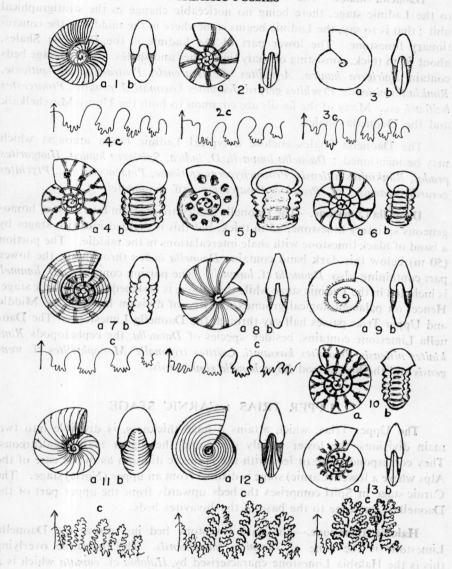

EXPLANATION OF PLATE XIII

1. *Meekoceras varaha* (0.3). 2. *Flemingites rohilla* (0.15). 3. *Hedenstroemia mojsisovicsi* (0.15). 5. *Ceratites subrobustus* (0.15). 5. *Stephanites superbus* (0.15). 6. *Celtites armatus* (0.3). 7. *Flemingites flemingianus* (0.07). 8. *Aspidites superbus* (0.1). 9. *Gyronites frequens* (0.2). 10. *Sibirites prahlada* (0.7). 11. *Ptychites gerardi* (0.4). 12. *Sturia sansovinii* (0.1). 13. *Gymnites jollyanus* (0.3).

LADINIC STAGE

Daonella Shales.—The Upper Muschelkalk bed shows a gradual passage to the Ladinic stage, there being no noticeable change in the stratigraphical unit; that is to say, the Ladinic begins somewhere in the middle of the concretionary limestone. The lower part of the Ladinic is the Daonella Shales, about 50 m thick, consisting of shaly limestones and shales. The passage beds contain *Spirigera hunica, Arpadites* cf. *lissarensis, Protrachyceras spitiense, Rimkinites nitiensis, Ptychites gerardi Joannites kossmati, J. proavus, Proarcestes balfouri*, etc. Many of the fossils are common to both the Upper Muschelkalk and the Daonella Shales.

The Daonella Shales enclose a typical Ladinic fauna amongst which may be mentioned: *Daonella lommeli, D. indica, Spirigera hunica, Hungarites pradoi, Rimkinites nitiensis, Protrachyceras spitiense, Pinacoceras* sp., *Ptychites gerardi, Joannites* cf. *Proavus, Proarcestes* aff. *bramantei*.

Daonella Limestone.—The Daonella Shales are overlain by a mass of homogeneous splintery limestone, 85 m thick, but this is divided into two stages by a band of black limestone with shale intercalations in the middle. The portion (50 m) below this dark band contains *Daonella indica* throughout, the lower part containing also *Daonella* cf. *lommeli*. The portion containing *D. lommeli* is included in the Ladinic stage while that above is assigned to the Carnic stage. Hence, on palaeontological grounds, the line of division between the Middle and Upper Trias passes halfway through the Daonella Limestone. The Daonella Limestone contains, besides species of *Daonella*, the cephalopods *Rimkinites nitiensis, Joannites kossamti, Celtites trigonalis, Monophyllites* cf. *wengensis* and the brachiopod *Rhynchonella rimkinensis*.

UPPER TRIAS : CARNIC STAGE

The Upper Trias, which attains a huge thickness, is divisible into two main divisions, the lower mainly shaly and the higher mainly calcareous. They correspond more or less with the faunistic division as in the case of the Alps where a lower (Carnic) stage is distinct from an upper (Noric) stage. The Carnic stage in Spiti comprises the beds upwards from the upper part of the Daonella Limestone to the base of the Juvavites beds.

Halobia Limestone.—The dark limestone bed just above the Daonella Limestone is the zone of *Joannites thanamensis*. The limestone overlying this is the Halobia Limestone characterised by *Halobia* cf. *comata* which is a typical European fossil of the Julic substage of the Carnic.

Grey Shales.—The Halobia Limestone is overlain by the Grey Shales which have a tickness of 150 m and consist of shales with intercalations of shaly limestone. They show fossil horizons a little above the base and again at 100 m above the base. The lower horizon contains *Trachyceras* aff. *ariae, Joannites* cf. *cymbiformis, Monophyllites* cf. *simonyi*, etc. The upper horizon yielded only one ill-preserved ammonoid (*Paratropites* sp.) and several brachiopods and bivalves:

Brachiopods	...	...	Rhynchonella laucana, R. himaica, Spiriferina shalshalensis, S. gregaria, Mentzelia mentzelii, Dielasma julicum.
Bivalves	...	...	Lilangina nobilis, Pomarangina haydeni, Lima sp.

Tropites Beds.—The Grey Shales are overlain by the Tropites Beds, the lower 200 m of which are calcareous shales with limestone intercalations. About 120 m above the base of these there is a nodular limestone containing a rich, but badly preserved, cephalopod fauna which includes *Tropites* cf. *subbullatus, T. discobullatus, Clydonautilus acutilobatus, Jovites spectabilis, Sandlingites* aff. *reyeri, Proarcestes* cf. *gaytani.*

The upper part of the Tropites Beds are dolomitic limestones, with a thickness of 100 m also containing Carnic fossils — *Dielasma julicum, Spiriferina* aff. *shalshalensis* ; *Lima* cf. *austriaca, Halobia* aff. *superba, Daonella* aff. *styriaca.*

The Carnic beds, which end with this limestone, have a total thickness of 500 m.

NORIC STAGE

Juvavites Beds.—The Tropites beds are followed conformably by brown-weathering limestone with shale and sandstone beds having a total thickness of 150 m. Their characteristic fossil is *Juvavites angulatus* wihch is associated with others, such as —

Juvavites aff. *ehrlichi, Anatomites* aff. *melchioris, Tibetites* cf. *ryalli, Pinacoceras* aff. *parma, Metacarnites footei, Dittmarites lilliformis, Atractites* cf. *alveolaris, Paranautilus arcesti-formis :*

Bivalves	...	...	*Lima* cf. *serraticosta, Pecten* aff. *monilifero, Halobia* aff. *fascigerae.*

Coral Limestone.—The base of the overlying stage, the Coral Limestone, is a calcareous sandstone with plant remains. The Coral Limestone, which is 30 m thick, abounds in crinoidal and coral remains and contains two brachiopods *Spiriferina griesbachi* and *Rhynchonella bambanagensis.*

Monotis hales.—Above the Coral Limestone are shaly limestones, black limestones, flaggy sandstones and sandy shales, which attain a thickness of 100 m. The sandy shales and sandstones especially contain abundant fossils which include —

Bivalves	...	...	*Monotis salinaria, Lima* cf. *serraticosta, Pecten* aff. *monilifero, P. margariticostatus, Pleuromya himaica.*
Brachiopods	...	...	*Spiriferina griesbachi, Spirigera dieneri,* ¡*Aulacothyris Joharensis, Rhynchonella bambanagensis.*

Quartzite Series.—Immediately above the Monotis Shales are white and brown quartzites, 90 m thick, which form a conspicuous horizon visible from a distance. Most of its fossils are also found in the Monotis Shales, but *Spirigera maniensis* is restricted to it.

Megalodon Limestone (= Kioto Limestone).—The topmost beds of the Triassic sequence are thick massive limestones and dolomites which Griesbach originally included in his Rhaetic system. Their total thickness is of the order of 800 m and they bear a striking resemblance to the Dachsteinkalk of the Alpine region both in lithology and stratigraphical position. They have a

PLATE XIV
TRIASSIC FOSSILS

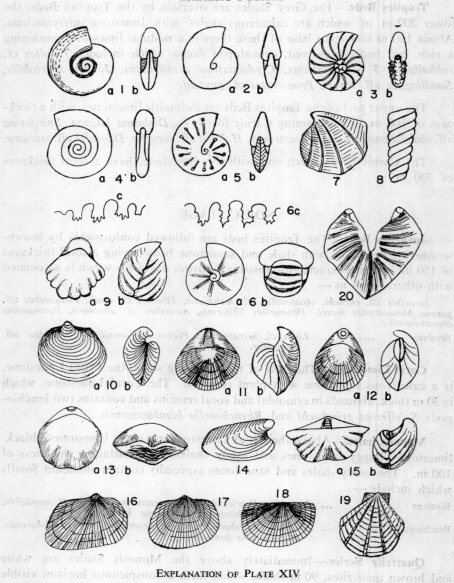

EXPLANATION OF PLATE XIV

1. *Prionolobus rotundatus* (0.2). 2. *Kingites lens* (0.2). 3. *Hollandites voiti* (0.2). 4. *Monophyllites confucii* (0.5). 5. *Hungarites* sp. (0.4). 6. *Isculites middlemissi* (0.4). 7. *Myophoria middlemissi* (0.7). 8. *Atractites smithi* (0.3). 9. *Rhynchonella trinodosi* (0.4). 10. *Pseudomonotis griesbachi* (0.4). 11. *Mentzelia mentzeli* (0.7). 12. *Coenothyris vulgaris* (0.5). 13. *Dielasma julicum* (0.5). 14. *Myophoria ovata* (0.5). 15. *Spiriferina stracheyi* (0.7). 16. *Daonella lommeli* (0.4). 17. *Daonella indica* (0.25). 18. *Halobia comata* (0.5). 19. *Lima serraticosta* (0.7). 20. *Diceracardium himalayensis* (0.15).

uniform appearance throughout their thickness and are for the most part unfossiliferous. Fossils have been found between 60 and 100 m from the base, amongst which are —

Megalodon ladakhensis, Entolium cf. *subdemissum, Pecten chabrangensis, Lima cumaunica ; Spirigera noetlingi, Spiriferina* cf. *haueri.*

Near Hansi, large numbers of *Megalodon ladakhensis* and *Diceracardium himalayense* are seen in the limestone about 15 m abve The Quartzite Series. This was called the *Para Limestone* by F. Stoliczka and referred to the Rhaetic. The fossil assemblage however shows this part of the limestone to be Noric. The upper part includes the Rhaetic and Liassic.

The Megalodon Limestone, to which Hayden has advocated the use of the name Kioto Limestone, is overlain by the *Spiti Shales* of Callovian age.

At Tera Gadh near Giumal, Von Krafft obtained an ammonite about 100 m below the top of the Kioto Limestone. This was thought to be *Stephanoceras coronatum,* but was later identified as a *Macrocephalites* by Uhling (*Pal. Ind.* Ser. XV, Vol. 4, 269, Pl. 77, fig. 5). There must therefore be some mistake about the locality or horizon of this fossil (Arkell: Jurassic Geology of the World, 1956, p. 408). Heim and Gansser found that only a small part of the Kioto Limestone may be of Liassic age. This limestone is overlain in the Niti Pass by a sandstone 50 m thick containing pelecypods and belemnites, and this by a thin ferruginous oolite bed containing Callovian fossils.

PAINKHANDA (KUMAON)

Excellent sections of the Trias are exposed in the Bambanag and Shalshal cliffs in the north-western part of Kumaon. Table 43 shows the general succession observed. The Painkhanda section, though resembling that of Spiti, has some peculiarities. In particular, the Ladinic stage dwindles down to insignificance and the Upper Trias is much less thick.

The lowest three beds of the Trias contain the *Otoceras -Ophiceras* fauna in which, besides species of these, there are *Episageceras dalailamae, Hungarites* sp., *Meekoceras hodgsoni, Xenodiscus himalayanus,* etc. The top of the lowest zone contains also abundant *Pseudomonotis griesbachi*. Dark shales having a thickness of 5.5 m intervene between the Ophiceras beds and the Meekoceras beds, which are succeeded by 1.5 m of unfossiliferous grey limestones and these again by the Hedenstroemia beds which contain *Flemingites rohilla, Xenodiscus nivalis* and *Pseudomonotis himaica*.

The Muschelkalk begins with a 1-metre limestone zone containing *Rhynchonella griesbachi* and *Sibirites prahlada*. The nodular limestone of Spiti is represented here by the Niti limestone having a thickness of 60 m. The *Spiriferina stracheyi* beds with *Keyserlingites dieneri, Monophyllites hara* and *Delmatitis ropini* overlie the Niti limestone. They are followed by the Upper Muschelkalk limestone (Ptychites beds) enclosing a very rich cephalopod fauna the important species in it being : *Hollandites voiti, H. ravana, Ceratites thuilleri,*

C. trinodosus, Beyrichites khanikoffi, Buddhaites rama, Gymnites jollyanus, G. vasantasena, Ptychites rugifer, P. gerardi.

The Ladinic stage is all but absent, being represented by 6 m of thin bedded limestones containing a few fossils, most of which are the same as those occurring in the overlying beds. It is only the presenc of *Joannites* cf. *proavus* that indicates the presence of the Ladinic. The characteristic fossil of this stage, *Daonella* cf. *lommeli*, is not found in Painkhanda.

The beds above them are the *Traumatocrinus beds* — well-bedded limestones containing abundant crinoid stems, cephalopods, etc., which indicate the Julic horizon of the Carnic stage. The fossil fauna includes :

Bivalves	... ...	*Daonella indica.*
Brachiopods	... ...	*Spirigera hunica, Retzia* aff. *ladina, Aulacothyris nilangensis, Rhynchonella rimkinensis.*
Cephalopods	... ...	*Proclydonautilus* cf. *buddhaicus, Joannites cymbiformis, J. kossmati, Grypoceras rimkinensis, Lobites delphinocephalus, Anatomites* sp., *Carnites* cf. *floridus, Rimkinites nitiensis, Arpadites rimkinensis. Dittmarites* cf. *circumscissus, Trachyceras austriacum, Sirenites cookei.*

Overlying the crinoidal beds are a thick series of dark carbonaceous shales and shaly limestones which constitute the *Halobia beds*. They are 200 m to 245 m thick and contain fossils throughout and especially in the lower part.

Brachiopods	... ...	*Speriferina shalshalensis, Retzia schwageri, Rhynchonella laucana.*
Lamellibranchs	... ...	*Halobia* cf. *comata, Avicula girthiana.*
Cephalopods	... ...	*Jovites* aff. *daci. Anatomites bambanagensis, Juvavites* cf. *tonkinensis, Placites polydactylus, Mojsvarites eugyrus, Discophyllites ebneri ;* also *Sagenites, Tibetites, Monophyllites, Proarcestes,* etc., which could not be determined specifically because of the poor state of preservation.

The lowest beds of the Noric stage are nodular and shaly limestones, 30 m thick, containing *Proclydonautilus griesbachi, Pinacoceras* aff. *imperator, Metacarnites* sp., *Arcestes, Sagenites, Juvavites,* etc.

The Halobia Beds are succeeded by the *Halorites beds*, consisting of dark shales with limestone bands. The fossiliferous Halorites zone, about 6 - 9 m above the base, is a rich cephalopod horizon containing numerous fossils of which mention may be made of —

Halorites procyon, H. sapphonis, Parajuvavites blanfordi, Tibetites ryalli, Paratibetites bertrandi, Helictites atalanta, Steinmannites desiderii. Clionites woodwardi, Sirenites richteri, Sandlingites nicolai, Pinacoceras metternichi, P. parma, Bambanagites dieneri, Placites sakuntala ; also a few brachiopods and lamellibranchs including *Rhynchonella bambanagensis, Anodontophora griesbachi, Lima serraticosta* and *Halobia* cf. *comata.*

The Halorites Beds pass upwards into earthy compact limestones, often dolomitic or micaceous, having a thickness of over 100 m. They contain abundant *Spiriferina griesbachi,* and also *Spirigera dieneri, Aulacothyris joharensis* and *Retzia schwageri,* and the bivalves *Lima cumaunica* and *Pecten interruptus.* Above them are the beds containing *Anodontophora griesbachi* in which a specimen of *Sagenites* was discovered. They are liver-coloured or brown limestones having a thickness of about 45 m. They are overlain by the *Quartzite Series* similar in constitution to that of Spiti and containing

TABLE 43—Triassic section in Painkhanda (after Diener)

			m
Lias		Megalodon limestone (in part) ...	650
	Rhaetic	Megalodon limestone (in part) ...	
Noric	Upper	Quartzite series with *Spirigera maniensis* ...	83
		Sagenite or Anodontophora beds : Brown limestones with *Anodontophora griesbachi* ...	53
	Lower	Earthy limestones with *Spiriferina griesbachi* passing down into calcareous shales ...	100
		Halorites beds : Massive grey limestones with numerous cephalopods, especially *Halorites procyon* and other species ...	60
		Nodular and slaty limestone with *Proclydonautilus griesbachi*	30
Carnic		Halobia beds : Black flaggy limestones, shales, massive earthy grey limestones and dolomite passing up into micaceous shales ; with *Halodia* cf. *comata* ...	250
		Traumatocrinus beds : Black flaggy limestones with shale partings, with *Traumatocrinus* and *Daonella indica* ...	3
Ladinic		Passage beds (Shalshal). Thin bedded concretionary Limestone with *Daonella indica*, *Spirigera hunica* ...	7
Muschelkalk		Upper Muschelkalk limestone with *Ptychites rugifer* ...	7
		Spiriferina strachyei beds with *Keyserlingites dieneri* ...	1
		Niti limestone—Hard nodular limestone ...	20
		Shaly limestone with *Rhynchonella griesbachi* and *Sibirites prahlada* ...	1
Lower Trias		Hedenstroemia beds : Thin bedded grey limestone with shale partings, with *Flemingites rohilla* and *Pseudomonotis himaica* near the top ...	8
		Grey limestone — no determinable fossils ...	2
		Meekoceras bed — Dark concretionary limestone with *Meekoceras varaha* and *M. markhami* ...	0.3
		Dark blue shales — Unfossiliferous ...	6
		Dark limestone with *Otoceras woodwardi* and *Ophiceras tibeticum* ...	0.16
		Dark hard clay with concretions containing *Episageceras dalailamae* and *Plychites scheibleri* ...	0.5
		Dark blue limestone with *Otoceras Woodwardi* and *Ophiceras Sakuntala* ...	0.3
Permian		Productus shales ...	

Spirigera maniensis and *S. dieneri*. The topmost beds are, as in Spiti, the Megalodon Limestones which here have a thickness of 550 to 600 m which apparently include Noric, Rhaetic and Liassic. In the Shalshal cliffs, the Kioto Limestones are overlain by the Belemnopsis sulcacutus Beds (about 6 m thick) followed by ferruginous oolites with Callovian fossils.

PLATE XV
TRIASSIC FOSSOLS

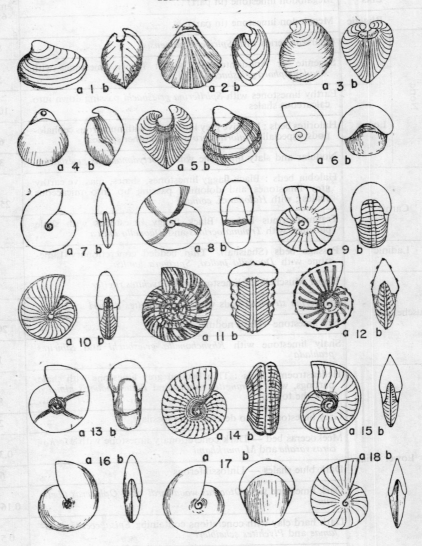

EXPLANATION OF PLATE XV

1. *Anodontophora griesbachi* (0.5). 2. *Retzia schwageri* (0.7). 3. *Megalodon ladakhensis* (0.4). 4. *Spirigera hunica* (0.7). 5. *Pomerangina haydeni* (0.4). 6. *Bellerophon veckei* (1.5). 7. *Dalmatites ropini* (0.7). 8. *Joannites cymbiformis* (0.4). 9. *Halorites procyon* (0.4). 10. *Dittmarites trailli* (0.4). 11. *Tropites subbulatus* (0.4). 12. *Tibetites ryalli* (0.5). 13. *Proarcestes gaytani* (0.5). 14. *Sirenites* aff. *argonautae* (0.4). 15. *Thisbites meleagri* (0.5). 16. *Placites polydactylus* (0.4). 17. *Didymites tectus* (0.7). 18. *Pinacoceras parma* (0.2).

TABLE 44 — TRIAS OF BYANS

		m
Noric	Megalodon Limestone (in part)	120
	Grey limestones with shales at the top, containing undeterminable ammonites	
	Black shales with *Arcestes*	300
Carnic	Tropites limestone very rich in fossils	1
? Ladinic	Light grey limestones, unfossiliferous	50
Muschelkalk	Do. Cephalopod bed with *Ceratites thuilleri, Buddhaites rama*, etc.	3
	Do. Brachiopod bed with *Spiriferina stracheyi* and *Rhynchonella griesbachi*	3
	Do. mainly unfossiliferous	20
Lower Trias	Chocolate limestone with some sahles — *Sibirites Spiniger* zone near top; *Meekoceras* and *Otoceras* fauna near bottom	50
Permian	Productus Shales	

BYANS

The Triassic succession in Byans in north-eastern Kumaon, close to the Nepal border, is less well developed than in the western areas and the facies is also generally different, limestones predominating to a very large extent.

The Lower Trias of Byans is composed of chocolate coloured limestone with shales in the lower part. The basal portion contains *Otoceras* and *Meekoceras* fauna, while the *Hedenstroemia* bed of Spiti is here represented by the *Sibirites spiniger* zone. The Muschelkalk is a light grey limestone without any shales, contrasting strongly with the shaly facies elsewhere. It contains a brachiopod bed 22 m above the base and a cephalopod bed a little above it. The latter contains, besides some familiar cephalopods such as *Ceratites thuilleri, Buddhaites rama, Gymnites jollyanus* and *Ptychites sahadeva*, also species not known elsewhere, e.g., *Smithoceras drummondi, Bukowskites colvini, Pinacoceras loomisii* and *Phillippites jolinkanus*. The Ladinic is apparently absent. At the top of the limestone, a 1-m zone constitutes the Tropites zone which is extraordinarily rich in fossils representing a mingling of Carnic and Noric types. This mixture is apparently due to the faunal remains accumulating more rapidly than the sediments at the sea bottom. The Noric is, however, well represented by a series of thick shales and limestones overlain by limestones of the Dachsteinkalk type (*i.e.*, Megalodon Limestone). The shaly beds and associated limestones are about 300 m thick and fossils found in them are all crushed and undeterminable, even generically. The Megalodon Limestone is 450 m thick, and as usual includes the Lower Jurassic.

Some 150 species of cephalopods have been identified from the Tropites Limestone besides which there are several which are not specifically determinable. Two-thirds of the species are peculiar to this region while the rest

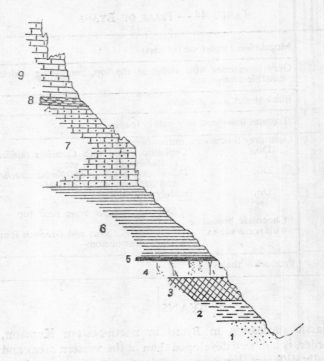

Fig 9—Section N.W. of Kalapani, Byans
(After A. von Krafft, as in Diener, *Mem.* 36, Pt. 3)

1. White quartzite. 2. Productus Shales. 3. Chocolate Limestone (L. Trias). 4. Grey Limestone (Muschelkalk, Ladinic and Carnic stages). 5. Tropites Limestone. 6. Black Shales with *Arcestes*. 7. Grey Limestone. 8. Shales with ammonites. 9. Megalodon Limestone.

are identical with species of the Hallstatt marble of the Alps. The fauna of the Tropites Limestone includes the following:—

Lamellibranchs	... *Halobia* cf. *comata*, *H.* cf. *fascigera*, *Avicula*, aff. *caudata*.
Cephalopods ...	... *Atractites* cf. *ellipticus*, *Orthoceras* cf. *triadicum*, *Proclydonautilus griesbachiformis*, *Pinacoceras parma*, *P. metternichi*, *Placites polydactylus* var. *oldhami* (*Placites* being numerically the most abundant fossil), *Bambanagites krafftii*, *Carnites* cf. *floridus*, *Megaphyllites jarbas*, *Discophyllites ebneri*, *Arcestes diceras*, *Proarestes* cf. *gaytani*, *Lobites* cf. *ellipticus*, *Helictites* cf. *geniculatus*, *Thisbites meleagri*, *Jellinekites barnardi*, *Arpadites tassilo*, *Dittmarites rawlinsoni*, *Drepanites schucherti*, *Tibetites* cf. *ryalli*, *Paratibetites adolphi*, *Himavatites watsoni*, *Clionites gracilis*, *Sandlingites* cf. *oribasus*, *Sirenites* cf. *argonautae*, *S. vredenburgi*, *Distichites sollasii*, *D. ectolcitiformis*, *Ectolcites hollandi*, *Isculites smithii*, *Halorites* aff. *procyon*, *Jovites spectabilis*, *Anatomites speciosus*, *Didymites tectus*, *Discotropites krafftii*, *Margarites acutus*, *Tropites subbullatus*, *T.* cf. *fusobullatus*, *Anatropites nihalensis*, etc.

NORTHERN KUMAON

The Triassic is well developed in northern Kumaon near the border of Tibet in Byans and further northwest, where it is seen as an almost continuous

PLATE XVI
TRIASSIC AND JURASSIC FOSSILS

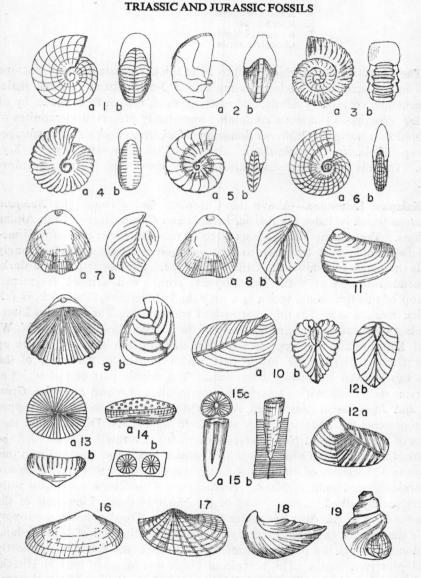

1. *Jovites daciformis* (0.2). 2. *Proclydonautilus griesbachiformis* (0.2). 3. *Thetidites huxleyi* (0.2). 4. *Parajuvavites jacquini* (0.2). 5. *Bambanagites dieneri* (0.4). 6. *Sagenites* sp. (0.4). JURASSIC FOSSILS. 7. *Terebratula acutiplicata*. 8. *Terebratula aurata*. 9. *Rhynchonella fornix* (0.7). 10. *Trigonia smeei* (0.5). 11. *Trigonia crassa* (0.3). 12. *Trigonia V-scripta* (0.4). 13. *Monlivaltia kachhensis* (0.5). 14. *Stylina kacchensis* (a, 0.2; b, 3.0). 15. *Belemnopsis gerardi* (0.4). 16. *Homoya tibetica* (0.4). 17. *Arca egertoniana* (0.5). 18. *Aucella spitiensis superba* (0.5). 19. *Pleurotomaria spitiensis* (0.4).

band for long distances. Heim and Gansser have suggested four major divisions as follows :—

4. Kioto Limestone
3. Kuti Shales
2. Kalapani Limestone
1. Chocolate Series

The Chocolate Series which is 30 to 50 m thick, consists of clay-ironstone layers intercalated with shales, passing into a 3-metre band of dark shale. It is generally covered by debris, and when exposed, is seen to be made up of nodular ferruginous limestone containing imperfectly preserved ammonites of the Ophiceras horizon (*Ophicersa demissum, Meekoceras hodgsoni, Pseudosageceras* sp., *Vishnuites* cf. *pralambha*). The age of the Chocolate Series is Scythian. There is generally a sharp discontinuity above this, but no disconformity.

Kalapani Limestone.—Above the Chocolate Series comes the *Kalapani Limestone* which includes the Anisic, Ladinic and Carnic stages of the Alpine divisions. The thickness varies from 20 to 60 m. The lower part of the limestone is characterised by rusty to orange-coloured patches which strongly recalls the facies of Schiltkalk of the Swiss Alps. The upper part is a dark, well-bedded limestone with brachiopods, corals and crinoid fragments. The top of this limestone, which is a grey sandy limestone, 2 m thick, is full of thick involute as well as thin sharp-edged ammonites. The Kalapani Limestone is a constant member in the whole of the Triassic development of N. W. Nepal, Kumaon and Spiti regions. It often contains thin layers of broken up shells (lumachelle) in the lower part which is Anisic, but none of the shells can be even generically determined. The middle part of this series at Kalapani shows a hematitic layer rich in ammonites of which *Ptychites, Gymnites* and *Japonites* are important Anisic (Muschelkalk) genera. The upper part represents the Ladinic, Carnic and Noric Stages, and Diener found that faunas of the Carnic and Noric Stages were mixed in a thin horizon called the *Tropites Limestone*, seen near Kalapani, Teragad, Nihal and Kuti. From this limestone 155 species of amm9nites have been described, most of which are of Carnic age and some of Noric. No less than 49 species are identical with, or very closely allied to, the species of the Noric Hallstatt Limestone or the Halorites Beds of the Alps (see under Byans). The Tropites horizon strongly recalls the similar, thin, but exceedingly rich, ammonite bed of Timor island in Indonesia, which is a pink limestone barely 2 metres thick, composed mostly of well preserved fossils. This horizon in Timor was found in 1904 by Hirschi and the fossils were described by J. Wanner and O. A. Weller. In both cases the stratigraphic condensation is exceedingly interesting, and it would appear that deposition consisted mainly of the dead animals which fell to the bottom with only a little other material which formed the scanty matrix around the fossils. At Tinkar Lipu in Nepal also, the sandy limestone horizon forming the top of the typical blue dense Kalapani Limestone yielded a very rich ammonite fauna containing *Halorites, Anatomites, Thisbites, Sirenites, Gymnites, Arcestes, Pinacoceras*, and many other genera and also *Halobia*. These indicate the Carnic and Noric horizons.

Kuti Shale.—The Kuti Shales which have a sharp boundary with the Kalapani Limestones, are micaceous shales with calcareous flags, 100 to 500 m thick. In the Jayanti Pass area, the upper part is characterised by well bedded grey limestone, the whole series being more than 500 m thick. There is generally a passage zone containing quartzite and oolite layers between this series and the succeeding Kioto Limestone, as for example near Kuti and Chidamu. Fossils are generally rare in the Kuti Shales but an excellent collection has been obtained on Tinkar Lipu from the black shales and the intercalated calcareous flags. These included *Halorites* cf. *procyon, Thetidites huxleyi, Placites sakuntala, Clionites woodwardi, Juvavites, Anatomites, Parajuvavites, Proarcestes, Discophyllites* and also *Orthoceras*, and several gastropods, lamellibranchs and brachiopods. *Monotis salinaria* is also found in the shales overlying the flags. The whole group of fossils is of Noric age. This facies is also of deep sea character.

Kioto Limestone.—After the passage zone comes the Kioto Limestone which is also well developed in the belt extending from northwest Nepal to Garhwal. It is a dark blue, well bedded, limestone with oolites in the lower part. Ripple marks are seen in the uppermost layers. Fossils are rather rare except for some bivalves. The thickness at Kuti is about 200 m but it increases to 500 m in the Garhwal and to 800 m in Spiti. A horizon about 150 m below the top of these limestones is of importance as it contains layers of lumachelle and impure bands one of which is a light grey limestone 10 m thick containing numerous irregularly lying spindles (which are 5 to 40 cm long and 1 to 2 cm thick). This has been called the *Horizon Problematica* by Heim and Gansser and little is known about the mode of formation of the spindles. Above this comes a zone of iron pisolites which definitely marks a discontinuity and which was overlooked by Griesbach. The Kioto Limestone is of upper Triassic and Rhaetic age (Dachsteinkalk). Heim and Gansser emphasize the fact that there is no gradual passage from the Upper Trias reaching into the Lias and Middle Juras, for the pisolite zone indicates a break about the end of the upper Triassic. The uppermost beds may be Lower Liassic. The ferruginous pisolite horizon represents a highly condensed Callovian.

JOHAR — HUNDES

Kiogad facies.—In the Kiogad and Amlang-La regions according to Heim and Gansser, there are rocks of two facies which are different from the Tethys Himalayan facies. They have been called the *Kiogad facies* and *Chitichun facies*. In the Kiogad facies the Lower Trias is represented by earthy limestones containing *Meekoceras joharense, Xenodiscus nivalis* and *Hedenstroemia byansica*, resembling the Tethyan Himalayan facies to some extent. The Upper Trias is represented by the Kiogad Limestone which is very similar to the Dachsteinkalk of the Alps. It is 200 to 300 m thick and is associated with basic Igneous rocks. It is a white, dense, very fine grained crystalline marble but contains no fossils. This is followed by red and violet marls, limestones and interbedded oolitic limestones with flaser structure. The oolitic limestone layers contain well preserved shells of *Calpionella alpina* and *C. elliptica*, which are of Portlandian age. These marls and limestones are succeeded by grey siliceous radiolarian limestone of Cretaceous age.

Chitichun facies.—The exotic or Chitichun facies, which is characteristic of the exotic blocks, is different from both the Tethyan and Kiogad facies. The Permian red limestones are overlain by red limestone containing Middle Triassic Ceratites and Upper Triassic (Carnic) ammonites. The limestones are of open sea facies and are extraordinarily like the red and white Hallstatt marble of the Alps both in lithology and fossil content. The Middle Trias was noticed near Chitichun I, containing *Ceratites (Danubites) kansa, Sibirites pandya, Monophyllites confucii, Xenapsis middlemissi*, etc., which can be correlated faunistically with the Spiriferina stracheyi beds of the Muschelkalk. The Carnic stage is represented by the red marble blocks in Balchdhura and Malla Johar containing *Cladiscites crassestriatus, Arcestes* cf. *richthofeni, Proarcestes gaytani, Pinacoceras* aff. *rex, Tropites* cf. *subbullatus, Juvavites kraffti, Tibetites bhotensis*, etc., which show greater affinities to the Julic and Tuvalic stages of the Alpine Hallstatt marble than to the corresponding stages of Spiti. *Arcestes* and *Cladiscites* which occur in great numbers in these blocks are characteristic of the Hallstatt marble and rare in Spiti. The Upper Noric is represented by white or grey dolomitic limestone different in appearance from the Megalodon Limestone, and containing no fossils. The overlying Liassic has also a striking resemblance to the Alpine facies. Arkell thinks that the Kiogad facies is the same as the Chitichun, as the Upper Triassic Kiogad Limestone may really be Upper Jurassic.

HAZARA

The Infra-Trias of Hazara (of Permo-Carboniferous age) are immediately overlain by 30 m of lavas of rhyolitic and felsitic character which are the equivalents of the Panjal Traps. The Triassics include thick, massive, grey limestones, up to 360 m thick, containing fossils of Upper Triassic age. It would appear that the Lower and Middle Trias are absent, being represe ted partly by the volcanics. One of the best sections of the Mesozoic rocks in Hazara is seen in Mount Sirban not far from Abbotabad.

KASHMIR

The Himalayan Triassic belt extends into Kashmir and occurs in the Sind and Lidar valleys, Wardwan, Gurais and Central Ladakh and also in northwest Kashmir and Pir Panjal. In the Pir Panjal the Trias occurs as a long thin band extending from Kishtwar on the east to Tosh Maidan beyond the Jhelum valley on the west. In northwestern Kashmir only the Upper Trias is seen. Good sections are observed in the Vihi district near Khunmu, Khrew and in the Guryal ravine. The Kashmir Triassic beds are easily accessible from the Kashmir valley, in contrast with the other Himalayan occurrences.

TABLE 45—TRIAS OF KASHMIR

Upper (several thousand m.)	Grey to dark, massive limestone with occasional fossils, mostly fragmentary. Zone of *Spiriferina stracheyii* and *S.* cf. *haueri*. Lamellibranch beds.	
Middle (over 300 m.)	Ptychites horizon : sandy shales with calcareous layers Ceratite beds ,, ,, ,, *Rhynchonella trinodosi* beds ,, ,, ,, Gymnites and Ceratite beds — Red and grey slabby limestone Lower nodular limestones and shales. Interbedded thin limestones, shales and sandy limestones	
Lower (over 100 m.)	Hungarites shales (position uncertain). Meekoceras beds — limestones and shales. Ophiceras limestone.	

LOWER TRIAS

The Lower Trias is well represented in the Sind and Lidar valleys. Good sections have been studied near Pastanna and other places but the lowest beds are much concealed by scree and vegetation. The horizons known in Spiti have been found here, except the basal Otoceras zone. The Ophiceras zone contains :

Ophiceras sakuntala, O. ptychodes, Xenodiscus himalayanus, X. cf. *ophioneus, Vishnuites pralambha ; Pseudomonotis griesbachi, P. painkhandana*, etc.

A slightly younger fauna, of the Hedenstroemia zone, is found in the Guryal ravine. It shows :

Flemingites sp., *Meekoceras* aff. *jolinkense, Prionites guryalensis, Sibirites kashmiricus, Kashmirites blaschkei, Stephanites superbo, Hungarites* sp.

The Lower Trias has a thickness of over 100 m and systematic excavations may reveal more fossiliferous zones.

MIDDLE TRIAS

Above the Hungarites zone there occurs a succession of thin bedded limestones with intercalated shale and sandstone layers. The lower 60 m consist of dark grey limestones with only occasional lamellibranchs. They are overlain by alternating beds of thin limestones and shales and these again by 30 m of grey, thin bedded sandy limestone containing a lamellibranch bed near its base. Above this is a bed 60 m thick, of pale nodular sandy limestone with hard shale partings, containing cephalopod horizons 6 m and 25 m respectively below the top. At the top of the last mentioned beds is a conspicuous horizon of red and grey slabby limestones rich in *Gymnites* and other fossils (Gymnites and Ceratites Beds). These have yielded :

Ceratites thuilleri, Hollandites voiti, H. ravana, Beyrichites khanikofi, Sibirites cf. *prahlada, Gymnites jollyanus, G. sankara, Acrochordiceras balarama, Buddhaites rama, Grypoceras vihianum*, and some lamellibranchs.

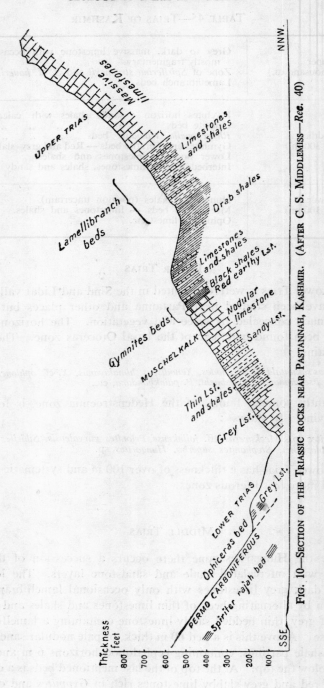

FIG. 10.—SECTION OF THE TRIASSIC ROCKS NEAR PASTANNAH, KASHMIR. (AFTER C. S. MIDDLEMISS—*Rec.* 40)

TABLE 46—CORRELATION OF THE TRIASSIC ROCKS OF THE HIMALAYA

	Divisions		Spiti	Ft.	Painkhanda	Ft.	Byans	Ft.	Kashmir	Ft.
Trias	Upper	Noric — Upper	Megalodon Limestone in part (total 2600)	800	Megalodon Limestone in part (2000)	500	Megalodon Limestone	300	Megalodon Limestone	—
		Noric — Middle	Quartzite Series (*Spirigera maniiensis*).	300	Quartzite Series (*Spirigera maniensis*).	250	Greenish black shales with sandy bands	1000	intercalated shales and limestone.	Several thousand m.
			Monotis beds (*Monotis salinaria*).	300	Sagenite beds (*Anodontophora griesbachi*).	160				
			Coral limestone (*Spiriferina griesbachi*).	100	Limestone with *Spirferina griesbachi*.	320				
		Noric — Lower	Juvavites beds (*juvavites angulatus*.)	500	Halorites beds (*Halorites procyon*),	200				
					Nodular limestones *Proclydonautilus griesbachi*).	100				
	Upper	Carnic — Tuvalic	Dolomitic limestone (*Lima cf. austriaca*)	300			Tropites Limestone		*Spiriferina stracheyi* zone	
			Tropites shales (*Tropites subbullatus*).	600						
		Carnic — Julic	Grey beds { Upper with brachiopods and bivalves. Lower with *Joannites cymbiformis*.	500	Beds with *Halobia cf. comata*.	800			Lamellibranch beds	
			Halobia limestone (*Halobia cf. comata*).		Traumatocrinus Limestone	25				

GEOLOGY OF INDIA AND BURMA

Divisions		Spiti	Ft.	Painkhanda	Ft.	Byans	Ft.	Kashmir	Ft.
Middle Trias	Ladinic	Daonella Limestone Upper with *D. indica.* Lower with *D. lommeli.* Daonella Shales (*D. lommeli*)	150, 160	Passage beds of Shalshal cliff	20				Over 900 feet
Middle Trias	Muschel-kalk	Concretionary limestone (*Ptychites rugifer*). Beds with *Spiriferina straceyi* and *Keyserlingite dieneri.* Nodular limestone. Shaly limestone (*Rhynchonella griesbachi.*)	100	Beds with *Ptychites rigifer.* Beds with *Spiriferina stracheyi* and *Keyserlingites dieneri.* Niti Limestone Shales with *Rhynchonella griesbachi*).	100	Cephalopod bed (*Ceratites thuilleri*). Brachiopod bed. *Spiriferina. stracheyi.* Limestone (*Rh. griesbachi*).		Ptychites beds Ceratites beds Rhynchonella trinodosi beds. Gymnites beds Nodular limestone Limestones and shales	Over 900 feet
Lower Trias	Campil beds	Hedenstroemia beds (*Flemingites rohilla*).	30	Hedenstroemia beds (*Flemingites rohilla*).		Sibirites spiniger zone		Hungarites shales	Over 300 feet
Lower Trias	Seis beds	Meekoceras bed (*Meekoceras varaha*). Ophiceras bed (*Ophiceras sakuntala*). *Otoceras* and (*Otoceras woodwardi*).	7	Meekoceras bed (*Meekoceras markhami*). Otoceras beds (*Ophiceras sakuntala* and *Otoceras woodwardi*).	51	Chocolate limestones	150	Meekoceras beds Ophiceras beds	Over 300 feet
Permian		Kling or Productus shales		Kuling or Products shales		Productus shales		Zewan beds	

Six metres above the main *Gymnites horizon* is another fossiliferous bed of the same character, the intervening beds being black shales. They are followed by 120 to 180 m of alternating limestones and shales. In the Khrew section, the *Ptychites horizon* occurs about 165 m above the Gymnites beds and contains the following :

Ceratites trinodosus, Buddhaites rama, Ptychites sahadeva, Ptychites sp.
Mojsvaroceras kagae, Grypoceras vihianum ; Myophoria, Lima, Pecten, etc.

These belong to the Upper Muschelkalk.

Upper Trias

The *Ptychites*-bearing beds pass gradually upwards into massive limestones. The lamellibranch bed and the zone of *Spiriferina griesbachi* occur in the lowest parts of the Upper Trias and belong to the Carnic stage. The lamellibranch bed contains only one brachiopod, *Dielsma julicum*, but is rich in lemellibranchs, including :—

Myophoria middlemissi, M. cf. *kefersteini, Hoernesia bhavani, Chlamys middlemissi, Pseudomonotis* sp., *Lima* cf. *subpunctata*.

The *Spiriferina stracheyi* bed contains, besides that fossil, *Spiriferina* aff. *lilangensis, Mentzelia mentzelii, Rhynchonella trinodosi,* etc.

The Upper Trias is well developed in the mountains of Vihi and Anantnag districts and in other places and forms conspicuous scarps. It lacks the Daonella and Halobia Beds which are developed in Spiti. But further search may reveal the equivalents of these. The upper portion resembles the Megalodon (or Kioto) Limestone but has not yielded any *Megalodon* in the Vihi district. Elsewhere in Kashmir it is known to contain some corals, crinoidal stems and bivalves.

SIKKIM

Tso Lhamo Series.—In the Lachi ridge, the calcareous sandstones containing Upper Permian fossils are overlain by several hundred metres of quartzites and shales which may be of Triassic age. East of the Lachi hill there are dark limestones and shales containing a rich brachiopod and ammonite fauna of Triassic age. They overlie 100 m of gritty flags containing plant remains. Similar flags associated with limestones and shales occur N.N.E. of Paunhari peak (7,065 m) which appear to be of Triassic age. They overlie shattered blue-grey limestones which may be the equivalents of the Everest Limestone (*Rec.* **69**, p. 152, 1935).

REVIEW OF THE WESTERN HIMALAYAN TRIAS

The Triassic rocks of the different areas in the western Himalayas may conveniently be reviewed at this atage.

The Lower Triassic is well represented in Kashmir, Spiti, Painkhanda and Byans. It is only 12 m thick in Spiti and slightly more in Painkhanda. Though it attains a thickness of over 100 m in Kashmir, it does not seem to be richer in fossils, but this is partly because it has not been investigated in detail. It is also quite thick in Byans (50 m) and contains a new element in the fauna, viz., *Sibirites*.

The Muschelkalk of Spiti falls into three divisions. The lower portion is a nodular limestone with a brachiopod horizon (*Rhynchonella griesbachi*) at the base ; the middle one contains a fauna with *Spiriferina stracheyi and Keyserlingites* (*Durgaites*) *dieneri*. The upper Muschelkalk is extraordinarily rich in fossils and characterised by *Ptychites rugifer*. The same characters are more or less recognised in Kashmir where the strata thicken to 300 m. In Painkhanda the Muschelkalk is only a little over 30 m thick, though the different zones are recognizable. In Byans it is a pure limestone facies but faunistically closely related to Spiti.

There is a great variation in the Ladinic stage of the different areas. It is 90 m thick in Spiti, dwindles down to about 6 m in Painkhanda and is apparently absent or represented by unfossiliferous limestone in Byans. Its equivalents in Kashmir should be looked for in the strata above the Ptychites beds. This marked difference continues also into the Carnic stage. The Carnic is thick in Spiti, consisting of a lower *Joannites* horizon and an upper *Tropites* horizon. It attains only half that thickness in Painkhanda and much less in Byans where the Tropites horizon is extremely rich in fossils, but contains also some Noric elements. The lamellibranch beds and *Spiriferina stracheyi* beds of Kashmir represent the Carnic stage.

The Noric stage is dominantly calcareous in all the areas except Byans where it is shaly and generally contains crushed fossils. In Kashmir the Noric seems to be practically unfossiliferous. The Upper Noric as well as Rhaetic is the Megalodon Limetone (Kioto Limestone) which attains large thickness in all areas. It is thickest in Kashmir (several thousand metres) and gradually becomes thinner southeastwards, being 800 m in Spiti, 600 m in Painkhanda and 450 m in Byans.

The correlation of the Triassic formations of these areas is given in Table 46.

SALT RANGE

Triassic rocks, known as *Ceratite beds*, are well developed in the Salt Range on either side of the Indus. In the Cis-Indus portion only the Lower Trias and the lower part of the Middle Trias are observed westward from Kathwai near Kundghat, their total thickness being 46 to 50 m. In the Trans-Indus region the Upper Trias is also seen, the thickness of the whole system being around 120 to 150 m.

It has already been mentioned that the Chidru stage of the Upper Productus beds becomes arenaceous at the top. A small unconformity intervenes between the Permian and Trias ; a conglomerate marking the break in deposition is found near Siram-ki-dhok, and a marked change occurs in the fauna with the earliest Triassic rocks. Table 47 shows the Triassic succession found in the Salt Range.

The Chidru Stage is overlain in several places by a bed of dolomite with thin limestone layers. The two together are 2 to 5 metres thick and constitute the Ophiceras zone with *Ophiceros connectans*, rhynchonellids, Crinoids and fishes. Productids occur rarely. These beds are overlain by the LOWER CERATITE LIMESTONE which is a hard thin-bedded light grey limestone containing numerous *Gyronites frequens*. The *Otoceras* horizon is probably represented by the unfossiliferous sands and clays which lie between this limestone and the Chidru beds of Upper Productus Limestone. The Lower Ceratite Limestone is overlain by thick greyish green marly beds with limestone bands, called the *Ceratite Marls*, containing abundant fossils. The marls constitute a very conspicuous and easily visible horizon because of their colour, and weather into rounded outcrops. The Ceratite Sandstones which succeed them are divided into an upper and a lower sandstone separated by a calcareous horizon rich in *Stachella*, a genus closely allied to *Bellerophon*. The Upper Ceratite Sandstone is characterised by the presence of *Flemingites flemingianus*, which can be correlated with the *Flemingites rohilla* zone of Painkhanda. Above the Ceratite Sandstone is the *Upper Ceratite Limestone* composed of hard limestones and intercalated grey marls with highly ornamented *Ceratites* of the genera *Sibirites*, *Stephanites*, etc. This corresponds to the *Sibirites spiniger* zone of Byans.

TABLE 47—TRIAS OF THE SALT RANGE

Division			Salt Range	Thickness Feet	Himalaya
Carnic			Crinoidal dolomite	250	? Halobia beds
Middle Trias			Sandoy limestones with bivalves	100	? Doanella beds and Muschelkalk
Trias	Lower	Beds	Upper Ceratte limestone with *Stephanites suberbus* and *Sibirites chidruensis*.	20	*Sibirites spiniger* zone (Byans)
			Ceratite Sandstone { Upper Ceratite Sandstone with *Flemingites flemingianus* / Stachella beds with *Stachella* sp. sp. and *Flemingites radiatus*. / Lower Ceratite Sandstone with *Celtites fallax*.	30	*Flemingites rohilla* zone / Hedenstrocemia beds
		Ceratite	Ceratite Marls with *Prinolobus rotundatus* and *Propytchites lawrencianus*	20 to 60	Meekoceras beds
			Lower Ceratite Limestone with *Gyronites frequens* / Unfossiliferous sands and clays.	10	Ophiceras zone / ? Otoceras zone
Permian			Chidru Stage / Upper Productus Limestones		Productus Shales

The Ceratite beds are succeeded by the *Bivalve beds* which range in composition from limestone to calcareous sandstone and contain abundant bivalves but very few ammonoids. The uppermost member consists of yellowish, somewhat brecciated or cavernous dolomites containing some obscure fossils. The top of these is sometimes marked by a thin limestone with bivalevs.

The succession varies in different places as to details ; limestones vary to dolomite and sometimes there are pisolitic limestone with glauconitic matter. Sandstones and limestones may vary to marls.

Fauna.—As mentioned already, fossils belonging to the Ceratite group of Ammonoids characterise the Ceratite beds. Some of the beds, and especially the bivalve bed, contain lamellibranchs. The chief fossils of the different sub-divisions are shown below :

Lower Ceratite Limestone	*Gyronites frequens, Prionolobus atavus, P. ophinoneus. Lecnites psilogyrus, Meekoceras varians, Dinarites cuccensis.*
Ceratite Marls	*Proptychites lawrencianus, Prionolobus rotundatus, Dinarites minutus, Koninckites ovalis, Clypites typicus, Kongites lens.*
Lower Ceratite sandstone	*Celtites fallax, Ceratites normalis, Gyronites rotula, G. radians.*
Stachella beds	*Stachella sp., Flemingites radiatus, F. rotula, Celtites acuteplicatus, Aspidites kingianus.*
Upper Ceratite sandstone	*Flemingites flemingianus, F. compressus, Aspidites superbus, Celtites armatus, Ceratites wynnei.*
Upper Ceratite Limestone	*Stephanites superbus, S. corona, Sibirites chidruensis S. kingianus, Dinarites dimorphus, Celtites patella, Prionites tuberculatus, Celtites dimorphus, Acrochordiceras coronatum.*

HAZARA

The southeastern parts of Hazara contain Mesozoic rocks. Excellent sections of a sequence ranging from the Permo-Carboniferous to Eocene are observed in Mount Sirban south of Abbottabad :

Murree Sandstones	Oligocene
Kuldana beds (passage beds)	
Nummulitic Series (limestones and shales with coal horizons near base, underlain by thin laterite)	Eocene
Grey Limestone	Up. Cretaceous
Giumal Sandstone	Cretaceous
Limestone and orange coloured Shales (Spiti Shales)	Jurassic
Limestones, grey, massive	Triassic
Lavas and Agglomerates	Panjal Trap
Infra-Trias Limestone	Permian
Boulder-bed	Up. Carboniferous

The Infra-Trias Limestones are overlain by about 30 m thickness of rhyolitic and felsitic lavas and these by thick-bedded grey limestones which vary in thickness from 150 to 360 m. The latter are similar to the Kioto Limestones and contain Upper Triassic fossils very closely related to those of Spiti.

ATTOCK DISTRICT, PUNJAB

Between Hazara and the Punjab Salt Range, and along the northern border of the Potwar plateau, lie the folded rocks of the Kala Chitta hills, the denuded anticlines of which expose a series of strata ranging in age from Upper Trias to Siwaliks, as shown below :

Siwalik System	...	Pliocene
Murree Series	...	Miocene
Nummulitic rocks	...	Eocene
Giumal Sandstones, etc. ...	...	Albian
Spiti Shales	...	Argovian — Tithonian
Kioto limestone	...	Upper Trias to Lias
Attock Slates	...	Pre-Cambrian to Cambrian

The Kioto Limestones of Spiti are represented by similar limestones which are grey or cream-coloured and include a few shaly bands. They are sparsely fossiliferous and range in age from Upper Triassic to Liassic as in Spiti. The Upper Triassic fossils found in them between Jhalar and Campbellpore include *Rhynchonella* cf. *bambanagensis, Terebratula* sp., *Velata velata, Lima serraticosta, Pecten* sp., etc.

BALUCHISTAN

The Upper Trias is represented in the Zhob and Pishin district of Baluchistan by a vast thickness, amounting to several thousand metres, of greenish slaty shales with intercalations of thin black limestone. They occupy an area 110 km long (east to west) and some 20 km broad. They are fairly rich in fossils which include abundant *Monotis salinaria* and a few species of ammonoids of the genera *Halorites, Didymites* and *Rhacophyllites*.

BURMA

Triassic rocks are found in the Shan States, Amherst district, the Arakan Yomas and possibly also in the Manipur hills.

NORTHERN SHAN STATES

Napeng Beds.—We find a hiatus in sedimentation between the Plateau Limestone and the Napeng beds of Rhaetic age, there being a well marked unconformity between the two. The Napeng beds occur in a series of patches and consist of a variety of sediments — yellow shales, clays, sandy marls, sandstones and limestones — deposited in very shallow and irregular basins. They contain a fauna including the following :

Corals	...	*Isastrea confracta*.
Lamellibranchs	...	*Palaeoneilo fibularis, P. nanimensis, Pinna* cf. *blanfordi, Conocardium superstes, Grammatodon lycetti, Gervillea shaniorum, G. rugosa, Avicula contorta, Myophoria napengensis, Plicatula carinata, Modiola frugi, Cardita singularis*.
Gastropods	...	*Turritella* sp., *Promathilda exilis*.

The fauna is of Rhaetic age and of peculiar character, being composed mainly of lamellibranchs and recalling the fauna of the Wetwin Shales. It is quite distinct from the fauna of the Himalayas or of the Salt Range.

AMHERST DISTRICT

Kamawkala Limestone.—In the eastern part of the Amherst district near the Thailand frontier, the Permo-Carboniferous beds are succeeded by the Kamawkala limestones which are referred to the Triassic, probably the Noric age. The fauna includes *Rhynchonella bambanagensis. Chlamys* aff. *valoniensis, Trachyceras* sp., and *Centrastraea cotteri.*

INDO-BURMA FRONTIER AND ARAKAN YOMA

The mountain belt of the Indo-Burma frontier is an inaccessible region on which very little geological information has been published. It is expected to contain a fairly full succession of Permian and Mesozoic rocks, as it is the counterpart of the Baluchistan Arc.

The oldest rocks exposed in the Arakan Yomas belong to the *Axial System* the lower part of which is referable to the Triassic. The rocks of this system are found in Prome and Thayetmyo districts, usually much folded and disturbed. They consist of dark shales and sandstones with some limestones. The only fossils found in them are *Daonella lommeli* and specimens of *Monotis* and *Avicula*. These seem to indicate a Triassic age.

Farther south in Minbu and Pakokku districts, the equivalents of the Lower Axials are the *Chin Shales*. Black slates, sandstones and quartzites found in the Manipur State on the Assam-Burma border have also been referred to the Axials and may probably be partly of Triassic age.

SELECTED BIBLIOGRAPHY

Bittner, A. Triassic brachiopoda and lamellibranchiata. *Pal. Ind. Ser. XV*, Vol. III, (2), 1899.

Diener, C. Trias of the Himalays. *Mem.* 36, Pt. 3, 1912.

Diener, C. Upper Triassic and Liassic faunae of the exotic blocks of Malla Johar, *Pal. Ind. Ser. XV*, Vol. I, (1), 1908.

Diener, C. The Cephalopoda of the Lower Trias. *Op. cit.* Vol. II(1), 1897.

Diener, C. The Cephalopoda of the Muschelkalk. *Op. cit.* Vol. II, (2), 1895.

Diener, C. Fauna of the Tropites limestone of Byans. *Op. cit.* Vol. V ,(1), 1906.

Diener, C. Fauna of the Himalayan Muschelkalk, *Op. cit.* Vol. V, (2), 1907.

Diener, C. Ladinic, Carnic and Noric faunae of Spiti. *Op. ct.* Vol. V, (3), 1908.

Diener, C. and von Krafft. Lower Triassic Cephalopoda from Spiti, Malla Johar and Byans. *Op. cit.* Vol. VI, (1), 1909.

Diener, C. Fauna of the Traumatocrinus limestone (Painkhanda). *Op. cit.* Vol. VI, (2), 1909.

Diener, C. Triassic faunae of Kashmir. *Pal .Ind. N.S.*, V. (1), 1913.

Gregory, J.W. *et al.* Upper Triassic fossils from Burma-Siam frontier. *Rec.* 63, 155-181, 1930.

Hayden, H.H. Geology of Spiti. *Mem.* 36, Pt. 1, 1904.

Healy, M. Fauna of the Napeng beds (Rhaetic) of Upper Burma. *Pal. Ind. N.S. II*, (4), 1908.

Krafft, A. von. Exotic blocks of Malla Johar. *Mem.* 32, Pt. 3, 1902.

Mojsisovics, E. von. Upper Triassic Cephalopod faunae of the Himalayas. *Pal. Ind. Ser. XV,* Vol. III, (1), 1899.

Waagen, W. Salt Range Fossils — Ceratite formation. *Pal. Ind. Ser. XIII*, Vol. II, 1895.

CHAPTER XIII

THE JURASSIC SYSTEM

GENERAL

Taking its name from the Jura mountains in Central Europe, this System was established by the work of Quensted and Von Buch. Three main divisions were recognised — Bthe lack, Brown and White Jura, which correspond respectively to the now accepted major divisions, the LIAS, DOGGER and MALM. At a later date the palaeontologist Oppel worked out a series of Stages, each characterised by the occurrence of particular fossils, in this case ammonites. Oppel's work laid the foundation for detailed stratigraphic zoning in other areas as well as in other geological formations. It also led to the clear recognition of variation in facies due to differences in the environment, climate and other factors. Numerous subdivisions of the Jurassic have been proposed by various authors at different times but those which have generally come into use are as below :

Stages.			Fossils.
Portlandian	{ Purbeckian	} Tithonian	Taramelliceras,
	Bononian		Craspedites, Zaraiskites
			Berriasella, Pavlovia
Kimmeridgian	{ Havrian		Strebiltes, Gravesia
	Sequanian		
	Rauracian		
	Argovian		Ringstedia, Cardioceras
Oxfordian			
	Divesian		Perisphinctes.
Callovian			Peltoceras, Macrocephalites
			Kosmoceras.
Bathonian	(Great Oolite)		Oppelia, Zigzagiceras,
Bajocian	(Inferior Ooolite)		{ Parkinsonia, Sonninia,
			Stephanoceras.
Aalenian			Leioceras, Ludwigia.
Toarcian			Lytoceras, Hildoceras.
Charmouthian (Domerian or Pliensbachian)			Amaltheus, Uptonia
Sinemurian			Arietites, Asteroceras.
Hettangian			Schlotheimia

NOTE :—A number of other stage names are also currect. Lusitanian is approximately Oxfordian-Sequanian. Corallian is Rauracian-Sequanian. Portlandian is applicable to Britain and N.W. France and Tithonian to the Mediterranean facies. The Volgian of Russia is approximately Tithonian.

Though mainly marine in character, the Jurassic formations show varied facies in different areas. The lower beds (*Lias*) are generally dark marls and limestones (Black Jura). The Middle Jurassic (*Dogger*) consist of hard limestones, often ooolitic, with crinoids and corals or ferruginous beds (Brown Jura). The Callovian and especially the Oxfordian consist of black clays. The Argovian-Sequanian are marly limestones with corals, sponges, etc. The Kimmeridgian and Portlandian are limestones (White Jura), while the Purbeckian shows shore facies deposits. The Upper Jurassic is also referred to as MALM.

Important mountain building movements took place, during the Jurassic period, in the circum-Pacific belt, in extreme southeast Asia (Shan States—Malaya — Borneo — Indo China) and in the Carpathians and Crimea. These are known as the Cimmerian movements. The Karakorum and its extension in Tibet may also have been affected. Though the Jurassic formations are extensively distributed in the Tethyan zone of the Himalayas and Tibet, they are not very thick nor do they show irregularities in deposition. But our knowledge of these regions is still very imperfect.

There was apparently fairly free interchange of fauna between the Mediterranean, Northwest European and Russo-arctic marine provinces in the Lower and Middle Jurassic. After the Middle Jurassic (Cimmerian) disturbances, the fauna of the Russian Province became distinct, especially in the Tithonian. The equivalent beds are called the Volgian, which are characterised by the ammonite species belonging to *Dorsoplanites, Zaraiskites, Virgatites, Craspedites,* etc.

Because of the great abundance and variety of ammonites in the Jurassic it has been possible to divide the formations into numerous zones and to carry out correlations between the marine strata in different parts of the world. There are at least 17 zones in the Lower, 21 in the Middle and more than 20 in the Upper Jurassic, according to Arkell.

The recommendations of an international committee of palaeontologists on the classification of the Jurassic System is given in Table 48. It has recommended also that the Stage names ARGOVIAN, RAURACIAN, SEQUANIAN, LUSITANIAN and PURBECKIAN be used for the local facies; that the upper boundary of KIMMERIDGIAN be taken as the base of the *gravesia* zone ; that the TITHONIAN (Mediterranean) and the VOLGIAN (boreal) be used to connote the strata above the Kimmeridgian and below the base of the Cretaceous ; that PORTLANDIAN be confined to Britain and Northwestern France with its base at the base of the *albani* zone in England and the *gravesia* zone in France.

FAUNA

In the Mediterranean region there are radiolarian cherts containing radiolaria and other protozoa. Sponges are abundant especially in the Upper Jurassic (Lusitanian) limestone sometimes forming flint pebbles. Corals belonging to the Hexacoral group are found widely distributed particularly in the Middle Jura, Rauracian (Corallian) and Kimmeridgian. Echinoderms form crinoidal limestones (*Pentacrinus, Apiocrinus, Millericrinus* etc.) while some of the echinoids such as *Dysaster, Clypeus* and *Cidaris* have species characteristic of certain stages. A large number of pelecypods and gastropods occur —*Gryphaea, Astarte, Pecten, Avicula, Mytilus, Aucella, Diceras ; Trochus, Pleurotomaria, Nerinea* — but they have no special stratigraphic significance. On the other hand, ammonites play a most important part because of their great variety and the short time-range of the species. The chief groups in the Jurassic are the Arietids, Hildoceratids, Stephanoceratids, Macrocephalitids,

TABLE 48.—STANDARD JURASSIC SUCCESSION SHOWING LIMITS OF STAGES

(B = Base, T = Top ; Internat. Geol. Rev. Vol. 7 (6), p. 845, 1965)

	Mediterranena	Boreal
VALANGINIAN	B. *Subthurmannia boisseri*	B. *Riasanites risaanensis*

TITHONIAN (= VOLGIAN)	T. *Virgatosphinctes transitorius*
	B. *Taramelliceras lithographicus*
		U { T. *Craspedites nodiger*
		    B. *Kaschpurites nikitini*
		L { T. *Epivirgatites nikitini*
		    B. *Subplanites sokolovi*

KIMMERIDGIAN	T. *Aulacostephanus pseudomutabilis* B. *Pictonia bayleyi*	
OXFORDIAN	T. *Epipeltoceras bimammatum* B. *Quenstedticeras mariae*	
CALLOVIAN	T. *Quenstedticeras lamberti* B. *Macrocephalites macrocephalus*	
BATHONIAN	T. *Clydoniceras discus* B. *Zizgagiceras zigzag*	
BAJOCIAN	T. *Parkinsonia parkinsoni* B. *Sonninia sowerbyi*	
AALENIAN	T. *Ludwigia concava* B. *Leioceras opalinum*	
TOARCIAN	T. *Pleydellia aalensis* B. *Dactylioceras tenuicostatum*	
PLIENSBACHIAN	DOMERIAN CARICIAN	T. *Pleuroceras spinatum* B. *Amaltheus margaritatus* T. *Prodactylioceras davoei* B. *Uptonia jamesoni*
SINEMURIAN	(UPPER) LOTHRINGIAN LOWER	T. *Echioceras raricostatum* B. *Asteroceras turneri* T. *Arnioceras semicostatum* B. *Arietites bucklandi*
HETTANGIAN	T. *Schlotheimla angulata* B. *Psiloceras planorbis*	

Oppelids and Perisphinctids. They are found in the Rhaetic (uppermost Trias) only in the Alpine region. *Cardioceras, Virgatites* and *Craspedites* characterise the Russo-arctic province while *Reineckeia, Oppelia* and *Gravesia* characterise the Mediterranean province. Different genera and species are characteristic of the formations in the different regions as well as in different facies. The Belemnites occur profusely in certain formations — *Belemnites* in Lias, *Belemnopsis* in middle Juras, and *Cyclindroteuthis* and *Duvalia* in Upper Juras. The brachiopods are already dwindling in importance, the major groups in these formations being *Spiriferina*, Terebratulids and Rhynchonellids.

In regard to the Vertebrates, the Mesozoic era supported a great variety of reptiles and fishes. The Ichthyosaurian, Plesiosaurian and Pterosaurian groups were the characteristic reptiles of the Jurassic.

Distribution and Facies.—Rocks belonging to the Jurassic System are developed in the Indian region in the Himalayas of Spiti, Kumaon, Nepal, Kashmir, Hazara; in Baluchistan and Salt Range; in Cutch and Rajasthan; in the Rajmahal hills; in the Puri district of Orissa; in the Ellore, Ongole, Madras and Trichinopoly regions on the eastern coast. In Baluchistan and the outer border of the Iranian arc, the rocks are mainly calcareous in this as well as in the succeeding Cretaceous System, for which reason the region of their development is called the Calcareous zone. To the interior (northwest) of this zone in Hazara as well as in the Himalayan area and a large part of Tibet the rocks are dominantly shaly. The Tibetan facies of the Lias, which is seen in the exotic blocks of the Kiogar region, consists of reddish earthy limestones which resemble rocks of the corresponding age in the Eastern Alps. The Rajasthan occurrence indicates a shelf facies, while those along the eastern coast are estuarine.

Unconformity and Marine Transgression.—The geosynclinal region shows a marked interruption of sedimentation commencing from the Callovian and lasting until the Oxfordian or later. In Spiti and northern Himalayas this interruption ranges up to the Oxfordian while in Baluchistan it extends to the Tithonian. In both cases the unconformity is marked by a bed of ferruginous laterite. In the coastal facies of Western India on the other hand, the Lower Jurassic rocks are absent and the deposition begins in the Bathonian, assuming a marine character in the Callovian and continuing on beyond the Jurassic times. It is interesting to note that the Callovian is often marked by a regression in the geosynclinal area but by a transgression along parts of the coasts of the Peninsula.

SPITI

Spiti Shales.—The beds succeeding the Triassic Kioto Limestones are the Spiti Shales which extend from Upper Oxfordian to Lower Neocomian, a part of the Kimeridgian being probably unrepresented. The type section is in Spiti but they are developed over a considerable length of the Himalayas from beyond the Kanchenjunga in the east through Nepal and Kumaon into Hazara in the west. They are generally 30 to 40 m thick consisting of micaceous shales with several intercalated layers of sandstones each of the latter being only a few centimetres thick. In other places, *e.g.*, at Kuti and north of Laptal, they may attain a thickness of several hundred metres, while in Eastern Tibet they occupy an area of several thousand sq. miles. Towards the top, the strata become marly and these pass into greenish glauconitic sandstones (Giumal Sandstones) of Neocomian age. Being soft, the shales are generally not well-exposed and are often hidden by talus and soil. The only indication of their presence is often provided by ovoid fossiliferous concretions weathered out of the shales after they have completely disintegrated. Table 48 shows the Jurassic succession of Spiti.

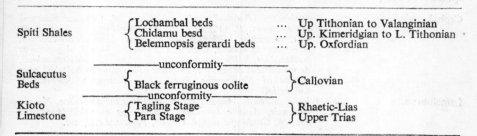

TABLE 49. JURASSIC SUCCESSION IN SPITI

The upper part of the Kioto Limestone, called the TAGLING STAGE by Diener, is considered by Heim and Gansser as mainly Rhaetic and possibly Lower Liassic. According to Arkell (1956, p. 408) the discovery of Liassic pelecypods in the Kioto Limestones of the Attock district in the Punjab makes it certain that the Tagling Stage includes some Liassic strata. There is some mistake about the *Stephanoceras coronatum* reported to have been found some 350 feet below the top of the Kioto Limestone at Tera Gadh near Giumal. This specimen was later identified as a *Macrocephalites*, which must have come from the Sulcacutus bed. In some other places (see below) the Laptal Series intervenes between the Kioto Limestone and the Spiti Shales. Arkell has suggested that the Laptal Series, which is of Liassic age, is a lateral variation of the Tagling Stage. At the top of the Kioto Limestone there is a break in deposition followed by thin bed of black ferruginous oolite of Callovian age which have been called the SULCACUTUS BEDS from the fact that the most characteristic fossil in them is *Belemnites sulcacutus*. They also contain other Callovian fossils like *Macrocephalites triangularis*, *Dolikephalites fluxuosus*, *Reineckeites waageni*, etc. at Laptal.

The Spiti Shales comprise three units. The BELEMNOPSIS GERARDI BEDS are grey shales with a few limestone bands ; they contain *Belemnopsis gerardi*, which occur also in the Chidamu beds, and some species of *Inoceramus* with coarse ribbing. The only ammonites found in these beds are *Mayaites* which confirm the age as Upper Oxfordian. There is apparently a gap above this, for the Lower and Middle Kimmeridgain are unrepresented. The succeeding black lustrous shales constitute the CHIDAMU and LOCHAMBAL BEDS, containing abundant calcareous concretions enclosing fossils. These constitute the typical Spiti Shales known over an extensive terrain from Baluchistan and Hazara in the west, through Kashmir, Spiti and Tibet to Eastern Himalayas. A more or less similar fauna and facies have been found in the islands east of Celebes in Indonesia, in New Guinea, Western Australia and New Zealand (Arkell 1956, pp. 445, 454, 458). The Chidamu beds are Lower Tithonian (Upper Kimmeridgian) in age while the Lochambal beds extend from Upper Tithonian through Berriasian to Valanginian. The ammonites of the Spiti Shales, known as *Saligrams* to the Hindus and considered by them as sacred objects, are brought down by the Gandak and other Nepal rivers. Some of these fossils have a coating of pyrites which gives them a golden colour. These and other Spiti

Shale fauna have been described in *Pal. Indica* (Ser. XV, vol. IV), some important species being :—

Cephalopods	... ...	*Perisphinctes (Paraboliceras) sabineanus, P. (Virgatosphinctes) denseplicatus, P. (V.) raja, P. (Aulacosphinctes) spitiensis, P. (A.), torquatus. Phylloceras plicatius Lytoceras exoticum, Hecticoceras kobelli, Oppelia (Strebilites) kraffti, Aspidoceras avellanoides, Spiticeras spitiensis, S. grotei, Himalayites seideli, Acanthodiscus octagonus, Hoplites (Thurmannia) boissieri,* also species of *Necoomites, Berriasella, Kossmatia, Sarasinella, Kilianella.*
Lamellibranchs	... ...	*Avicula spitiensis, Aucella spitiensis, Lima melancholica, Nucula spitiensis, Astarte hermanni, Cosmomya egregia, Homomya tibetica, Goniomya uhligi, Ostrea, Pecten, Leda,* etc.

As already mentioned, the Spiti Shales pass upwards into the Giumal Sandstones of Neocomian age, these being mainly of the flysch facies.

NORTHERN AND NOTH-WESTERN KUMAON

Laptal Series.—This series is known only in North-west Kumaon from Kungri-Bingri to Laptal, but it may extend towards Spiti as the strata show uninterrupted deposition from the ioto limestone upwards. Where the Laptal Series is absent, there is always a definite stratigraphical break as in the Kuti Valley. It is 60 to 80 m thick at Laptal and is characterised by several layers of *lumachelle* (which was called shell limestone by Griesbach) containing small oysters, *Trigonia, Pecten, Lima* and some *Belemnites*. Though they do not contain ammonites in this region, Liassic ammonites have been found by Stoliczka farther to the northwest where they are seen in several places — (*e.g.* Shalshal and Bambanag hills). The section at Chidamu above the dense blue Kioto Limestone is as follows, the beds marked *a* to *f* forming the Laptal Series :—

Spiti Shales

f. Brown sandy limestone and lumachelle with *Belemnites* and bivalves, *Cardiaum, Arca,* etc. (5 m.)
e. Well-bedded limestone (15 m.)
d. Brown lumachelle limestones with *Belemnites* and *Trigonia* (8 m.)
c. Dark marly limestone (5 m.)
b. Impure dense limestone with shaly layers (45 m.)
a. Yellow and red spotted limestone with lumachelle layers, *Belemnites* and bivalves (10 m.)

Thin bedded limestone with some bivalves, oolite and microbreccia recalling the Alpine Urgonian facies (20 m.)
Dark blue dense Limestone ; upper part with a layer of smooth bivalves (120 to 150 m.)

The Laptal Series appears to have been included in the Tagling Stage by Diener. It shows Liassic molluscs. Its equivalent has been found in the Attock district (Punjab) where it is part of the Kioto Limestone. It is therefore a lateral variation of the Upper Kioto Limestone. Wherever the Laptal Series is missing, for instance between the Kali and Garhwal, the upper beds of the Koto Limestone show ripple marks indicating an interruption in deposition. At the top of the Laptal series, there are layers of shaly ferruginous oolite, 3 to 4 metres thick on the whole, interbedded with thin bedded limestone. In one of these layers *Belemnites* and *Reineckeites* were found. These

TABLE XVII
JURASSIC FOSSILS

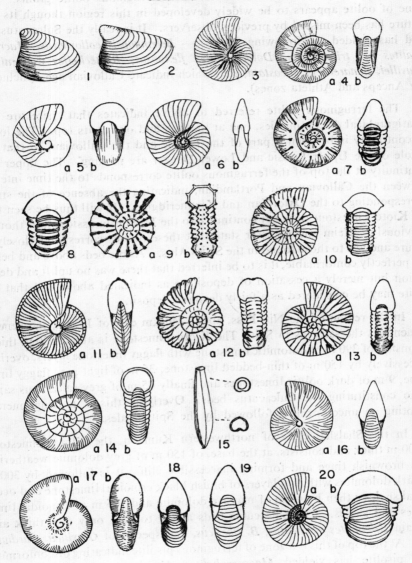

EXPLANATION OF PLATE XVII

1. *Astarte hermanni* (0.5). 2. *Nucula spitiensis* (0.8). 3. *Phylloceras plicatius* (0.2).
4. *Lytoceras exoticum* (0.2). 5. *Aspidoceras avellanoides* (0.4). 6. *Oppelia acucincta* (0.5).
7. *Spiticeras spitiensis* (0.4). 8. *Astieria schencki* (0.5). 9. *Hoplites (Acanthodicus) octagonus* (0.2) 10. *Himalalayaites Seideli* (0.2). 11. *Oppelia (Streblites) kraffti* (0.2).
12. *Hoplites (Blanfordia) wallichi* (0.2). 13. *Perisphinctes (Paraboliceras) sabineanus* (0.2).
14. *Perisphinctes (Virgatosphinctes) raja* (0.2). 15. *Belemnopsis grantana* (0.5) 16. *Macrocephalites triangularis* (0.2). 17. *Macrocephalites macrocephalus* (0.2). 18. *Macrocephalites chariensis* (0.2). 19. *Epimayaites transiense* (0.2). 20. *Indocephalites diadematum* (0.2).

are succeeded by typical Spiti Shales. The thin shaly ferruginous oolite bed is megascopically mistaken for a shale which, on careful examination with a magnifying lens, is seen to consist of black ferruginous oolite grains. This zone of oolite appears to be widely developed in this region though its true nature has been missed by previous observers. It is really the Sulcacutus bed and has yielded the following fossils :— *Belemnopsis calloviensis, Macrocephalites* cf. *triangularis, Dolikephalites flexuosus, Reineckeites waageni, R. douvillei, Bonarella* cf. *bicostata*, etc., which indicate Callovian age (including the Anceps and Athleta zones).

The ferruginous oolite referred to above indicates that there are two stratigraphical discontinuities, one at its base and one at its top. The lower discontinuity is between a part of the Liassic and the Callovian, so that the whole of the Upper Liassic and Lower Dogger are missing. The upper discontinuity at the top of the ferruginous oolite corresponds to the time interval between the Callovian and Portlandian, indicating the absence of the strata corresponding to the Oxfordian and Kimmeridgian. It will thus be seen that the Kioto Limestone does not continue into the Middle Jurassic as was thought previously. Heim and Gansser state that these oolites correspond closely in nature and age to those found in the Swiss Alps. As the beds above and below are perfectly conformable, it is to be inferred that there was no uplift and denudation but merely a cessation of deposition as indicated above and that the oolite may be considered as a fairly deep sea deposit.

In the region of the Niti Pass, about 240 km east of Lilang, the general sequence is the same as in Spiti. The Kioto Limestone is about 630 m thick; it consists of 390 m of dolomitic limestone with flaggy limestone layers, overlain successively by 150 m of thin-bedded limestone, 35 m of light grey flaggy limestone, 9 m of dark oolitic limestones and finally 45 m of grey calcareous sandstone constituting the Sulcacutus beds. Overlying this is a conglomerate denoting an unconformity, followed by the Spiti Shales.

In the Shalshal cliffs of northwestern Kumaon, the Kioto Limestone is 600 m thick and consists, at the base, of 150 m of grey dolomite weathering to a brownish tinge and forming inaccessible cliffs; it is followed by 300 m of dark dolomites with thin layers of bluish black crinoidal limestone and occasional shales ; then by 45 m of massive dolomite and 105 m of crinoidal limestones and shales. The Sulcacutus beds at the top are only 6 m thick and contain *Belemnites sulcacutus, B. tibeticus*, and species of *Gervillea, Cardium*, etc. At the top of this is a zone of ferruginous pisolite indicating disconformity. This pisolite has yielded *Macrocephalites, Kepplerites, Sphaeroceras*, etc., and though of small thickness represents the full Callovian Stage. It is overlain by the Spiti Shales.

Farther southeast, in Byans, near the Nepal border, the Kioto Limestone is less massive and only about 360 m thick, composed of 70 m of flaggy limestone with shaly bands, succeeded by 210 m of shaly limestone and shales, then by 60 m of massive grey limestone and 10 m of dark shales and finally by the Sulcacutus beds with a thin laterite bed at the top.

JOHAR — HUNDES

The exotic blocks of the Kiogar region show red limestones of Lower Jurassic age. They are earthy, thin-bedded and sometimes nodular, with layers of marls and shale. Though the dominant colour is red, there are also pink, grey and black coloured layers. The more important elements in the enclosed fossils are Phylloceratids and Lytoceratids such as *Phylloceras, Schistophylloceras, Juraphyllites, Analytoceras, Euphyllites, Schlotheimia, Arietites* etc. They have a remarkable resemblance to the Adneth Limestone (Adnetherkalk) of Salzburg and to the Hallstatt marble of the Alps and to the similar limestones of Armenia. It is probably the only full fossiliferous succession of the Liassic known in the Tethyan Himalayan basin. According to Arkell, the white Kiogar Limestone (200 to 300 m thick) which is unfossiliferous and is overlain by pink oolites and shales of Tithonian age, is most likely to be of Upper Jurassic age, and not Upper Triassic as supposed by Heim and Gansser because of their lithological resemblance to the Triassic Dachsteinkalk of the Alps.

MOUNT EVEREST REGION AND TIBET

The Spiti Shale facies occupies a large area east and southeast of Spiti, in central Tibet, Sikkim and Tsangpo valley. Hayden summarised the stratigraphy of the region as below :—

Spiti Shales ...	Upper Jurassic
Lungma Limestone	Bajocian
Shales and Quartzites	
Crinoid Limestones	
Slates and Quartzites	⎫ Liassic
Brachiopod Limestones	⎭

Near Kampa Dzong the Lungma Limestones and Spiti Shales are readily identifiable. The Lungma Limestones have yielded *Sonninia, Witchellia* and *Dorsetensia*, indicating a condensed Bajocian. The strata below, down to the Brachiopod Limestone, are Liassic. To the south of these are Triassic and Permo-Carboniferous strata, the latter occurring at the summit of Mount Everest as already mentioned.

SUB-HIMALAYA OF GARHWAL

The Jaunsars are overlain by the KROL BEDS which are probably of Permian (? Permo-Trias) age. East of the Ganges in Garhwal, the Krols are overlain by the TAL SERIES consisting of shales in the lower portion and quartzites and limestones in the upper portion. The Tal Series contains fragmentary molluscs and corals and may probably represent the Jurassic System or even the Cretaceous. It is overlain by Eocene rocks.

KASHMIR

We have already seen that the Megalodon Limestone is present in the Vihi district and in Ladakh. The upper portion of this is referable, as in

Spiti, to the Lias. This limestone is known to be overlain, in parts of Ladakh and the Zanskar range, by the Spiti Shales. Fossils are found in concretions in the shales but have not been studied in detail. They include ammonites (*Macrocephalites* among them), *Belemnites,* lamellibranchs and brachiopods.

Jurassic rocks are also found in a small area north of the Banihal Pass in the Pir Panjal mountains. It is likely that Jurassic rocks are associated with bands of Trias in parts of this range.

In the area of the syntaxial bend of northwestern Kashmir, there is a band of orange coloured strata, some 50 m thick, which it is thought may be Jurassic to Cretaceous in age.

HAZARA

Hazara is a region of transition from the Himalayan to the Baluchistan type of development. Here there are two neighbouring zones running N.E.-S.W. and parallel to each other, the northwestern one containing massive limestones overlain by the Spiti Shales as in the Himalaya, and the southeastern showing limestones overlain by Neocomian strata as in Baluchistan.

The massive limestone (of the northwestern belt) is dark grey in colour and varies in thickness from 150 to 360 m. It contains *Megalodon* and *Dieceracardium* in the lower part and is therefore similar to the Kioto Limestone of Spiti. The upper portion is presumed to be Lower Jurassic in age. It is succeeded by shales of the Spiti Shale type about 60 m thick containing *Gymnodiscoceras acucinctum, Virgatosphinctes frequens, Belemnites gerardi, Inoceramus* sp., and *Trigonia ventricosa*.. These are followed by the Giumal Sandstones of Neocomian to Albian age.

ATTOCK DISTRICT

Jurassic rocks are present in the folded strata of the Kala Chitta hills of this district. The lower portion is included in the *Kioto Limestone* and the upper in *Spiti shales.* The latter are however not easily separable from the overlying Giumal Series because the lithology is continuous and shows very gradual change. The Spiti Shales consist of sandy and carbonaceous shales Belemnite-bearing beds and olive clays. The fossils found in the Upper Kioto Limestones are :

Sphaeroidothyris attockensis, Burmirhynchia cf. *namyauensis, Lophothyris euryptycha.*

The fossils of the Spiti Shales are :

Epimayaites polyphemus, Peltoceratoides sp., *Perisphinctes orientalis, P.* cf. *indogermanus, Prososphinctes virguloides, Blanfordiceras wallichi, Spiticeras, Himalayites, Belemnopsis tanganensis, Hibolites budhaicus, H. tibeticus,* etc.

These indicate an age similar to that of the Chari beds of Kutch and younger. The horizons above these do not show any fossils except one which has yielded *Oxytropidoceras,* indicating an Albian (Gault) age. It may be mentioned here that there is a fairly close resemblance between the Gault horizons of Attock, Hazara and Kohat.

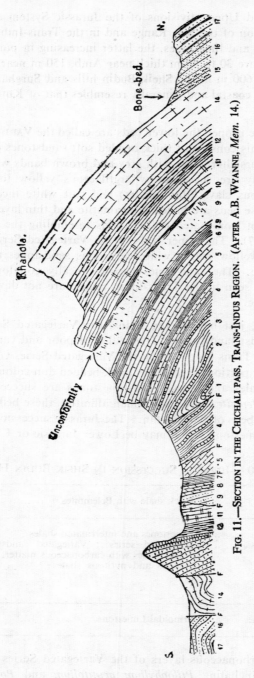

Fig. 11.—Section in the Chichali Pass, Trans-Indus Region. (After A.B. Wynne, Mem. 14.)

1. Limestones and shales. 2. Golden oolite limestone. 3. Shales and sandstones. 4. Limestones and shaly sands. 5. White limestone. 6. Olive sandstone and clay. 7–9. Sandy beds. 10. Alum shales. 11. Lower limestone. 12. Dark shales. 13. Nummulitic marls. 14. Compact grey limestone. 15. Bed with bone fragments. 16. Sandstones. 17. Red clays.

SALT RANGE

The Middle and Upper divisions of the Jurassic System are developed in the western portion of the Salt Range and in the Trans-Indus region as a series of sandstones and limestones, the latter increasing in porportion westwards. The strata are 30 to 60 m thick near Amb, 150 m near Kalabagh on the Indus and over 600 m in the Sheik Budin hills and Surghar range. The facies is dominantly coastal and generally resembles that of Kutch in Western India.

In the Salt Range proper, the lower beds are called the VARIEGATED SERIES. The lower part of this consists of thick bedded soft sandstones of red, yellow and variegated colours alternating with grey and brown bands which are often ripple-marked. They are succeded by argillaceous yellow limestone, grey gypseous and pyritous clays (alum shales) and soft white incoherent sandstones. Amidst these beds are bands of heamatite and thin layers of ' golden oolite ', composed of iron-coated oolite grains resembling the golden oolite of Kutch (*see* p. 367). The upper beds of the Variegated Series are coarse brown sandstones, yellow marls, white cavernous sandstones and bands of hard grey limestone. The sandstones become locally conglomeratic. The Upper Jurassic beds seen in the Trans-Indus region are not developed in the Cis-Indus Salt Range.

Near Kalabagh, just beyond the Indus, the Variegated Series contains thin coal seams amidst sandstones. The coal is of poor and variable quality, high in ash. In the Trans-Indus region the Variegated Series consists mainly of shales in the lower portion, overlain by thick-bedded dun-coloured limestone intercalated with shales and clays. The limestones are succeeded by black shales containing *Belemnites* with Neocomian affinities, these beds corresponding to the Belemnite beds of Baluchistan. The Jurassic succession is underlain by white crinoidal limestone which may be Lower Jurassic or Upper Triassic.

TABLE 50.—JURASSIC SUCCESSION IN SHEIK BUDIN HILLS

		Feet
Neocomian	Black shale with Belemnites	60
Upper Jurassic	Limestones and intercalated shales... ...	800
Middle to Lower Jurassic ...	Variegated series : Variegated sandstones and shales with carbonaceous matter, gypceous and pyritous shales	1,000 to 1,400
? Lower Jurassic	Crinoidal Limestone	200

Fossils.—The carbonaceous layers of the Variegated Series have yielded some plant fossils including *Ptilophyllum acutifolium* and *Podozamites* sp. which indicate Upper Gondwana affinities. The associated limestones contain marine fossils including lamellibranchs, gastropods, terebratulids and echinoids.

PLATE XVIII
JURASSIC FOSSILS

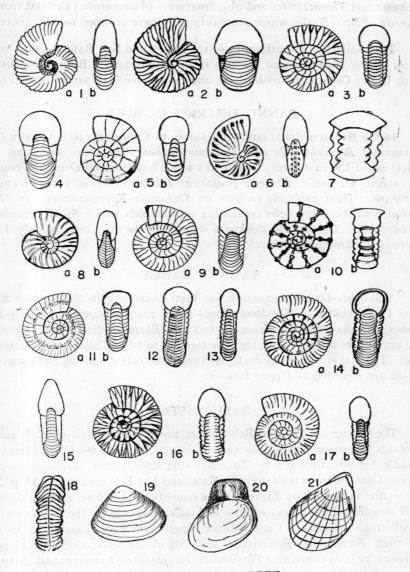

EXPLANATION OF PLATE XVIII

1. *Mayaites maya* (0.3). 2. *Kamptokephalites dimerum* (0.4). 3. *Torquatisphinctes torquatus* (0.2). 4. *Epimayaites polyphemus* (0.06). 5. *Hemilytoceras rex*. (0.1). 6. *Oppelia (Taramelliceras) kackhensis* (0.1). 7. *Peltoceras ponderosum* (0.2). 8. *Harpoceras (Sublunuloceras) lairense* (0.2). 9. *Peltoceratoides semirugosum* (0.15). 10. *Peltoceras athleta* (0.2). 11. *Perisphinctes orion (Orionoides indicus)* (0.25). 12. *Perisphinctes indogermanus* (0.4). 13. *Perisphinctes (Ataxioceras) leiocymon* (0.25). 14. *Katroliceras katrolensis* (0.1). 15. *Procerites hyans* (0.1). 16. *Reineckeia anceps* (0.2). 17. *Indosphinctes calvus* (0.2). 18. *Perisphinctes rehmanni (Reineckeites waageni)* (0.5). 19. *Corbula lyrata* (0.7). 20. *Gryphaea balli* (0.5). 21. *Lima roberti* (0.7).

The ammonites include *Indocephalites* aff. *transitorius, Pleurocephalites habyensis, Kamptokephalites* cf. *magnumbilicatus*, etc., which are fossils of the Macrocephalites Beds of Kutch. The Neocomian strata above contain *Holcostephanites* and *Thurmannites* and also fragments of ammonites derived from the erosion of Spiti Shales which must originally have cropped out in this region.

The similarity of stratigraphy of the Kutch and Salt Rang regions would show that a sea connected the two through Jaisalmer. Both are associated with Upper Gondwana fossil plants and the marine beds are of coastal facies.

BANNU DISTRICT (N.-W.F.P.)

In the Bannu district, sandy limestones of Callovian age have been found containing *Burmirhynchia* and *Daghanirhynchia*. These are separated by a thickness of 150 m of strata from rocks which have yielded *Ornithella coulsoni, O. indica, O. ovalis, Kingena punjabica, Daghanirhynchia coulsoni* and *D. pezuensis*. These probably indicate an Oxfordian-Kimmeridgian age. Associated with these are beds containing cephalopods which Spath regarded as Neocomian. Though the thickness separating the two beds is only 1.3 m, there is no stratigraphic break.

SAMANA RANGE

A Jurassic-Eocene succession has been discovered in the Samana Range near Fort Lockhart. The lowest beds noted are sandstones and shaly limestones with shale bands. From the fact that *Rhynchonella arenata* is found in the upper part of these beds, they are thought to be of Callovian to Bathonian age. They are succeeded by dark grey limestones called *Samana Suk Limestones* which are regarded as Upper Jurassic.

BALUCHISTAN

The Calcareous Zone of Baluchistan, which is composed of N.-S. ranges, contains limestones and shales varying in age from Permo-Carboniferous to Middle Jurassic, followed by Tithonian and later strata. Productus-bearing Permo-Carboniferous beds occur in Kalat and Las Bela areas (*Rec.* 38, p. 201). These are succeeded by Liassic strata consisting of 900 to 1200 m of dark, well stratified limestones, oolites and calcareous shales, the last containing fossiliferous horizons. These indicate Sinemurian and Toarcian stages with *Arietites, Fuciniceras, Bouleiceras,* and *Spiriferina* and other brachiopods, followed by beds containing *Phylloceras, Juraphyllites, Lytoceras* and *Dactylioceras*. Overlying these are thick, massive, grey limestones, some 1,200 m thick, well exposed near Quetta, which are referred to the Middle Jurassic but have not yielded any ammonites. The lower part of the succeeding limestones called MAZAR DRIK LIMESTONES, contain the Upper Bathonian ammonites *Bullatimorphus bullatus,* and *Clydoniceras* (originally described as *Harpoceras*). The upper part, consisting of thin bedded limestones and shales, constitutes the *Polyphemus Beds* so named by Noetling as they contain the large ammonites identified by him as *Macrocephalites polyphemus*. This identification is mistaken, for the large ammonites are *Indocephalites* ; the Upper Mazar Drik limestones should therefore be called INDOCEPHALITES BEDS. They contain also *Macrocephalites macrocephalus* and other species, *Dolikephalites, Pleurocephalites, Indosphinctes, Choffatia, Terebratula* and *Rhynchonella*. These beds are of Lower Callovian age. After an unconformity, they are followed by the black Belemnite Shales of Neocomian age (up to 400 m thick) with *Hibolites fusiformis*, and *H. pistilliformis* which occur also in similar beds in the Salt Range. Overlying these are bright red and white banded limestones of Lower Cretaceous age which form circular outcrops around the Jurassic domes.

The Liassic and Middle Jurassic limestones form prominent hill masses in the Calcareous Zone of Jhalawan and Sarawan. Amongst the conspicuous hills may be mentioned Zaradak, Anjiro and Sumbaji in Jhalawan (Liassic) and Chehiltan and Koh-i-Maran in Sarawan (Middle Jurassic.)

WESTERN INDIA

General.—Peninsular India, which was practically devoid of marine sediments since the Vindhyan times, witnessed marine transgressions in the coastal regions in the Jurassic, Lower Cretaceous and Miocene times. A sea invaded the area north of Kathiawar during the Jurassic.

The Jurassic rocks occupy large areas in Kutch and Rajasthan which seem to have formed parts of a large sedimentary basin, which extended northwards from Kutch as far as the Salt Range in the Punjab. The outcrops are now isolated by intervening large stretches of desert sands and alluvia. The Salt Range deposits, though folded, compressed and much disturbed, are similar to those of Kutch and show much less affinity with those of the Himalayan and Baluchistan geosynclinal facies of the same age. Compared to the Tibetan facies, these rocks are of shallower origin and laid down not far from the land, as evidenced by the presence of intercalated plant remains.

RAJASTHAN

TABLE 51—POST-PALAEOZOIC SUCCESSION IN WESTERN RAJASTHAN

BANDAH & KHUIALA BEDS (? 50 m.)—Dense siliceous limestones with fuller's earth and clays. Contain Nummulites. Pelecypods, gastropods, echinoids, etc.   —Kirthar & Laki
————————————————Unconformity————————————————
ABUR BEDS (60-90 m.)—Limestones, subordinate grits, clays and shales containing Pseudohaploceras and other ammonites   —Aptian
PARIHAR SANDSTONES (? 600-850 m.)—Unfossiliferous ferruginous and feldspathic sandstones and quartzites. Overlapped in places by Eocene   —L. Cretaceous
BEDESIR BEDS (110-300 m.)—Ferruginous sandstones and grits with ammonites
    —Up. Jurassic
BAISAKHI BEDS (600 m.)—Sandstones and shales, gypseous. Lower beds are fossiliferous
    —Oxfordian to Kimmeridgian
JAISALMER LIMESTONE (160 m.)—Buff to yellow Oolitic and shell limestones and Calcareous sandstones   —Callovian-Oxfordian
LATHI BEDS (330 m.)—Sandstones and some layers of limestones in the upper part ; with molluscan fossils and pieces of gymnospermous wood   —L. Jurassic
————————————————Unconformity————————————————
BADHAURA FORMATION   —Permo-Carboniferous.

There is a very good development of Jurassic rocks in Jaisalmer (Western Rajasthan) which links Kutch with the Salt Range. Table 51 gives the general stratigraphical succession found in Jaisalmer by the geologists of the Geological Survey of India and the Oil & Natural Gas Commission. The earliest sedimentary rocks overlying the basement are of Vindhyan or Permo-Carboniferous age, the latter being partly marine (Badhaura formation). There are no Triassic strata. The Jurassics are transgressive and are followed by the Lower Cretaceous after which there is a regression, followed again by a transgression in the Lower Eocene.

The Mesozoic and Tertiary rocks form a gentle arch plunging northwest towards Mari in Sind. The beds dip in the same direction at very low angles. The sedimentary basin greatly deepens to the west and northwest where the total sedimentary thickness may be of the order of 6000 m. The succession is also fuller in the northwest. For instance, the Upper Cretaceous strata which attain a large thickness at Mari in Sind, are not present in Jaisalmer. The Eocene at Sui, Mari and Khandkot in the Indus basin is thicker than in Jaisalmer and contains reservoirs of natural gas. The strata are cut by NW-SE faults near Jaisalmer. A large N-S fault occurs on the western side of the arch, as indicated by geophysical data. From the nature and distribution of the strata it is inferred that this region formed the shelf of an arm of the Tethys which extended from the Salt Range to Kutch.

The Jurassic succession in Jaisalmer commences with the LATHI BEDS which are sandstones in the lower part with molluscs and pieces of fossil wood, indicating a shallow facies. The upper part contains some limestone layers. These are succeeded by the JAISALMER LIMESTONES which are richly fossiliferous and have yielded *Stephanoceras fissum* (=*Idiocycloceras singulare*), *Sindeites sindensis*, *Reineckeites* aff. *reissi*, *Grossouvria steinmanni*, etc. These indicate the same age as the Upper Chari beds of Kutch (M. Callovian to Oxfordian). The Jaisalmer Limestones are grey to bluish grey in colour when unexposed but become orange to brown on weathering. There are some shell limestones also of the same colour. Both the compact orange-brown limestone and the yellow shell limestone with white shells are good decorative building stones and are quarried for that purpose. The succeeding BAISAKHI BEDS are deposits of rather arid environment and lagoonal in character, as shown by their content of Gpysum. The BEDESIR BEDS are again of open sea character and contain Tithonian ammonites like *Pachysphinctes* aff. *bathyplocus*. The overlying PARIHAR SANDSTONES are unfossiliferous and correspond to a part of the Umia Series, part of which would be of Neocomian age. The ABUR BEDS at the top of the sequence consist of limestones and subordinate grits and shales whose age is clearly Lower Aptian as shown by the contained ammonites.

KUTCH

Jurassic rocks occupy a large area in Kutch and are the oldest rocks except for some patches of Pre-Cambrians. They are bordered to the south by the Deccan traps while on the north lies the saline marsh of the Rann of Kutch.

TABLE 52—JURASSIC SUCCESSION IN KUTCH

	Age	Sub-divisions	Leading fossils
UMIA (1,000 m.)	Post—Aptian	Bhuj beds (Umia Plant beds) Sandsontes and shales	*Palmoxylon* in upper beds *Ptylophyllum* flora, similar to Jabalpur flora in lower beds.
	Aptian	Ukra beds — Marine calcareous shales	*Australiceras, Colombiceras, Cheloniceras, Tropaeum,,* etc.
	U. Neocomian	Umia beds :— Barren sandstones and shales	Unfossiliferous
	Valanginian	Trigonia beds Barren sandstones	*Trigonia ventricosa, T. crassa* Unfossiliferous
	U. Tithonian	Umia ammonite bed	*Virgatosphinctes, Ptychophylloceras, Micracanthoceras, Hemilytoceras, Umiaites,* etc.
KATROL (300 m.)	M. Tithonian	Up. Katrol Shales	*Hildoglochiceras kobelli, H. propinquum, Dorsoplanites mirabilis, Haploceras elimatum.*
	M. Tithonian	Gajansar beds.	*Belemnopsis gerardi, Streblites gajinsarensis, Phylloceras* cf. *Plicatius, Hildoglochiceras* spp.
	L. Tithonian	Upper Katrol (barren) Sandstone	No fossils. *Aulacosphinctoides meridionalis, Virgatosphinctes indosphinctoides.*
	M. Kimeridgian	Middle Katrol (red sandstones).	*Waagenia kachhensis, Katroliceras katrolense, Katroliceras pottingeri, Pachysphinctes* spp. *Aspidoceras lerense.*
	M. Kimeridgian	Lower Katrol (Sandstones, shales, marls).	*Torquatisphinctes similis, Aspidoceras asymmetricum, Ptychophylloceras ptychoicum, Taramelliceras kacchhense, Streblites plicodiscus, Waagenia,* spp., *Hybonoticeras* spp.
	Up. Oxfordian	Kantkote sandstone. (Bimammatum zone)	*Epimayaites* spp. *Prograyiceras grayi, Ataxioceras leiocymon, Biplices wagurensis, Prososphintes virguloides, Torquatisphinctes torquatus. Trigonia smeei.*
CHARI (360 m.)	Up. to L. Oxfordian	Dhosa oolite (green and brown oolites). (Transversarium zone)	*Taramelliceras jumarense, Discosphinctes* aff. *kreutzi, Perisphictes indogermanus, Mayaites maya, Epimayaites polyphemus, Paracenoceras kumagunense, Peltoceratoides semirugosus.*
	U. Callovian	Athleta beds (marls and gypseous shales).	*Peltoceras athleta, P. ponderosum, P. metamorphum, Orionoides indicus, O. purpurus.*
	M. Callovian	Anceps beds (limestones and shales).	*Perisphinctes anceps, Indosphinctes calvus, Reineckeia ravana, Kinkeliniceras* sp., *Hubertoceras mutans.*

Age		Sub-divisions	Leading fossils
CHARI	M. Callovian	Rehmanni beds (yellow limestone).	*Reineckeia rehmanni, R. tyranniformis, Sivajiceras kleidos, Idiocycloceras singuare, Kellawaysites greppini.*
	L. Callovian	Macrocephalus beds (shales with calcareous bands, with golder oolite—diadematus zone — in the uppr part)	*Macrocephalites macrocephalus, M. chariensis, Dolichocephalites subcompressus, Indocephalites diadematus, Kamptokephalites dimerus, Pleurocephalites habyensis, Belemnites.*
PATCHAM (300 m)	L. Callovian	Patcham coral bed	*Macrocephalites triangularis, Sivajiceras congener, Procerites hyans, Thamnastrea, Stylina, Montlivaltia.*
	L. Callovian to Bathonian	Patcham shell limestone	*Macrocephalites triangularis, Trigonia pullus, Corbula Iyrata.*
		Patcham basal beds (Kuar Bet Beds).	*Corbula, Eomiodon, Trigonia,* etc.

The Jurassic rocks have an estimated thickness of some 1950 m and crop out in three anticlinal ridges trending E.-W. Owing to an E.-W. fault the whole sequence is repeated. The northern range is about 160 km long and is broken up into four islands (Patcham, Karrir, Bela and Chorar) in the Rann of Kutch. The middle ridge is 190 km long, trending E.S.E. from Lakhpat on the west. The southern ridge, south of Bhuj, is 65 km long and forms the Charwar and Katrol hills. The Jurassic rocks are repeated in these two ridges as the main outcrop, of which they form parts, is cut by an E.-W. strike fault. An isolated but large outcrop, on which Wagur and Kantkote stand, is about 80 km long, in N.E. Kutch. These anticlines show transverse undulations so that the dome-like parts have been separated from each other by denudation. Some distance to the east, in Dharngadhra in Kathiawar, there is a large outcrop of Jurassic rocks which forms part of the same sedimentary basin.

The Jurassic sequence in Kutch consists of four main divisions, as shown below :

UMIA SERIES (1000 m.) ... ... Neocomian and Up. Tithonian
KATROL SERIES (400 m.) ... ... Mid. and L. Tithonian and L. Kimmeridgian
CHARI SERIES (450 m.) ... ... Oxfordian and Up. Callovian
PATCHAM SERIES ⎫
  including   ⎬ (300 m.) ... L. and Mid. Callovian to
KUAR BET BEDS ⎭   ... Up. Bathonian

Table 53—Section in the Jumara dome, Kutch

(*Pal. Ind. N. S.* IX, II, Part 6)   (age modified after Arkell)

Age	Stage and zone	Characteristic ammonites
Up. Argovian	Upper Dhosa oolite (green oolite).	*Taramelliceras jumarense.*
L. Argovian	Lower Dhosa oolite (brown oolite).	*Mayaites maya, Peltoceratoides semirugosus.*
Divesian	Lower Dhosa oolite (borwn oolite).	*Euaspidoceras* sp.
U. Callovian	Upper Athleta beds (shales and yellow marls)	*Metapeltoceras* spp., *Peltoceras ponderosum, Orionoides indicus.*
Do.	Middle Athleta beds (shales and yellow marls).	*Peltoceras metamorphum, Orionoides purpurus.*
Do.	Lower Athleta beds (gypseous shales).	*Petloceras* sp., *Reineckeites* sp.
M. Callovian	Upper Anceps beds (yellow limestone).	*Kinkeliniceras* sp. *Hubertoceras mutans.*
	Lower Anceps beds (yellow limestone).	*Indosphinctes calvus, Sivajivceras fissum.*
M. Callovian	Upper Rehmanni beds (yellow limestone).	*Reineckeia tyranniformis, Sivajiceras kleidos, Ideocycloceras singulare.*
	Lower Rehmanni beds (yellow limestone).	*Reineckeia rehmanni, Kellawaysites greppini.*
L. Callovian	Upper Macrocephalus beds (limestone).	*Dolicephalites subcompressus, Macrocephalites,* spp. *Kamptokephalites* aff. *magnumbilicatus.*
Do.	Midlde Macrocephalus beds (Shales with ferrugenous nodules).	*Dolicephalites subcompressus, Nothocephalites semilaevis, Kamptocephalites* aff. *magnumbilicatus.*
Do.	Do.	*Macrocephalites chariensis, Kamptokephalites dimerus, Indosphinctes* sp.
Do.	Do.	*Macrocephalites chariensis, Alcidia* sp., *Parapetoceras* sp.
Do.	Do.	*Macrocepha ites chariensis, Kamptokephalites dimerus, Pleurocephalites habyensis.*
Do.	Lower Macrocephalus beds (White limestones and shales).	*Macrocephalites triangularis, M. madagascariensis, Sivajiceras* aff. *congener.*
L. Callovian	Upper Patcham (coral limestone).	*Macrocephalites triangularis, Sivajiceras congener, Procerites hians.*
Do.	Lower Patcham (shelly limestone).	*Macrocephalites triangularis* and other species.

Patcham Series.—Deposition began in the Kutch region with transgressive clastic sediments, named after the island of Patcham in the Rann. The lowest beds, seen near Khera, known as the KUAR BET BEDS, are clastic Sandstones and yellow limestones which yield a rich pelecypod fauna (*Corbula lyrata,* and other species, *Eomiodon, Protocardia, Trigonia,* etc.) and corals. The only ammonite found in these beds was a Stephanoceratid. Similar beds are found in Madagascar and in the Attock district in N.W. Punjab. They are of Upper Bathonian age.

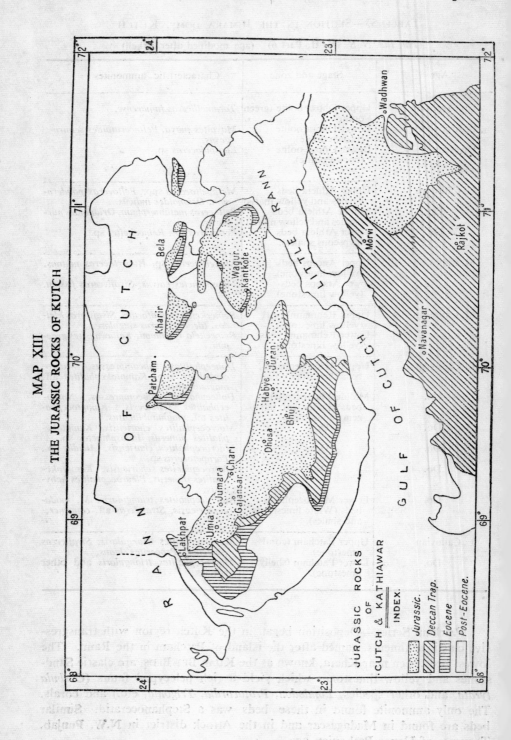

MAP XIII
THE JURASSIC ROCKS OF KUTCH

Overlying the Shell and Coral limestones are the Patcham Beds proper. They are characterised by the presence of *Macrocephalites triangularis, Sivajiceras congener*, etc. The total thickness is of the order of 300m. Arkell (1956, p. 391-2) states that Spath was mistaken in including the Patcham as well as the Chari beds in the Bathonian, for the ammonite fauna indicates an unmistakable Callovian age. It is only the Kuar Bet beds which are Bathonian.

Chari Series.—This takes its name from a village 50 km northwest of Bhuj and contains five main stages. The lowest is the MACROCEPHALUS BEDS (so called from the common fossil — *Macrocephalites macrocephalus* and other species) which can be sub-divided into several zones by means of the ammonite fauna. The upper part of the middle division of this stage contains a few layers of 'golden oolite' constituting the DIADEMATUS ZONE (*Indocephalites diadematus*) which is a calcareous oolite, the grains of which are coated with thin films of ferric oxide giving them a golden colour. Species of *Nucula* and *Astarte* are also very common in the Macrocephalus Stage. Above this occur dark shales and sandy shales with calcareous and ferruginous nodules. The lower portion of these beds is the REHMANNI ZONE (*Rieneckeia rehmanni*) in which *Macrocephalites* persists and *Phylloceras* and *Lytoceras* begin to appear ; *Idiocycloceras* and *Subkossmatia* are characteristic, as also some terebratulids and Trigonias. The ANCEPS BEDS (*Perisphinctes (Reineckeia) anceps*) succeeding the Rehmanni zone contain ammonites, brachiopods and lamellibranchs. The ATHLETA BEDS (*Petloceras athleta*) above them are composed of white limestones with a similar fauna. The topmost beds of the Chari Series are the DHOSA OOLITES, composed of green and brown oolitic limestones very rich in fossils, among which may be mentioned *Mayaites maya, Epimayaites polyphemus, Perisphinctes indogermanus, Peltoceratoides semirugosus, Paracenoceras kumagunense, Euaspidoceras waageni* and some terebratulids. The Chari beds range from Callovian to Lower Oxfordian. The Patcham and Chari beds constitute an exceedingly well developed Callovian, characterised by numerous Macrocephalitids. The Dhosa oolites contain the late Macrocephalitids such as *Mayaites, Epimayaites* and *Dhosaites*.

Katrol Series.—This series, composed of different types of sandstones and shales, includes the KANTKOTE SANDSTONES, KATROL BEDS proper and GAJANSAR BEDS in the ascending order, and range in age from upper Argovian to Portlandian. The Kanktote sandstone is found in the neighbourhood of Wagur and indicates a horizon below the Katrol beds proper, the chief fossils being *Epimayaites transiense, Prososphinctes virguloides* and *Torquatisphinctes torquatus* of Upper Oxfordian age. The lower and middle Katrol beds proper contain Oppelids, and Perisphinctids (*Taramelliceras, Glochiceras, Streblites, Aspidoceras ; Torquatisphinctes, Pachysphinctes, Katroliceras, Subplanites*, etc. but the upper Katrol beds are barren. Interstratified with these strata are horizons containing plant remains. The Gajansar beds contain a fauna in which appear species of *Glochiceras, Phylloceras*, etc.

Umia Series.—The Katrol Series is overlain by the Umia Series which is over 900 m thick and consists of sandstones, conglomerates and shales, part of the earlier beds being of continental origin and considered by Arkell as similar to the Nubian Sandstone of Egypt, which is partly of Lower Cretaceous age. The sub-divisions of the Umia Series are :—

BHUJ STAGE (450 m.)	...	Palmoxylon Beds Ptylophyllum Beds Zamia Beds	} Post-Aptian
UKRA STAGE (30 m.)	...	Calcareous shales with *Australiceras* and other ammonites	} Aptian
UMIA STAGE (500 m.)	...	Barren sanstones and shales Trigonia beds Barren sandstones Umia ammonite beds Barren sandstones	} Neocomian to Up. Tithonian

The shales at the top of the Katrol Series, called formerly the Zamia Shales, really form part of the Gajansar beds. They are better called the UPPER KATROL SHALES. They contain *Haploceras elimatum, Hildoglochiceras* spp., *Phylloceras* etc. and are of Portlandian age. The Umia Stage consists mainly of barren sandstones, with fossliferous intercalations. The AMMONITE BEDS which are barely 15 m thick, are green oolites with abundant ammonites, some brachiopods and corals. Large *Virgatosphinctes* are characteristic, but *Aulacosphinctes, Ptychophylloceras, Umiaites* and *Micracanthoceras* also occur; other fossils are *Acanthorhynchia, Lobothyris,* and *Trigonia retrorsa*. The TRIGONIA BEDS are characterised by *T. ventricosa* and *T. crassa (Trigonia smeei* is Upper Jurassic and does not belong here). The Umia Stage is therefore Upper Tithonian and Neocomian. The Barren Sandstones at the top may be Upper Neocomian.

The succeeding UKRA BEDS, mainly calcareous shales, are of Aptian age. They contain *Australiceras, Cheloniceras,* and *Tropaeum*. The BHUJ BEDS are probably estuarine. They contain *Ptilophyllum* and other plants closely related to the Jabalpur Series, and are presumably of post-Aptian age. The regression which was taking place here seems to be contemporaneous with the transgression on the Trichinopoly coast and elsewhere.

The neighbouring area of Kathiawar (Saurashtra) contains a large thickness of sandstones and some shales of Lower Cretaceous age (Arkell, 1956, p. 392). They are overlain by marine beds resembling the Bagh Beds of the Narmada valley.

This sequence is overlapped by the Deccen trap lava flows or by Nummulitic rocks, with a distinct unconformity.

MADRAS COAST

Marine fossils are associated with the Upper Gondwana beds of the East coast of the Peninsula in the Godavari-Krishna-Guntur tract. Ammonites and other marine fossils were found in some of the beds. The Ammonites, which are however in a bad state of preservation, have been examined by L.F. Spath who assigns an Upper Neocomian age for them. The plant fossils in these beds are thought to be of Upper Jurassic age. Further details about them will be found in the chapter on the Gondwana System.

BURMA

Namyau Series.—The Napeng beds of Northern Shan States are overlain by red sandstones, conglomerates and shales with occasional lime-stone bands. They are well developed in the Namyau valley and show evidences of shallow water origin. The outcrops are in wooded country, characterised by labyrinthine depressions, and the strata have been folded and tilted. Fossils are found in the limestones and calcareous shales, comprising only brachiopods and lamellibranchs but no cephalopoas, differing in this respect from other areas. The fossils include species of *Burmirhynchia, Holcothyris, Terebratula, Modiola, Pecten*, etc. which are considered to indicate a Bathonian age. The lower part contains more than 35 species (including *Eomiodon*) found in Kuar Bet beds at the base of Patcham beds in Kutch. The uppermost beds are Namyau shales whose age may be Cretaceous. The Namyau beds extend into Yunnan and Szechuan in China.

The LOI-AN SERIES of Southern Shan States is considered to be of Jurassic age. It is composed of shales and sandstones, with coal seams in the upper portion. The plant remains found in them include *Ginkgoites digitata, Cladophlebis denticulata, Pagiophyllum divaricatum, Brachyphyllum expansum* and *Podozamites distans*.

There are certain red sandstones in Amherst and Mergui districts which continue into Thailand and Tonkin. Though fossils are rare in them and of little value for determining the age, they are assigned to the Jurassic from their lithological similarity to beds in Northern Shan States and Siam (Thailand).

The Jurassic strata of Burma (Shan-Tenasserim belt) and Malaya are associated with granites and pegmatites containing tin and tungsten ores. These continue into the Tin Islands of Sumatra and into Borneo.

SELECTED BIBLIOGRAPHY

Arkell, W.J.   Jurassic Geology of the Wrold.   London (Oliver & Boyd) 1956.
Buckman, S.S.   The brachiopods of Namyau beds, Burma, *Rec.* **45**, 75-81, 1915.
Buckman, S.S.   The brachiopods of Namyau beds, *Pal. Ind. N.S. III*, (2), 1917.
Cotter, G. de P.   Geology of a part of Attock district.   *Mem.* **75**(2), 1933.
Cox, L.R.   The Triassic, Jurassic and Cretaceous gastropoda and lamellibranchiata of the Attock district.   *Pal. Ind. N.S. XX*, (5), 1935.
Holdhaus, K.   Fauna of the Spiti shales — Lamellibranchiata and gastropoda.   *Pal. Ind. Ser. XV*, Vol. IV, Fasc. 4, 1913.
Koken, E.   Kreide und Jura in der Salt Range.   *Z. Bl. Min. Geol.* IV, p. 437, 1903.
Middlemiss, C.S.   Geology of Hazara.   *Mem.* **26**, 1886.
Muir-Wood, H.M.   Mesozoic brachiopoda of the Attock district.   *Pal. Ind. N.S. XX*, (6), 1937.
Raj Nath.   A contribution to the stratigraphy of Cutch.   *Q.J.G.M.M.S. India*, **4**, 161-174, 1933.
Spath, L.F.   Revision of the Jurassic cephalopod fauna of Cutch.   *Pal. Ind. N.S. IX*, 1924-33.
Sath, L.F.   Jurassic and Cretaceous ammonites and Belemnites of the Attock district.   *Pal. Ind. N.S. XX*, (4), 1934.
Uhlig, V. and Steiger, P.   Fauna of the Spiti shales — Cephalopoda.   *Pal. Ind. Ser. XV*, Vol. VI, (1, 2, 3, 5), 1903-14.
Waagen, W. Gergory, J.W. and Kitchin, F.L.   Jurassic fauna of Cutch.   *Pal. Ind. Ser. IX*,
Wynne, A.B. Geology of Kutch, *Mem.* 9, Pt. I, 1872.

CHAPTER XIV

THE CRETACEOUS SYSTEM
GENERAL

This System takes its name from Chalk (Craie or Kreide) which is one of its important constituents in Western Europe. It is generaly classified into two sub-systems, the lower stretching upto and including the Albian, and the upper comprising the rest. Sometimes a Middle Cretaceous is also recognised, in which case the strata from the base of the Albian to the end of the Turonian are included in it. The chief sub-divisions of this System, named mainly after places in France, are shown below.

DANIAN	...	Bryozoan limestone of Denmark. (Now placed in the Eocene).
MAESTRICHTIAN	...	Siliceous chalk of Maestricht (Holland) with *Belemnitella mucronata*.
SENONIAN	...	CAMPANIAN — Chalk of Champagne (near Paris) with *Belemnitella quadrata*.
		SANTONIAN — Chalk of Saintonge with *Micraster coranguinum*.
		CONIACIAN — Chalk of Cognac (W. France) with *Micraster cortestudinarium*.)
TURONIAN	...	Siliceous marly chalk of Touraine.
CENOMANIAN	...	Glauconitic sands of Maine (W. of Paris).
ALBIAN	...	Greensand and blue clays of Aube (S.E. of Paris) = Gault clay of England.
APTIAN	...	Limestones and marls of Apt (Vaucluse, France).
BARREMIAN	...	Marls of Barreme (Basse Alps, France).
HAUTERIVIAN	...	Marly limestone of Hauterive, Switzerland.
VALANGINIAN	...	Marly limestone of Valangin, Switzerland.

Note:—The team *Neocomian* (from Neuchatel Switzerland) is used as the equivalent of Valanginian and Hauterivian together. Similarly *Emscherian* is used for coniacian and Santonian together. The basal part of the Valanginian is referred to as the *Berriasian*.

The Cretaceous is largely marine in most places though a continental facies of the Lower Cretaceous with plant fossils and reptiles, fishes etc. is also found in some regions, especially in Gondwanaland. There are two distinct marine facies in Europe, one being that of N.W. Europe with characteristic chalk deposits, the other being the Mediterranean one in which the aberrant group of bivalves called Rudistids developed. It is now generally agreed that the Chalk formations were deposited in medium deep seas as they contain abundant remains of benthonic and pelagic foraminifera and of sponges.

A remarkable geologic event in the Cretaceous is the universal marine transgression usually known as the *Cenomanian transgression*, perhaps the most conspicuous of all marine floodings during Phanerozoic geologic times. The transgression is often noticed in the uppermost Aptian or Albian, though fossiliferous marine strata become important only from the Cenomanian onwards.

The Alpine-Himalayan mountain building movements began in the Upper Cretaceous. It is most conspicuous in the belt stretching from New Guinea, along the Indonesian archipelago, the Himalayan arcs, Southern Persia, Oman, Iraq and Turkey into the Alps and Mediterranean regions. Some regions of the circum-pacific belt were also affected. The dismemberment of Gondwanaland was completed in this period, some parts having already begun to drift from each other. The margins of these lands, bordering the South Atlantic and Indian Oceans, are conspicuously faulted. The Indian Ocean which seems to have originated with the rifts formed initially in Permo-Carboniferous times, had considerably widened in the Cretaceous, for we find marine Cretaceous deposits along the coasts of East and South Africa, India and Western Australia, with an Indo-Pacific fauna which differs appreciably from the Tethyan Mediterranean fauna. The fragmentation of Gondwanaland was also attended by extensive eruption of lava flows from the Rhaetic or Lower Jurassic to early Tertiary in different countries like Brazil, South and East Africa, and Antarctica, India.

FAUNA

Foraminifera play an important part in Cretaceous stratigraphy. Orbitolines, which are characteristic of coastal and reef facies, are found in the Lower Cretaceous up to the Cenomanian, after which they disappear. Orbitoids continue from the Campanian into the Tertiary, the genus *Orbitoides* being characteristic of Upper Cretaceous. Globigerinids, miliolids, lagenids and rosalines are abundant in the chalk formations. Several species of foraminifera have been used as zone fossils.

Both calcareous and siliceous sponges occur, the latter in the chalk formations. Though a number of corals are known, they are not of stratigraphic importance. Amongst the echinoderms, crinoids often form limestones (*Uintacrinus, Marsupites*) while genera of echinoids such as *Discoidea, Cidaris, Toxaster, Holaster, Micraster, Hemipneustes* etc. are useful in zoning. The brachiopods are less important than in the earlier Systems, but sometimes enormous numbers of individuals of *Pygope*, Terebratulids, Rhynchonellids, *Thecidia*, etc. are to be found. Gastropods and Pelecypods are of limited importance. The more important genera of the former are *Actaeon, Actaeonella, Turritella, Cerithium, Physa, Natica*, etc. *Inoceramus* is an important genus as several of its species mark certain zones. The Rudistids occurring in the Mediterranean region are also useful ; *Velletia* in Neocomian ; *Requienia, Toucasia* and *Polyconites* in Barremian-Aptian ; *Hippurites* and *Radiolites* in Cenomanian to Danian. The ammonites still play an important part but only in the Lower Cretaceous, though several genera continue to the end of the Cretaceous. The important genera are *Berriasella, Hoplites, Crioceras, Desmoceras, Acanthoceras, Douvilleiceras, Schloenbachia, Mortoniceras, Placenticeras*, etc.

Amongst the Belemites, the genera *Cylindroteuthis* and *Belemnopsis* continue from the Jurassic into the Cretaceous. *Belemnitella, Gonioteuthis* and *Actinocamus* occur in the Upper Cretaceous.

Northern Europe.—The northern facies is developed in northern Europe. In England the Neocomian-Barremian strata are of continental nature known as the WEALDEN, with plant, fish and vertebrate remains. Both continental and marine strata occur in France. The Aptian forms the LOWER GREENSAND while the Albian is a clay formation called the GAULT. The Cenomanian shows greensands in the shallower facies and chalk in the deeper facies. Ammonites, echinoderms, Inoceramus and foraminifera serve as zone fossils.

In Poland and Germany the Valanginian is estuarine but the beds above are marine. As many as 30 cephalopod horizons have been identified in the Lower Cretaceous.

Mediterranean Region.—Cretaceous strata have been studied in the Dinarides, the Eastern Alps, in Southern France, and in N.W. Africa, in all of which the Mediterranean facies is developed. In addition to ammonites, echinoderms, foraminifera, there are also corals and Rudistids. The distinguishing faunal characters of the two provinces (northern and mediterranean) trend to disappear in the Upper Cretaceous when the faunae assume a more cosmopolitan character.

The Cretaceous is one of the most widely distributed sedimentary systems in India and is, moreover, represented by a variety of facies. The Himalayan area, as in the earlier systems, shows the Tethyan geosynclinal facies. A large area of Tibet and northern Himalayas is covered by these rocks. In Baluchistan there is an eastern calcareous fossiliferous facies and a western arenaceous unfossiliferous facies consisting of sandy strata of the European flysch type. Similar rocks are developed also in the Burma and Arakan ranges at the other border of India. Marine incursions of this period have left their deposits in the Narmada valley, in the Trichinopoly-South Arcot area of southern Madras and in Assam. The latter two have a faunal assemblage somewhat different from that the Tethyan geosynclinal deposits. Estuarine and lacustrine deposits are developed in various parts of the centre of the Peninsula underlying the great group of lava flows called the Deccan Traps.

Towards its end, the Cretaceous was a period of extensive igneous activity. Granites and particularly basic rocks belonging to this period are found in the Himalayan area. Baluchistan and the Arakan region of Burma likewise witnessed igneous phenomena. The Peninsula was the scene of stupendous volcanic outbursts which were responsible for a large thickness of lava flows and fragmental products which at one time probably covered close upon half a million square miles of the land surface.

The eastern coast of India seems already to have taken shape at the beginning of the Cretaceous, but on the west there was still land to the west of Bombay, stretching westwards into what is now part of the Arabian Sea. An arm of the Tethys extended into Baluchistan and farther west on the one hand and into the Arakan region on the other. A sea possibly connected with this arm, transgressed into the Narmada valley in Lower Cretaceous times and about the same time a southern sea invaded the eastern shores of the Peninsula in Southern Madras and Assam.

There is evidence of local regression or shallowing of the sea in some parts of the Tethyan basin in the upper part of the Cretaceous, though the Cenomanian and Maestrichtian are characterised by transgression in most parts of the world.

SPITI

Giumal Series.—At Giumal, Kibber and Chikkim, the Lochambal beds gradually merge into yellow and brown sandstones and slaty quartzites. These constitute the Giumal Sandstones and have a thickness of about 100 m. The uppermost Lochambal beds contain the cephalopods *Hoplites* (*Neocomites*) *neocomensis* and *Acanthodiscus subradiatus* which clearly point to a Neocomian age. The Giumal sandstones enclose a fauna which indicates that they cannot be older than Upper Velanginian or Lower Hauterivian nor younger than the Albian :

Cephalopods ... ... *Holostephanus* (*Astieria*) aff. *atherstoni*, *Stephanoceras* sp., *Perisphinctes* sp.
Lamellibranchs ... *Cardium gieumalense* (abundant), *Pseudomonotis superstes* (abundant), *Ostrea* sp., *Gryphaea* aff. *baylei*, *Pecten* sp. etc.

Chikkim Series.—The Giumal Sandstones are overlain by grey or whitish limestones, having a thickness of 30 m or more. These are the Chikkim Limestones which contain *Belemnites, Hippurites,* and some foraminifera including *Cristellaria, Textularia, Nodosaria, Dentalina,* etc. and also *Globotruncana*. These are probably of the same age as the Hemipneustes beds of Baluchistan, which are Maestrichtian. Overlying the Chikkim Limestones are the Chikkim Shales (45-60 m) consisting of calcareous, sandy, grey-green shales. (*Jour. Pal. Soc. Ind.* I, 199-201, 1956). Similar rocks are found also in Hazara, Afghanistan and Iran.

The Chikkim Series is overlain by a group of sandtsones and arenaceous shales of the *flysch* facies, which are entirely unfossiliferous. It appears that after the deposition of the Spiti Shales the sea gradually became shallow and became unsuitable for supporting a rich fauna. There is also a distinct difference between the Spiti Shale fauna and the Cretaceous fauna and the two have very few common elements.

TETHYS HIMALAYA

Near the Niti Pass the Chikkim Limestones are well developed. Beyond it in Tibet, there are volcanic rocks (of Cretaceous age) overlain by sandstones and shales of Eocene age.

In North Kumaon the Cretaceous is represented by the Giumal Sandstones (Lower flysch) of Lower Cretaceous age and Upper flysch sediments of Upper Cretaceous age. The Spiti Shales show a gradual passage into the Giumal Sandstones. The basal portion of these sandstones is mainly shales with calgareous sandstone layers containing glauconite. They pass up into hard green clauconitic sandstones with shaly layers. The total thickness of the Giumal Sandstones in the Kiogar region is 500 to 700 m. They weather into steep-sided black crags with fantastic shapes, some of the peaks rising to over 5,000 m in altitude. Fossils are rare, but several bivalves and some ammonites have been found in the Spiti region. These indicate that the lower part of the Giumal Sandstone is of Upper Valanginian age, while the uppermost part may be slightly younger.

The upper part of the Cretaceous, called the Upper Flysch by Heim and Gansser, is estimated to have a maximum thickness of 1000 m., but varies a great deal. It shows the following sub-divisions :—

- d. 300 to 400 m. red and green siliceous sandstones and dense radiolarian cherts alternating with siliceous shales.
- c. 500 to 600 m. black slaty shales with clay ironstone layers and flaggy limestones with Fucoids.
- b. 100 m. of purple marly shales containing foraminifera (Globigerinids and Rotalids)
- a. 50 m. greenish shale with sandstone layers.

The thickness of each of the divisions is very variable. The division (c) bears a great resemblance to the similar flysch of the Alps. It also resembles the Spiti and Kuti shales from a distance but does not contain the characteristic ammonites of those formations. Heim and Gansser state that subdivision (d), which was described as being a tuff in the upper part by Von Kraft, is not correct as it really consists of an alternation of red and green chert with thin shaly layers, all being rich in radiolarians. What was mistaken for tuff was really radiolarian chert which is a deep-sea sedimentary deposit. It shows under the microscope, an opalescent ground mass without any volcanic material and is rich in numerous small globular radiolarians and nassellarians. The shells of these are often replaced by bluish glauconite or filled with opaque ferruginous silica or chalcedony. The Upper Flysch indicates that after the deposition of the Giumal Sandstone there was a deepening of the sea. The Upper Flysch is considered to be of Turonian to Senonian age.

JOHAR — HUNDES

In the area of the exotic blocks of Koigar on the Tibetan frontier, the Spiti Shales are overlain by the Lower flysch (Giumal Sandstones of Lower Cretaceous age) which attain a thickness of 600 to 700 m. Their exact age is Upper Valanginian to Albian as they contain *Olcostephanus* (*Rogersites*) cf. *atherstoni* and other fossils. Above them are the Upper flysch consisting of siliceous shales and radiolarian cherts overlain by the igneous sheets associated with the exotics. Some 80 km to the east of this area, to the south and southwest of Rakshas lake, similar Cretaceous rocks are seen with associated igneous rocks at Jangbwa and Amlang-La. Cretaceous flysch is also seen in the Darchen region to the south of Mount Kailas, where they are thrust towards the north (counter-thrust) and are overlain by the Kailas Conglomerate of Eocene age.

According to Von Krafft, the following succession is seen in the Chitichun area, which was studied by him :—

		Meters
4f(b).	Red Tuffs, thin bedded	15
4f(a).	Green tuffs, thin bedded	45
4e.	Greenish and grey sanastones alternating with shales, passing through tuffaceous sandstones upwards into 4f	90
4d.	Hard black siliceous shales and crumbling shales ...	10

			Metres
4c.	Brown weathering sandstones alternating with shales.	...	3
4b.	Black crumbling shales	...	60—90
4a.	Red and greenish shales and reds shaly limestones	...	30
3.	Giumal Sandstones (Lower Cretaceous)	... ...	120—150
2.	Spiti Shales (Portlandian)	...	—
1.	Grey limestone passing down from the Lower Jurassic into the Dachsteinkalk (Up. Trias).		—

All the strata under the sub-division 4 belong to the Upper Flysch, of Upper Cretaceous age. Of these, 4a is remarkable as it consists of red siliceous shales of a dark terracotta colour intercalated with a few bands of red hornstone and splintery greenish shales. These pass upwards into earthy calcareous shales with greenish grey limestone intercalations. Because of their colour, they are very conspicuous from a distance and are well seen north of Talla Sangcha and also in the Chitichun area. The so-called tuffs were all found to be radiolarian cherts.

The Giumal Sandstones in this region contain grains of glauconite, their total thickness being 500 to 700 m., much thicker than Von Krafft thought. The upper part of this formation forms a huge anticline on the north side of the Kiogad river. Its age is Lower Cretaceous, ranging from Valanginian to the Upper Gault. According to Heim and Gansser the Cretaceous section in this area is as follows :—

		Metres
d.	Green and red siliceous sandstones and dense radiolarian hornstone alternating with siliceous shales. These are overlain by thrust sheets of basic igneous rocks	300—400
c.	Black shales and slates with intercalated layers of fucoid limestones and grey ironstone layers. The upper part contains some brown sandstones	500—600
b.	Purple marly shale and earthy limestone with foraminifera	100
a.	Green shale with sandstone layers, (Giumal Sandstone) glauconitic with shaly bands in the middle and greenish sandstones in the upper part	500—700

It will be noticed that the major difference found by Heim and Gansser relates to the sub-division (d) (4e and 4f of Von Krafft), the so-called tuff being really siliceous radiolarian material.

According to Heim and Gansser, there are three different facies of Mesozoic rocks in the Kiogar region. One is the *Tethys Himalayan facies* characterised by the Kioto Limestone, Spiti Shale and Cretaceous flysch. The second is the *Kiogar facies* consisting of white Kiogar Limestone (lithologically similar to the Dachsteinkalk but unfossiliferous), pink-violet oolite with *Calpionella* of Tithonian to Lower Neocomian age, Cretaceous limestones and radiolarian siliceous limestones closely associated with basic igneous rocks. The third is the *Exotic Chitichun facies* comprising the conspicuously red limestones (with subordinate white and grey bands) of Permian, Triassic Jurassic and Cretaceous ages. This facies is unknown in the region escept in the exotic nappes. Table 54 shows these facies :—

TABLE 54—DIFFERENT FACIES IN THE HIMALAYAN CRETACEOUS

	Tethys Himalaya	Kiogar	Chitichun
Cretaceous	Upper Flysch Giumal Sst.	Basic igneous rocks Shales, limestones & radiolarian cherts	Basic igneous rocks
Jurassic	Spiti Shale Iron oolite Laptal Series	Pink coolite and shale with *Calipionella*	Adneth marble and radiolarian chert
Triassic	Kioto Limestone Kuti Shale Kalapani Limestone Chocolate Shale	Kiogar limestone (Dachsteinkalk)	Red Carnic limestone Red limestone with *Ceratites*
Permian	Kuling Shales		Red Crinoidal limestone

Reviewing the evidence, Arkell (1956, p. 411-412) suggests that the Kiogar and Chitichun facies are really one, as the unfossiliferous white Kiogar Limestone which has been assigned an Upper Triassic age by Heim and Gansser on purely lithological resemblance, may really be Upper Jurassic as it underlies the Tithonian pink oolite. This simplifies the correlation as only two facies can then be recognised, as shown in the Table 55.

The Exotic Blocks, though apparently strewn pell-mell, and merely held together by the igneous rocks on which they lie, can often be traced from one to the other by the continuity of the bedding planes. The limestones are penetrated by the igneous rocks and generally metamorphosed and serpentinised. Some of them are seen to be incorporated in the flysch sediments into which they have been thrust. Their structure and position suggest that they represent blocks detached from the bottom of a thrust sheet and mechanically thrust into and over the Upper Cretaceous flysch (E.B. Bailey, 1944). It is not unlikely that they represent the remnants of an earlier thrust sheet in which the Upper flysch and some unexposed stratigraphic horizons are involved in a later thrust. If so, the earlier thrust may be considered to have been semi-contemporaneous with the sub-marine eruptions of basic lavas (including pillow lavas) and intrusives whose age appears to be Upper Cretaceous.

The igneous rocks in the exotic thrust mass consist mainly of dolerites, greenstones, spilites (pillow lavas), amygdaloidal porphyrites, enstatite-peridotites and pyroxenites which have been considerably serpentinised. Where they are intimately associated with limestones they have produced serpentine-marble. It is interseting that the peridotites are closely associated with deep sea radiolarian cherts and siliceous shales. All these intrusions are of late

THE CRETACEOUS SYSTEM

TABLE 55—CORRELATION OF HIMALAYAN AND TIBETAN FACIES

(AFTER ARKELL)

Age	Tethys Himalayan facies	Tibetan facies
Cretaceous	Basic igneous rocks and Upper Flysch Lower Flysch (Giumal)	Basic igneous rocks Grey radiolarian limestone
	Spiti Shales	Pink oolite with *Calpionella* White Kiogar Limestone
Jurassic	Callovian ferruginous oolite	?
	Laptal Series	Adneth Limestone with Hettangian and Sinemurian ammonites
Triassic	Kioto Limestone Kuti Shale Kalapani Limestone Chocolate Shale	Red Limestone with Carnic ammonites Red Ceratite Limestone
Permian	Kuling Shales	Red crinoidal limestone with brachiopods

cretaceous or post-Cretaceous age, but pre-thrusting. The sills of greenstone may however, be contemporaneous with, or slightly earlier than, the flysch sediments, at least in the Amlangla region. The gabbro and peridotite intrusive masses appear to be post-flysch, perhaps late Cretaceous, in age.

CENTRAL TIBET

Cretaceous rocks, consisting of Giumal sandstones and Cenomanian limestones, occupy a large area in Western and Central Tibet. They are overlain by late Cretaceous and Eocene rocks towards the southeast. In the Phari plain are exposed a thick series of limestones, shales, slates and quartzites which include Triassic and possibly older beds. The upper limestones of this group contain Liassic brachiopods. Jurassic rocks occupy a large area in the Provinces of Tsang and U, and fossiliferous Middle Jurassic limestones are known. Typical Spiti Shales have been found near Kampa Dzong and in the hills east and southeast of Gyantse.

A full Cretaceous sequence, though the whole of it is not fossiliferous, is noted near Kampa Dzong in a series of faulted folds. Together with the associated early Tertiary rocks it constitutes the *Kampa System*. The succession near Kampa Dzong is shown in Table 56. The lower Cretaceous is represented by the Giri Limestone which is unfossiliferous. The overlying Kampa Shale and Hemiaster beds are Cenomanian as shown by their fossil content. The Scarp Limestones and the Tuna Limestones range in age from Campanian to Maestrichtian and are apparently the equivalents of the Hemipneustes beds and Cardita subcomplanata beds of Baluchistan. The ferruginous sandstones at the top may be partly Cretaceous and partly Eocene.

In this region there are altered basic intrusives and serpentines which appear to be late Cretaceous or Eocene in age. There is also an unfoliated hornblende granite intrusive into the Mesozoic rocks.

KASHMIR

Indus Flysch and Dras Volcanics.—Cretaceous limestones with *Gryphaea vesiculosa* and other fossils have been encountered in Rupshu and along the Leh-Yarkand road in Sinkiang. Cretaceous deposits are known in the upper Hunza valley and in Eastern Karakorum and also along the Indus valley north of the Himalayas, the last being connected with a great tectonic zone.

At the western end of the Himalayas rises the solitary high mountain Nanga Parbat (8113 m). It was originally formed of the Salkhala metamorphics which have since been intruded and partly replaced by late Tertiary tourmaline granite. This granite forms the central mass, surrounded by migmatised gneiss. All stages of replacement can be seen from the outer zone of Sulkhalas at the base and on the lower flanks of the mountain, increasing progressively towards the top. The Salkhalas swing around the syntaxis and form a zone which widens out to the south as the higher part of the mountain has an elliptical outline trending NNE-SSW. They form a zone 8 to 10 km wide at the base and are composed of slate, phyllites (including carbonaceous ones) garnetiferous mica-schists with bands of crystalline limestones, all showing tight isoclinal folding. Just beyond this zone occurs the Upper Cretaceous flysch accompanied by contemporary volcanics and slightly younger basic and ultrabasic intrusives.

The Indus River north of the Himalayas flows along the flysch zone which marks a major tectonic zone which Gansser (1964) calls the *Indus Suture*. This zone can be followed from the western side of the syntaxis around the northern side of the Nanga Parbat to its east and southeast. The volcanics accompanying the flysch zone are named after Dras, where they were studied by Wadia in the middle thirties. Here, both Lower and Upper Cretaceous rocks enclosing *Orbitolina* and other fossils are found. The volcanics consist of thin-bedded tuffs and ash beds of purple and greenish colours with layers of agglomerate, shale, cherty jasper and lava flows. The earlier beds are volcanic breccias containing fragments of Cretaceous sediments and igneous rocks, grading into lava flows, the assemblage attaining a thickness of 2000 m. The lavas are basaltic and andesitic but are generally chloritised and epidotised. The intrusives are partly contempraneous and partly slightly later and comprise norites, peridotites, pyroxenites, hypersthene-diorites. Granites also occur, accompanied by porphyries and dolerite dykes and sills. This ophiolite suite marks a deep thrust zone which has, in Northern Kumaon, brought up the rocks of exotic facies from a deeply buried zone. Exposures of the flysch and associated ophiolites are seen more or less all along the Indus valley from the Nanga Parbat to Spiti valley and south of Mount Kailas, where they disappear. They probably underlie the alluvial valley of the Tsangpo which follows the trend of the Eastern Himalayas some distance to their north. It is of interest to note that ophiolites are found along the eastern base of the main axis of the

TABLE 56—THE KAMPA SYSTEM OF TIBET

Age	Sub-divisions	Thickness Metres
LAKI	Dzongbuk Shales	30
	Alveolina limestones	120
	Shales and limestones, unfossiliferous	45
	Sandy micaceous shales and flaggy sandstones	15
	Orbitolites limestone with *Orbitolites* and *Miliolina*	
PALEOCENE	Spondylus shales — Varicoloured and needle shales	50
	Operculina limestone — Shaly nodular foraminiferal limestone with *Nummulites*	—
	Gastropod limestone — Limestones, thin-bedded below but hard, massive and dark above with gastropods, especially *Velates schmideliana*	?
	Ferruginous sandstones — no deterriminable fossils	60
DANIAN	Grey Limestone with echiniods and brachiopods	15
	Sandstone (unfossiliferous)	15
MAESTRICHTIAN — TUNA LIMESTONE	Lithothamnion Limestone with *Omphalocyclus macropora*, *Orbitoides*, *Cyclolites regularis*.	—
	Red arenaceous Limestone with grey and brown limestone below, with *Hemiaster*, *Plicatula*, *Neithea quadricostala* and *Omphalocyclus*.	20
	Thin-bedded limestone with *Orbitoides media*.	—
MAESTRICHTIAN	Third Scarp Limestone — massive limestone, calcareous shale and flaggy limestone with *Radiolites* and *Orbitoides media*	66
	Second Scarp Limestone — with Rudistae, *Orbitoides media*, echinoids and lamellibranchs	45
	Dark shales and limestones with *Gryphaea*	45
	First Scarp Limestone, hard and splintery, no determinable fossils	45
CAMPANIAN TO CENOMANIAN	Hemiaster beds — Pale grey shales with *Hemiaster grossouvrei*, *H. cenomanense*, *Gryphaea*, *Inoceramus*...	80
	Kampa Shale — Brown shaly limestone and black needle shales with *Acanthoceras rhotomagense* and other species and *Turrilites costatus*	110
LOWER CRETACEOUS	Giri limestone — hard thin-bedded limestone, unfossiliferous	120
JURASSIC	Spiti Shales	...

Burmese Arc down to the Arakan on one side, and at several places in the Baluchistan Arc on the other.

HAZARA

Mention has already been made of the fact that there are two parallel zones in Hazara in which different facies of rocks are developed side by side. This applies not only to the Jurassic but also to the Cretaceous.

In the northwestern zone, the Spiti Shales of Jurassic age pass conformably upwards into the Giumal Sandstones, succeeded by 3 to 6 m of orange-yellow sandstone with cherty and ferruginous patches containing Albian ammonites. Amongst the fossils are *Lyelliceras lyelli* (abundant), *L. cotteri, Acanthoceras rhotomagnesis, Eutrephoceras* sp., *Douvilleiceras mammillatum, D.* aff. *monile, Brancoceras, Diploceras* aff. *bouchardianum, Hamites* cf. *attenuatus, Metahamites* aff. *elegans, Oxytropidoceras multifidum, O. roissyanum, Mojsisovicsia* aff. *delaruei* and also *Echinoconus* and *Micraster*. This bed is followed by 120 m of well-bedded grey limestone with a laterite layer on top, the latter indicating subaerial weathering before the Eocene was laid down.

In the southeastern zone, the massive limestone of Upper Jurassic age is succeeded by Giumal sandstones which are 30 m thick, and dark green to nearly black in colour. Calcareous intercalations in these sandstones contain abundant *Trigonia*. They are succeeded by 30 m of sandstone and shell limestone, overlain by 30 m of buff, thin-bedded shaly limestone. All these form a conformable series and appear to be of Neocomian to Albian age. Their junction with the overlying Eocene is marked by a bed of laterite. In this southeastern zone, therefore, the beds between the Albian and Eocene are missing whereas in the northwestern zone the unconformity is much smaller *i.e.*, between the Middle Albian and some part of the Upper Cretaceous.

ATTOCK DISTRICT

The Spiti Shales of this region pass imperceptibly upwards into the Giumal horizon containing *Oxytropidoceras* aff. *roissyanum* in the uppermost beds which are of Albian age. There are no strata belonging to an age between these and the Eocene.

SAMANA RANGE

Above the Samana Suk Limestone of Jurassic age there occur dark grey glauconitic sandstones with limestone bands containing the ammonite *Astieria* and *Belemnites* of Neocomian age. The glauconitic sandstones are unconformably succeeded by the MAIN SANDSTONE SERIES of Albian age consisting of white quartzitic sandstones with haematitic bands and having a thickness of about 220 m. Fossils occur in the topmost grits.

Cephalopods	*Douvilleiceras mammillatum* (very abundant), *Pictetia* cf. *astieriana, Cleoniceras daviesi, Desmoceras latidorsatum. Brancoceras indicum, Hamites* cf. *attenuatus.*
Brachiopods	*Rhynchonella samanensis, R. reticulata, Terabratula samanensis, T. daviesi, Kingena spinulosa.*
Gastropods	*Metacerithium* cf. *ornatissimum, Semisolarium moniliferum.*
Lamellibranchs ...	*Pecten, Neithea, Venus, Cyprina* cf. *quadrata.*
Echinoderms	*Discoidea* sff. *decorata, Conulus* sp., *Holaster* sp.

The Main Sandstones Series is succeeded, after a slight unconformity, by 45 m of flaggy limestone called the LOWER LITHOGRAPHIC LIMESTONE, of probably Chikkim age. Above this there are variegated quartzitic sandstones and the UPPER LITHOGRAPHIC LIMESTONE, which also appear to be of Upper Cretaceous age.

CHITRAL

Middle to Upper Cretaceous strata, consisting of limestones containing *Orbitolina* and *Hippurites*, are found n a series of outcrops in Chitral, often faulted against older strata. They are succeeded by the Reshun conglomerate beds whose age may be Upper Cretaceous or Eocene.

BALUCHISTAN-SIND

In the eastern or Calcareous zone of Baluchistan, the Cretaceous System is well developed as a series of shales and limestones, whereas in the western or Flysch zone it consists mostly of dark greenish grey sandstones and sandy shales.

The Calcareous Zone

Belemnite Shales.—In the calcareous zone the Neocomian succeeds the Callovian strata unconformably. Above the lower BELEMNITE SHALES are the PARH LIMESTONES of Cenomanian age. These formations are abundantly intercalated with agglomerates and tuffs (especially abundant in the lower part) and pillow lavas of basic composition. The Belemnite Shales are dark splintery shales containing abundant belemnites of the genus *Duvalia* (a flatted form) and other fossils including *Duvalia dilatatus, Belemnites latus, B. subfusiformis, Gryphaea oldhami,* etc. At times they swell into a thick formation of gree n shales with thin limestone intercalations ; by the accession of arenaceous materials these pass into the flysch facies developed on a large scale in Lower Zhob. The Belemnite Shales are younger than the Spiti Shales. Their age may extend into the Lower Cenomanian.

Parh Limestones.—Conformably overlying the Belemnite Shales are the Parh Limestones of white, red and purple colours and often of porcellanoid texture in the upper part. They occasionally reach a thickness of 500 m. They contain *Inoceramus* and the aberrant genus *Hippurites* characteristic of the Mediterranean province and also abundant tiny pelagic foraminifera — *Globotruncana, Rotalipora,* etc.

The Parh Limestones and the Belemnite bedsare seen forming the peripheral portions of the domes and anticlines whose cores are of dark-coloured Jurassic limestone. The age of the Parh Limestones is Cenomanian to Campanian.

There are intrusions of gabbroid and peridotitic rocks in the Parh Limestones, which may be of Upper Cretaceous age. With these are associated the chromite deposits of Zhob-Pishin areas.

Hemipneustes Beds.—The Upper Cretaceous is developed in the Laki range of Sind and in a large area in the Calcareous zone of Baluchistan. The lowest strata, called the Hemipneustes Beds, are limestones of Maestrichtrian age, which have yielded a rich fauna :—

Foraminifera	*Siderolites calcitropoides, Orbitoiaes media, Omphalocyclus macropora.*
Cephalopods	*Nautilus sublaevigatus, Parapachydiscus dulmensis, Bostrychoceras polyplocum.*
Echinoids	*Hemipneustes pyrenaicus, H. Leymeriei, H. compressus, Echinoconus gigas, E. helios, Holetypus baluchistanensis, Pyrina ataxensis, Hemiaster blanfordi.*

Lamellibranchs ...	*Alectryonia pectinata, Gryphaea vesicularis Vola quadricostata, Spondylus santoniensis, Pecten (Chlamys) dujardini, Corbula harpa.*
Gastropods ...	*Ovula expansa*

Cardita subcomplanata beds.—The Hemipneustes beds are overlain by shales containing occasional fossiliferous bands, one of these containing a rich ammonite fauna, though only 15 cm. thick. The fossils from this zone include *Cardita (Venericardia) subcomplanata* (by which name the beds are sometimes called), and the ammonites *Indoceras baluchistanensis, Sphenodiscus acutodorsatus, Parapachydiscus dulmensis* and *Gaudryceras* sp.

Pab Sandstones.—The above-mentioned shales are followed by a thick series of flysch-like sandstones, called the Pab Sandstones. Though mainly unfossiliferous, they contain rare fossiliferous horizons of shaly constitution, especially in the upper part, in which *Cardita (Venericardia) beaumonti* has been found. Other fossils of these beds include :—

Lamellibranchs ...	*Venericardia vredenburgi, Ostrea acutirostris, O. flemingi, Liostrea orientalis, Corbula sulcifera, C. harpa, Crassatella austriaca, Cardium inaequiconvexum, Glycimeris vredenburgi.*
Gastropods ...	*Bellardia indica, Nerita haliotis, Morgania fusiformis, Turritella praelonga, Potamides chaprensis, Procerithium triplex, Campanile breve.*
Cephalopods ...	*Nautilus forbesi, N. lebechi.*

The Jurassic and Cretaceous succession of Baluchistan is shown in Table 57, based on the work of E. Vredenburg. The ages have been modified in accordance with information now available as a result of the work of the oil Companies.

TABLE 57—MESOZOIC SUCCESSION IN BALUCHISTAN
(Avter Vredenburg — *Rec.* 38, p. 199—200)

Age	Sub-divisions.
Eocene	Ghazij beds — Gypseous clays etc. with *Assilina granulosa*
Danian	Cardita beaumonti beds with volcanic agglomerates, basalts, dolerites and serpentines.
Maestrichtian	Pab Sandstones — Massive, rather coarse, sandstones sometimes of large thickness, accompanied by volcanic materials.
,,	Olive Shales — with numerous ammonites, occasionally interbedded with volcanic ash.
,,	Hemipneustes beds — Limestones and calcareous shales with *Parapachydiscus, Bostrychoceras*, and several echinoids.
	———unconformity———
Senonian	Lituola beds — Flaggy porcellanic limestones and shales of buff and pale green colours, containing numerous small foraminifera principally of the genus *Lituola*.
Cenomanian	Parh Limestone — white, procellanic, well stratified limestones with ferruginous basal beds.
Neocomian	Belemnite beds — Black splintery shales with abundant *Belemnites (Duvalia)* and *Hoplites*.
	———unconformity———
Callovian	Indocephalites beds — thin bedded limestones and shales containing the large ammonite *Indocephalites* sp.
Bathonian and Bajocian	Massive grey limestones, several thousand m thick.
Lias	Dark well-bedded limestones, several thousand m thick, sometimes with fossiliferous horizons.
Trias	Thick, greenish grey slaty shales with thin black limestone layers. *Monotis salinaria, Didymites, Halorites,* etc.
Permo-Carboniferous	Limestones with *Productus* and other Permo-Carboniferous fossils.

The Belemnite Shales extend into the Cenomanian only in the Sulaiman Range. In the more southern areas the Parh limestones commence in the Upper Albian and continue through to the Campanian. A low ridge seems to have been formed in the Upper Parh times in the Sibi-Jacobabad area which separated the northern part of the sedimentary basin (Sulaiman and farther north) from the southern (the region of Quetta and the Kirthar range). In the northern area, the Parh Limestones extend up to the middle of the Campanian and are overlain by the NISHPA FORMATION whose range is from the Upper Campanian to Lower Maestrichtian. In the southern area of Kirthar range and the Quetta-Lorelai region, the Nishpa Formation is not developed but its place is taken by the Orbitoid Limestone which is continuous with the Parh Limestones. The PAB SANDSTONE succeeds the Nishpa Formation or the Orbitoid Limestone, as the case may be. In the north there is a disconformity between it and the succeeding DUNGHAN SHALES of Palaeocene age, while in the south the sedimentation is continuous and does not show a break. The Orbitoid Limestone corresponds to the Hemipneustes Beds. The bottom part of the Dunghan Shales has been called VENERICARDIA SHALES as they contain *Cardita*(*Venericardia*) *beaumonti*, and species of *Globigerina* (but no *Globotruncana* or *Globorotalia*), indicating the age as DANIAN which is now included in the Paleocene. The general succession in the Sulaiman Range area, as revealed in a borehole at Giandari and in surface exposures, is given in Table 58 (Gigon, 1963).

TABLE 58.—CRETACEOUS SUCCESSION IN THE SULAIMAN RANGE AND IN THE NORTHERN AREA.

PALEOCENE	DUNGHAN	Dark calcareous sandy shales with *Globigerina velascoensis, G. compressa* etc.
	PAB SANDSTONE	Grey, greenish, brownish sandstone with silt intercalations and streaks of carbonaceous matter Oolitic in the upper part. Lower and upper parts marine, middle continental. Fossils not common. *Globotrucana* present.
MAESTRICHTIAN	NISHPA	Calcareous sandy shale at top. Hard grey argillaceous compact limestone mainly. *Omphalocyclus macrocarpa, Orbitoides media, O. minor, Globotruneana elevata, G. stuarti,* etc.
CAMPANIAN	PARH LIMESTONE	Top — light grey, hard, porcellanous limestone —*Globotruncana ventricosa, G. lapparenti* etc. Main part — Argillaceous limestone with intercalations of calcareous shale. *Globotruncana imbricata, G. helvetica, Rotalipora turonica, R. apenninica,* etc.
SANTONIAN		
CONIACIAN		
TURONIAN		
CENOMANIAN	BELEMNITE SHALE	Dark, hard, micaceous, calcareous silts and claystones — *Planomelina buxtorfi, Rotalipora ticinensis*
ALBIAN		

SALT RANGE

Belemnite Beds.—Only the lower portion of the Cretaceous is represented here, consisting of 3 m thick grey sandstone supposed to be glauconitic, overlain by white sandstones. At Kalabagh, just beyond the Indus, they contain *Belemnites* of Neocomian age. They continue on to Chichali, Makarwal and Sheik Budin hills where the lower Belemnite-bearing beds are shaly and associated with grey marls overlain by white and yellow sandstones. The Belemnite beds contain a fairly rich fauna belonging to the Valanginian age, amongst which are the following :

Olcostephanus salinarius(abundant), *O. sublaevis, O. fascigerus, O.* (*Rogersites*) *schenki, Blanfordiceras* aff. *wallichi, B.* cf. *boehmi, Subthurmannia media, S. fermori, S. patella, Himalayaites* cf. *seideli, Neocomites similis, Parandiceras* sp., *Kilianella asiatica, K. besairiei, Neocosmoceras hoplophorum, Sarasinella uhligi, Neohaploceras submartini, Neolissoceras grasianum, Hibolites subfusiformis* (abundant) :

Pleurotomaria blancheti, Astarte herzogi, Exogyra imbricata.

Also fish and reptilian remains.

It was believed that there are no Mesozoic beds in this region younger than the Albian (Gault). S.R.N. Rao has recorded the foraminifer *Globotruncana rosetta* of Maestrichtian age from limestone band in the Nammal gorge. They are succeeded by ferruginous marls having a thickness of about 2 m., and these by Eocene limestone. The marls may be Cretaceous or Eocene.

PENINSULAR AREAS

BOMBAY

Ahmednagar (Himmatnagar) Sandstone.—The Ahmednagar Sandstones, discovered by C. S. Middlemiss in the Idar State, consist of thick, horizontally bedded sandstones, shales and conglomerates, having pink, red and brown colours. Amongst the plant fossils found in these, *Matonidium indicum* and *Weichselia reticulata* have been identified, which point to a Lower Cretaceous (Wealden) age. These sandstones are therefore older than the Bagh beds and may be contemporaneous with the Nimar Sandstones underlying the Bagh beds.

The Ahmednagar Sandstone is similar to the Dhrangadhra Sandstone (*see* Upper Gondwanas, p. 253) the Songir Sandstones of Baroda, the Barmer Sandstone of Western Rajputana and the Nimar Sandstone of the Narmada Valley, all of which appear to be of about the same age. These indicate a regressive estuarine facies. The Cretaceous transgression is seen in the Bagh Beds, Wadhwan Sandstones, Ukra Beds of Kutch and the Abur Beds of Rajasthan.

KUTCH

The Jurassic rocks of Kutch pass upwards into the Umia plant beds which overlain by sandstones containing marine fossils like *Colombiceras* and *Cheloniceras.* The age of these beds is Aptian. They are followed by the Deccan Traps.

NARMADA VALLEY

Bagh Beds.—Strata of Cretaceous age are observed in a series of outcrops between Barwah east of Bagh in Gwalior and Wadhwan in Kathiawar, the best exposures being found around Chirakhan and in the Alirajpur-Jhabua area in Gujarat. They rest on the ancient metamorphic rocks and attain a thickness of 20 m or more in Gwalior but are believed to be much thicker in Rajpipla. The lower part of the beds is arenaceous while the upper is mainly calcareous.

Near Bagh, the lower beds consist of a basal conglomerate with sandstone and shale layers above. These are called the Nimar Sandstones. The upper portion consists of three divisions, a lower nodular limestone, a middle marly bed (Deola marls) and an upper coralline limestone.

		Deccan Traps
Bagh beds	Upper ...	Coralline limestone Deola Marls Nodular argillaceous limestone
	Lower ...	Nimar Sandstones
		Metamorphics

The Nimar Sandstones thicken when followed westwards. They yield good and durable building stones, quarries for the extraction of which exist near Songir in Baroda. Thick veins and lenses of fluorite have been formed in them as a result of the action of volcanic emanations at Amba Donger in Gujarat.

The Nodular Limestone is compact, argillaceous and light coloured. The Deola (or Chirakhan) marl is only 3 m thick but richly fossiliferous. The Coralline Limestone is red to yellow in colour and contains abundant fragments of Bryozoa. The Deola marls are the chief fossiliferous beds, but closely related fossils occur also in the limestones above and below them :

Brachiopods	... ...	*Malwirhynchia* (several species)
Cephalopods	... ...	*Knemiceras mintoi, Namadoceras scindiae, N. bosei.*
Lamellibranchs	... ...	*Crassinella trigonoides, C. planissima, Neithea morrisi, Ostrea arcotensis, O. leymeriei, Plicatula multicostata, Protocardium pondicherriense, Cardium (Trachycardium) incomptum, Macrocallista* cf. *sculpturata, Grotriana* cf. *jugosa, Inoceramus concentricus, I. multiplicatus, I. lamarcki.*
Gastropods	... ...	*Lyria granulosa, Fulguraria elongata, Fasciolaria rigida, Turritella multistriata.*
Echinoids	... ...	*Cidaris namadica, Salenia keatingei, Cyphosoma namadicum, Orthopsis indica, Echinobrissus haydeni, E. malwaensis, Hemiaster cenomanensis, H. chirakhanensis.*
Fishes	... ...	*Lamna, Otodus, Oxyrhina, Prionodon, Odontospis*

This fauna is referable mainly to the Turonian, perhaps extending from Cenomanian to Senonian, according to G.W. Chiplonkar.

In his examination of the Bagh beds of the Dhar Forest area, Vredenburg came to the conclusion that they were contemporaneous with, and represented the marine facies of, the Lametas. The fossil evidence points to the age of these beds as Cenomanian to Upper Senonian.

The fauna of the Bagh beds has a close resemblance to that of the Cretaceous of Arabia and Southern Europe. There are, some elements which show affinities with the South Indian Cretaceous — *Protocardium pondicherriense, Trachyacardium incomptum, Macrocallista* cf. *sculpturata, Turritella* (*Zaria*) *multistriata.* and the Uttattur forms *Grotriana* cf. *jugosa* and *Crassinella* cf. *planissima,* It would appear that the two areas became connected after the Cenomanian, by which time India moved off from Madagascar, leaving an open sea-way by Cape Comorin.

SOUTHERN MADRAS

The great Cenomanian transgression covered a large area of the Coromandel coast from near Pondicherry to south of the Cauvery valley. The Cretaceous rocks are found in three areas separated by alluvium of the Pennar and Vellar rivers. There is, besides, a small patch to the south of the Cauvery, a few km west of Tanjore. The largest of these is in the northeast of the Trichinopoly district and occupies about 650 sq. km

Trichinopoly (Tiruchirapalli) district

Here the rocks lie on a platform of granitic gneisses and charnockites, a thin fringe of Upper Gondwanas intervening between the sediments and the Archaeans in a few places along the western margin. On the north they are covered by alluvium and on the east by the Cuddalore sandstones of Mio-Pliocene age. The Cretaceous rocks are divided into four stages as shown in Table 59. The oldest stage is found on the west and is successively overlain, and in places overlapped, by the younger ones which occur to the east. The general dip of the formations is to the east or E.S.E. at low angles. Slight unconformities exist between the three lower stages, though these are not prominent.

Uttattur Stage.—The lowest beds are named after the village of Uttattur (about 32 km N.E. of Trichinopoly) which lies on the western margin of the Cretaceous strata. This stage lies on charnockites for the greater part, and on gneiss and granite at its northern and southern ends. Near its junction with the Cretaceous rocks, the charnockite is heavily weathered and replaced by or veined with tufaceous limestone (kankary matter). Upper Gondwana shales

MAP XIV

THE CRETACEOUS ROCKS OF TRICHINOPOLY

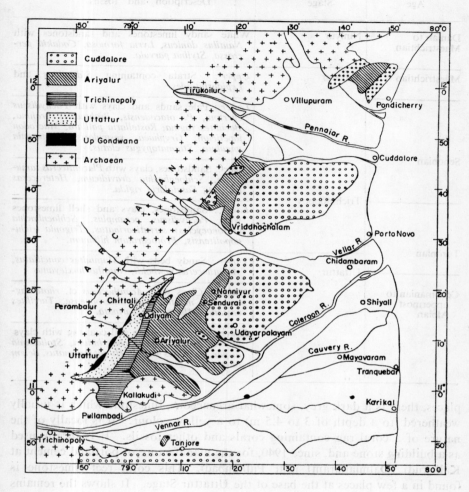

and sandstones intervene between this stage and the charnockites in a few places, forming four or five separate patches. The junction between the Gondwanas and basal Cretaceous rocks does not show any disconformity but a thin ferruginous bed may be seen intervening between the two. The outcrop of this stage is 6 to 8 km wide near the southern end but narrows down to less than 2 km in the north, being overlapped at both ends by the next higher stage.

The Uttattur beds consist of fine silts, calcareous shales, and sandy clays containing ferruginous, phosphatic and calcareous nodules. The clays are often streaked with yellow and brown ferruginous stains. At the base, in a few

TABLE 59—CRETACEOUS SUCCESSION IN TRICHINOPOLY

Age	Stage	Description and fossils
Danian to Maestrichtian	Niniyur	White sandy limestones and sandstones with *Nautilus danicus, Lyria formosa, Codakia percrassa, Stylina parvula*.
Maestrichtian	Ariyalur	*Upper*: Strata containing *Siderolites* and *Lepidorbitoides*.
Senonian		*Lower*: Pale sands and clays with *Pachydiscus egertoni, P. otacodensis, Brahmaites brahma, Baculites vagina, Rostellaria palliata, Macrodon japeticum, Gryphaea vesicularis, Aletryonia ungulata, Stigmatopygus elatus*.
	Trichinopooy	*Upper*: Sandstones, clays with *Placenticeras tamulicum, Schloenbachia dravidicum, Heteroceras indicum, Fasciolaria rigida*.
		Lower: Sandstones, clays and shell limestones with *Pachydiscus peramplus, Schloenbachia (Prionocyclus) serraticarinatus, Trigonia trichinopolitensis, Protocardium hillanum*.
Turonian	Uttattur	*Upper*: Sandy beds with *Mammites conciliatus, Acanthoceras newboldi, Nautilus huxleyanus*
Cenomanian to uppermost Albian		*Middle*: Clays with *Acanthoceras* cf. *rhotomagense, A. mantelli, A. coleroonense, Turrilites costatus, Alectryonia carinata*.
		Lower: Basal limestone and coal rag with clays above, with *Schloenbachia inflata, Stoliczkaia dispar, Turrilites bergeri, Hamites armatus, Belemnites*.

places, there is a dark grey, somewhat arenaceous, limestone which is usually weathered (to a depth of 3 to 4.5 m) to a yellow colour. It is locally of the nature of a coral rag containing corals and other fossils. It is locally used as a building stone and, since 1940, for the manufacture of portland cement at Kallakudi (Dalmiapuram) near Pullambadi. This coral reef limestone is found in a few places at the base of the Uttattur Stage. It shows the remains of corals, several foraminifera (*Nodosaria, Textularia, Rotalia,* etc.) and algae (*Solenopora, Marinella, Pseudolithothamnium* etc.). At Kallakkudi (Cullygoody in Blanford's Memoir) this is said to contain some pebbly limestone at the base, interpreted by Prof. S. R. Narayan Rao as possibly derived from an earlier Jurassic limestone which is said to explain the presence of characteristic Jurassic algae in them. The succeeding beds are clays of buff colour streaked with red and yellow and containing thin intercalations of sandstone. They are traversed by veins of gypsum which are up to five inches thick. In certain areas similar but sparsely distributed veins of fibrous celestite, barite and calcite also occur as also calcareous and phosphatic nodules up to six inches long. The upper beds are more arenaceous and show current-bedding. There must have been an interval before the close of the Uttattur times when the sea became desiccated, impregnating the sediments with gypsum and salt.

PLATE XIX

CRETACEOUS FOSSILS

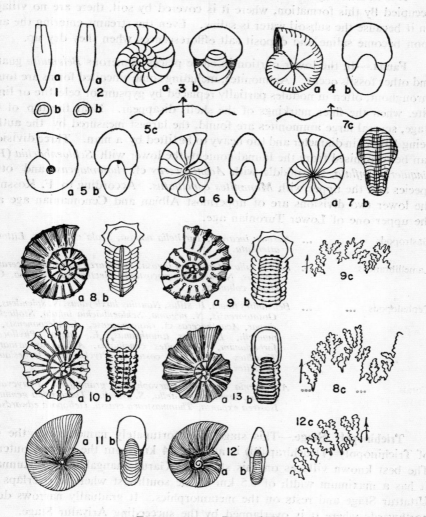

EXPLANATION OF PLATE XIX

1. *Belemnites stylus* (0.5). 2. *Belemnites fibula* (0.5). 3. *Nautilus bouchardianus* (0.25).
4. *Nautilus huxleyanus* (0.25). 5. *Nautilus danicus* (0.2). 6. *Nautilus kayeanus* (0.2).
7. *Nautilus trichinopolitensis* (0.1). 8. *Schloenbachia inflata* (0.3). 9. *Acanthoceras rhotomagense* (0.3). 10. *Acanthoceras mantelli* (0.3). 11. *Stoliczkaia disper* (0.4). 11. *Phylloceras velledae* (0.4). 12. *Phylloceras surya* (0.4).

The dips are low and irregular, from a few degrees to as much as 25 degrees towards the east or southeast, the average being about 10 degrees. The thickness of the Uttattur Stage should however be more than the 300 m allowed for by Blanford, and probably about 600 m.

It is noteworthy that though there is some cultivation in parts of the area occupied by this formation, where it is covered by soil, there are no villages on it because the subsoil water is saline. Even the streams entering the area soon become saline and deposit salt efflorescences when they dry up.

Fauna.—In the lower portion in some places numerous *Belemnite* guards and other fossils occur. Ammonites, including some uncoiled forms are found throughout, often in nodules partially replaced by gypsum or celestite or limonite, when the finer markings of the shells disappear. Near the top of the stage, several large ammonites are found, the largest measured by the author being 107 cm in diameter and too heavy to be lifted by a man. Three divisions can be recognised from the faunal content, the lower with *Schloenbachia* (*Pervinquieria*) *inflata*, the middle with *Acanthoceras* cf. *rhotomagense* and other species and the upper with *Mammites conciliatus*. According to F. Kossmat, the lower two divisions are of uppermost Albian and Cenomanian age and the upper one of Lower Turonian age.

Gastropods	*Nerinea incavata, Turritella nodosa, Scala clementina, Littorina attenuata.*
Lamellibranchs	*Lucina fallax, Grotriana jugosa, Trigonarca gamana Inoceramus labiatus, Neithea quinquecostata, Alectryonia diluviana, Gryphaea columba.*
Cephalopods	*Belemnites fibula, B. stilus, Nautilus huxleyanus, N. splendens, N. Ootatoorensis, N. negama, Schloenbachia inflata, Stoliczkaia dispar, Acanthoceras* cf. *rhotomagense, A. coleroonensis, A. mantelli, Lytoceras timotheanus, L. sacya, Phylloceras forbesianum, Mammites conciliatus, Anisoceras armatum, Turrilites bergeri, T. costatus, Ptychoceras forbesianum, Baculites vagina.*
Corals	*Astrocoenia retifera, Caryophyllia granulifera, Platycyathus indicus, Stylina multistella, S. grandis. Thecosmilia geminata, Isastrea expansa, Thamnastraea crassa, Heliopora edwardsana.*

Trichinopoly Stage.—This stage is unfortunately named since the city of Trichinopoly (Tiruchirapalli) is at least 24 km from the nearest outcrop. The best known villages on this stage are Garudamangalam and Kunnam. It has a maximum width of 6.5 km in the southwest where it overlaps the Uttattur Stage and rests on the metamorphics. It gradually narrows down northwards where it is overlapped by the succeeding Ariyalur Stage.

The Trichinopoly Stage seems to have been laid down after the Uttattur Stage was slightly uplifted and denuded. A light unconformity, sometimes marked by a bed of conglomerate or coarse pebbles, is seen at the base of the Trichinopoly Stage, but this is not always noticeable since the upper Uttattur beds are arenaceous and similar in constitution to the succeeding beds and angular unconformity is not easy to detect in the irregularly dipping strata. Some of the conglomerate is full of granite, quartz and feldspar pebbles derived from the granitic area to the south and southwest.

PLATE XX

CRETACEOUS FOSSILS

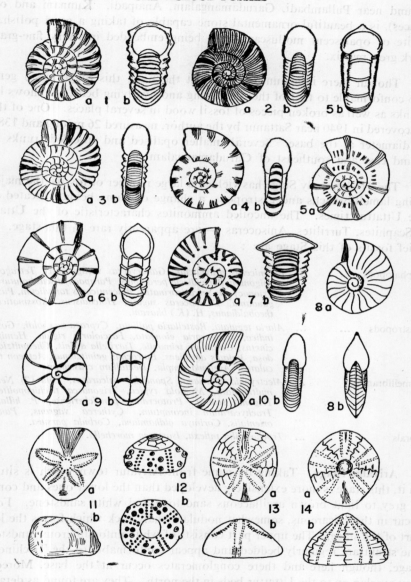

EXPLANATION OF PLATE XX

1. *Pachydiscus peramplus* (0.3). 2. *Lytoceras sakya* (0.4). 3. *Holcodiscus (Kossmaticeras) theobaldianus* (0.2). 4. *Pachydiscus gollevillensis* (0.3). 5. *Brahmaïtes brahma* (0.3).
6. *Brahmaïtes vishnu* (0.3). 7. *Olcostephanus superstes* (0.3). 8. *Oxytropidoceras royssianum* (0.3). 9. *Desmoceras latidorsatum* (0.4). 10. *Puzosia gaudama* (0.3). 11. *Hemiaster similis* (0.4). 12. *Salenia arcotensis* (0.7). 13. *Echinoconus conicus* (0.7). 14. *Stigmatopygus elatus* (0.3).

The Trichinopoly Stage is a littoral, shallow marine formation consisting of sandstones, grits, calcareous grits, occasional shales and bands of shell limestone full of gastropod and lamellibranch shells. This sehll limestone (found near Pullambadi, Garudamangalam, Anaipadi, Kunnam and other places), is a beautiful ornamental stone capable of taking a good polish, the white or opalescent molluscan shells being embedded in a very fine-grained dark grey matrix.

Though there is an unconformity at the base, this Stage has a general dip conformable to that of the underlying and overlying beds. It shows large trunks as well as broken pieces of fossil wood in several places. One of these, discovered in 1940 near Sattanur by the author, measured 26 m long and 138 cm in diameter at the base. Several smaller opalised and silicified trunks were found east and southeast of Garudamangalam.

The Trichinopoly Stage has yielded a large number of fosslis, the majority being lamellibranchs and gastropods, a change of fauna being indicated after the Uttattur times. The uncoiled ammonites characteristic of the Uttatturs —Scaphites, Turrilites, Anisoceras — are apparently rare in this stage. The chief fossils of this Stage are :

Cephalopods	...	*Peroniceras dravidicum, Gaudryceras varagurensis, Tetragonites epigonum, Pachydiscus peramplus, Parapachydiscus koluturensis Prionocyclus serraticarinatus, Placenticeras tamulicum, Puzosia gaudama, Desmoceras sugata, Holcodiscus (Kossmaticeras) theobaldianus, H. (K.) bhavani.*
Gastropods	...	*Alaria tegulata, Rostellaria palliata, Cypraea newboldi, Gosavia indica, Fulguraria elongata, Fasciolaria rigida, Hemifusus cinctus, Cerithium trimonile, Turritella affinis, Chemnitzia undosa, Velates decipiens, Eutrochus geinitzianus, Actaeon turriculatus, Avellana ampla, Dentalium crassulum.*
Lamellibranchs	...	*Alectryonia diluviana, Spondylus calcaratus, Vola (?) Neithea multicostata, Modiola typica, Trigonarca trichinopolitensis, Trigonia scabra, Protocrdium pondicheriense, P. hillanum, Trachycardium incomptum, Cytherea vagrans, Panopea orientalis, Corimya oldhamiana, Corbula parsura.*
Corals	...	*Trochosmilia inflexa, Isastraea morchella.*

Ariyalur Stage.—Taking its name from Ariyalur town which is situated on it, this stage is more extensively developed than the lower ones and consists of grey to light brown argillaceous sandstone and white sandstone. Fossils occur in the calcareous, somewhat nodular, shaly rock which forms the lower part of the stage. The upper part consists of white unfossiliferous sandstones. The strata are regularly bedded and appear conformable to the Trichinopoly Stage, though here and there conglomerates occur at the base. Moreover, they overlap on to the Uttattur beds in the north. They are found as detached outcrops in the Vridhachalam and Pondicherry areas and as small patches amidst the alluvium a few km west of Tanjore. To the east they are covered by the Cuddalore Sandstones. They have a gentle easterly or northeasterly dip (averaging 3 to 5 degrees) and attain a thickness of around 300m. Though shallow water deposits, they are not coastal deposits like the Trichinopoly Stage.

THE CRETACEOUS SYSTEM

The fauna of the Airyalur Stage resembles that of the underlying Trichinopoly Stage though many new fossils appear. There is therefore no faunal break between these two as is evident between the Uttattur and Trichinopoly Stages. The fauna includes reptiles, fishes, mollusca, echinoderms, brachiopods, corals and bryozoa, amongst which gastropods and lamellibranchs are the most numerous. Reptilian remains were found in the Upper beds east of Otakovil. The more important fossils are:

Reptiles	Megalosaurian and Titanosaurian bones.
Fishes	Otodus cf. semiplicatus, Oxyrhina sp., Ptychodus latissimus.
Cephalopods	Nautilus bouchardianus, N. clementinus, N. valudayurensis, N. trichinopolitensis : Schoenbachia blanfordiana, Kossmaticeras bhavani, K. pacificum, K. aemilianum, Baculites vagina, Sphenodiscus siva, Desmoceras sugata, Brahmaites brahma, Parapachydiscus otacodensis, P. egertonianus, P. grossouvrei, Pachydiscus crishna, Hauericeras gardeni, H. rembda, Puzosia varuna.
Lamellibranchs	Alectryonia ungulata, Pholadomya caudata, Cyprina cristata, Trigonia orientalis, Trigonarca galdrina, Macrodon japeticum, Yoldia striatula, Radiolites mutabilis, Inoceramus balticus.
Gastropods	Pugnellus uncatus, Rostellaria palilata, Alaria tegulata, Cypraea kayei, Fulguraria elongata, Volutolithes septemcostata, Neptunea rhomboidalis, Cancellaria breviplicata, Cerithium arcotense, C. karasurense, Turritella pondicheeriensis, T. dispassa, T. (Zaria) multistriata, Euspira pagoda, Gyrodes pansus, Helcion corrugatum, Actaeon pugilis.
Echinoderms	Stygmatopygus elatus, Hemiaster cristatus, Epiaster nobilis, Cidaris sceptrifera.
Corals	Cyclolites filamentosa, C. conoidea, Stylina parvula

It may be added that the uppermost beds of Ariyalur Stage have yielded *Siderolites calcitropoides* and *Lepidorbitoides* which indicate Maestrichtian age. This stage is found east of Ariyalur and extends into Sendurai, Vridhachalam and Pondicherry areas.

Niniyur Stage.—The fossil-bearing beds overlying the Ariyalur Stage are regarded as a separate stage. They are well exposed around Niniyur (Nanniyur) and Sendurai, northeast of Ariyalur. The rocks are grey, brown, ochreous and calcareous sands and shales, in which there are fragments of flint and chert containing algae. There are no ammonites in this stage, but the presence of *Nautilus (Hercoglossa) danicus*, and *Orbitoides minor* fixes its age as extending from Maestrichtian to Danian. Alternatively, the beds containing *Orbitoides* can be relegated to the Maestrichtian while those with *Nautilus danicus* would be Danian. The fossils found in the Niniyur stage include :

Cephalopods	Nautilus (Hercoglossa) danicus N. (H.) tamulicus.
Lamellibranchs	Tellina arcotensis, Lucina (Codakia) percrassa, Cardita jaquinoti.
Gastropods	Pseudoliva percrassa, Euspira hirata, Solarium arcotense, Lyria formosa, Turritella elicita.
Corals	Caryophyllia arcotensis, Stylina parvula, Thamnastraea brevipes.
Protozoa	Orbitoides minor, O. fauiasi.
Algae	Dissocladella savitriae, Indopolia satyavanti, Acicularia dyumatsenae, Parachaetetes asvapattii, Archaeolithothamnium sp., Orioporella malaviae.

Prof. L. Rama Rao and collaborators have in recent years studied the fossil content of the Niniyur beds. The nodular limestone at Athankurichi (Authencoorchy) contains a rich algal flora in which *Parachaetetes asvapatii, Archaeolithothamnium lugeoni, Dasicladella savitriae, Acicularia dyumatsenae, Indopolia satyavanti* have been identified. The flints and cherts in the Niniyur group are also rich in algae and foraminifera. The find of *Globorotalia* in the Niniyur Beds recently fixes the age as Danian.

Vriddhachalam Area (South Arcot)

Beyond the Vellar river, near Vridhachalam, the Ariyalur beds emerge out of the alluvium and are overlapped on the east by the Cuddalore Sandstone. They are exposed in a strip 24 km long (N.N.E.—S.S.W.) and 5 to 8 km wide between the two rivers Manimukta and Gadilam. The strata are of the same nature as around Ariyalur and have yielded *Trigonia semiculta, Neithea quinquescostata, Pecten verdachellensis,* etc.

Valudavur — Pondicherry Area

Farther north, around Valudavur (16 km W. by N. of Pondicherry) and in the erstwhile French territory of Pondicherry, the upper stages of the Cretaceous are exposed in a strip 15 km long (N.N.E.—S.S.W.) and 7 km broad. A large part of the area is under cultivation or under water (ponds or tanks). Dr. H. Warth, who investigated this area in 1895, distinguished six lithological horizons designated by letters A to F, but the investigation of the fossils by Dr. F. Kossmat proved that only three faunal horizons could be distinguished, viz., the Anisoceras, Trigonarca and Nerinea Beds, from below upwards.

The area was restudied by N. Rajagopalan (1965) leading to a new classification on the basis of foraminiferal fauna. It is also recognised that the Cretaceous strata continue without interruption into the Lower Eocene (Ypresian) The Cretaceous and Lower Eocene sections in these beds are 190 m and 60 m thick respectively. Table 60 shows the division of the strata based on the new data.

The Valudavur Formation is Campanian to Maestrichtian. It has yielded a rich fauna described by Kossmat, amongst which are :— *Anisoceras* (common), *Baculites, Gaudryceras kayei, Pseudophyllites indra, Sphenodiscus siva, Parapachydiscus egertoni, Brahmaites brahma* ; *Pholadomya, Rostellaria, Athleta, Turritella, Trochus* etc. The Trigonarca Beds are now designated Mettuveli Formation. They contain *Cyclolites filamentosa* (abundant), *Nautilus sublaevigatus, Brahmaites, Baculites* and numerous gastropods. The top of this formation marks the top of the Gretaceous and there is a distinct faunal break here. The beds above this, which include the former Nerinea Beds, are now called the Pondicherry Formation. They are divided into four units extending in age from Danian to the Ypresian. *Globigerina* occurs in the Danian and several species of *Globorotalia* occurring in this and the Upper beds serve as zone fossils. The Algal Limestone is a hard greyish limestone encrusted with algal remains, while the Discocyclinid limestone is compact and brown, yielding *Nummulites* and *Discocyclina*. The two Marlstone horizons are sandy and comparatively soft.

THE CRETACEOUS SYSTEM

TABLE 60—CRETACEOUS SUCCESSION IN THE PONDICHERRY AREA

Kossmat (1897)	Rajagopalan (1965)		Standard
	Units	Foram. Zones	Stages
Nerinea Beds	PONDICHERRY FORMATION — Marlstone	Globorotalia Pseudoscitula / G. formosa formosa	Ypresian
	Discocyclinid Limestone	G. rex / G. Velascoensis	Sparnacian
	Algal Limestone	G. Whitei	
	Lower Marlstone	G. uncinata / G. trinidadensis	Danian
Trigonarca Beds	METTUVELI FORMATION	Globotrucana gansseri	Maestrichtian to Upper Campanian
Anisoceras (Valudavur) Beds	VALUDAVUR FORMATION	Globotruncana tricarinata	

Some limestones from the Nerinea beds in this area have yielded *Discocyclina* and *Nummulites* (*Ranikothalia*) which appear to have been mistaken for *Orbitoides* and *Amphistegina* by Kossmat (L. Rama Rao, *Curr. Sci.* 7, No. 5) and which prove that the strata are of Paleocene age. The presence of the Middle Eocene in the Pondicherry area has also been pointed out by Furon and Lemoine (*Compt. Rend. Ac. Sci.* 207, 1424-1426, 1939) who discovered undoubted *Nummulites* in bore-hole cores obtained from a depth of about 150 metres. It is obvious therefore that the Danian beds of the Pondicherry area continue uninterruptedly into the Eocene.

Rajamahendri (Godavari District)

In the vicinity of Rajamahendri (Rajahmundry) at the head of the Godavari delta, the Upper Gondwana rocks are seen to be superposed by estuarine sandstones and limestones of a total thickness of 15 m and these by the Deccan Traps. The limestones, which form the upper portion, contain bivalves and gastropods among which *Turritella* is very common. The fauna, though not examined in detail, is thought to be related to that of the Ariyalur stage. From these beds H. C. Das Gupta obtained *Nautilus danicus*, wihch indicates Danian age. The fossil algae contained in them include *Neomeris*, *Acicularia* and some *Charophyta* which, according to S. R. Narayana Rao and K. Sripada Rao indicate early Eocene age. Some early Eocene ostracods have also been found here.

ASSAM

Marine Cretaceous rocks occur in the Garo, Khasi and Jaintia Hills, the beds being dominantly arenaceous, with occasional shales and carbonaceous

PLATE XXI
CRETACEOUS FOSSILS

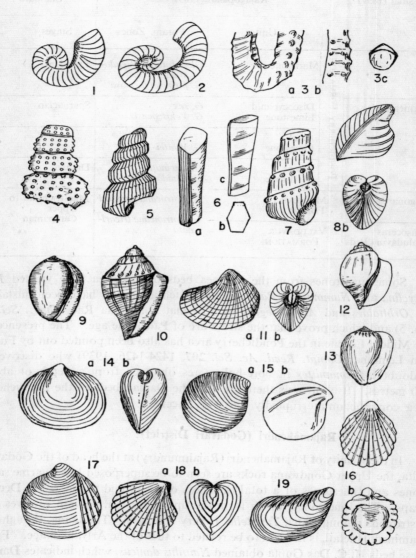

EXPLANATION OF PLATE XXI

1. *Scaphites obliquus* (0.7). 2. *Scaphites kingianus* (0.7). 3. *Anisoceras armatum* (0.4). 4. *Turrilites* (*Heteroceras*) *bergeri* (0.4). 5. *Turrilites* (*Heteroceras*) *indicus* (0.3). 6. *Baculites vagina* (0.3). 7. *Cerithium karasurense* (0.5). 8. *Cyprina cristata* (0.7). 9. *Cypraea kayei* (0.7). 10. *Gosavia indica* (0.7). 11. *Pholadomya candida* (0.7). 12. *Pseudoliva elegans*. 13. *Ovula expansa* (0.5). 14. *Lucina* (*codakia*) *percrassa* (0.4). 15. *Grotriana jugosa* (0.7). 16. *Trachycardium incomptum* (0.5). 17. *Protocardium hillanum* (0.4). 18. *Cardita jaquinoti* (0.7). 19. *Incoceramus simplex* (0.4).

material. In the Garo Hills they rest on a gneissic platform and are composed of sandstones with some coaly layers. In the Khasi and Jaintia Hills the basal beds are conglomeratic, reaching a thickness of up to 30 m. The conglomerates are succeeded by the Mahadek and Langpur Stages. The Mahadek Stage consists of hard, gritty coarse glauconitic sandstones with a fossiliferous horizon near the top. The maximum thickness is 230 m. The succeeding Langpur Stage, which is up to 90 m thick, consists of impure limestones, calcareous shales and sandstones. Some impure limestones found in Manipur are probably of this age. The best exposed section near Therria Ghat is shown below :

		Feet
Langpar Stage (305 ft.)	Sandstones and sandy shales with thin sandy limestone	100
	Yellow-brown impure limestone with bands of sandy shales	135
	Shales with thin limestone and argillaceous sandstone bands	70
Mahadek Stage (770 ft.)	Sandy shales and sandstones	40
	Shales mainly (not exposde)	40
	Greenish sandstones	40
	Hard gritty massive sandstones with a thin fossiliferous sandstone band near top	650

The Cretaceous beds are 180 to 300 m thick near the edge of the Assam plateau, having gentle dips on the plateau but plunging steeply down the southern flanks under the alluvial valley to the south. There are several isolated but small outcrops on the plateau ; they are interesting as indicating the large original spread of the Cretaceous rocks on the plateau region.

Fossils have been obtained from the hard galuconitic sandstone band near Mahadek and from a locality about 2 km northeast of Therria Ghat on the Cherrapunji road. They include the following :

Foraminifera ...	... *Orbitoides* sp.
Echinoderms ...	... *Echinoconus douvillei, Stigmatopygus elatus, Hemiaster.*
Brachiopods ...	... *Rhynchonella depressa ,Terebratula carnea.*
Lamellibranchs ...	... *Neithea faujasi, Vola propinqua, Chlamys* cf. *dujardini, Ostrea (Alectryonia) arcotensis, O. (Alectryonia) angulata, Gryphaea vesicularis, Lima ovata, Inceramus* sp.
Gastropods ...	... *Helcion corrugatum, Nerita divaricata, Natica (Lunatia) larteti, Turritella pondicherriensis, T. multistriata, Gyrodes pansus, Campanile turritelloides, Aporrhais tegulata,Rostellaria palliata, Cypraea* cf. *globulina, Lyria crassicostata.*
Cephalopods ...	... *Nautilus baluchistanensis, Baculites vagina, Puzosia planulatus, Stoliczkaia dispar, Anisoceras indicum.*

The Assam fauna is very closely related to that of the Trichinopoly Cretaceous, especially the Ariyalur Stage. According to E. Spengler, it is of Senonian age. Very similar fauna is found in Madagascar and in Natal in South Africa and all apparently lived in one zoo-geographical province — the Indo-Pacific. As already mentioned, these faunas have only a small degree of resemblance to the Bagh fauna which latter is related to the Baluchistan-Arabian (Mediterranean) Cretaceous fauna. A fairly effective barrier seems to have existed, which prevented the free intermingling of the two faunal groups, though there was connection by a circuitous route.

Recent work by the Assam Oil Co. has shown the presence of *Gumbelina, Globotruncana, Orbitoides, Siderolites* in calcareous bands near the top of

PLATE XXII
CRETACEOUS FOSSILS

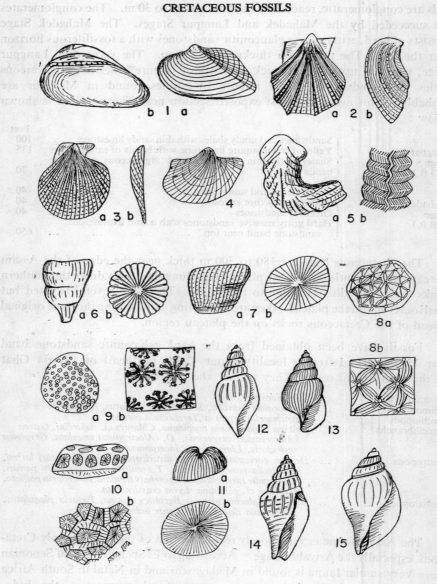

EXPLANATIONS OF PLATE XXII

1. *Trigonarca galdrina* (;0.7). 2. *Neithea quinquecostata* (0.7). 3. *Pecten verdachellensis* (0.4). 4. *Macrodon japeticum* (0.7). 5. *Alectryonia pectinata* (0.5). 6. *Caryophyllia arcotensis*. 7. *Trochosmilia tuba*. 8. *Stylina parvula*. 9. *Astrocoenia decaphylla* (b-enlarged) 10. *Isastraea morchella* (b-enlarged). 11. *Cyclolites filamentosa* (2). 12. *Alaria tegulata* (0.7). 13. *Rostellaria palliata* (0.7). 14. *Fulguraria elongata* (0.7). 15. *Athleta purpuriformis* (0 3).

Mahadek Stage, these indicating Maestrichtian age. The Langpar is Danian as it has yielded *Globigerina pseudobulloides* and *G. triloculinoides*.

The occurrence of Cretaceous rocks in other parts of Assam has not been definitely established. It is however probable that the lower part of the Disang Series is of this age. Some limestone found around Akral in Manipur may possibly belong to the Cretaceous. No Cretaceous rocks have so far been reported from the little-explored sub-Himalayan region of Assam.

BURMA

In the vicinity of Kalaw in Southern Shan States there are soft friable purple sandstones, known as the ' Red Beds, ' overlying the Loi-An series. They contain *Turrilites* and *Baculites* which indicate an Upper-Cretaceous age. The coal measures associated with these are also of Cretaceous age. In Upper Burma, Cretaceous rocks are known to exist in the Jade Mines tract, in the region of the first defile of the Irrawaddy between Sinbo and Bhamo, and in the second defile in the Bhamo district. The rocks are mostly limestones and calcareous shales containing *Orbitolina* and other foraminifera and molluscan shells. Strata of Danian or pre-Danian age are known in Pakokku and Henzada districts. The serpentines in this region are slightly earlier in age than the Danian. Resting on the Axials and serpentines there are conglomerates, slaty shales and fine sandstones which are unfossiliferous and may be Cretaceo-Eocene.

ARAKAN YOMA BELT

A wide belt of Cretaceous and Eocene rocks stretches from Assam-Burma border west of the upper reaches of the Chindwin river, along the mountainous tracts of the Arakan Yoma down to Cape Negrais. Our knowledge of this belt is fragmentary, since much of the area is covered by mountains and forests difficult of access.

In the Sandoway district of the Atakan coast there are some argillaceous limestones of creamy colour near Mai-i, from which *Schloenbachia inflata*, a Cenomanian fossil, has been obtained. In the Ramri island near the coast of this region an ammonite identified as *Acanthoceras daviesi* was found, which, according to G. de P. Cotter (*Rec.* **66**, 225, 1932) belongs to the *Acanthoceras coleroonensis* group which is characteristic of the Uttattur stage of Trichinopoly Cretaceous.

The foothills of the Arakan Yoma in the Thayetmyo district have yielded *Cardita beaumonti, Orbitoides* and some lamellibranchs. The rocks of the Yomas in Arakan are known to contain *Placenticeras* and *Mortoniceras*.

The NEGRAIS SERIES, consisting of grey sandstones, shales and limestones, occurring on the flanks of the Yoma north of the Cape Negrais, and extending to the Prome District, is considered to be of Cretaceous age. It is intruded by peridotites and serpentines.

PLATE XXIII
CRETACEOUS FOSSILS

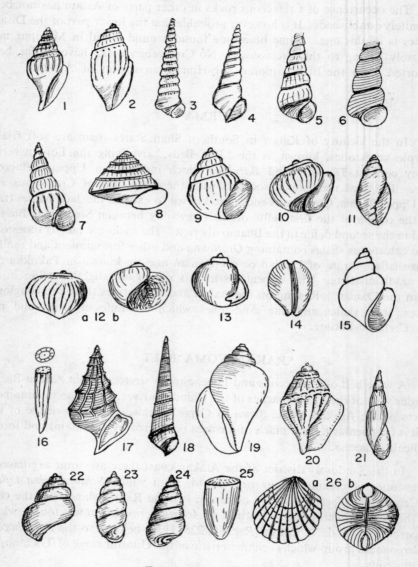

EXPLANATION OF PLATE XXIII

1. *Fasciolaria rigida* (0.4). 2. *Actaeonella cylindrica* (0.5). 3. *Nerinea incavata* (0.4). 4. *Cerithium arcotense* (0.4). 5. *Turritella dispassa* (0.4). 6. *Turritella (Zaria) multistriata* (0.7). 7. *Chemnitzia undosa* (0.4). 8. *Solarium karapaudiense*. 9. *Euspira pagoda*. 10. *Gyrodes pansus* (0.6). 11. *Actaeon pugilis* (0.7). 12. *Nerita divaricata* (0.7). 13. *Avellana ampla* (0.7). 14. *Cypraea cunliffi* (0.7). 15. *Phasianella globoides* (0.7). 16. *Dentalium crassulum* (0.7). 17. *Cerithium stoddardi* (0.7). 18. *Turritella praelonga* (0.5). 19. *Physa prinsepii* (0.5). 20. *Diploconus elegans* (0.5). 21. *Limnaea subulata*. 22. *Paludina normalis* (0.7). 23. *Vicarya fusiformis* (0.7). 24. *Morgania fusiformis* (0.7). 25. *Radiolites muschketoffi* (0.5). 26. *Cardita (Venericardia) beaumonti* (0.7).

The Upper Axial Group contains *Cardita beaumonti* and other lamellibranchs and gastropods and is therefore of Cretaceous age in part.

The Andaman Islands expose Cretaceous rocks in several places. Porcellanic limestones similar to the Parh Limestones are known to be present. In the Middle Andamans there are Chalky limestones and grey calcareous clays which enclose a rich microfauna including *Bathysiphon, Marginulina, Robulus, Nodosaria, Globiogerina, Gumbelina, Bolivina, Globotruncana* etc. which indicate Maestrichtian age. The Danian may also be present. There are also Radiolarian Cherts associated with ultramafic intrusives of late Cretaceous age in the Andaman Islands.

Cretaceous rocks are found in Western Sumatra. There is a slight break in sedimentation in the early Eocene which indicates a period of uplift during the first stage of Himalayan orogeny.

IGNEOUS ROCKS IN THE CRETACEOUS

Reference has already been made to the presence, in parts of India, of igneous rocks in close association with Cretaceous formations. They comprise acid, basic and ultrabasic types both in the intrusive and extrusive phases. They occur in the northern zones of the Himalaya, in Kashmir, Baluchistan, Burma and in the Peninsula, the last being the Deccan Trap which is dealt with in the next chapter.

Volcanic breccias and lavas of basaltic, spilitic and andesitic composition are intimately associated with the 'exotic blocks' of Johar, Bhalchdhura and their neighbourhood. They are considered to be of late Cretaceous or early Eocene age. In the Burzil-Dras region of Kashmir marine Cretaceous limestones are seen in association with basic lava flows, ash-beds and agglomerates of Lower Cretaceous age. The volcanics have a bedded character and are intercalated with the Cretaceous *Orbitolina*-bearing limestones and shales which have undergone folding. They occupy a large area in the Burzil valley and have a thickness amounting to a few hundred meters. The sediments and volcanics have been invaded by bosses and veins of hornblende-granite, gabbro and serpentine which are therefore younger than the volcanic rocks. The hornblende-granite of this region, which appears to be identical with similar granite in other parts of the Himalaya, is therefore of post-Cretaceous age. The ultramafics are of Upper Cretaceous age.

A short distance to the west of the folded ranges forming the Sind-Baluchistan frontier, is a belt of basic and ultra-basic rocks of Cretaceous and Nummulitic ages which probably formed an island arc in front of the advancing Indian shield. According to Crookshank (*Gondwana Symposium*, p. 178, International Geological Congress, Algiers, 1952) igneous activity commenced during the deposition of Belemnite Shales of early Cretaceous age. The earliest manifestations were violent eruptions producing agglomerates which are intercalated with red and green shales. These are followed by pillow lavas with shale intercalations in the Parh Limestone (Barremian to Aptian).

Later basic and ultra-basic rocks were also intruded into these. Then followed more pillow lavas associated with the Nummulitics. As the agglomerates contain rolled pebbles, they are thought to have been derived from volcanic islands.

This igneous belt can be traced northward from Cape Monze, first proceeding N.N.W. to Wad, and thence slightly east of north to near Quetta and then through Hindu Bagh to Fort Sandeman and Wana to west of Bannu. This zone lies along the border of the Calcareous and Flysch zones of Baluchistan and contains the chromite deposits of the Quetta-Zhob region. The intrusive phase may be correlated with the earliest phase of Himalayan movement. All the rocks of this zone have been involved in the subsequent folding and thrusting.

Late Cretaceous or early Eocene igneous activity is known in the Arakan Andaman belt. The rocks are intrusive into the Axials and comprise peridotites and serpentines which contain chromite in places. Gabbros, serpentines and enstatite-peridotites cover large areas in the Andaman and Nicobar islands where they are associated with radiolarian cherts. This zone continues into the Nias-Mentawei ridge and the southern border of Sumatra and Java which lie on the same tectonic belt. At the northern end of Burma, in the Kamaing sub-division of Myitkyina district, there occur numerous outcrops of serpentinised peridotites, dunites, pyroxenites and amphibolites amidst crystalline schists. They contain masses of chromite and jadeite, the latter being commercially exploited. The deposits of jadeite occur at Tawmaw, Meinmaw, Pangmaw and other places.

EARTH MOVEMENTS IN THE CRETACEOUS

The Cretaceous period witnessed the initiation of the important physiographic changes which became accentuated during the Tertiary. The southern arm of the Tethys *i.e.*, the portion which occupied the Baluchistan arc continuing through the Mekran into the Oman region of Arabia, experienced the earliest phase of mountain building. From the work of G.M. Lees and others in Arabia, the Oman mountains are known to have been formed during Pre Gosau times (Upper Cretaceous). It is also known that the igneous belt of eastern Baluchistan occurring along the border of the Calcareous zone and the Khojak flysch zone was formed during the Cretaceous as basic lavas are interbedded with Lower as well as Upper Cretaceous rocks while ultra-basic plutonic rocks are intrusive into the Upper Cretaceous in this region. In the Himalayan region the Lower Flysch (Giumal Sandstones) indicats a considerable shallowing of the Tethys in Lower Cretaceous times. There is a gap between the Lower Flysch and the Upper Flysch, which probably implies a period of disturbance. The Upper Flysch in northern Kumaon contains much radiolarian chert indicating the local deepening of the sea due to compression of the strata. The Upper Flysch was intruded by ultra-basic and basic rocks part of which are contemporaneous with them and part somewhat later. In the Burmese arc, there were earth movements of Upper Cretaceous or Laramie age which brought into being a central ridge intruded by peridotites and serpentines. This ridge separated the Burmese region into a Burma Gulf and Assam Gulf, the deposits in the two areas showing noticeable differences from the early Tertiary onwards.

On the other hand, the Indian ocean region shows evidences of marked marine transgression in the Albian and Cenomanian. Upper Cretaceous strata, containing practically identical fauna are found on the Madras coast, on the southern part of the Assam plateau, in West Australia and in East Africa. There seems to have been a barrier between the Narmada valley and the main area of the Indian ocean, as evidenced by the differences in the fauna of the Bagh beds and the Trichinopoly and Assam Cretaceous, and this barrier was Madagascar which was attached to Southwestern India until the Middle Cretaceous. Towards the end of the Cretaceous, however, there seems to have been a free intermingling of the fauna of the Mediterranean and the Indian Oceans.

In the Upper Cretaceous, while there were compressive movements throughout the whole of the Tethyan region, tension fractures developed in the Peninsula of India resulting in the outpouring of the Deccan Traps. In other parts of Gondwana-land, however, major fracturing seems to have already taken place in the Jurassic with the eruption of the great floods of Stormberg lavas and their equivalents in South Africa, South America and Antarctica.

SELECTED BIBLIOGRAPHY

Bailey, E. B. Mountains that travelled over volcanoes. *Nature* 154, 752, 1944.

Blanford, H.F. Cretaceous and other rocks of the S. Arcot and Trichinopoly districts. *Mem.* 4, 1862.

Blanford, W.T. Geology of the Taptee and Lower Nerbudda valleys. *Mem.* 6, Pt. 3, 1869 (Bagh beds, p. 45-57).

Chiplonkar, G.W. Fauna of the Bagh Beds. *Proc. Ind. Ac. Sci.* 7B, 300-316 ; 9B, 235-246; 10B, 98-109, 255-274 ; 14B, 271-276.

Clegg, E.L.G. The Cretaceous and associated rocks of Burma. *Mem.* 74, Pt. I, 1941.

Davies, L.M. *et al.* Fossil fauna of the Samana Range. *Pal. Ind. N.S.* XV, (1-5), 1930.

Douville, H. Les couches a *Cardita beaumonti*. *Pal Ind. N.S.X.*, (3), 1928-29.

Forbes, E. Report on the fossil invertebrates from South India collected by Kaye and Cunliffe. *Trans. Geol. Soc.* (London) 2nd Ser. Vol. VII, Art. 5, 97-174, 1845.

Gigon, W.D. Upper Cretaceous stratigraphy of well Giandari-I and its correlation with Sulaiman and Kirthar Ranges. ECAFE *Min. Res. Dev. Ser.* 18(1), 282-284, 1963.

Griesbach, C.L. On the exotic blocks of the Himalayas. *C.R.* IX, *Int. Geol. Congr.* Vienna, 1904.

Gupta, B.C. and Mukerjee, P.N. Geology of Gujarat and Southern Rajputana. *Rec.* 73, 164—205, 1938.

Heyden, H.H. Geology of the Provinces of Tsang and U in Tibet. *Mem.* 36, Pt. 2, 1907.

Heim, A. and Gansser, A. Central Himalaya. *Denkschr. Schweiz, Naturf. Gesell.*, 73(1), 1939.

Heune, F. Von, and Matley, C.A. The Cretaceous Saurischia and Ornithischia of the Central Provinces, *Pal. Ind.* N.S. XXI, (I), 1933.

Ghosh, A.M.N. The stratigraphical position of the Cherra Sandstone. *Rec.* 75, Paper 4, 1941.

Koken, E. Kreide und Jura in der Salt Range. *Zbl.* Min. Geol. IV, 437, 1903.

Kossmat, F. Importance of the Cretaceous rocks of S. India in estimating the geographical conditions during later Cretaceous times. *Rec.* 28, 39-54, 1895.

Kossmat, F.   Cretaceous deposits of Pondicherry.   *Rec.* **30**, 51-110, 1897.
Kossmat, F.   Die Beduetung der Sud-Indische Kreide-Formation.   *J.B.K.K. Geol. Reichanstalt*, (Wien) **44**,(3), 459, 478, 1894.
Kossmat, F.   Untersuchungen uber die sud-Indische Kriede-Formation.   *Beitr. z. Pal. Geol. Oster.—Ungarns*, IX, 97-203, 1895 ; XI, 1-46, 89-152, 1898.
Kraft, A. Von.   Exotic blocks of Malla Johar in the Bhot Mahals of Kumaon.   *Mem.* **32**, Part 3.

Matley, C.A.   The Cretaceous Dinosaurs of the Trichinopoly district.   *Rec.* **61**, 337-349, 1929.
Matley, C.A.   Stratigraphy, fossils and geological relationship of the Lameta beds of Jubbalpore.   *Rec.* **53**, 142-164, 1921

Nagappa, Y.   Foraminiferal biostratigraphy of the Cretaceous—Eocene succession in the India-Pakistan-Burma region.   *Micropalaeontology*, **5** (2), 145-192, 1959.
Noetling, F.   Fauna of Upper Cretaceous beds of Mari hills.   *Pal. Ind. Ser.* XVI, Vol. I, (2), 1897.

Rao, L.R. The Cretaceous of South India.   *Lucknow Univ. Studies.*   No.XVII, 1942.
Rao, L.R. and Pia J.   Fossil algae from the Niniyur group of the Trichinopoly district.   *Pal. Ind.* N.S. XXI (4), 1936.
Rajagopalan, N.   Late Cretaceous and early Tertiary stratigraphy of Pondicherry, South India. *Jour. Geol. Soc. India*, **6**, p. 108-121, 1965.
Rao, L.R. The Maestrichtian Stage in the Cretaceous of South India, In Wadia Commemorative volume; *Min. Geol. Met. Inst. India*, p. 126-137, 1965.

Spengler, E.   Contributions to the Palaeontology of Assam.   *Pal. Ind.* N.S. VIII, (I), 1923.
Spitz, A.   Lower Cretaceous fauna of the Himalayan Giumal sandstone.   *Rec.* **44**, 197-217, 1914.
Stoliczka, F.   Cretaceous fauna of S. India.   *Pal. Ind. Ser.* I, III, Vol. I ; Ser. V, Vol. II, Ser. VI, Vol. III, Ser. VIII, Vol. IV ; 1861-73.

Wadia, D. N. Cretaceous volcanics of Astor-Deossi, Kashmir.   *Rec.* **72**, 151-161, 1937.
Warth, H.   Cretaceous formations of Pondicherry, *Rec.* **28**, 15-21, 1895.

Vredenburg, E.   Geology of Sarawan, Jhalawan, Mekran, etc.   *Rec.* **38**, 189-215, 1910.

CHAPTER XV

THE DECCAN TRAPS
GENERAL

The close of the Mesozoic era was marked by the outpouring of enormous lava flows which spread over vast areas of Western, Central and Southern India. They issued through long narrow fissures or cracks in the earth's crust, from a large magma basin and are therefore called *fissure eruptions*. The lavas spread out far and wide as nearly horizontal sheets, the earliest flows filling up the irregularites of the pre-existing topography. They appear to have been erupted sub-aerially. At a few places, *e.g.*, Girnar hills, Ranpur, Dhank, Chogat-Chamardi, and in crater-like exposures in Gujarat and near Bombay, the eruptions were of the 'Central' type showing differentiated rocks of varying characters. These and certain associated flows appear to be younger than the main flows. Because of their tendency to form flat-topped plateau-like features and their dominantly basaltic composition, such lavas are called *plateau basalts*. The flows are called *traps* because of the step-like or terraced appearance of their outcrops, the term being of Scandinavian origin.

DISTRIBUTION AND EXTENT

The area now occupied by the Deccan Traps is about 200,000 square miles, including Bombay, Kathiawar, Kutch, Madhya Pradesh, Central India and parts of the Deccan. They are found as far as Belgaum in the south, Rajahmundry* in the southeast, Amarkantak, Sirguja and Jashpur in the east and Kutch in the northwest. Rocks of the same age and characters found in the hills of Sind are also considered to belong to them but this is subject to doubt. The present distribution shows that the traps may have occupied some of the areas intervening between the main mass and the outlying patches, and that the original extent may well have been over 1.5 million sq. km including the segment of unknown extent which has foundered in the Arabian sea to the west of Bombay. The Deccan Traps are thus the most extensive geological formation of Peninsular India at present, with the exception of the metamorphic and igneous complex of Archaean age.

The Traps have been divided into three groups,--Upper, Middle and Lower, with the Infra-trappean beds or Lametas at their base.

Upper traps (450 m.) thick	... ...	Bombay and Kathiawar; with numerous inter-trappean beds and layers of volcanic ash.
Middle traps (1,200 m.)	... ...	Central India and Malwa; with numerous ash-beds in the upper portion and practically devoid of intertrappeans.
Lower traps (150 m)	... ...	Central Provinces and eastern areas; with inter-trappen beds, but rare ash-beds.

---

* Rajamahendri, a contraction of Rajamahendravaram, is the correct spelling, which has however been mutilated to the present Rajahmundry. The former spelling will be found in Oldham's Manual.

STRUCTURAL FEATURES

The Trap country is characterised by flat-topped hills and step-like terraces. This topography is a result of the variation in hardness of the different flows and of parts of the flows, the hard portions forming the tops of the terraces and plateaux. In the amygdular flows the top is usually highly vesicular, the middle fairly compact and the bottom showing cylindrical pipes filled with secondary minerals; while in the ordinary flows the top is fine-grained and the lower portion coarser with sometimes a concentration of basic minerals like pyroxene and olivine. Vesicular and nonvesicular flows may alternate with each other, or the flows may be separated by thin beds of volcanic ash or scoriae and by lacustrine sediments known as Inter-trappean beds.

Ash beds are particularly well seen in the upper part of the traps, for instance around Bombay, Poona and in the Western Ghats. They usually reveal a brecciated structure, fragments of the trap being found in a matrix of dusty or fine-grained material. Columnar jointing in the traps may be seen in a few places, for instance in the Salsette Island (Bombay) near Hoshangabad and in some places in Malwa. Fine columnar basalt 40 m high, is exposed at Andheri in Bombay city, but it is disappearing as a result of quarrying operations.

The traps attain their maximum thickness near the Bombay coast where they are estimated to be well over 2,000 m thick. They are very much less thick farther east; at Amarkantak and in Sirguja they are about 150 m thick, while near Belgaum south of Bombay they are only about 60 m. The bauxite and laterite occurrence of Lohardaga in the Ranchi district, Bihar, is the lateritised remnant of the Deccan Trap which must originally have extended there.

In the boring at Lakhra in Sind by the Burmah Oil Co., a basalt flow 15 m thick was encountered below the Lower Ranikot bed at a depth of 755 m. This was followed below by 45 m of sandstones and shales; 75 m of basalt; 30 m of shale of *Cardita beaumonti* horizon; 27 m of basalt; and sediments with traces of basalt for a further 75 m. A boring near Tatta (24° 45' : 67° 54') met with Deccan Traps at a depth of about 508 m and went through a thickness of 144 m, consisting of 11 flows. Nearly a sixth of this thickness was Intertrappean sediments. The Deccan Trap has not been encountered in the borings at Sukkur and Khairpur and it is probable that it did not extend so far out. (Crookshank: Gondwana Symposium. *Int. Geol. Congress*, Algiers, 1952, p. 177).

A boring at Dhandhuka in Ahmedabad district penetrated a thickness of 464 m of Trap; at Jamnagar the thickness was more than 311 m and at Khambalia more than 215 m; while in the Girnar area Fedden estimated the thickness at about 1,070 m.

The individual flows vary in thickness, from a few feet to as much as 36 m. A borehole at Bhusawal, 370 m deep, revealed 29 flows, the average thickness being 12 m. In the Chindwara district of the Madhya Pradesh,

16 flows have been identified with an average thickness of 21 m. In the Sausar tahsil of the same district, the average for 7 flows has been found to be about 17 m.

The flows have a great superficial extent in comparison with their thickness. Individual flows have been traced for distances of 100 km and more, as for example, between Chhindwara and Nagpur. This extraordinary spread is explained by Fermor (1935) as due to a high degree of super-heat in the erupted mass which is believed to have been derived from the basaltic or eclogitic shell of the earth, the heat being due probably to exothermic mineral transformations.

The lavas are generally horizontal in disposition. Near the Bombay coast as well as north of Bombay they dip towards the sea at an angle of about $10°$. In Rajpipla, dips varying from $5°$ to $20°$ are known; in Betul in Madhya Pradesh, $5°$-$10°$; and in the western part of the Narmada valley $10°$-$15°$. Gentle warping has been noted in parts of Madhya Pradesh e.g., in the Satpura region. The traps have also been faulted as in the Sausar tract in Chhindwara and at the southern foot of the Gawilgarh hill in Berar.

DYKES AND SILLS

Sills of the Deccan Trap are noted in the Satpura area of Madhya Pradesh particularly in the Upper Gondwana formations, and also in the Gondwana basin of Rewa. A few sills are also found penetrating the Jurassic strata in Kutch. They are composed of fairly coarse-grained dolerite which is occasionally porphyritic.

Dykes are very numerous in the Traps but they are not evenly distributed, some areas being devoid of them and others closely crowded with them. In Saurashtra they show a more or less radial disposition around Amreli and Jasdan and also around the plutonic mass of Girnar and Chogat-Chamardi. Dykes are very common in Eastern Saurashtra, in Gujarat, in the Narmada Valley, in the Satpura region of Madhya Pradesh and also in Western Hyderabad.

The dyke system in Gujarat has a general E.N.E.-W.S.W. direction but a subsidiary W.N.W.-E.S.E. trend is also noticed. The dyke system in Gujarat and in the Narmada Valley follows a general E.N.E.-W.S.W. trend. The dyke system in Southern Gujarat has a major N.-S. trend and a less prominent N.W.-S.E. trend. A triangular pattern is seen in the Lake Tansa-Asangaon area. As pointed out by Blanford and Auden, the major direction favoured by the dykes appears to be conditioned by the direction of folding and fracture to which the Deccan Traps have been subjected. It is known that the Narmada Valley has suffered dislocation along a E.N.E.-W.S.W. direction, continuing into Saurashtra. Gentle folds with the same axial trends are seen in the Tertiary rocks of the Surat area. The Bombay coastal region shows a monoclinal flexure (Panvel flexure) which follows a N.-S. direction upto Kalyan, to the north of which it follows a N.N.W.-S.S.E. direction; the western limb of this is bent down at a low angle.

The dykes vary considerably in their dimensions. They may be from a few to as much as 60 m thick (*e.g.*, Khokri dyke near Gondal and some dykes in the Satpura mountain region). Many of them are several hundred metres long, while a few have been traced over a length of 30 to 50 km.

PETROLOGICAL CHARACTERS

The Deccan traps belong to the type called *plateau basalt* by H.S. Washington. They are extra-ordinarily uniform in composition over much the greater part of the area and correspond to dolerite or basalt, with an average specific gravity of 2.9. The minimum value of specific gravity, in the more acidic lavas, is 2.58 while the maximum found in ultrabasic types is 3.03. They are generally dark-grey to dark-greenish grey, but brownish to purplish tints are also met with. The more acidic (trachytic) types found near Bombay have a buff to creamy colour. The non-vesicular basaltic types are hard, tough, compact and medium to fine-grained, breaking with a conchoidal fracture. The vesicular types are comparatively soft and break more easily.

Tachylite or basalt-glass is distinctly rare and may be found only as a thin selvage where the hot lava encountered a cold surface and suddenly became chilled thereby. In parts of Western India including Kathiawar and Kutch, however, the traps are associated with acid, intermediate and ultrabasic rock types derived through differentiation of the original magma.

In the Girnar and Osham hills of Junagarh (Kathiawar) there are, besides the usual dolerite and basalt, lamprophyre, limburgite, monchiquite, olivine-gabbro, porphyrite, andesite, monzonite, nepheline-syenite, granophyre, rhyolite, obsidian and pitchstone, which have been studied and described by M.S. Krishnan.

K. K. Mathur and others who mapped the Girnar hills have expressed opinion that the basaltic flows of this area were domed up by later intrusives representing the result of differentiation through progressive crystallisation. The domed up portion of the flows has been eroded away in the centre, exposing an intrusive mass of diorite-maonzonite below. This is surrounded by olivine gabbro and a mass of granophyre intrusive into basalt. The monzonite contains intrusions of nepheline-syenite associated with lamprophyre. The Chamardi-Chogat mass is roughly of the same dimensions as Girnar and shows some differentiated types. The Barda hills are made up of granophyre.

In the Gir Forest of Kathiawar, there occur some dykes of olivine-dolerite and masses of granophyre and rhyolite. The coastal region of Bombay has been shown to contain a large variety of types — rhyolite, granophyre, trachyte, andesite, ankaramite and oceanite, the acid types being considered to belong to a late phase of igneous activity. From the Pavagad hill in Gujarat, Fermor has described pumice, pitchstone, rhyolite, felsite, quartz-andesite, etc. The rhyolite which caps this hill is considered by Fermor and Heron to be a flow and not a plug-like intrusion as advocated by Mathur and Dubey. (*Rec.* **68**, p. 17, 1934). S.C. Chatterjee has studied this hill more recently (1961, 1964) and has described mugearite, Hawaiite, picritic and alkali-olivine basalt (ankaramitie) etc. The mugearite is composed of oligoclase, subordinate orthoclase and pyroxene with abundant laths of feldspar showing trachytic texture. Augite, olivine and magnetite occur as phenocrysts. The rock occurs above the Machi Rest House and on the spur of the Bhadrakali temple at an altitude above 300 m and continues towards the top. The Hawaiite type is an andesine-andesite (or basalt) with normative andesine and olivine. Soda is more abundant than potash in the analysis. The groundmass shows abundant pyroxene and magnetite grains, much of the pyroxene being a highly pleochroic titanaugite liable to be mistaken for hypersthene. The flow occupying the top is a rhyodacite. There is also another rhyolite flow. The rhyolite flows are genetically unconnected with the basaltic rocks and are younger.

In parts of Kathiawar where borings have been put down through the traps, very basic types like limburgite and ankaramite are found inter-bedded with normal dolerites and basalts. The ultra-basic types seem to be more or less restricted to the western edge of the trap country in Bombay, Gujarat and Kathiawar as dykes and sills, while the acid types are found mainly along two zones, one running from Pavagad hill to Bombay and the other from the Narmada valley to Porbandar in Kathiawar.

The Barda, Dhank, Osham and Girnar hills lie on a line which appears to strike into the region of the Narmada dislocation, while the Pavagad and Chamardi-Chogat line is parallel to and north of it. It would, therefore, appear that the loci of the central eruption type are located on important fault zones in Western India.

It is of interest to note here that the traps at Worli Hill, Bombay Island, are associated with bituminous matter, whose presence in cavities is explained by Dr. C. S. Fox as due to the distillation of the organic matter present in the associated sedimentary beds, by the heat of the traps.

PETROGRAPHY

The common type of trap is composed of abundant labradorite of the composition $Ab_2 An_2$ and clinopyroxene, the two forming the bulk of the rock. The clinopyroxene is always abundant in the holocrystalline types, its amount decreasing with increase in the content of interstitial glass. It is usually greenish or brownish grey in colour in hand-specimens but practically colourless in thin section. A little pigeonite is found in a few places.

Ophitic texture is common, the labradorite laths lying in a mass of more or less anhedral enstatite-augite. The labradorite is almost always the earlier mineral to crystallise, though contemporaneous crystalliation of the two may have taken place to some extent.

The rocks often show phenocrysts of feldspar in the doleritic types and interstitial glassy matter in the basaltic types. The proportion of the glassy matter varies a good deal. The glass is often highly corroded and contains abundant dust-like inclusions which are presumably magnetite. The glass is liable to alteration into palagonite, chlorophaeite, celadonite and delessite.

Since the amount of augite and magnetite bear an inverse ratio to that of the glass in thin section, it is obvious that these two minerals are together represented in the glass.

Magnetite is quite a common, though a rather minor constituent. It occurs usually as discrete grains amidst the other minerals and as grains, dust and skeletal growths in the glassy groundmass. Some ilmenite and leucoxene are generally always present in the rocks and this is confirmed by the appreciable amount of titanium shown by the analyses, though some titanium may be present in the augite also. In some varieties a fair amount of olivine is present but biotite and hornblende are generally absent from normal types. Quartz is rare or absent, but there is generally an excess of silica in the C.I.P.W. norm. Sodic plagioclase and orthoclase are absent, but interstitial granophyric or micrographic patches are sometimes seen.

Investigation by Fermor of the flows enountered in a boring at Bhusawal showed that olivine, and to some extent labradorite, had settled down to the bottom of the flows, apparently through gravitative settling. In the Satpura region of Jabalpur-Chhindwara, thick sills often show the phenomenon of crystal settling, the lower portions being coarse and holocrystalline and the upper portions finely crystalline and containing quartz and micropegmatite. A dyke in this region, which is over 13 km long, consists of porphyrite containing oligoclase, enstatite-augite, hornblende, quartz and micro-pegmatite. There are, however, other areas where there is no evidence of crystal settling even in thick flows, as mentioned on a previous age.

W.D. West (1958) has published the results of his petrographic examination of some flows representatives of which were recovered from the cores of boreholes put down in Western India. He has recognised three types of basalts in the materials examined, including types containing phenocrysts of labradorite, bytownite and olivine in different proportions. Amongst them are also picrite-basalts including such types as Oceanite, Ankaramite and Limburgite. The olivine phenocrysts in these rocks vary in composition from Fo_{60} in the average basalt to Fo_{75} in some olivine-basalts, and Fo_{90} in some picrite-basalts. The pyroxene pheoncrysts are diopsidic augites in the olivine and picrite-basalts. The groundmass pyroxenes in the normal basalts are in general sub-calcic augites. West has shown that the composition of the mineral phases present in the various types of basalts is closely related to the composition of the host rocks.

Secondary minerals are often developed in the traps, either as fillings in the amygdular cavities or as products of alteration and replacement. The minerals of late hydrothermal activity are the zeolites — stilbite, apophyllite, heulandite, laumontite, analcite and prehnite ; also calcite, chalcedony and its varieties (agate, jasper, carnelian, chrysoprase, heliotrope, onyx etc.) opal and sometimes quartz and amethyst crystals in drusy cavities. These minerals are generally found in amygdular cavities which may be lined with chlorophaeite and delessite. Amongst the zeolites, radiating and sheaf-like aggregates of stilbite are the most common, though prismatic crystals also occur. Next in importance are apophyllite, heulandite, laumontite and scolecite, excellent crystals of which are not infrequently found. The alteration products are chlorophaeite, palagonite, delessite, celadonite, iddingsite and serpentine, the last two formed from olivine. Several of these are collectively spoken of as " green earth, " which is sometimes used as a pigment material.

CHEMICAL CHARACTERS

As might be expected from the uniformity in mineralogical composition, the chemical composition of the traps also tends to be uniform. The plateau basalts (of the fissure eruption type) when compared with the normal cone basalts (central eruption type) show a higher iron and titanium content, the iron being dominantly in the ferrous state. Magnesia and the alkalies are lower. The pyroxene is a sub-calcic augite intermediate in composition between diopside and pigeonite.

Taking all the types found in and associated with the Deccan Traps, the silica percentage varies from 43 to 73 per cent. The rocks fall into groups ; basic and ultra-basic types with 45 to 51 per cent silica ; intermediate types with 52 to 61 per cent silica ; and acid types with over 61 per cent silica. In the under-saturated types, olivine and nepheline appear in the norm. As the silica increases, it gives rise to more pyroxene at first and then more feldspars. In the over-saturated types the pyroxene is subordinate, the felspars become dominant and silica appears as free molecules.

The average chemical composition of Lower and Upper Traps as well of some types are presented in Table 61. A number of differentiated types are present near the western coast and in Kathiawar, all these belonging to the Upper Traps. Differentiated types are also conspicuous at all the loci of central eruptions — Girnar, Osham, Choghat, Pavagad etc.

Fermor (1934) investigated the flows at Linga in Madhya Pradesh and the analyses showed that even though the composition of the different flows were extraordinarily similar, normative composition of the Upper flows indicated a diminution in the silica, total iron and water and an increase in alkali felsdspars, total amount of feldspars and pyroxenes, as compared to the lower flows. West (1958) found that phenocrysts of pyroxene are more common in the Upper Traps. The feldspar phenocrysts show anorthite content of 55 to 57% in the lower traps and 60 to 60% in the upper traps ; in the groundmass the pyroxenes and feldspars are more basic in the upper traps than in the lower. The value of the *differentiation index* (*i.e.*, the sum of normative Q, Or, Ab, Ne and Lc) devised by Thornton and Tuttle (1960), is around 29 for the Lower Traps and around 39 for the Upper Traps (Saxena, 1962). The prevalence of the differentiated types in the Upper Traps is due to tectonic disturbances in Western India including normal and trough faulting along the coast and elsewhere which brought up differentiated materials from the deeper crust.

The petrochemical studies made by Vemban (1947) with the help of CIPW norms and Niggli values led him to conclude that the original magma was picritic in composition and that both the calc-alkali trend and alkali-olivine trend were present. As a result of the study of some flows encountered in borings in Western India, West suggested that the original magma was tholeiitic but more basic than the normal type ; that initial crystal settling of olivine, pyroxene and calcic plagioclase may have taken place leaving a tholeiite liquid which would be erupted ; that at a later stage there was remelting of the earlier crystallised material yielding an alkali-olivine type of magma.

Experimental studies by Yoder and Tilley (1962) on the melting and crystallisation of eclogitic material has thrown light on the possible mechanism of the origin of tholeiitic and alkali type of basaltic rocks. Primary mantle material presumed to be of the composition of garnet-peridotite may, on partial melting, yield two types of magmas. The increase of the garnet constituent of the original material at high pressure by the removal of the omphacite material (or by a shift of the omphacite-garnet " boundary-surface") will give rise to theoleiite at low pressure. On the other hand, the increase of the omphacite constituent at high pressure and elimination of the garnet constituent will lead to the generation of the alkali-olivine type of magma. It would appear that in general the alkali-basalt type would be generated at a greater depth than the tholeiite type. Differentiation of either type of magma may take place during and after emplacement through one or more of the wellknown processes — oxidation and reduction, gas flaxing, crystal settling and contamination. The presence of much water would produce rocks rich in hornblende or other amphiboles instead of gabbro, basalt or eclogite. Hornblendic basalts are however rare in the Deccan Traps. The presence of rocks of overwhelmingly tholeiitic and andesitic composition in the continents probably indicates that contamination by sialic material has played a very important part. The major part of the Oceanic crust may be of the alkali-basalt composition, though theoleiitic material is also present.

ALTERATION AND WEATHERING OF THE TRAPS

The traps weather with characteristic spheroidal ex-foliation which gives rise to large rounded boulders on the outcrops. The weathering starts along the well-developed joints, first rounding off the angles and corners and then producing thin concentric shells or layers which become soft and fall off gradually. The interiors of the spheroidal masses are however, quite fresh.

The traps give rise to either a deep brown to rich red soil or to *regur* (black cotton soil) which can be seen in many parts of the Deccan. The regur is rich in plant nutrients such as lime, magnesia, iron and alkalies, on which cotton and certain of the ' dry ' crops flourish. It has the property of swelling greatly and becoming very sticky when wetted by rain ; on drying it contracts again with the production of numerous cracks. Another product of weathering is *laterite*, a material from which silica, alkalies and alkaline earths have been leached away, leaing behind alumina, iron, manganese and titanium. It has a vermicular or pisolitic structure and contains much water. Some laterites

THE DECCAN TRAPS

TABLE 61—ANALYSES OF DECCAN TRAPS (Calculated to 100, *water-free* basis)

	1	2	3	4	5	6	7	8	9	10	11	12	13
SiO_2	51.71	51.50	50.75	50.51	51.50	50.30	47.36	44.05	44.40	46.94	45.29	46.50	45.16
TiO_2	1.91	2.37	2.40	2.85	1.77	3.29	2.24	1.76	2.18	2.08	2.90	2.77	2.55
Al_2O_3	13.87	13.47	13.37	12.65	14.93	11.92	16.69	10.49	11.14	9.78	11.09	16.43	12.16
Fe_2O_3	3.27	3.10	3.15	3.12	3.41	3.10	3.10	1.81	3.09	4.40	1.66	4.97	6.40
FeO	10.14	10.51	10.64	11.23	9.77	11.80	7.82	10.55	8.01	7.00	9.65	7.85	6.86
MnO	0.16	0.18	0.22	0.23	0.09	0.24	0.34	Trace	...	0.22	...	0.28	0.01
MgO	5.58	5.28	5.85	5.45	4.54	5.08	7.61	18.44	18.80	14.98	16.11	6.53	13.06
CaO	9.64	9.79	10.43	10.57	9.51	10.70	11.92	11.25	10.58	12.08	11.06	9.86	10.73
Na_2O	2.65	2.68	2.30	2.43	3.24	2.55	2.09	1.08	1.26	1.40	1.48	2.62	1.20
K_2O	0.72	0.81	0.52	0.61	1.02	0.69	0.60	0.40	0.54	0.91	0.76	1.11	0.75
P_2O_5	0.35	0.31	0.37	0.35	0.22	0.32	0.23	0.17	...	0.21	...	0.38	0.30
										Carb matter		0·71	0·82

(1) Average of 11 anal., H.S.W.   (2) Average of 18 superior anal., S. & P.   (3) Average of 4 anal. of Lower Traps.   (4) Average of 8 superior anal. of Lower Traps, S., & P.   (5) Average of 21 anal. of Upper Traps., S. & P.   (6) Average of 4 anal. Lower Traps of Linga, L.L.F.   (7) Three-Phenocryst basalt from Wadhwah, W.D.W.   (8) Oceanite from Botad., W.D.W.   (9) Oceanite from Bombay area, S. & P.   (10) Ankaramite from Dhandhuka, W.D.W.   (11) Ankaramite, Bombay area, S. & P.   (12) Titanaugite-alkali-olivine basalt, Main hill quarry, Pavagad hill, S.C.C.; Olivine-basalt, from 640 m. elevation, Main Pavagad hill, S.C.C.

Abreviations: H.S.W.—H.S. Washington; S.& P.—Sukheswala and Poldervaart; W.D.W.—W.D. West; S.C.C.—S.C. Chatterjee. For references see at the end of Chapter.

which are highly aluminous, form deposits of bauxite. Laterite plateaux capping the traps are present in Maharashtra and Madhya Pradesh, some of these containing good deposits of high grade bauxite.

THE LAMETA BEDS

The Lameta beds, named after the Lameta Ghat near Jabalpur in Madhya Pradesh, are fluviatile or estuarine beds occurring below the traps at about the same horizon or slightly above that of the Bagh beds of the Narmada valley. They are found to rest on various older formations such as the Archaeans, the Upper Gondwanas or the Bagh beds.

They are fairly extensively developed though not found everywhere underneath the traps. They usually occur as a narrow fringe around the trap country, particularly in Madhya Pradesh (*e.g.*, around Nagpur and Jabalpur), and along the Godavari Valley to up Bhopal and Indore, and in the western part of the Narmada valley. The chief rock types found in them are limestone, with subordinate sandstones and clays. The limestones are gnerally arenaceous and gritty, though occasionally pure, but a cherty type containing lumps of chert and jasper may be said to be characteristic. Earthy greenish sandstones are common, while clays which are usually sandy and red or green in colour are also found. The Lameta beds vary in thickness from 6 to 35 m, the individual beds frequently varying in character when followed horizontally.

In the type area at Lameta Ghat the following section can be made out, though all the members are not present in the same individual section.

4. Sandstone similar to No. 1, containing bands of flint or thin ilmestone. The sandstone may occasionally be composed of grains of glassy quartz with a white powdery cementing medium.

3. Pale green or purplish mudstone, often finely laminated, sometimes arenaceous or calcareous.

2. Limestone or indurated marl often earthy and drab to bluish in colour. Has a tufaceous appearance because of vein-like cavities which may be filled with chalcedony or calcium carbonate by infiltration, this marly limestone being the characteristic member.

1. Greenish, poorly compacted, sandstone, sometimes hard and cherty. These are brownish near Jabalpur.

The Lametas are shallow marine formations. In the section north of Jabalpur, the contact between the Jabalpurs and Lametas is apparently conformable but close examination reveals the presence of a layer of red clay, which is an oxidised surface indicating a period of uplift, exposure and erosion before the sea encroached over the area and deposited the Lameta Beds. The Greensand layer in Chui Hill has a thickness varying from 1.5 to 6 metres, due to deposition on a sloping surface. This formation also filled the irregularities on the surface on which it was laid down. The bed above this, the marly formation, is of uniform thickness.

Occasionally, the rocks underlying the Deccan traps are found to be calcified by solutions descending down from the traps, the original rocks being Archaean gneisses and schists. The occurrence of fragmentary fossils remains in the true Lameta shelps us to distinguish them from the calcified rocks.

The Lametas only rarely contain good determinable fossils, though small fragmentary fossils are common. They include mollusca, fishes and dinosaurian reptiles.

Mollusca	... ...	*Melania, Physa (Bullinus), Paludina* and *Corbicula*.
Fishes	... ...	*Lepidosteus, Eoserranus, Pycnodus*.
Dinosaurs	...	*Antarctosauras septentrionalis, Titanosaurus indicus, Indosaurus matleyi, Lametasaurus indicus, Jubbulporia tenius, Laplatasaurus madagascariensis*.

The Dinosaurian remains have been found mainly at Jabalpur and at Pisdura 12 km north of Warora in Madhya Pradesh. Coprolites are also found in these places. *Physa (Bullinus) Prinsepii* is associated with these reptilian remains at Pisdura, while the other mollusca have been found in a bed at the base of the traps at Nagpur and Ellichpur. According to Von Huene, the age indicated by the Dinosaurs found at Jabalpur and Pisdura, which are allied to forms recorded from Madagascar, Brazil and Patagonia, is Turonian.

INFRA-TRAPPEAN AND INTER-TRAPPEAN BEDS

Allied to the Lametas are the beds occurring below the traps in the Rajahmundry area, called the Infra-trappeans. They occur only on the right bank of the Godavari river. At Dudukuru, a few miles N.W. of Rajahmundry, they are composed of yellowish, whitish and greenish sandstones overlying the Upper Gondwanas. They are about 15 m thick, the upper portion being calcareous and containing a fossiliferous limestone, 30 to 60 cm in thickness, at the top. The fauna is undoubtedly marine and comprises a nautilus, several lamellibranchs and gastropods, the latter including a *Turritella* which seems to be identical with *T. dispassa* of the Ariyalur Stage of the Trichinopoly Cretaceous.

There seems to be a slight unconformity between the Infra-trappeans and the basal basalt flow since the former appear to have been partially denuded before the traps were erupted. The Infra-trappeans do not contain, so far as known at present, fossils identical with any in the Bagh beds or the Trichinopoly Cretaceous except the *Turritella* mentioned above. There is, moreover, an absence of any characteristic genera. Oldham states that, on the whole, the Infra-trappeans have perhaps more affinity with the Cretaceous than with the Tertiary. H. C. Das Gupta found *Cardita (Venericardia) beaumonti* from the Dudkur (Dudukuru) beds.

During the considerable intervals of time which elapsed between succesive eruptions of lava, there came into existence some rivers and fresh water lakes in the depressions and in places where there was obstruction to drainage. The fluviatile and lacustrine deposits formed in them are intercalated with the lava flows, and are of small horizontal extent and generally 0.5 to 3 m thick, though occasionally only 15 cm thick. They contain, in several places, animal and plant remains which should prove to be valuable in the determination of the age of these beds and incidentally of the associated traps. They comprise cherts, impure limestones and pyroclastic materials, and have been recorded from the Godavari, Chhindwara, Nagpur and Jabalpur districts and parts of Maharashtra.

The traps of the Rajahmundry area appear on either bank of the Godavari river, with a length of some 55 km in an E.N.E.-W.S.W. direction, and a thickness varying from 30 to 60 m. Exposures are found at Kateru on the Rajahmundry side of the river, resting on Archaean rocks, and on the other side (right bank) near Pangadi and Dudukuru, resting often on the Upper Gondwanas. The traps are overlain by the Rajahmundry Sandstones (Cuddalore Sandstones) all the formations having a general gentle dip towards the south or southeast.

The lower flow of trap is about 15 m thick. It is overlain by a fossiliferous bed which is 4 to 4.5 m thick near Kateru and only 0.6 to 1.2 m thick near Pangadi. The fossiliferous inter-trappean bed is exposed for about 1 km near Kateru and for over 16 km on the other side of the river. The fauna is unmistakably estuarine and comprises *Cerithium, Potamides, Pirenella, Cytherea*, etc., some of the characteristic species being *Corbicula ingens, Cerithium stoddardi, C. leithi, Cytherea meretrix, Physa* (*Bullinus*) *prinsepii, Paludina normalis* and *Lymnaea subulata*. There are no corals, cephalopods or echinoderms to indicate any marine affinities. The fauna is said to have more affinity with the Ariyalurs of the South Indian Cretaceous than with the Eocene though no cases of identical fossil species have yet been established.

Fossil algae found in these beds have been studied by S. R. Narayan Rao and K. Sripdaa Rao. They state that some of these, like *Neomeris* and *Acicularia*, and the *Charophyta* are Tertiary in age and that the associated traps are therefore early Eocene in age. Some ostracods found in these beds also show Eocene affinities.

The Inter-trappean beds of Bombay are high up in the Upper Traps excellent sections of which can be seen on the Malabar hill and Worli hill at Bombay. Here they are about 30 m. thick and consist of brown, grey or dark shales the last being carbonaceous and showing plant impressions and remains of frogs with occasional pockets of bitumen and coaly material. They contain also the fresh water tortoise *Hydraspis* (*Platemys*) *leithi*, the frog *Rana pusilla* (=*Indobatrachus pusillus*) and three species of Cyprides (Crustacea) the common one being *Cypris submarginata*.

The Inter trappeans and Infra-trappeans occur also in some parts of Madhya Pradesh ; in Chhindwara they have yielded plant remains, among which are palms with distinct Eocene affinities. In Berar and the Narmada valley, the beds are found 90 to 150 m above the base of the traps and contain plant and animal fossils in some places.

AGE OF THE DECCAN TRAPS

In the previous sections are stated the facts which should enable us to gain an idea of the age of the Inter-trappeans and of the associated traps.

At Dudukuru in West Godavari district, the traps are underlain by the Infra-trappeans containing gastropods, lamellibranchs and a nautilus. The fossils, though not identical with any found in the Trichinipoly Cretaceous beds, seem to indicate some general affinities with them. Recent work here has revealed the presence of several algae including *Holosporella, Dissocladella, Neomeris, Torquemella,* and *Acicularia*, the last of which has not been recorded from any beds older than the Paleocene. Several foraminifera have also been recorded —*Triloculina, Nodosaria, Textularia, Spheroidinella, Nonion, Globotruncana.*

In the Narmada valley the traps are underlain by the Bagh beds of Upper Cretaceous age, possibly in part equivalent to the Lametas. Between the traps and the Bagh beds there is a slight but distinct unconformity.

In Surat and Broach there is a distinct erosional unconformity between the top of the traps and the Nummulitic strata, for the basal Eocene contains materials derived from the denudation of the traps. This is also the case in Cambay where borings have been put down to reach the Traps through the Tertiaries. In Kutch the traps overlie unconformably the Jurassic and Lower Cretaceous beds and are overlain by the Nummulitics (Ranikot). Here also there seems to be an unconformity between the traps and the Nummulitics though this is not very clear.

In Sind, the Bor hill near Ranikot shows a bed of calcareous, gritty to conglomeratic, sandstone overlying the Hippuritic limestone (Upper Cretaceous) in which occurs an inter-stratified bed of basalt 12 m thick, some 100 to 120 m above the base of the sandstone. This sandstone is overlain by Olives Shale and sandstones, the latter containing some volcanic ash or decomposed fragments of basalt. *Cardita (Venericardia) beaumonti* occurs in several horizons in the Olive Shale, but especially abundantly in a bed 60 to 75 m below the top of the series. In addition to some corals, echinoids and gastropods, *Nautilus bouchardianus*, (which occurs in the Ariyalur beds of the Trichinopoly district), is also to be found here. The faunal assemblage indicates an age in the uppermost Cretaceous.

The *Cardita beaumonti* bed is overlain by another bed or flow of basalt, the thickness of strata separating this and the lower flow being about 180 m. The upper one is much more extensive than the lower and has been traced for 35 km from Ranikot to Jakhmari, at the base of the Ranikot beds. This upper flow (lying on the *Cardita beaumonti* bed) has a thickness varying from 12 to 27 m, but is itself composed of two individual flows each of which is vesicular at the top. There is no doubt, according to R. D. Oldham, that the basalt is a flow and not an intrusive (Manual, Second Edition, P. 289), and it is conformable both to the underlying *Cardita beaumonti* beds and the overlying Ranikot beds. Though separated by a good distance from the main Deccan Trap area, these trap beds in Sind are considered to belong to the Deccan Traps.

In a boring put down in 1948 by the Burmah Oil Company at Lakhra, about 40 km northwest of Hyderabad, Sind, the *Cardita beaumonti* bed was encountered at a depth of 2,925 ft. (891·5 m). Deccan Traps overlie as well as underlie this bed which is 30 m thick. Traces of traps were encountered also in the sediments below the lower trap horizon. This indicates that the eruptions began in the Maestrichtian.

Recent work on the Inter-trappean fossils, especially by B. Sahni and collaborators, lends support to a Lower Eocene age for the beds from which the fossils were obtained. The chief points in the evidence are : There is a large proportion of palms (*Palmoxylon* predominating) amongst the angiospermous flora ; the palms are said to be much more abundant here than in any Cretaceous flora so far studied. The gerns *Azolla* found in these beds has not been recorded from any beds earlier in age than the Tertiary. *Nipadites*, a characteristic Eocene genus, occurs in the Inter-trappeans.

Smith Woodward's work on the fish remains from the Lametas has shown that perhaps they are more allied to Eocene than to Cretaceous forms. S.L. Hora who studied the fossil fishes found at Takli, Paharsingha and other places in the Madhya Pradesh, has identified *Lepidosteus indicus, Pycnodus lametae, Eoserranus hislopi, Nandus, Pristolepis, Scleropages* and some percoid fishes. Scales of *Musperia* and species of *Clupea* have also been recovered from Infra-trappean beds. These have led him to favour a Lower Eocene age for the lavas associated with the Inter-trappeans.

The fact still remains that much of the stratigraphical field work on the traps and the associated sedimentaries is old, and it is very desirable to restudy them and make fresh fossil collections. The evidence of age, as it stood in the early nineties of the last century, has been admirably summed up by R. D. Oldham in the second edition of the Manual of the Geology of India (pp. 280-281, 289). The base of the traps lies, in various places, on the Bagh Beds (Cenomanian to Senonian), the Lametas (roughly Turonian according to Von Huene and Matley) and the Infra-trappeans or *Cardita beaumonti* beds (Danian). In the Lakhra borings mentioned above, the traps occur both above and below the Cardita Beaumonti beds. It is therefore reasonable to conclude, as Oldham did in the Manual, that the traps commenced to be poured out in the Uppermost Cretaceous and that they continued through the gap of time marked in Europe by the unconformity between the Mesozoic and the Tertiary, which in North America is represented by the Laramie formation, and perhaps well into the Eocene. How far into Eocene the activity continued is not known.

Recent work on Inter-trappean fossils would seem to place a part of the traps in the Eocene. This is reasonable since the traps have a very large thickness (perhaps 2,000 m in Western India) and some of the products of the latest phases of the activity found in Bombay and Saurashtra are distinctly later than the main mass. This however leaves the question of the age of the base of the traps practically where it was. It is necessary to know by field work where exactly the lowest traps exist and whether they may not be of different ages in different places — e.g., in Madhya Pradesh and in Rajamahendri. It has already been mentioned that the Lower Traps occur in Madras and Madhya Pradesh and the Middle and Upper traps progressively westward. There is also an opinion current, supported by Dr. C.S. Fox, that the Rajmahal traps and the basic dykes in the coalfields of Bengal and Bihar represent an early manifestation of the Deccan Trap activity.

Some work on the radiometric age of the Deccan traps seems to show that the traps range in age from Upper Cretaceous to Eocene. Radiometric age determinations at the Carnegie Institution of Washington (Year book, 1964) on a few specimens of basalt flows from Bombay and Pavagad Hill gave 60-65 M.Y. and 42-45 M.Y. respectively. The palaeomagnetic positions of the same specimens were in South Latitudes for the first group, and equatorial to very low Northern Latitudes for the others. As the drift of India should have commenced sometime after the Middle Cretaceous, rapid movement is indicated during the Eocene. From these data it may be inferred that the central eruption type of activity (probably also some of the flows of the alkali-olivine rock types) took place in the late Eocene, long after the main fissure eruptions took place.

Summing up, it may be stated that the available evidence shows that a good part of the traps may have been erupted in early Eocene. Though some of the genera of plants occurring in the Inter-trappeans are identical with those in the Eocene of other parts of the world, there are apparently no identical species. Such being the case, there is no reason why a flora allied to the Eocene of Europe should not have flourished in India in the uppermost Cretaceous and Laramie times. At any rate, the evidence of marine animal fossils in Sind is clear that the earliest traps are of late Upper Cretaceous age. Radiometric age data support this conclusion and indicate also later activity in Upper Eocene to Oligocence.

ECONOMIC GEOLOGY

Being dense, hard and durable, the Deccan traps are used fairly extensively as building stones in the areas in which they occur in large masses. But, being dark in colour, they are not used to the extent to which their durability will entitle them. The light buff and cream-coloured trachytic rocks found in the Salsette island and the neighbourhood of Bombay are generally more preferred than the dark traps. The commemoration arch called the "Gateway of India,' at the Bombay harbour, is constructed of such trachytic rock obtained from Kharodi and Malad near Bombay. This rock, however, frequently contains some calcite and pyrite which are liable to produce unsightly brown stains and weakness, on weathering.

As road metal, the Deccan Traps are excellent for macadam and tarred roads and are among the best stones obtainable in India. They are hard, tough, wear-resisting and have good binding properties. They are also excellent for use as aggregates in cement concrete.

The Deccan Traps of Western India are a great store-house of quartz, amethyst, agate, carnelian, onyx and other varieties of chalcedony which occur as geodes and are used as gem stones. These are made into trinkets, beads, ring stones and ornamental objects. There is a small agate-cutting industry at Ratnagiri, Rajpipla and Cambay, the necessary raw stone being collected from streams and from the debris on weathered outcrops. The supply for Cambay used to come from a Tertiary Conglomerate, the pebbles of which were derived from the traps.

The traps are often capped by ferruginous and aluminous laterite. The latter is in several places — e.g., Saurashtra, Kolhapur, Belgaum, Katni, Jabalpur, Mandla, Surguja — rich enough in alumina to be high grade bauxite. Indian bauxite has been used in petroleum filtration for many years and for the manufacture of alumina and aluminium since about 1945. It is however generally rich in titania, as much as 10 per cent. or more of this constituent being present. The ferruginous laterite forms a good building stone and has also been used formerly for iron smelting in indigenous furnaces. There seem to be possibilities also for smelting the ferruginous laterite and obtaining pig iron and cement by a suitable process.

The black soil or *regur* formed over the Deccan trap is a rich soil particularly suitable for raising cotton. It is similar to the Russian *Chernozem*, but is by no means confined to the trap areas, for we find it on gneisses, charnockites, Cretaceous sediments, etc., in South India. This would suggest that it is not only the chemical composition of the parent rock but also climatic factors that play an important part in its formation.

SELECTED BIBLIOGRAPHY

Auden, J.B. Dykes in Western India. *Trans. N.I.S.* 1949.

Blanford, W.T. Traps and Inter-trappean beds of Western and Central Asia. *Mem.* VI, Pt. 2, 1867.

Crookshank, H. Geology of the northern slopes of the Satpuras between Morand and Sher rivers. *Mem.* 66 (2), 1936.
Chanda, S.K. Petrography and origin of Lamata Beds, Lambata Ghat. *Proc. Nat. Inst. Sci. India.*, 29-A, 273-287, 1963.
Chatterjee, S.C. Petrochemistry of the lavas of Pavagad Hill. Advancing Frontiers in Geology and Geophysics. *Ind. Geophys. Union* (Hyderabad) p. 385-397. 1964.
Chatterjee, S.K. Petrology of igneous rocks from the West Gir Forest. *Jour. Geol.* 40, 154-170, 1932.

Fedden, F. Geology of Kathiawar. *Mem.* 21(2), 1884.
Fermor, L.L. Lavas of Pavagad Hill. *Rec.* 34, 144-166, 1906.
Fermor L.L. and Fox C.S., Deccan Trap flows of Linga, C.P. *Rec.* 47, 81-136 1916.
Fermor, L.L. Basaltic lavas penetrated by deep boring for coal at Bhusawal. *Rec.* 58, 93-240, 1926.
Fermor, L.L. Deccan Traps of Linga, C.P. *Rec.* 68, 344-358, 1934.
Fermor, L.L. The Role of garnet in nature. *Ind. Assoc. Cult. Sci.*, Calcutta, 1935.

Hora, S.L. Fossil fish scales from the Inter-trappeans of Deothan and Kheri. *Rec.* 73, 267-292, 1938.

Karkare, S.G. Geochemical studies of the Deccan Traps. Banaras Hindu University, Banaras, 128 pp. 1965.

Krishnan, M.S. Petrography of rocks from the Girnar and Osham Hills, Kathiawar. *Rec.* 58, 380-424, 1926.

Krishnan, M.S. Deccan Trap Volcanism. *Bull. Volc.* 26, 387-399, 1963. (Also *Bull. Nat. Geophys. Res. Inst. India*, 1, 4-19, 1963).

Mathur, K.K. Problems of petrogenesis in the Deccan traps. *Proc. 21st Ind Sci. Cong.* 329-344, 1934.
Mathur, K.K. *et al.* Magmatic differentiation in the Girnar Hills. *Jour. Geol.* 34, 298-307, 1926.

Rao, S.R.N. and Rao, K. Sripada. Foraminifera of the Inter-trappean beds near Rajahmundry. *Rec.* 71, 389-396, 1936.

Rao, S.R.N. and Rao, K. Sripada. Fossil Charophyta from the Kateru Inter-trappeans. *Pal. Ind.* N.S. XXIX (2), 1940.

Sukheswala, R.N. and Poldervaart, A. Deccan besalts of the Bombay area. *Bull. Geol. Soc. Amer.* **69**, 1475-1494, 1958.

Sukheswala, R.N. and Sethna, S.F. Deccan Traps and associated rocks of the Bassein area. *Jour. Geol. Soc. India*, 3, 125-146, 1962.

Vemban, N.A. Differentiation trends in the Deccan Traps. *Proc. Ind. Acad. Sci.* **25-A**, 75-118 1947.

Washington, H.S. Deccan Traps and other plateau basalts. *Bull. Geol. Soc. Amer.* 33, 765-804, 1922.

West, W.D. The petrography and petrogeneisi of 48 flows of Deccan Trap penetrated by borings in Western India. *Trans. Nat. Inst. Sci.* India, IV (1), 1-56, 1958.

Yoder, H.S. and Tilley, C.E. Origin of basalt magmas — an experimental study of natural and synthetic rock systems. *Jour. Petrology*, 3(3), 342-532, 1962.

CHAPTER XVI

THE TERTIARY GROUP

The great Cenomanian transgression during the Cretaceous was worldwide and in the middle Cretaceous there were fairly distinct faunas in the Mediterranean region, in the Atlantic, in the lands bordering the Indian Ocean (South Africa, Eastern Coast of India and W. Australia), and in the Russo-Arctic regions. The end of the Cretaceous was marked by important changes in the fauna and flora. The great reptiles amongst the land animals and the ammonites amongst the marine ones disappeared. Many changes occurred also in the generic composition of the surviving fauna. The angiosperms established themselves firmly as the most important group of plants. These changes were roughly contemporaneous with the first phase of the compression of the Tethys in the Upper Cretaceous, resulting in the formation of mountain ranges in the Oman region of Arabia and in the shallowing of the Tethys and ridging up of its bottom in the Himalayan region. In addition to the fragmentation and drifting of the component parts of Gondwanaland, evidences of the effects of tension are found in the formation of the Rift Valleys of East Africa and the connected Red Sea rift, as also in the outpouring of the enormous lava flows in India and at a later time in Ethiopea.

The Tertiary (Kainozoic or Cenozoic) deposits of Europe have been classified as shown below.

Pliocene	Astian Plaisancian Pontian	
Miocene	Sarmatian Tortonian Helvetian Burdigalian Aquitanian	}Vindobonian
Oligocene	Chattian Stampian (Rupelian) Sannoisian (Tongrian or Lattorfian)	
Eocene	Ludian (Wemmelian) Bartonian Auversian Lutetian Cuisian Ypresian Sparnacian Thanetian Montian	}Priabonaian }Paleocene

There are numerous local names in the different countries for the different subdivisions but the standard nomenclature given above is followed for purposes of correlation. The Eocene and Oligocene are characterised by numerous foraminifera which are very useful for correlation. Nummulites occur in both these; Assilines and Orthophragmina are confined to the Eocene while the Lepidocyclines occur in the Oligocene.

In the Western Mediterranean there was an important marine transgression in Miocene times over the pre-Alpine depression. The marine deposits laid down then constitute the BURDIGALIAN, the type area being near Bordeaux. The type area of the succeeding VINDOBONIAN is the Vienna basin, where the lower part of the deposits is sandy and of the HELVETIAN type (*i.e.*, marine MOLASSE of Switzerland somewhat similar to the Siwaliks in India) ; while the upper part consists of deep marine blue marls of the TORTONIAN type found in the Piedmont. The normal marine Tortonian in the Eastern Mediterranean is overlain and is in part replaced by the SARMATIAN (named after Southern Russia) which consists of near-shore deposits. The deposits of the Upper Tertiary are of different types in the Eastern and Western Mediterranean. The Eastern Mediterranean covered also the Black Sea and Caspian Sea which later became isolated into inland seas in the Pliocene. After the Sarmatian, the fauna of the Eastern Mediterranean acquired characters typical of the Caspian area, characterised by abnormal *Cardium* and *Congeria*. The PONTIAN consists of such inland-sea deposits, the name being derived from the Black Sea (= Pont Euxin). There are apparently no well established marine equivalents of the Pontian in the Mediterranean region. The Alpine chains were uplifted during the Oligocene, when the Mediterranean was cut off from the Indian Ocean. Most of the species of the Miocene fauna are extinct, but some genera survived into the Pliocene. The Pontian deposits are now assigned to the Pliocene for they are characterised by the appearance of Hipparion and a few other Vertebrates. The PLAISANCIAN strata are marine with large Pleurotomes. The ASTIAN are partly the littoral representatives of the Plaisancian and partly younger.

At the end of the Pliocene there was a slight uplift in some of the Mediterranean areas. The Pliocene deposits are followed by the Pleistocene whose marine representatives are the CALABRIAN (with *Cyprina Islandica*) and continental representatives the VILLAFRANCHIAN. The latter contain the first true horse *Equus* and also the elephant genera *Elephas* and *Mastodon*. Four important periods of glaciation have been identified, known successively as the GUNZ, MINDEL, RISS and WURM. Many species of the rich mammalian fauna perished during the glaciation and only a few survive into the present age, these being a meagre fraction of the prolific Upper Tertiary fauna.

The Rise of the Himalayas.—The great Mediterranean ocean, the Tethys, was first shallowed in the Upper Cretaceous as a result of the tangential compression to which the crust was subjected. The mountain building activity continued intermittently throughout the Tertiary and brought into being the great equatorial mountain systems which include the Atlas, the Pyrensee, the Alps, the Caucasus, the Himalayas and the Burma Andaman Arc.

The rise of the Himalayas took place in a series of five or more stupendous movements punctuated by intervals of comparative quiescence. The first movement took place during the Upper Cretaceous when the Tethys was furrowed into a series of ridges and basins running longitudinally. It is likely that in some places 'island arcs' were formed though it would be difficult now to identify them, particularly in the Himalayan region. Simultaneously with this, certain portions of the Tethys became deeper and Radiolarian cherts were formed at great depths, while in the shallow areas flysch-like sediments were deposited e.g., the Giumal and Chikkim Series. This movement was also responsible for the separation of the northern extremity of the Bay of Bengal into two gulfs, one extending into Upper Burma and the other into Upper Assam. Similarly also a ridge appeared in the northwest, which separated the Sind Gulf from the Baluchistan Gulf.

Then followed a period of comparative rest after which another upheaval took place during the Upper Eocene i.e., after the deposition of the Kirthar beds. After this the Nari, Gaj and Murree strata were laid down which were marine in the south and brackish water in the north. The third movement, which was probably the most powerful of all, took place during the Middle Miocene times. It was probably during this time that the Himalayas acquired their major features and the Tethys disappeared more or less completely, being replaced by mountain ranges with intervening shallow lakes, marshes and large river valleys. At the same time a long narrow trough seems to have been formed between the rising Himalayas and the Peninsular mass. In the trough, which is often referred to as the *foredeep*, were deposited sediments from both sides and especially from the newly risen mountain ranges on the north. These sediments constitute the Siwalik System of the Himalayan foothills and their counterparts in Sind and Burma which are called respectively the Manchhar and Irrawaddy formations. These are largely of fresh water origin, but even though their thickness is large, they are shallow-water formations.

At the end of Siwalik sedimentation i.e., towards the close of the Pliocene, a fourth upheaval took place. This heralded the incoming of the Pleistocene Ice Age which contributed to the virtual extinction of the spectacularly rich mammalian fauna of the Siwalik times, though a small part of the fauna managed to migrate to other areas where they survived. The final phase of Himalayan movements took place in early Pleistocene when the Pir Panjal was raised upto its present height and possibly also some other ranges in the Lesser Himalayas ; for we find Pleistocene deposits on the flanks of the Pir Panjal elevated to a height of several thousand feet above the level of the lakes in which they were originally deposited. Before this movement, man had already appeared on the globe and must have witnessed the final phases of the rise of the Himalayas. Minor adjustments have been taking place since then and the movements in the Himalayan region cannot yet be considered to have completely died down.

Fluviatile and marine facies.—The Tertiary rocks have an early marine facies and a later fluviatile facies, not only in the Himalayas but also in the Burmese and Baluchistan arcs. The original marine basins of deposition were filled up and shallowed and later became estuarine and deltaic. In the northwestern Himalaya, for instance, the Eocene was marine, the succeeding Murree sediments estuarine while the Siwaliks were distinctly fluviatile (i.e., fresh-water) in nature.

Absolute Ages.—A large number of determinations of absolute age by radiometric methods are now available. As a result of the symposium held by the Geological Society of London on the Phanerozoic time scale (Q.J.G.S., London, Vol. 120, 1964) the following limits are assigned to the different Systems in the Tertiary era.

		Million Years
Base of	Pleistocene	1.5
,, ,,	Pliocene	7
,, ,,	Miocene	26
,, ,,	Oligocene	38
,, ,,	Eocene	54
,, ,,	Paleocene	65

The division of the Tertiary is based essentially on palaeontological data, *i.e.*, the disappearance or appearance of certain species of animals and on the percentages of marine molluscan populations. These are related to some extent to the phases of the Alpine-Himalayan orogeny.

Distribution.—The Extra-Peninsular region shows a great development of Tertiary rocks continuously from the Mekran coast of Baluchistan, through the mountainous frontier tracts of Sind and N.W.F.P., to Kashmir and thence along the Himalayan foot-hills to the Brahmaputra gorge in the extreme north-east of Assam. They are probably continuous, underneath the Brahmaputra alluvium, with the broad Tertiary belt of Eastern Assam and Arakan. This is separated from the Burmese Tertiary belt by a zone of Cretaceous (and older) rocks forming the central parts of the Arakan Yoma. In Peninsular India, Tertiary rocks are developed in comparatively small areas in Cutch, Gujarat and Travancore on the western coast and in several places along the eastern coast from Southern Madras to Orissa and the Ganges-Brahmaputra delta.

To enable the reader to form a comprehensive idea of the Tertiary succession in the different areas, a summarised account is first presented before proceeding with the detailed descriptions in later chapters.

Sind and Baluchistan

This region may be considered the type area of the Tertiaries, not only because of the excellent development of the various divisions, but also because it was one of the earliest areas to be studied in detail. It is divided into two regions by the mountain ridges which form the boundary between Sind and Baluchistan Provinces. In the Punjab-Sind region, as one proceeds from south to north, the strata change their marine character to brackish water and fresh water. The Lower Eocene and Oligocene are largely marine, while the Lower Miocene shows two facies and the succeeding Manchhar Series (the equivalent of the Siwalik) is almost entirely of fresh-water origin.

In Baluchistan, however, the formations are mostly marine. This region consists of three parallel provinces or zones adjoining each other; the eastern or *Calcareous zone* shows mainly calcareous and argillaceous sediments of Mesozoic and Eocene ages followed by Upper Tertiary sediments. To the west of the calcareous zone is the *Khojak Mekran zone* composed mainly of flysch-like sediments laid down in a marine basin, typified by the Khojak shales forming the Khwaja Amran ranges. The argillaceous sediments are overlain

TABLE 62—TERTIARY SUCCESSION IN SIND AND BALUCHISTAN

Upper Manchhar (1,500 m)	Sandstones, conglomerates and clays	Pliocene
Lower Manchhar (1 000–1,500 m)	Conglomerates and sandstones with mammalian fossils	Upper to Middle miocene
Upper Gaj (150–300 m)	Red and green shales, occasionally gypseous	
Lower Gaj (150–300 m)	Limestones and shales with marine fossils (represented by fluviatile Bugti beds in Baluchistan)	Lower Miocene
Upper Nari (1,200–1,800 m)	Thick unfossiliferous sandstones and shales	Up. Oligocene
Lower Nari	Fossiliferous marine limestones	Lr. Oligocene
Kirthar (1,500–2,700 m)	Upper—(Spintangi limestone) massive limestones, poorly developed in Sind Middle—Limestone Lower — Shales and sandstones, practically absent from Sind	Middle to Upper Eocene
Laki (150–800 m)	Ghazji shales / Shales and limestones Dunghan limestone / with coal seams and Meting shales and / sometimes oil seepages limestones	Lower Eocene
Upper Ranikot (240 m)	Buff to brown Nummulitic limestone and shales	Up. Pale9cene.
Lower Ranikot (300–450 m)	Gypseous shales and sandstones with lignite lignite and coal	L. Paleocene

Cardita beaumonti beds	Danian
Pab Sandstone	Maestrichtian.

successively by higher Tertiary strata in a southerly and southwesterly direction. To the west of the Khojak zone is the northwestern *Chagai zone* in which are found rocks of various ages including Cretaceous and Tertiary sediments with interbedded lavas as well as intrusives. In this zone lie the volcanoes Koh-i-Sultan and Koh-i-Taftan which have been active in recent geological times.

The calcareous zone includes the hill tracts of the Sind-Baluchistan border composed largely of limestone and the overlying sandstone-clay beds of the Baluchistan foothills, the boundary of this Province with the next adjoining one running a few km to the west of Quetta. The oldest rocks exposed in this area the Productus Limestones which crop out as small inliers amongst Triassic shales at the boundary of the Quetta and Zhob districts. The Triassic rocks are succeeded by a great thickness of calcareous Jurassic sediments which are well seen in the Chiltan and Takatu mountains. These are followed by calcareous and argillaceous rocks of Cretaceous age, which are intercalated with pyroclastics and volcanic flows. Slight sub-aerial weathering of the rocks at the end of the Mesozoic is indicated by the presence of a thin layer of laterite on them. Marine conditions were re-established during the Eocene when the Dunghan limestones were deposited. The younger Eocene strata, partly of shallow water or estuarine origin, contain some coal-bearing rocks as also some gypsum bearing (Spintangi) beds. The post-Eocene deposits, represented by the Gaj and Manchhar beds, ar eof fluviatile origin in the north but of marine character in the south in Lower Sind.

In the Khojak zone which lies to the west of the calcareous zone, the sequence is mainly argillaceous and composed of thick olive green slaty shales and sandstones with thin limestone layers. The beds are sharply folded and thrust over those to the east. Some fossils have been found in the Khojak Pass north of Quetta in these rocks, indicating their Oligocene age. On the western side these are faulted against igneous and metamorphic rocks to the south of Chaman and west of the Khwaja Amran range. The Khojak shales continue northeast towards Fort Sandemen. Continued to the south, they are found in the Mekran ranges where they are overlain by Middle and Upper Tertiary silty clays and sandstones which are of marine origin.

In the northwestern or Chagai zone there are igneous and metamorphic rocks which continue southwards to Nushki and then turn southwestwards. The sedimentary sequence here consists of Hippuritic limestones, Eocene limestones and shales and Upper Tertiary sandstones, clays,a nd conglomerates. These contain some interbedded Tertiary volcanics also. This zone is considered to be part of the Iranian median mass.

The Salt Range

The Punjab Salt Range shows a fine development of Tertiary rocks. The top of the scarp over the greater part of the range is formed of Eocene limestone, mainly of Laki age, while the Ranikot Series is seen as a shaly facies in the eastern part. The limestones are intercalated with marls and are overlain, with a pronounced unconformity, by the Murree Series of Lower Miocene age and this in turn by rocks of the Siwalik System.

TABLE 63—TERTIARIES OF THE SALT RANGE

Siwalik system	Conglomerates, grits, sandstones and shales.	Pliocene to Upper Miocene
Murree series (600 m)	Pseudo-conglomerates, sandstones and purple shales.	Lower Miocene
Laki	*Bhadrar beds* (30–90 m) shales, limestones and marls. *Sakesar limestone.* (60–150 m) Scarp limestone. *Nammal limestone and shales* (30–60 m) Limestones, shales and thin mrals.	Lower Eocene
Ranikot	*Patala shales** (30–75 m) Shales with thin limestones and sandstones and a coal seam at the base. *Khairabad limestone* (15-150 m) Nummulitic limestones and calcareous shales. *Dhak Pass beds* (5–30 m) Sandstones and shales and haematitic beds.	Paleocene

* Part of the Patala shales is of Laki age.

The Potwar Plateau

The northern slopes of the Salt Range merge into the Potwar plateau which forms the type area of the Siwalik formations. The Siwaliks are divided into several stages on lithological and faunal characters since they enclose a rich mammalian fauna. The succession is shown in the accompanying table. (Table 64).

TABLE 64—TERTIARIES OF THE POTWAR REGION

Siwaliks	Upper (1,800 m)	Bouler conglomerate.	Conglomerates, sandstones and clays.	Lr. Pleistocene
		Pinjor stage	Coarase sandstones	Up. Pliocene
		Tatrot stage	Sandstones	
	Middle (1,800 m)	Dhok Pathan stage.	Sandstones and shales	Lr. Pliocene
		Nagri stage	Sandstones and shales	Up. Miocene
	Lower (1,500 m)	Chinji stage	Psudo-conglomerates, red shales and grey sandstones.	Mid. Miocene
		Kamlial stage	Pseudo-conglomerates, grey sandstones and shales.	Mid. Miocene
Murree Series.			Sandstones and purple shales.	Lr. Miocene

Outer Himalaya of Jammu and the Punjab

Tertiary rocks are developed all along the Himalaya, the Siwalik strata forming a practically constant zone of outer hills. Older Tertiary rocks are also known in the Western Himalaya, where they are best developed in Jammu and the neighbourhood. Table 65 shows the sequence here:

TABLE 65—TERTIARIES OF JAMMU

Siwaliks		Upper (1,800 m) Middle (1800 m) Lower (1,500 m)	
Murrees		Upper (900 m) Lower (1,500 m) Basal—Fatehjang zone of ossiferous conglomerates.	
		—————— Unconformity ——————	
Chharat stage	...	Nummulitic shales limestones and marls.	Middle Eocene.
Hill Limestone (500 m)		Massive Nummulitic limestones with coaly layers.	Lower Eocene.

In the foot-hill region of the Simla-Garhwal Himalaya, the Eocene is represented by the Subathu beds consisting of grey to red shales, often gypseous, and some limestones. The Lower and Upper Murrees are represented by the Dagshai and Kasauli beds respectively which are brackish or lagoonal deposits having a total thickness of 2,100 to 2,500 m. The Tertiary rocks of the Eastern Himalayas have been visited by geologists only in a few places and our knowledge of them is meagre. Similar strata are known to be present through almost the entire length of the Himalayas.

Assam

Eastern and southeastern Assam show excellent development of Tertiary rocks but there is a good deal of variation in the succession in different areas. In Upper Assam the Disang Series represents part of the Upper Cretaceous and the Lower and Middle Eocene. The Jaintia Series of southern Assam appears to cover the greater part of Eocene including Paleocene. The succeeding Barail Series is of Upper Eocene and Oligocene age and contains coal seams and petroliferous beds. There is a widespread unconformity in the Oligocene between the Barails and the Surma Series. A minor unconformity is known between the Tipam Series and the Dupi Tila Series, while the latter is generally separated from the Dihings by another unconformity in the Pliocene.

The Tipams are Miocene, while the Dihings are mainly Pliocene, probably extending into the Pleistocene and resembling the Upper Siwaliks in general. The general succession in Assam is shown in Table 66.

There is some variation in the lithology of the different systems developed in Upper Assam and in the Surma Valley and Khasi and Jaintia hills. Owing to paucity of fossils and their being of little use for precise age determination when present, the geologists of the Assam Oil Company have used parallelism in lithology and especially sedimentary petrology (heavy mineral residues) extensively or purposes of correlation. For most of our knowledge of the Assam Tertiaries we are indebted to the excellent work carried out by that organisation.

Burma

The succession in Burma resembles that of Assam in some measure. Eocene beds are developed in the mountainous region of the Arakan Yoma, closely following the Cretaceous rocks. The Oligocene and Lower Miocene are represented by the Pegu Series, corresponding to the Murrees and to the Nari and Gaj beds of northwestern India. The beds above these constitute the Irrawaddy System corresponding to the Siwalik System.

A marine facies is observed in the greater part of the succession in the south, but when the same beds are followed northward, they show estuarine and fresh-water facies. This is due to the fact that a Tertiary gulf existed in the region between the Arakan Yoma on the west and the Shan plateau on the east. This gulf was gradually filled up, the waters receding southward as deposition proceeded.

TABLE 66—TERTIARY SUCCESSION IN ASSAM

(After P. Evans—Figures in metres)

Age	Series	Geosynclinal Facies		Shelf facies
		Surma Valley	Up. Assam	
PLIOPLEISTOCENE	DIHING	Dihing (400)	Dihing (400)	Dhekiajuli (1800)
		—— Unconformity ——		
MIOPLIOCENE	DUPITILA	Duptila (3600)	Namsang (800)	Namsang (600)
		—— Unconformity ——		
MIOCENE	TIPAM	Girujan Clay (1500) / Tipam Sst. (1600)	Girujan Clay (1800) / Tipam Sst. (2300)	Girujan Clay (600) / Tipam Sst. (900)
	SURMA	Bokabil (1500) / Bhuban (4000)	Surma (900)	Surma (200)
		—— Unconformity ——		
OLIGOCENE	BARAIL	Renji (1,000) / Jenam (1200) / Laisong (2400)	Tikak Parbat (600) / Baragoloi (3,300) / Naogaon (2200)	Barail (1200)
EOCENE	DISANG	Disang (1500)	Disang (3000)	JAINTIA SERIES Kopili (500) / Sylhet Lst. (500) / Therria (100)
		—— Unconformity ——		
CRETACEOUS				

TABLE 67—TERTIARY SUCCESSION IN BURMA

Irrawaddy System (1,500 m)	Fluviatile deposits with mammalian fossils and fossil wood.	Pliocene to Upper Miocene
	—Unconformity—	
Upper Pegu Series	Obogon sands and clays (900 m) Kyaukkok sandstone (900 m) Pyawbwe clays (1,500 m)	Vindobonian Burdigalian Aquitanian
	—Palaeontological break—	
Lower Pegu Series	Okhmintaung sandstone (900 m) Padaung clays (750 m) Shwezetaw sandstones (600–1.200 m)	Chattian Stampian Lattorfian
Eocene and Paleocene	Yaw shales (600 m) Pondaung sandstone (1,800 m) Tabyin clays (1,500 m) Tilin sandstone (1,200 m) Laungshe shales (3,000 m) Paunggyi conglomerate (900 m)	Bartonian-Ludian Auversian Upper Lutetian Lower Lutetian Ypresian Londinian

Bengal and Ganges Delta

Western and Eastern Bengal formed the shelf area of a marine basin which stretched from the eastern coast of Orissa to Upper Assam and the geosynclinal facies lay to the southeast of the shelf. The succession of strata in West Bengal (Table 68) is now known from information provided by boreholes put down in connection with exploration for petroleum by the Indo-Stanvac project in the early fifties (B. Biswas, 1959, 1962). The strata dip gently in a S.S.E. direction. It is found that the Rajmahal Traps which are exposed in the Rajmahal hills have been stepfaulted and are encountered in boreholes at increasing depths to more than 3,500 m in the south. The Dauki fault more or less terminates the continuity of the beds into the Assam plateau, while in the Surma valley and Upper Assam the older strata are thrust over the younger in a series of thrusts directed northwestward. In Table 68 the maximum thickness of the formations is given.

TABLE 68—TERTIARY SUCCESSION IN BENGAL BASIN

Shelf Facies	Geosynclinal Facies	
DEBAGRAM (900 m.). Clays and silts showing lateral variation ; shallow marine	RANAGHAT (2,530 m.) below 200m. of alluvium. Estuarine to marine	Plio-Pleistocene
PANDUA (700 m.). Open shelf deposits of sands, silts and shales similar to Durgapur Beds	MATLA (1,280 m.) Alternating marine and deltaic	Mio-Pliocene
All of Shelf facies below this		
MEMARI (60 m.). Estuarine to marine, Calcareous shales and glauconitic sandstones		Upper Oligocene to L. Miocene
BURDWAN (160 m.). Fresh to brackish water sandstones and lignitic shales. Rich in fossils including brachyhaline forams and spore-pollen. Unconformable at bottom, but pass up conformably into the Memari formation		Oligocene

———————————————— Unconformity ————————————————

KOPILI (20-25 m.). Brackish water grey shales and calcareous shales. ⎫ Upper Eocene
Resemble the Kopili Alternations of Assam ⎭

SYLHET (320 m.). Marine clastics with limestone bands. Rich in forma- ⎫ Mid to Upper
minifera. ⎭ Eocene

JALANGI (700 m.). Mainly estuarine sands and shales with lignitic matter ; ⎫ L. Eocene to
basal portion marine in the south ⎬ Upper
⎭ Cretaceous

GHATAL (120 m.). Lagoonal black shales, limestone bands and calcareous ⎫ Upper
sandstones similar to Mahadeo formation of Assam plateau ⎭ Cretaceous

BOLPUR (160 m.). Estuarine to fresh water ; mainly Trap-wash, some white ⎫ Lower
sandstone ⎬ Cretaceous
⎬ Upper
⎭ Jurassic ?

———————————————— Erosional Unconformity ————————————————

RAJMAHAL TRAPS (Over 600 m.). One borehole penetrated 287 m. of the Traps.

The Tertiary section becomes thicker in a southerly direction and the strata which are generally brackish and estuarine in the north, become progressively more marine in the south. The total thickness of the shelf facies is about 5,000 m. near Port Canning and thicker farther to the south. The Rajmahal Traps also extend to the south of Port Canning at the bottom of the section ; they have been met with in borings in East Bengal to 89° 30′ E. and are apparently continuous with Sylhet Traps exposed on the Assam plateau.

Andaman Islands

In these islands a fairly full Tertiary section is seen. The Archipelago is the emergent portion of a great submarine mountain system connecting the Arakan Yomas with the southern strip of Sumatra-Java and the island chain lying immediately to their south. The earliest rocks are Upper Cretaceous clastics and radiolarites (Jacob, 1951) associated with ultramafic and mafic intrusives. The Tertiary succession begins with the Eocene PORT BLAIR SERIES containing Nummulites, Assilina, etc. These are overlain by the GREYWACKE SERIES which are unfossiliferous and are Upper Eocene to Upper Oligocene. Over these are found the ARCHIPELAGO SERIES, the lower part of which is arenaceous, containing Aquitanian fossils such as *Miogypsina, Nephrolepidina, Orbulina, Cycloclypens,* etc. The middle and upper parts contain Burdigalian fossils including *Orbulina universa, Globigerina menardi, Nephrolepidina sumatrensis* and *Operculinoides.* The strong mountain building movements in the Miocene have produced an unconformity, above which appear the WHITE SHALE SERIES of mainly Pliocene age, but which may possibly contain some Miocene. Pleistocene sands clays and coral banks are the youngest formations.

Eastern Coast

Orissa.—The BARIPADA BEDS in Orissa are mainly Upper Burdigalian and they may be underlain by Burdigalian and Aquitanian. These should extend towards the coast.

Godavari Delta.—The seaward part of the Godavari delta should be expected to contain Miocene to Pleistocene strata, but no published data are available to show what lies below the peat-bearing Pleistocene sands and clays.

Pondicherry.—The Cretaceous strata exposed here are known to be overlain by the Eocene. Younger strata should be expected below the alluvium in the Cuddalore-Porto Novo area.

Cauvery Basin.—The Tertiary formations occupy the coastal strip of South Arcot, Tanjore and Ramanathapuram (Ramnad) districts attaining a maximum width of about 75 km in the basin of the Cauvery River. A borehole put down at Karikal about the beginning of the present century indicated the presence of fossiliferous Mio-Pliocene strata which are called the KARIKAL BEDS. From data provided very recently by borings for petroleum by the O.N.G.C. in the Cauvery delta the following Tertiary succession (Table 69) is now known. Further borings are now in progress from which more details will become available.

TABLE 69—TERTIARY SUCCESSION IN THE CAUVERY BASIN

PLEISTOCENE & RECENT—Sands and Clays with peat beds.

PLIOCENE (0–70 m.) (Sands and clays with molluscan fossils, mainly the equivalents of Karikal Beds.

BURDIGALIAN (70-425 m.). Grey shales and silts with *Austrotrillina, Miogypsina, Operculina, Bolivina, Elphidium, Taberina,* etc.

AQUITANIAN (425-550 m.). Sandstones and clays with *Spiroclypeus, Operculina, Bolivina, Textularia, Nodosaria, Quinqueloculina* etc.

OLIGOCENE (550-710 m.). Poorly fossiliferous sandstones. Reticulate Nummulites absent.

U. to MID. EOCENE (710-1670 m.). Grey shales and pyritous silts with *Uvigerina, Hantkenina* (*Aragonella* etc.), *Globorotalia,* etc.

LOWER EOCENE (1,670-1760) m.). Shales and sandstones with *Alveeolina* and arenaceous forams.

This boring was located on a gravity 'high' and the Eocene was underlain by crystalline Archaeans. But the Cretaceous and Upper Gondwana (estuarine and terrigenous) sediments should be present in the depressions in the basement.

The CUDDALORE SANDSTONES of Mio-Pliocene age are exposed from Cuddalore to near Ramnad to the west of the alluvium which covers the greater part of the Cauvery delta and the coastal fringe. Beyond Ramnad, in the Tirunelveli district there is a soft calcareous sandstone of Pliocene to Pleistotocene age while some small islands off the Tirunelveli Coast expose Sub-Recent Coral limestone.

The Tertiary belt of South Arcot-Tanjore-Ramnad shows three interconnected basins separated by two basement ridges, one of them running E.N.E. just to the north of the Coleroon River and the other northeast from near Manamadurai to Kattumavadi on the coast, as the general strike of the

Archaean rocks in the region is E.N.E. to N.E. The basement slopes towards the sea gently at first and more steeply near the coast, where the sedimentary column reaches a thickness of about 2500 to 3000 m. There is a N.E.-trending trough fault, about 16 km wide, passing between Karaikudi and Devakottai, the trough being more than 160 m deep. The basin has a larger thickness in the Palk Strait and shaols up again towards N.W. Ceylon where the Miocene and Upper Gondwana sediments crop out. The depth to basement at Karikal is 1650 m on the ridge, and 2750 m in the furrows on either side; it is 2000 m near Devipatnam and 1000 m at Tondi. Seismic work indicates the presence of a full Tertiary succession all along the basin near the sea coast.

Kerala and Mysore Coast

Along the western coast, from near Nagarcoil to Cannanore and also in Ratnagiri, there are exposures of the VARKALA SANDSTONES (Warkilli Beds) which are of the same age as the Cuddalore Sandstones. They are underlain in Travancore by the Quilon limestones exposed at Padappakara near Quilon. The same strata are known from a borehole at Sakthi near Quilon and also from near Chavara. They have yielded rich Burdigalian forams and Ostracods. Borings put down in the Varkala Sandstones in Travancore for lignite found Archaean crystallines about 50 m below the sandstones, but no Eocene.

Gujarat

Tertiary rocks are exposed all along the coastal region of Surat-Broach-Cambay and also in southern and eastern Kathiawar and in Kutch. In these areas the Deccan Traps seem to have been erupted subaerially and partially submerged at a later date. A trough fault was formed in the Cambay area in early Eocene. This trough is known to extend beyond N. Latitude 24° north of Cambay and along the Bombay coast to the south. The Traps are overlain by a full Tertiary succession in these areas. The Eocene is well developed, followed by regressive Oliogecne and transgressive Miocene and younger beds. The regression in the Oligocene may have been caused by the strong earth movements during that period in the Mediterranean region. The extensive uplift during the Miocene phase of orogeny in the Himalayan, Burmese and Indonesian area and the consequent withdrawal of waters from these areas into the Indian Ocean would explain the Miocene transgression all around this Ocean.

Following the deposition, along the coast of the Peninsula, of Mio-Pliocene strata, there was the final faulting along the western coast of India and along the Makran coast in late Pliocene times. The Owen fracture zone in the Western Arabian Sea may also have been extended towards Karachi at the same time. The down-faulting in the Arabian Sea may have been accompanied by an upward tilt of the Peninsula along the western coastal region, producing the steep westward slope and the more gentle eastward slope. Further movements are inferred from the Pleistocene uplift of the Charnockite massifs in South India, and of Kathiawar and Kutch.

Cambay-Broach Region.—Good Tertiary sections are found both to the north and south of the Narmada valley. In the middle of the Cambay trough the sediments attain a thickness of nearly 2,000 m and are gently folded in a N-S to NNW-SSE axial direction. To the south of the Narmada, in the Ankleswar-Olpad oilfields the Tertiaries are somewhat thinner and the fold axes run N.E.-S.W. to E.N.E.-W.S.W. Table 70 gives the section established in oil wells in the Cambay area.

TABLE 70—TERTIARY SUCCESSION IN THE CAMBAY TROUGH

PLEISTOCENE (100 m.)—Sands and Clays.

PLIOCENE (700 m.)—Brown to greenish sandy clays, sands and gravels.

MIOCENE (600-850 m.)—Grey and greenish silts, calcareous claystone bands, pyritic and Carbonaceous Clays.

OLIGOCENE (60-150 mm.)—Grey and black shales with limestone bands.

KIRTHAR (300 m.).—Black shales, Carbonaceous and pyritic shales with thin limestone layers.

LAKI (200 m.)—Grey to black carbonaceous shales with sandstone layers.

PALEOCENE (40 m.)—Lateritic and carbonaceous variegated clays and Trap-wash.

DECCAN TRAP (Over 1,200 m.)—Bottom not reached.

The same strata continue northward to beyond 24° N. latitude, but become thinner and estuarine. The northern limit of the trough may be in line with the E-W fault along the northern border of the Rann. The Paleocene is largely trap-wash and carbonaceous silty clays indicating very shallow deposition.

Kathiawar

Tertiary strata are exposed in the southwest, south and east of Kathiawar along the coasts. Table 71 shows the strata exposed. They generally dip at a low angle towards the sea.

TABLE 71.—TERTIARY SUCCESSION IN KATHIAWAR

PORBANDAR LIMESTONE —Oolitic Miliolite limestone	PLEISTOCENE
DWARKA BEDS—Clays, silts, cherty limestone	(?) PLIOCENE
PIRAM BEDS—Conglomerates, grits and clays with mammalian fossils of Siwalika age; exposed on the east coast and Piram Island	PLIOCENE
GOGHA BEDS—Thin bedded grits and sandstones exposed on the east coast	MIO-PLIOCENE
GAJ BEDS—Variegated clays, marls, impure limestones	L. MIOCENE
DECCAN TRAPS	CRETACIO-EOCENE
Light coloured and variegated sandstones	MID. CRETACEOUS
WADHWAN SANDSTONES—Light sandstones with marine intercalations	CENOMANIAN TO ALBIAN
Dark grey marls with some plant remains *Ptilophyllum, Araucarites, Cladophlebis*, etc.	NEOCOMIAN-TITHONIAN

Near Dharangadhra, boreholes penetrated about 420 m of horizontal sandy strata of Upper Jurassic age without reaching Lower Jurassic Strata.

Kutch

The Tertiary formations of Kutch are of marine shelf facies and attain a thickness of 600 m especially in the southwest. They overlie some 600 m of of Deccan Traps, 900 m of Cretaceous Umia Series and 1500 m of Jurassic strata. Table 72 shows the succession of Strata.

Rajasthan.—In Bikaner and Jaisalmer there are Eocene strata consisting of Nummulitic limestones associated with beds containing lignite and fuller's earth. The Palana lignite field of Bikaner is situated in these rocks. They are underlain by Cretaceous and Jurassic strata.

This short summary of the Tertiary group will now be followed by more detailed and systematic descriptions of the different systems which form its constituent parts. The inter-relationship of the strata of the different areas will be apparent from Table 73 which gives at a glance the correlation of the Tertiary rocks.

TABLE 72.—TERTIARY SUCCESSION IN KUTCH

PORBANDAR SERIES (15-20 m.)—Calcareous sandstones and Oolitic limestones with Miliolites	PLEISTOCENE
KANKAWATI SERIES (370 m.)—Pink grits with fossil wood. Grey sandstones and clays with *Textularia, Spiroloculina, Triloculina, Rotalia, Nonian* etc. Resemble Machhar Series of Sind. Show transgressive overlap.	PLIOCENE
KHARI SERIES :—	
KHARI STAGE (Upper, 340 m.)—Grey and khaki shales and variegated siltstones with *Lepidocyclina, Miogypsina, Operculina, Austrotrillina howchini, Taberina malabarica ; Ostrea gajensis, Breynia carinata* etc.	BURDIGALIAN
WAIOR STAGE (Lower, 10m.)—Creamy marls and Oolites with *Miogypsina, Spiroclypeus, Nephrolepidina* etc. of typical Aquitanian aspect	AQUITANIAN
LAKHPAT SERIES (10-12 m.)—Yellow and white marls and chalky limestones with *Nummulites intermedius, N. clypens, Operculina* etc. of Nari age	OLIGOCENE
BERWALI SERIES (130 m.)—The upper 60 m. of buff massive limestones and lower 25 m. of gypseous Ochreous and greenish clays, with *Assilina, granulosa, A. exponens, Nummulites atacicus, Lockhartia* etc. Best developed near Bar	KIRTHAR
KAKDI SERIES (45 m.)—Upper 20 m. of grey shales and limestones and lower 26 m. ferruginous, gypseous, and dark lignitic shales ; *Assilina spinosa, A. granulosa, Nummulites atacicus, Operculina, Cibicides, Ostrea, Turritella* etc.	LAKI
MADH SERIES (30-40 m.) of laterite, tuffaceous clays bentonitic clays, pyritic and gypsiferous sandstones ; *Venericardia, Neomeris*,etc.	PALEOCENE

TABLE 73—CORRELATION OF THE TERTIARY FORMATIONS.

	Sind-Baluchistan	Salt Range	Potwar	Simla Hills	Burma	Assam	E. Coast
Pleistocene						Dihing	
Astian	U. Manchhar	U. Siwalik	U. Siwalik				Karikal
Plaisancian						Dupi Tila	Cuddalore
Pontian	L. Manchhar	M. Siwalik	M. Siwalik				Warkalli
Sarmatian		L. Siwalik	L. Siwalik			Tipam	Quilon
Tortonian						Surma	Jaffna
Helvetian	U. Gaj	Murree	Murree	Kasauli	Break		
Burdigalian	L. Gaj		Fatehjang	Dagshai	U. Pegu	Barail	
Aquitanian	U. Nari						
Chattian	L. Nari	Break	Break	Break	L. Pegu		
Stampian							
Lattorfian							
Ludian	Kirthar		Chharat	Subathu	Eocene	Jaintia	
Bartonian	Laki	Laki	Hill				
Auversian	U. Ranikot	U. Ranikot	Limestone				
Lutetian							
Londinian							
Thanetian	L. Ranikot	L. Ranikot					
Montian							

SELECTED BIBLIOGRAPHY

Anderson, R. V. Tertiary startigraphy and orogeny of Northern Punjab. *Bull. Geol. Soc. Amer.* **38**, 665-720, 1927.

Barber, C. T. Tertiary igneous rocks of the Pokokku district. *Mem.* **68**(2), 1936.
Biswas, B. Sub-surface geology of West Bengal. ECAFE *Min. Res. Dev. Ser.* **10**, 159-161, 1959.
Biswas, B. Exploration for petroleum in the western part of the Bengal Basin. ECAFE *Min. Res. Dev. Ser.* **18**, 241-244, 1963.
Biswas, S.K. A new classification of the Tertiary rocks of Kutch, Western India. *GMMSI Bull.* **35**, 1965.
Blanford, W.T. Geology of Western Sind. *Mem.* **17**(1), 1880.

Clegg, E. L. G. Geology of parts of Minbu and Thayetmyo. *Mem.* **72**(2), 1938.
Cotter, G. de P. Geology of parts of Minbu, Myingyan, Pokokku and Lower Chindwin districts. *Mem* **72**(1), 1938.

Duncan, P.M. and Sladen, W.P. Tertiary and Upper Cretaceous fauna of Western India. *Pal. Ind. Ser.* VII and XIV, 1871-1885.

Evans, P. Tertiary Succession in Assam. *T.M.G.I.I.* **27**(3), 1932.

Guha, D.K. Madan Mohan, et al. Marine Neogene Microfauna from Karikal. *GMMSI Bull.* **34**, 1965.

Kailasam, L.N. On the structure of the sedimentary basin of the coastal area of Madras. *Bull. Ind. Geophys. Union*, **1**, 103-108, 1964.
Kailasam, L. N. Geophysical studies of the Ramnad Basin, Madras State. *Current Science*, **35**(8), 197-200, 1966.
Khan, A.H. and Azad, J. Geology of the Pakistan gas fields. ECAFE *Min. Res. Dev. Ser.* **18**, 275-282, 1963.
Krishnan, M.S. The Tertiary Basin of Tanjore, South India. ECAFE *Min. Res. Dev. Ser.* **10**, 151-153, 1959.

Lepper, G. W. Outline of the geology of the oil-bearing region of the Chindwin — Irrawaddy, valley and of the Assam-Arakan. *World Petr. Cong. Proc. I*, 15-23, 1933.

Mathur, L.P. and Kohli, G. Exploration and development of the oil resources of India. *Proc. 6th World Petrol Cong.* Sec. I, 1963.
Mathur, L.P. and Evans, P. Oil in India. XXII *Int. Geol. Cong.* (New Delhi) Special brochure, 86 pp., 1964.
Medlicott, J. G. Tertiary and alluvial deposits of the central portion of the Narbada valley. *Mem.* **2**, Pt. 2, 1860.
Medlicott, H. B. Sub-Himalayan ranges between the Ganges and the Ravi. *Mem.* **3**, Pt. 2, 1864.
Middlemiss, C. S. Geology of the Sub-Himalaya of Garhwal and Kumaon. *Mem.* **24**, Pt. 2, 1890.

Pinfold, E.S. Structure and stratigraphy of N.W. Punjab. *Rec.* **49**, 137-160, 1918.

Sale, H.M. and Evans, P. Geology of the Assam-Arakan region. *Geol. Mag.* **77**(5), 337-362, 1937.
Stamp, L. D. Outline of the Tertiary geology of Burma. *Geol. Mag.* **59**, 481-501, 1922.

Theobald, W. Geology of Pegu. *Mem.* **10**, Pt. 2, 1873.

Vredenburg, E. Geological sketch of the Baluchistan desert and part of Eastern Persia. *Mem.* **31**, Pt. 2, 1901.
Vredenburg, E. Considerations regarding the age of the Cuddalore Series. *Rec.* **36**, 321-323, 1908.
Vredenburg, E. Geology of Sarawan, Jhalawan, Mekram and Las Bela. *Rec.* **38**, 189-215, 1910.
Vredenburg, E. Classification of the Tertiary System of Sind. *Rec.* **34**, Pt, 3. 1906.

Wadia, D. N. Geology of Poonch State, Kashmir. *Mem.* **51**, Pt. 2, 1928.
Wadia, D. N. Tertiary geosyncline of N.W. Punjab. *Q.J.G. M. M. S.* **4**(3), 1932.
Williams, M.D. Stratigraphy of the Lower Indus Basin, West Pakistan, *5th World Petrol. Cong.* Sec. I, paper 19.
Wynne, A.B. The salt region of Kohat. *Mem.* **11**, Pt. 2, 1875.

CHAPTER XVII
THE EOCENE SYSTEM

General.—The end of the Cretaceous period was marked by a widespread marine regression which was, to a large extent, responsible for the destruction of the specialised groups of animals like the ammonites and the coralloid lemallibranchs — the *Rudistae*. This change was similar to that at the close of the Palaeozoic era when the *Goniatites* and specialised brachiopods disappeared from the scene of life. The changes which happened on the surface of the land were similarly responsible for the sudden end of many of the Mesozoic reptiles.

This marine regression accounts for the stratigraphical gap, with erosion unconformity, which separates the Cretaceous from the Tertiary formations in many parts of the world. In India the Eocene begins with the Ranikot stage (Lower Eocene) which is developed in Sind and farther north. The overlying Laki and Kirthar stages (Middle to Upper Eocene) are developed much more extensively in northwestern India. The Uppermost part of the Eocene coincided with the second Himalayan upheaval, so that it is unrepresented by deposits in many parts of the Tertiary belt. The Eocene underwent some uplift and disturbance before the deposition of the Oligocene began.

Distribution.—The Eocene comprises three facies — deep sea, coastal and fluviatile. The first is well-developed in Western Sind and adjoining parts of Baluchistan, parts of the N.-W. Frontier Province, Hazara, Kashmir and presumably along the northern zone of the Himalaya up to the meridian of Lhasa ; and also in the Arakan Yomas on the borders of Burma. The coastal facies is developed in south Kashmir, and the sub-Himalaya from Jammu to near Naini Tal ; in Gujarat, Kutch, Rajputana and to the south of the Shillong Plateau. The fresh-water facies is seen in Upper Burma and in north-western Punjab.

SIND AND BALUCHISTAN

The Kirthar, Laki, Sulaiman and other ranges of the Sind-Baluchistan border show an excellent development of Eocene rocks. The upper part of the Kirthar range exposes upper Eocene rocks which are appropriately named after the range. The eastern flanks expose successively younger beds, *viz.*, Nari, Gaj and Manchhar, dipping towards the Indus plains. To the west, in Kalat, older Eocene rocks are seen, which attain a thickness of 3,000 m. The deposition of the strata in the Laki range also is similar.

RANIKOT SERIES

The lowest division of the Eocene is called the Ranikot series, after Ranikot in Sind. It rests on the Deccan Traps or the *Cardita beaumonti* beds, being conformable to the latter. The lower Ranikot beds, which are 300-450 m thick, comprise soft sandstones, shales and variegated clays. Gypsum and carbonaceous matter frequently occur in them, while in one place there is a coal seam 2 m thick. The fossils found in them are dicotyledonous leaf impressions and oysters in an oyster bed at the base.

The Upper Ranikots, which have a thickness of 210-240 m, consist of fossiliferous brown limestones interstratified with sandstones and clays. Nummulites first appear in the upper part of the upper division, the most characteristic species being *Nummulites nuttalli* and *Miscellanea miscella*. These indicate a Paleocene age.

The fauna of the Upper Ranikot comprises foraminifera, corals, echinoids and molluses, the earliest Nummulites occurring together with the last Belemnites — *Styracoteuthis orientalis*. The Eocene genus *Belosepia* which forms a link between Belemnites and the modern cuttle-fish is also present. A large species of *Calyptrophorus* (gastropod), which is characteristic of the uppermost Cretaceous and lowermost Eocene is found in the lowest bed of the Upper Ranikot. The following are the chief Ranikot fossils :

Foraminifera	*Nummulites nuttalli, Miscellanea miscella, Lockhartia haimei, Assilina dandotica, Operculina* cf. *canalifera, O. sindensis.*
Corals	*Montlivaltia, Isastraea, Thamnastraea, Feddenia, Cyclolites, Trochosmilia, Stylina.*
Echinoids	*Phyllacanthus sindensis, Cyphosoma abnormale, Salenia blanfordi, Dictyopleurus haimei, Conoclypeus sindensis, Plesiolampas placenta, P. ovalis, Eurhodia morrisi, Hemiaster elongatus, Schizaster alveolatus.*
Lamellibranchs	*Ostrea* cf. *multicostata, O. bellovacensis, O. talpur, Flemingostrea haydeni, Spondylus roxanae, Venericardia hollandi, Cardium sharpei, Meretrix morgani, Corbula vredenburgi.*
Gastropods	*Surcula polycesta, S. vredenburgi, Pleurotoma jhirakensis, Calyptrophorus indicus, Conus blagrovei, Athleta noetlingi, Volutocorbis eugeniae, Lyria feddeni, Clavilithes leilanensis, Strepsidura cossmanni, Murex sindiensis, Gisortia jhirakensis, Rostellaria morgani, Rhinoclavis subnuda, Turritella halaensis, Natica adela, Crommium dolium, Velates affinis.*
Cephalopods	*Nautilus subfleuriausianus, N. deluci, N. cossmanni, N. sindiensis.*

LAKI SERIES

The Laki beds are well developed in the calcareous zone of Baluchistan and Sind and also in South Waziristan, Kohat, Salt Range, Attock district, Jammu, Bikaner, Cutch and Assam. They succeed the Ranikot beds and may sometimes be found directly overlying the Cretaceous. The base is sometimes marked by a zone of ferruginous laterite indicating sub-aerial weathering of the underlying beds. The Lakis are the chief oil-bearing beds of Northwestern India. They were divided into three divisions by Vredenburg :—

Upper	Ghazij beds (2,000 ft.)	Gypseous clays, greenish sandstones and clays of the flysch facies, with occasional limestones and coal seams.
Middle	Meting (Dunghan) limestone (500-800 ft.)	White or pale, massive, nodular limestone in Sind (dark coloured in Baluchistan).
Lower	Meting Shales and limestones (50-250 ft.)	White chalky limestones and shales.
	Basal laterite	Thin crust of ferruginous laterite.

At a later date, W. L. F. Nuttall (*Q.J. G. S.* **81**, 417-453, 1925) found that the Laki Limestone had been mistaken by Vredenburg for the Meting Limestone, and proposed the following classification :

	Thickness (Feet)
Laki Limestones	200-600
Meting Shales	95
Meting Limestones z	140
Basal Laki Limestone	25

The name Meting has been taken from Lower Sind where the formations are well developed in the Laki Range. There are well marked unconformities at the base and at the top of the Laki succession which is followed by -Middle Kirthar beds. In Baluchistan, the Meting Shales are absent and the Laki Limestones (Bolan Limestones) are overlain by the Ghazij Shales which appear to be partly Laki and partly Lower Kirthar.

The name DUNGHAN LIMESTONE was proposed by R. D. Oldham (*Rec.* **23**, p. 94, 1890) after the Dunghan hill (29° 52' : 68° 22') where the formation is well exposed. It was considered to be entirely of Laki age, but later work by the geologists of the Burma Oil Co. has shown that the lowest part contains *Orbitoides media* and *Omphalocyclus macropora* which indicate Upper Cretaceous age. The middle part has yielded typical Paleocene fossils such as *Nummulites nuttalli, N. thalicus, Miscellanea miscella, Lockhartia haimei,* etc. The upper part (about 75 m) is of Laki age and corresponds to the BOLAN LIMESTONE of the Bolan Pass area. L. M. Davies has therefore suggested that the term Bolan Limestone be adopted for the Laki, as that formation occurs over a distance of more than 400 km, extending from Baluchistan through the Sulaiman mountains to Warizistan. The Bolan Limestone is a massive, well bedded, tough and somewhat nodular limestone attaining a thickness of several hundred feet.

The Lakis are represented in the Kohat district by the SHEKHAN LIMESTONES, PANOBA SHALES and TARKHOBA SHALES ; in the Salt Range by the SAKESAR LIMESTONES which form the scarps at the top, and by the BHADRAR BEDS ; while in Waziristan the equivalents are mainly shales of estuarine facies.

The Laki Series is characterised by *Nummulites atacicus, Assilina granulosa* and *Alveolina oblonga.* The Ghazij Shales are rarely fossiliferous but the Laki Limestones contain a rich echinoid fauna. There are also plant fossils including seeds and leaf impressions.

Foraminifera	...	...	*Assilina granulosa, A. spinosa, Nummulites atacicus, N. irregularis, Lackhartia conditi, Orbitolites complanata,* etc.
Echinoids	...	...	*Leiocidaris canaliculata, Porocidaris anomala, Cyphosoma macrostoma, Micropsis venustula, Conoclypeus alveolatus, Echinocyamus nummuliticus, Amblypygus sub-rotundus, Eolampas excentricus, Echinolampas rotunda, E. obesa, Hemiaster nobilis, H. carinatus, Metalia sowerbyi, M.depressa, Schizaster symmetricus, Macropneustes specious, Euspatangus* sp.

PLATE XXIV
CRETACEOUS AND TERTIARY FORAMINIFERA

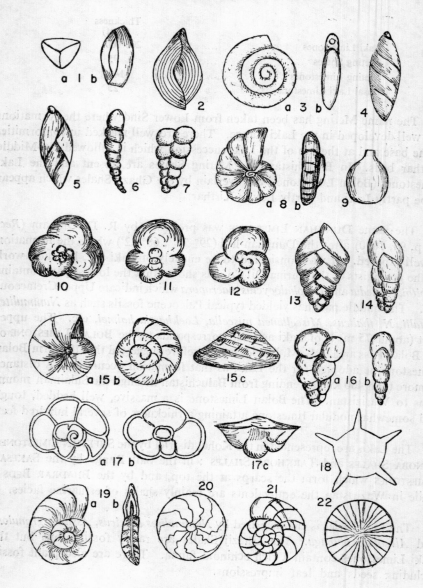

EXPLANATION OF PLATE XXIV

1. *Triloculina.* 2. *Spiroloculina.* 3. *Ammodiscus.* 4. *Virgulina.* 5. *Pyrulina.* 6. *Dentalina.* 7. *Textularia.* 8. *Anomalina.* 9. *Lagena.* 10. *Globotruncana appenninica.* 11. *Globigerina bulloides.* 12. *Globigerina triloculoides.* 13. *Bolivina incrassata.* 14. *Bulimina.* 15. *Rotalia.* 16. *Gumbelina globosa.* 17. *Globorotalia.* 18. *Calcarina,.* 19. *Operculina.* 20. *Amphistegina,* 21. *Heterostegina.* 22. *Discocycliua.*

The following mollusca are found in the Laki and Kirthar beds and may probably be common to both ;

Lamellibranchs ... ... *Ostrea vesicularis, pholadomya halaensis, Vulsella legumen.*
Gastropods ... ... *Turritella angulata, Nerita schmideliana, Natica longispira, Terebellum plicatum, Rostellaria angistoma, R. prestwichi, Ovulum murchisoni.*

KIRTHAR SERIES

The Lakis and Kirthars are exposed in the hilly tract of Northwestern Sind, both containing similar-looking massive limestone. The Kirthars are exposed in the Kirthar, Dumbar and Kimbu ranges while the Lakis are seen to their southeast in the Laki, Sumbak, Surjana and Kara ranges.

The Kirthar Series consists of two divisions, the lower mainly shaly, and the upper mainly calcareous. The lower shaly division is chiefly of the flysch facies, consisting of thin-bedded greenish shales and some sandstones and limestones, attaining a tihckness of several thousand metres.

There is a distinct stratigraphical and faunal break between Laki and Kirthar, marked by brecciated limestone strata. The basal beds in Baluchistan are locally the GHAZABAND LIMESTONES, named after a hill 25 Km. from Quetta. They contain *Nummulites laevigatus* and *Assilina exponens.*

The Middle Kirthars form a passage zone to the massive Upper Kirthar limestones, and may in fact be considered as their lower part. They are characterised by *Nummulites gizehensis, N. beaumonti, Disscocyclina javana, D. undulata* and *Assilina spira*, the last extending into the Upper Kirthar. Only the middle portion of the Kirthars is well developed in Sind, the lower beds and the uppermost beds being often missing. But all the divisions are well developed in Baluchistan.

The Upper Kirthars, also known as the SPINTANGI LIMESTONES in Baluchistan, attain locally a thickness of 1,000 m. Their characteristic foraminifer is the large form *Nummulites complanatus.*

The equivalents of the Kirthars are found in many places in the Tertiary Alpine belt of Western Asia and the Mediterranean region. The Kirthars are rich in fossils. Amongst the more important of them are :

Foraminifera ... ... (Lower) *Nummulites laevigatus, N. obtusus, N. atacicus, Assilina exponens.*
(Middle) *Nummulites gizehensis, N. acutus, N. beaumonti, N. murchisoni, N. discorbina, N. perforatus, Dictyoconoides cooki, Discocyclina javana, D. dispansa, D. undulata, Assilina exponens, A. spira, A.papillata, Alveolina elliptica.*
(Upper) *Nummulites perforatus, N. complanatus, N. biarritzensis.*
Echinoids ... ... *Cyphosoma undatum, Conoclypeus rostratus, Sismondia polymorpha, Amblypygus tumidus, A. latus, Echinolampas sindensis, Echinanthus intermedius, Micraster tumidus, Schizaster simulans.*

Since the foraminifera are of great importance in the zonal sub-division of the Eocene, Table 74 gives the chief forms and their distribution in Sind-Baluchistan.

SALT RANGE

The Eocene strata of the Salt Range are intermediate in character between the deep sea facies of Sind and the coastal facies of the Sub-Himalaya. The greater part of the Eocene is well developed here, the beds generally thickening towards the west. The beds overlap the older formations and are overlain by the Murrees or Siwaliks.

TABLE 74—FORAMINIFERA OF THE EOCENE OF WESTERN INDIA
(AFTER W. L. F. NUTTALL)

v.a. = very abundant a = abundant c = common f = frequent p = present or rare Species	Upper Ranikiot	Laki — Meting and Dunghan Limestone	Laki — Middle Laki Limestone	Laki — Ghazij shales	Kirthar — Lower	Kirthar — Lr. Middle	Kirthar — Up.-Middle
Miscellanea miscella (D'Arch & Haime)	f	–	–	–	–	–	–
Assilina ranikoti Nuttall	a	–	–	–	–	–	–
Operculina canalifera D'Arch	f	–	–	–	–	–	–
O. hardiei D'Arch & Haime	c	–	–	–	–	–	–
Lokhartia conditi Nuttall	P	–	–	–	–	–	–
Alveolina oblonga D'orb.	–	c	c	–	–	–	–
Flosculina globosa (Leym.)	–	v.a.	–	–	–	–	–
Alveolina subpyrenaica Leym.	–	v.a.	v.a.	–	–	–	–
Nummulites irregularis Desh.	–	P	–	f	–	–	–
Assilina granulosa D'Arch	–	a	a	a	–	–	–
Nummulites atacicus Leym.	–	a	a	f	a	a	–
Orbitolites complanata Lmk.	–	a	c	P	P	P	–
Assilina exponens (Sow.)	–	–	–	–	c	c	–
Nummulites obtusus (Sow.)	–	–	–	–	p	a	c
Nummulites acutus (Sow.)	–	–	–	–	–	f	–
N. beaumonti D'Arch and Haime	–	–	–	–	–	f	–
N. stamineus Nuttall	–	–	–	–	–	f	–
Dictyoconoides cooki (Carter)	–	–	–	–	–	a	–
Discocyolina dispansa (Sow.)	–	–	–	–	–	a	–
D. javana var. *indica* Nuttall	–	–	–	–	–	v.a.	–
D. Undulata Nuttal	–	–	–	–	–	a	–
D. sowerbyi Nuttall	–	–	–	–	–	p	–
Alveolina elliptica (Sow.)	–	–	–	–	–	c	p
Nummulites leavigatus (Brug.)	–	–	–	–	–	p	a
N. aff. *scaber* Lmk.	–	–	–	*	–	–	a
N. carteri D'Arch & Haime	–	–	–	–	–	–	p
N. gizehensis (Forksal)	–	–	–	–	–	–	p
Assilina cancellata Nuttall	–	–	–	–	–	–	f
A. papilata Nuttall	–	–	–	–	–	–	c
A. spira de Roissy	–	–	–	–	–	–	f

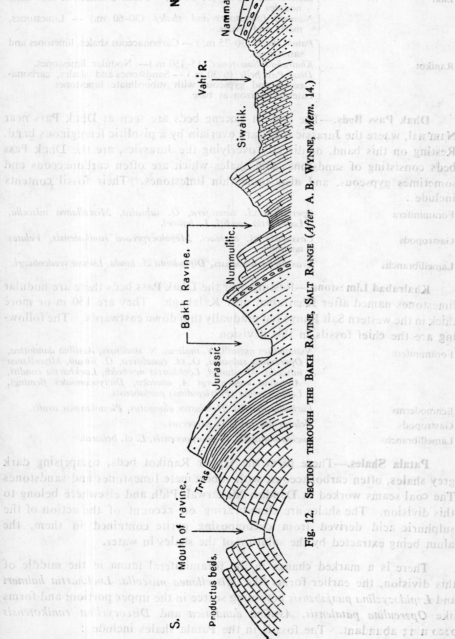

Fig. 12. Section through the Bakh Ravine, Salt Range (*After* A. B. Wynne, *Mem.* 14.)

446   GEOLOGY OF INDIA AND BURMA   [CHAP.

The following succession (in the descending order) has been established

Laki
- *Bhadrar beds* (30–90 m.) — Sandstones, limestones, marls and clays.
- *Sakesar Limestones* (60–90 m.) — Massive limestone with chert nodules.
- *Nammal Limestones and shales* (30–60 m.) — Limestones, marls and shales.

Ranikot
- *Patala shales* (30–75 m.) — Carbonaceous shales, limetones and sandstones.
- *Khairabad limestones* (15–150 m.) — Nodular limestones.
- *Dhak Pass beds* (6–30 m.) — Sandstones and shales, carbonaceous and gypseous with subordinate limestones.
- Laterite horizon at base.

Dhak Pass Beds.—The earliest Eocene beds are seen at Dhak Pass near Nammal, where the Jurassic rocks are overlain by a pisolitic ferruginous band. Resting on this band, or directly overlying the Jurassics, are the Dhak Pass beds consisting of sandstones and shales which are often carbonaceous and sometimes gypseous, and also some thin limestones. Their fossil contents include :

Foraminifera	*Operculina* cf. *canalifera, O. subsalsa, Miscellanea miscella, Lockhartia conditi, L. haimei.*
Gastropods	*Cassidaria* cf. *archiaci, Megalocypreaea ranikotensis, Velates noetlingi.*
Lamellibranchs	*Crassatella salesensis, Diplodonta* cf. *hindu, Lucina vredenburgi.*

Khairabad Limestone.—Resting over the Dhak Pass beds there are nodular limestones named after Khairabad near Kalabagh. They are 150 m or more thick in the western Salt Range but gradually thin down eastwards. The following are the chief fossils in this division :

Foraminifera	*Nummulites nuttalli, N. thalicus, N. sindensis, Assilina dandotica, Operculina subsalsa, O.* cf. *canalifera, O. jiwani, Miscellanea miscella, M. stampi, Lockhartia newboldi, Lockhartia conditi, Alveolina vredenburgi A. ovoidea, Dictyoconoides flemingi, Lepidocyclina (Polylepidina) punjabensis.*
Echinoderms	*Eurhodia morrisi, Hemiaster elongatus, Plesiolampas ovalis.*
Gastropods	*Velates noetlingi, V. perversus.*
Lamellibranchs	*Lucina mutabliis, L. noorpoorensis, L.* cf. *bellardi.*

Patala Shales.—These form the Upper Ranikot beds, comprising dark grey shales, often carbonaceous, with subordinate limestones and sandstones The coal seams worked at Dandot, Makarwal, Pidh and elsewhere belong to this division. The shales are alum-bearing on account of the action of the sulphuric acid derived from decomposing pyrite contained in them, the alum being extracted by the solution of the shales in water.

There is a marked change in the foraminiferal fauna in the middle of this division, the earlier forms like *Miscellanea miscella, Lockhartia haimeri* and *Lepidocyclina punjabensis* becoming scarce in the upper portion, and forms like *Operculina patalensis, Assilina dandotica* and *Discocyclina ranikotensis* becoming abundant. The fossils in the Patala shales include :

Foraminifera	*Operculina patalensis, O.* cf. *canalifera, O. salsa, O. subsalsa, Assilina dandotica, A. spinosa, Discocyclina ranikotensis, Nummulites globulus, N.* cf. *mamilla, Alveolina globosa, A. vredenburgi, A. ovoidea.*
Corals	*Astroceonia blanfordi, Trochocyathus,* cf. *epithecata.*

| Gastropods | ... | ... | *Turritella ranikoti, T. hollandi, T. halaensis, Mesalia fasciata, Rimella jamesoni.* |
| Lamellibranchs | | ... | *Crassatella salsaensis, Ostrea pharaonum* var. *aviculina.* |

Nammal Limestones and Shales.—These consist of limestones, marls and shales, a fine section being seen in the Nammal gorge. The characteristic fossils are *Nummulites atacicus, N.* cf. *mamilla, N. irregularis, Assilina granulosa*. Some long-range forms like *N. lahirii, Lockhartia tipperi, Discocyclina ranikotensis, Assilina subspinosa* and *Ostrea flemingi* are also found in them.

Sakesar Limestone.—This is a massive limestone, 60 to 120 m thick, containing numerous chert nodules in places. It is the characteristic member of the Laki series, forming high cliffs like the Sakesar hill, towering above the scarp. Occasionally it is seen to pass into gypsum, as near Kalabagh. On account of its massiveness and well-developed joints, it weathers into steep and irregular masses having the appearance of ruined fortress walls. The weathered surface shows numerous Nummulites. The chief fossils found are : *Nummulites atacicus, N.* cf. *mamilla, Assilina granulosa, A. spinosa, Lockhartia tipperi, L. conditi, Alveolina oblonga, A. ovoidea, A. globosa.*

Bhadrar Beds.—These constitute the uppermost Laki division, overlying the Sakesar limestone. They consist of sandstones, limestones, clays and marls varying in thickness in different places from 60 m to a few metres. In some places they are associated with the red clays characteristic of the Chharat beds of the Kala Chitta hills. Massive gypsum, regarded as derived from Laki limestone, occurs at the base of this division near Mari-Indus. The characteristic foraminifera are *Orbitoides complanatus* and *Assilina* cf. *pustulosa*, together with longer range forms like *Nummulites* cf. *mamilla, N. atacicus, Assilina subspinosa, Lokhartia conditi, L. tipperi, Alveolina ovoidea, A.oblonga.*

KOHAT DISTRICT

In the Kohat district, northwest of the Salt Range, there occur beds of rock salt overlain by massive gypsum, the latter being altered limestones of Laki age. The salt of this region is grey or dark owing to inclusions of bituminous matter. The Laki limestone and gypsum are intercalated with greenish shales and are succeeded by Kirthar rocks which consist mainly of limestones and red clays.

Near Kohat itself the Lakis consist of greenish shales at the base, succeeded by the SHEKHAN LIMESTONE (Upper Laki) and by red gypseous clays. The overlying Kirthars comprise a lower division of KOHAT SHALES and limestones and an upper division including Nummulitic shales and ALVEOLINA LIMESTONES.

SAMANA RANGE

Quartzites and Hangu Shales.—In the Samana Range, which lies some distance to the northwest of the Kohat area, the lowermost Eocene consists of about 45 m of white quartzite sandstones followed by the Hangu Shales which are 4.5 m thick and full of fossils. The Hangu Shales form a useful marker horizon in this region. The fossils show affinities with those of the

TABLE 75—EOCENE FORAMINIFERA OF THE SALT RANGE (AFTER DAVIES AND PINFOLD, *Pal. Ind.* XXIV, I, p. 67, 1937).

*present only in a small part of the strata ×present throughout or in greater part Species.	Ranikot			Laki		
	L	M	U	L	M	U
Nummulites nuttalli Davies	—	×	×	—	—	—
„ *thalicus* Davies	—	×	×	—	—	—
„ *sidensis* (Davies)	—	*	×	—	—	—
„ *globulus* Leym.	—	*	×	×	×	×
„ cf. *mamilla* (Fich. & Moll.)	—	—	×	×	×	×
„ *atacicus* Leym.	—	—	×	×	×	*
„ *lahiri* Davies	—	—	×	—	—	—
„ *irregularis* Desh.	—	—	×	×	×	×
„ *subirregularis* De la Harpe	—	*	×	—	—	—
Assilina dandotica Davies	—	—	×	—	—	—
„ *granulosa* D'Arch	—	—	—	×	×	—
„ *spinosa* Davies	—	*	×	×	×	×
„ *subspinosa* Davies	—	—	—	×	×	×
„ cf. *pustulosa* Don.	—	—	—	×	×	×
Operculina cf. *canalifera* D'Arch.	—	—	—	—	—	—
„ *salsa* Davies	—	×	×	—	—	—
„ *subsalsa* Davies	—	×	×	—	—	—
„ *patalensis* Davies	—	×	×	—	—	—
„ *jiwani* Davies	—	—	×	—	—	—
Miscellanea stampi (Davies)	—	×	×	—	—	—
„ *miscella* (D'A. & H.)	×	×	×	—	—	—
Lockhartia haimei Davies	—	×	×	×	×	×
„ *newboldi* (D'A. & H.)	×	×	×	×	×	×
„ *conditi* (Nuttall)	—	×	×	×	×	×
„ *tipperi* (Davies)	—	—	×	*	*	*
Sakesaria cotteri Davies	—	×	×	—	—	—
Dictyoconoides flemingi Davies	—	—	—	—	*	*
Heterostegina cf. *ruida* Schw.	—	—	—	—	—	—
Lepidocyclina (Polylepidina) punjabensis Davies	—	×	×	×	×	×
Discocyclina ranikotensis Davies	—	×	×	—	—	—
Orbitolites complanatus Lamk.	—	*	×	—	—	—
Alveolina vredenburgi Davis	—	—	—	—	×	×
„ *oblonga* D'Orb.	—	*	*	—	×	×
„ *ovoidea* D'Orb.	—	—	—	—	×	×
„ *globosa* Leym.	—	×	×	—	—	—

Caraita beaumonti beds, but the absence of cephalopods and of the larger foraminifera shows that they belong to early Eocene age, *i.e.*, Lower Ranikot. All the fossils found in them are new except a few which have a long time range.

Corals ... ... *Blagrovia simplex, Placotrochus tipperi, Euphyllia thalensis, Astrocoenia (Platastrocoenia) ranikoti, A. blanfordi, Cyclolites vicaryi, C. striata, Placosmilia wadiai.*

Gastropods ... ... *Companile brookmani, Turritella daviesi, T. ranikoti, Mesalia fasciata, Tibia samanensis, Rimella levis, Euspira roei, Globularia brevispria, Architectonica mainwaringi, Hemifusus montensis, Murex wadiai, Strepsidura tipperi, Voluta vredenburgi, Athleta (Volutocorbis) deviesi, Lyria samanensis.*

Lamellibranchs ... ... *Cardita hanguensis, Cardium inaequiconvexum, Meretrix indica, Trapezium daviesi, Crassetellites exiguus, Corbula samanensis,*

Lockhart Limestone and Hangu breccia.—The overlying rocks show two facies, one being a massive grey limestone of 60 m thickness (Lockhart Limestone) and the other a limestone breccia. The larger foraminifera make their first appearance here, species of *Lockhartia* (*L. haimei*, *L. newboldi* and *L. conditi*) being common.

Upper Ranikot.—Above the Lockhart Limestones are clays, shales and impure limestones having a thickness of 20 m capped by a limestone-breccia which is 10 m thick. The most important fossils in these are *Nummulites nuttalli*, *N. thalieus*, *Operculina* cf. *canalifera* and *Discocyclina* sp. The Upper Ranikot contains several corals including the following :

Paracyathus altus, Feddenia jacquemonti, Astrocoenia blanfordi, A. ramosa, Thamnastrea balli, Diploria flexuosissima, Pachyseris murchisoni, Trochoseris obliquatus, Isis ramikoti.

POTWAR PLATEAU

Hill Limestone.—The lowest zone in the Tertiaries of the Kawagarh and Kala Chitta hills is a ferruginous pisolite associated with unfossiliferous shales of Lower Ranikot age. This is overlain by the Hill Limestone, a massive limestone with shale intercalations, including both the Ranikot and Laki series. The shaly beds in the Hill Limestone occasionally contain layers of coal. They attain a thickness of several hundred m but vary from place to place, the upper portion containing the Laki fossil *Assilina granulosa*. The Hill Limestone is therefore Paleocene to L. Eocene in age.

Chharat Series.—The Hill Limestone is succeeded by the Chharat Series in the Kala Chitta hills, where the following divisions have been recognised:

3. Nummulitic shales (15-60 m).
2. Thin bedded limestones and green shales (30-60 m).
1. Variegated Shales and Limestones (100 - 150 m).

The passage bed between the Hill Limestone and the Variegated Shales is a chalky limestone with gypsum, showing oil seepages near Chharat. The Variegated Shales and Limestones show fragments of reptilian and mammalian fossils and shells of *Planorbis*. The middle division contains *Nummulites* and molluscs including *Cardita* (*Venericardia*) *subcomplanata*. The Nummulitic shales contain numerous *Assilina papillata* and *Discocyclina javana*. It is therefore considered to represent the lower part of the Middle Kirthar. The lower part in the Chharat may represent a part of the Laki and basal Middle Eocene while the upper part is Middle Eocene.

Kuldana Beds.—Some calcareous conglomerates and red shales which are found between the Nummulitics and the Murrees were described by Wynne as the Kuldana series and regarded as the equivalents of the Subathus. Pinfold showed later that they are approximately of the same age as the Chharat series.

HAZARA

The southeastern border of the mountains of Hazara shows a well developed zone of Eocene rocks. At the base is a band of pisolitic laterite followed by beds of variegated sandstones and clays, about 6 m thick, containing seams of inferior coal. These are overlain by 60 m of greyweathering, well bedded, massive limestones which emit foetid smell when broken. The limestones contain small *Nummulites* of the size of barley grains and a zone with *Echinolampas* near the base. These may be of Laki age. They are overlain by shales, marls and nodular limestone containing Montlivaltia and large *Nummulites*. These beds rather resemble the Chharats and are of Kirthar age.

Overlying these with an unconformity is a band of shales, clays and marls 4-6 m thick, known as the Kuldana beds. They are purple to deep brown in colour and contain *Nummulites* derived from the denudation of the older beds. The beds are succeeded by the Murree System.

KASHMIR

Eocene rocks similar to those of Hazara are developed on the southern flanks of the Pir Panjal. They consist of limestones resembling the Hill Limestone, followed by a large thickness of variegated shales containing a few coal seams in the lower part. The limestones are thin-bedded, pale grey and cherty containing a few *Nummulites* of Ranikot age and gastropods. They attain a thickness of 100 to 150 m. The overlying beds are pyritous, coaly and ferruginous shales with thin carbonaceous beds. These are succeeded by thin-bedded dark limestones containing *Nummulites*, *Assilina* and *Ostrea*, and these in turn, by variegated shales of several hundred m thickness with sandstone intercalations. This shale and limestone formation is similar to the Chharats in characters.

Chharat	Variegated red and green shales (800 ft.). Dark thin-bedded lenticular nummulitic limestone (100 ft.). Pyritous and carbonaceous shales with ironstone (50 ft.).
Ranikot	Thin bedded, pale grey, cherty limestones with a few *Nummulites* and gastropods (400 ft.)

To the south of the Pir Panjal there is a series of outcrops of Eocene rocks near Riasi and Jammu. These contain a basal zone of laterite succeeded by grey and green pyritous and carbonaceous shales and Nummulitic limestones. They attain a thickness of 180 m or more and are similar to the Subathu beds of Simla foothills farther east. The laterite is often highly aluminous and may therefore be useful as an ore of aluminium. The shales overlying them contain seams of coal which are workable but are more or less crushed and graphitic. The Nummulitic limestone is dark and thin-bedded, but when followed westwards becomes paler, more massive and thicker and contains *Nummulites atacicus* and *Assilina granulosa*.

PLATE XXV
TERTIARY FOSSILS

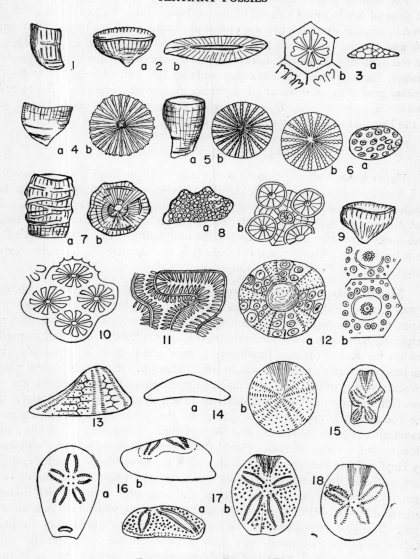

Explanation of Plate XXV

1. *Caryophyllia compress*. 2. *Trochosmilia medlicotti*. 3. *Astrocoenia numisma*. 4. *Montlivaltia ranikoti* (1a, 0.7; b, 1.5). 5. *Montlivaltia vignei* (a, 0.3; b, 0.5). 6. *Porites indica* (a, 0.8; b, 8). 7. *Cyclolites orientalis* (a, 0.25; b, 0.4). 8. *Stylina reussi* (a, 0.2; b, 8). 9. *Blagrovia simplex*. 10. *Astrocoenia blanfordi* (6). 11. *Meandrina medlicotti* (4). 12. *Salenia blandfordi* (b, magnified). 13. *Conoclypeus sindensis* (0.25). 14. *Plesiolampas placenta* (0.4). 15. *Hemiaster elongatus* (0'7). 16. *Eurhodia morrisi* (0.4). 17. *Macropneustes speciosus* (0.4). 18. *Schizaster alveolatus* (0.4).

SUB-HIMALAYA OF SIMLA

The Jammu belt of Eocene rocks continues southeastwards along the foot-hill zone of Simla and Garhwal as far as Naini Tal. The deposits gradually thin down in this direction and are of lagoonal nature. They are called *Subathu beds* and consist of a basal bed of pisolitic laterite overlain by greenish grey and red gypseous shales with occasional sandstones and a few impure limestone bands. The Subathus are the equivalents of the Lakis as they have yielded *Assilina granulosa, A. spinosa, A. leymerei, Nummulites atacicus, N. mamilla, Lockhartia ,Globorotalia, Operculina* etc. They are succeeded, after a gap, by the Dagshai beds of Lower Miocene age.

CENTRAL HIMALAYA AND TIBET

Upper Indus Valley.—Eocene rocks are found in the Upper Indus valley in Ladakh along a zone parallel to the Himalayan axis from Kargil to Leh, Hanle and beyond. They consist of feldspathic grits, green and purple shales and limestones containing badly preserved *Nummulites* and other fossils. The rocks have been subjected to folding and crushing and igneous intrusions on a large scale. From fossil evidence it is known that the sediments extend in age from the Cretaceous to Oligocene.

Mount Kailas.—In the southwestern part of Tibet is Mount Kailas which is made up entirely of arkose, sandstones and conglomerates. According to Heim and Gansser, the conglomerates are at least 2,000 m thick and may originally have been more than 4,000 m thick. On the north the conglomerates rest on the Kailas (Trans-Himalayan) granite and contain large boulders near the base which gradually diminish in size further up. The boulders and pebbles consist not only of this granite, but also of granophyre, lipartite, dacite, andesite, and tuff of intermediate composition. These must have been derived from the rocks which are known to occur in the Trans-Himalaya mountain of Bongthol as shown by Hennig and Sven Hedin. The Kailas granite is a hornblendic type without tourmaline and is similar to the Kyi-Chu granite described by Hayden from the region of Lhasa. As it has contributed pebbles and boulders to the Kailas conglomerates, it should be of Upper Cretaceous or earlier age. It is different from the tourmaline and muscovite granite which is characteristic of the main Himalayan ranges farther south and which may be of different ages. The great thickness of the conglomerates and sandstones in Mount Kailas indicates that these must have been deposited in a steadily sinking but shallow furrow formed in Cretaceous times. There are no rocks in this region younger than the Eocene.

Southern Tibet.—Eocene rocks occur over large areas in Southern Tibet and form part of the KAMPA SYSTEM. The sub-divisions recognised by Hayden are given in Table 69. The ferruginous sandstone is similar to the Dhak Pass beds of the Salt Range. The succeeding three beds are the equivalents of the Khairabad Limestone, and all contain foraminifera, especially the *Operculina* limestone. The chief foraminifera in these are *Miscellanea miscella, Nummulites sindensis, N. thalicus, Operculina subsalsa, Lockhartia haimei, L. newboldi,*

L. conditi, Dictyoconoides cf. *flemingi, Lepidocyclina (Polylepidina) punjabensis Verneuilia* sp., and these bear a striking resemblance to the fauna of the Salt Range. The mollusca found in these beds include *Megalocypraea ranikotensis, Gosavia humberti, Hippochrenes* cf. *amplus, Campanile brevis, Velates perversus Vulsella legumen, Ostrea (Liostrea) flemingi.*

Table 76—Eocene Succession in Kampa Dzong

6. Dzong-buk shales (150 ft.). Sandy micaceous shales with thin sandstone layers	Laki
5. Orbitolites limestone (50 ft.). Limestone full of *Orbitolites* and *Alveolina*	Laki
4. Spondylus shales (150 ft.). Fine-grained, greenish grey and black shales.	
3. Operculina limestone (150 ft.). Shaly nodular limestone full of foraminifera.	Upper Ranikot
2. Gastropod limestone (300 ft.). Hard, dark, massive limestone, thin bedded at base, with a shale band 40 ft. thick just above the middle.	
1. Ferruginous sandstones (200 ft.)	Lower Rainkot

ASSAM

The Eocene is well developed in Assam. In the southern and eastern parts of the Shillong plateau it is represented by the *Jaintia Series*. A different geosynclinal lithological facies, called the *Disang Series*, is found in Upper Assam extending from the Brahmaputra valley southwards into Manipur and beyond. These two facies have been brought into juxtaposition by the Haflong-Disang thrust fault. While the Jaintia Series covers almost the entire Eocene System it is thought that the Disangs extend from Upper Cretaceous to the upper part of the Eocene.

Jaintia Series.—This can be divided into several formations by reason of their lithological units and fossil content. The three main lithological divisions are the Therria, Sylhet and Kopili. The different units now recognised by the geologists of the Assam Oil Co. Ltd. are shown below :

Table 77—Eocene Succession in Assam

Kopili Stage	Alternations of shales and sandstones with bands of calcareous sandstones and shales (1,500 ft.)	Upper Eocene
Sylhet Limestone Stage.	Prang Limestone. Fossiliferous limestone (400-900 ft.)	Middle Eocene
	Nurpuh Sandstone. Sandstone with subordinate calcareous bands (60 ft.)	
	Umlatodoh Limestone. Limestones with occasional sandstone bands (200 ft.)	Lower Eocene
	Lakadong sandstone. Coal-bearing sandstones (80 ft.)	Upper Paleocene
	Lakadong Limestone. Fossiliferous limestone (500 ft.)	
Therria Stage	Upper Therria. Hard sandstones (upto 100 ft.)	Lower Paleocene
	Lower Therria. Limestones and calcareous sandstones (upto 225 ft.)	

PLATE XXVI
TERTIARY FOSSILS

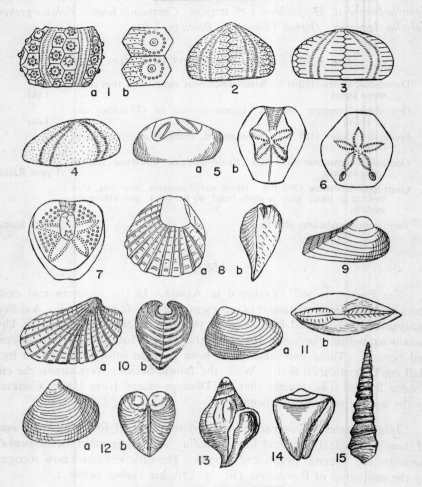

EXPLANATION OF PLATE XXVI

1. *Leiocidaris canaliculata* (b-enlarged). 2. *Conoclypeus alveolatus* (0.2). 3. *Amblypygus subrotundus* (0.3). 4. *Echinolampas desori* (0.7). 7. *Breynia carinata* (0.4). 8. *Ostrea promensis* (0.25). 9. *Corbula socialis* (2). 10. *Arca burnesi* (0.4). 11. *Nucula alcocki* (2). 12. *Cyrena crawfurdi* (0.4). 13. *Cancellaria martiniana* (1.5). 14. *Conus literatus* (0.7). 15. *Turritella angulata* (0.7).

The rocks shown under the Therria Stage were originally included in the Cretaceous as Cherra Sandstone. Fox suggested the name Tura Sandstone for these rocks to avoid confusion. The Assam Oil Co. geologists point out that the Tura Sandstone near Tura is younger than the sandstone of Therriaghat in Khasi hills and therefore advocate the term Therria Sandstone to the Palaeocne division. The Therria Stage has not yielded fossils useful for fixing their precise age but they are undoubtedly Paleocene. The overlying Lakadong Limestone has yielded typical Ranikot fossils such as *Nummulites thalicus*, *N. sindensis*, *Lockhartia haimei*, *Miscellanea miscella*, *M. meandrina*, *Operculina* cf. *canalifera*, *Alveolina*, *Orbitosiphon tibetica*, *Discocyclina ranikotensis*, *Gypsina* sp. and some calcareous algae. The Umlatdoh Limestone contains some *Nummulites*, *Alveolina*, *Discocyclina*, *Miliolidae* and calcareous algae and are thought to be of Laki age. The Prang Limestone encloses *Nummulites obtusus*, *N. beaumonti*, *Assilina papillata*, *Discocyclina omphalus*, *D. sowerbyi*, *Eodictyoconus*, *Linderina*, *Orbitolites complanatus*, *Alveolina*, *Calcarina*, etc. which indicate Kirthar age. The Kopilis contain *Discocyclina*, *Nummulites*, *Heterostegina* and *Pellatispira*, of Upper Eocene age. The Sylhet Limestone Stage therefore includes the Ranikot, Laki and Kirthar Series. The Sylhet Limestones in the Khasi and Jaintia hills are detrital limestones with abundant foraminifera and algae. Both the Sylhet Limestones and Kopilis become more arenaceous when followed to the east as well as to the west.

The Lakadong Sandstone which contains only poor coal seams in the south becomes thicker on the plateau and shows workable and better quality coal seams, as for instance near Cherrapunji and Laitryngew.

The Prang Limestone attains a thickness of about 200 m to the south of the Jaintia hills and is the 'Sylhet Limestone' used in the manufacture of cement.

The Kopili Series (Kopili alternations), consisting of an alternating series of shales and sandstones contain several horizons of fossiliferous limestones which indicate Upper Eocene age.

In the Garo Hills, the Tura Sandstones consist of sandstones, shales and coal seams. They have now been shown to be equivalent to the Middle Sylhet Limestones of the Khasi hills. They form an anticline to the south of the Tura range and one limb dips steeply down towards the plains in the south.

The Siju Limestones overlie the Turas and are correlated with the Upper Sylhet Limestones on fossil evidence. They are overlain by the Rewak Series which are apparently the equivalents of the Kopilis farther east. The succession in the Garo hills is shown below :

REWAK	...	Marine sandstones and shales	— 4,000 ft.
SIJU	...	Marine shales and limestones	— 500 ft.
TURA	...	⎧ Upper sandstone	— 200 ft.
		⎪ Upper coal seam	— 2 to 4 ft.
		⎨ Middle sandstone	— 180 ft.
		⎪ Lower coal seam	— 5 to 6 ft.
		⎩ Lower sandstone	— 210 ft.

Disang Series.—This facies of the Eocene is well developed to the east and southeast of the Haflong-Disang thrust fault in the Naga Hills, Manipur etc. The rocks consist of splintery dark grey shales intercalated with fine grained sandstones, passing upwards into well-bedded sandstones. In the Naga hills the shales are often found to have been metamorphosed to slates. Along the western edge of the Barail range, the Disangs are thrust over to the northwest against the Kopili series. The Disangs attain a thickness of about 3,000 m but are mostly devoid of fossils. They seem to represent the whole of the Eocene, though the lowest part may possibly extend down into the Upper Cretaceous.

BENGAL BASIN

The present deltas of the Ganges and Brahmaputra and also the Shillong plateau appear to have been under the Sea in the Upper Cretaceous and to have formed a shelf area trending in a northeast direction. The geosynclinal part of the basin lay to the southeast, covering the southeastern part of East Bengal and eastern and southeastern Assam. As revealed in boreholes put down in this basin by the Indo-Stanvac project for petroleum, the sedimentary deposits here commence from the Lower Cretaceous and consist of the estuarine to fresh-water BOLPUR FORMATION and lagoonal black shales and pink shaly limestone with calcareous sandstone intercalations forming the GHATAL FORMATION. The Eocene comprises the JALANGI, SYLHET and KOPILI FORMATIONS. The JALANGI, about 700 m thick, consists of estuarine carbonaceous shales containing pyrites, lignite and resin; the strata become marine in the dip direction and enclose Lower Eocene formanifera. It is possible that the lowermost beds of this formation are of late Cretaceous age. The SYLHET FORMATION overlying the Jalangi can easily be correlated with the SYLHET LIMESTONE which is exposed on the southern side of the Assam plateau. In the boreholes this formation is marine, and 320 m thick. It consists of three limestone bands separated by clastic beds, which have yielded a rich fauna including *Assilina daviesi, A. papillata, Nummulites obtusus, Alveolina elliptica, Discocyclina javana*, etc. the age being Middle to Upper Eocene. The KOPILI FORMATION which is of Upper Eocene age, is only 20-25 m thick and consists of brackish grey shales and calcareous shales, poor in fossils. They are easily correlated with the Kopili Alternations on the Assam plateau. There is an unconformity above this formation.

BURMA

Eocene rocks are found in a belt stretching from the Indonesia through the Nicobar and Andaman Islands and the Arakan Yoma to Upper Burma.

The Andaman and Nicobar Islands are composed, for the most part, of Eocene rocks. The lower beds are conglomerates and sandstones resting on rocks resembling the Axials. They contain *Nummulites atacicus* and *Assilina* and are of Laki to Kirthar age. The Eocene is represented by limestones in Sumatra and Java where only the Kirthars seem to be present.

PLATE XXVII
TERTIARY FORAMINIFERA

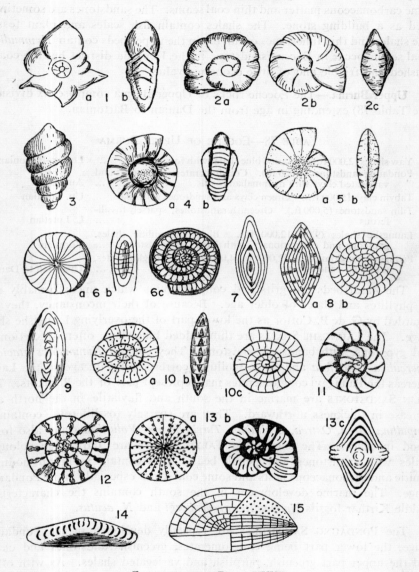

EXPLANATION OF PLATE XXVII

1. Hantkenina. 2. Cibicides. 3. Uvigerina. 4. Elphidium. 5. Camerina. 6. Nummulites nuttali. 7. Nummulites lahirii. 8. Nummulites atacicus. 9. Assilina dandotica. 10. Assilina granulosa. 11. Operculina salsa. 12. Operculina patalensis. 13. Miscellania miscella. 14. Lockhartia tipperi. 15. Lepidocyclina.

Lower Burma.—The eastern foothills of the Arakan Yoma in Lower Burma show Eocene rocks apparently faulted aginst the pre-Tertiary rocks. They comprise alternating sandstones, sandy shales and bluish shales containing some carbonaceous matter and thin coal seams. The sandstones are sometimes used as a building stone. The shales contain fish scales and plant fossils. The shales and thin limestones occurring in the upper beds contain *Nummulites*. Coal seams occur in Eocene sandstone in the Henzada district but the coal is crushed and friable and of little economic value.

Upper Burma.—The Eocene rocks of Upper Burma comprise six divisions (see Table 78) extending in age from the Danian to Bartonian.

TABLE 78—EOCENE OF UPPER BURMA

6.	Yaw shales (2,000 ft.). Marine blue shales with fossils ...	Ludian-Bartonian.
5.	Pondaung sandstones (6,000 ft.). Conglomerates, sandstones and variegated clays with mammalian fossils ...	Auversian
4.	Tabyin clays (5,000 ft.). Green clays and sandtones with coal ...	Up. Lutetian
3.	Tilin sandstones (5,000 ft.). Greenish sandstones, sparsely fossiliferous ...	L. Lutetian
2.	Laungshe shales (9,000-12,000 ft.). Blue Nummulitic shales, gypseous and concretionary, with bands of sandstone ...	Ypresian
1.	Paunggyi conglomerates (2,000 to 4,000 ft.). Conglomerates, grits and sandstones ...	Thanetian to Danian

The basal beds are grits and conglomerates which unconformably rest on phyllites and slates of older age. Because of their inconstancy, they are regarded by G. de P. Cottor as the lower part of the overlying Laungshe shale stage. The LAUNGSHE SHALES are thin-bedded blue clays, often concretionary and gypseous, with bands of sandstone. They contain *Nummulites atacicus*, *Operculina canalifera* and some mollusca corresponding in age to the Lakis, whereas the Paunggyi conglomerates may represent part of the Ranikots. The TILIN SANDSTONES are marine in the south and fluviatile in the north and increase in thickness northward. They are sparsely fossiliferous, containing *Ampullina, Arca, Ostrea, Cerithium, Turritella* and *Volutilithes* and also fossil wood in places. The succeeding TABYIN CLAYS are dark blue coloured shales with sandstones and pebble beds. They contain septarian nodules, lignitic and carbonaceous nests and some coal seams especially in the Pondaung range. The marine development in the south contains the characteristic middle Kirthar fossils *Nummulites vredenburgi* and *N. acutus*.

The PONDAUNG SANDSTONES are typically developed in the Pondaung range, the lower part being conglomerates, greenish sandstones and clays, and the upper part greenish, purplish and variegated shales. As with other formations they are marine in the south and brackish to fresh-water in the north. They enclose fossil wood which is usually carbonised in the lower part of the stage, and partly carbonised and partly silicified in the upper part. The conglomerate bed at the base contains *Cardita mutabilis, Arca pondaungensis, Alectryonia newtoni, Corbula daltoni* and some gastropods. The fresh-water facies consists of red, buff, and cream coloured earths interstratified with sandstones. The earthy beds contain reptilian and mammalian remains in the Pakokku district, the chief mammalian fossils being:

Primates	... ...	*Pondaungia cotteri, Amphipithecus mogaungensis.*
Brontotheridae	...	*Stvatitanops birmanicum, S. cotteri.*
Tapiridae	...	*Indolophus guptai, Deperetella birmanicum.*
Anthracotheridae	...	*Anthracothema pangan, A. crassum, A.rubricae, Anthracokeryx hospes, A. birmanicus, A. tenuis, A. bambusae, Anthracohyus choeroides.*
Tragulidae ...	...	*Indomeryx cotteri, I. arenae.*

The fossils indicate an upper Eocene (Auversian) age.

The YAW SHALES rest on the Pondaung sandstones and comprise bluish grey shales of essentially marine character, though the fluviatile representatives with coal seams are developed in the Minbu district. They often show thin bands of impure calcareous matter, septarian nodules, phosphatised coprolites and fish remains, foraminifera and molluscs. The chief fossils are:

Foraminifera	... ...	*Nummulites yawensis, Discocyclina omphala, D. sella, Gypsina globulus, Operculina canalifera.*
Lamellibranchs	...	*Solen manensis, Corbula subexarata, C. pauhensis, Meretrix (Callista) yawensis, Venus pasohensis, Tellina salinensis, Cardium hanleanum, Lucina yawensis, Ostrea minbuensis, Leda silvestris.*
Gastropods	...	*Velates schmideliana, Cypraedia birmanica, Gosavia birmanica, Lithoconus gracilispira, Athleta rosalindae, A. archiaci, Volutilithes arakanensis, Clavilithes cossmanni, Velates perversus, Ampullina* cf. *grossa.*

This fauna bears a distinct resemblance to that of the Upper Eocene of Java and to the Upper Kirthar of Western India.

RAJASTHAN

The comparatively lowlying tracts of Bikaner and Jaisalmer in southwestern Rajputana were under the sea in Eocene times. The strata exposed here belong to the Laki Series and especially its middle division, and comprise a considerable thickness of white or pale buff limestone with *Nummulites atacicus* and *Assilina granulosa.*

The Eocene beds contain lignite beds which are successfully worked at Palana in Bikaner, and also an earthy brown shale used as fuller's earth. The fuller's earth has yielded the typical Laki foraminifera *Assilina leymerei* and also species of *Rotalia, Cibicides, Nonion,* etc.

KUTCH

The Eocene marine invasion has left deposits in Kutch belonging to the Paleocene, Laki and Kirthar Series. One of the two bands is in the interior and rests upon the Deccan Trap, while the other, nearer the coast, overlaps on to the Jurassic rocks.

The lowest Tertiary beds lying over the Deccan Traps are the MADH SERIES of Paleocene age, composed of 30 to 40 m lateritic or tuffaceous or bentonitic clays of bright colour, some probably derived from the erosion of Deccan Traps. Pyritous and gypsiferous sandstones follow. These contain *Venericardia, Neomeris* etc. The Paleocene is overlain by KAKDI SERIES of Laki age, well seen near Umarsar. They are 40 to 50 m thick, the lower half being carbonaceous shales with lignite and ferruginous and gypsiferous clays. The upper half consists of grey shales and limestones. The fossils in these strata include *Nummilites atacicus, Assilina Spinosa, A. Granulosa, Cibicides, Operculina* and also *Ostrea, Turritella* and other molluscs and some echinoderms. The overlying strata of Kirthar age (BERWALI SERIES) have a maximum thickness of 130 m, whose lower portion (30 m) consists of ochreous, greenish and gypsiferous clays, and the upper portion of massive buff limestones. The fossils in these beds are *Assilina exponens, A. granulosa, Nummulites atacicus* and also *Hantkenina, Lockhartia, Discocyclina,* etc. The Eocene beds are overlain by the Oligocene LAKHPAT BEDS.

GUJARAT

There are two exposures of Eocene rocks in the area between Surat and Broach, separated by the alluvium of the Kim river. The smaller southern exposure extends for 16 km northward from the Tapti and is 24 km at its widest. The larger exposure, between the Kim and the Narmada, is 50 km long (N.E.-S.W.) and 20 km wide. The basal beds are impure limestones and some laterite and contain such characteristic Ranikot Nummiulites as *Nummulites thalicus, N. globosus* and *Discocyclina* aff. *ranikotensis*. The beds above these contain *Assilina exponens, Nummulites ramondi, Ostrea flemingi, Rostellaria prestwichi, Natica longispira* and *Vulsella legumen* which are regarded as indicating a Kirthar age. The uppermost Eocene beds here, of Priabonian age, have yielded *Discocyclina javana, D.* cf. *dispansa, Pellatispira indica,* etc. Narayana Rao has suggested naming them Pellatispira beds. A large thickness (1,300 - 1,500 m) of gravel, conglomerate, sandstones and shales of Miocene age overlies the Eocene limestones near Ratanpur east of Broach.

Drilling for oil at Ankleshwar near Broach has shown that the Eocene is about 1000 m thick, resting on the Deccan Traps. The strata are pyritic shales, silts and sandstones with some limestone intercalations. There are three oil producing horizons in the Eocene. Besides the fossils mentioned above, there are also *Actinocyclina, Asterocyclina, Bulimina, Hantkenina* and several ostracods.

Eocene beds occur in a trough-faulted area running N-S through Cambay for nearly 400 km to latitude 24° N and beyond, gradually shaoling up northward. It is 80 to 100 km wide. In this trough is found a full Tertiary succession 1,500 to 2,000 m thick. The Eocene, which is 500 m thick, begins with clastics derived from erosion of the Deccan Trap, overlain by carbonaceous and pyritic clays and sands, the more northern areas (Kalol etc.) containing more carbonaceous sediments. Gas and oil bearing structures occur in the Eocene and in the overlying, but much thinner, Oligocene. In this area, which lies in the more or less isolated and narrow Cambay Gulf, there are no important members of the Indo-Pacific fauna or larger foraminifera. The chief fossils are Ostracods, *Bulimina, Discocyclina, Hantkenina* and *Operculinoides. Pellatispira* is not present.

PONDICHERRY AREA

The discovery of Lower Eocene foraminifera (*Nummulites* and *Discocyclina*) was announced by L. Rama Rao in 1939 in some limestones in the Pondicherry area which was hitherto known to contain only Cretaceous rocks. Upper Eocene rocks with fossils of Lutetian to Bartonian age have also been found in some borings near Pondicherry. It may therefore be expected that a full Eocene sequence will be found in this area overlying the Cretaceous rocks, and below the Cuddalore Sandstones of Miocene age.

RAJAMAHENDRI (RAJAHMUNDRY)

The Infra-trappean sandy limestone of the Rajamahendri area contains a fauna whose age is probably Upper Cretaceous according to Medlicott and Blanford. The Inter-trappeans have in recent years yielded a rich algal flora containing *Acicularia, Neomeris, Chara,* etc. which have a decided Eocene aspect. Some Paleocene Ostracods have also been found. The age of these beds may be taken as Paleocene and Lower Eocene.

CAUVERY BASIN

Geophysical work in the South Arcot-Tanjore-Ramanathapuram Tertiary basin has shown that the basin slopes sea-ward and contains about 3,000 m of sediments near the coast, overlying the basement which shows two low ridges, one roughly parallel to the Coleroon in South Arcot and the other running from east of Karaikudi to Karikal. One borehole at Karikal which reached a depth of 1,760 m shows the Eocene resting on the crystallines, but the neighbouring depression which is 2,700 m deep must contain some Cretaceous and probably Upper Gondwana sediments. The Lower Eocene consists of 100 m of sandstones and shales enclosing Alveolines and some arenaceous foraminifera. The Middle and Upper Eocene consist of pyritous silts and grey shales with *Globorotalia, Hantekenina, Uvigerina* and other fossils. The Eocene is likely to contain petroliferous horizons in the deeper parts of the basin along the coast and in the off-shore area. The Eocene is succeeded by Oligocene which shows regressive characters.

ECONOMIC MINERALS

Amongst the more important resources of the Eocene strata are coal, limestones and clays. In some cases they are also the source rock of petroleum, though the petroleum may have migrated to later formations having suitable constitution and structure to act as reservoir rock. Notes on petroleum will be found in the next chapter.

Coal.—Coal seams appear to have been formed both in the Ranikot and Laki times, the latter being more extensive than the former. In the Palana field in Bikaner, lignite is found as a bed 6 to 15 m thick and is of Laki age. It is associated with marine shales including beds of fuller's earth. In the Shillong plateau of Assam (Garo, Khasi and Jaintia hills) there are several coalfields in the Sylhet Limestone Stage. Some coal is also found in the Paleocene of Khasi hills. These coals are generally rather high in volatiles and in sulphur. The reserves of coal in this region have yet to be investigated but they may be of the order of 300 million tons.

Coal seams occur in W. Pakistan in two horizons in the Eocene, *viz.*, Ranikot and Laki. Coal seams in the Makerwal area are in Lower Ranikot while those in Dandot in Eastern Salt Range are in the Upper Ranikot. The Dandot seam is horizontal and crops out on the Salt Range scarp. It is worked behind the scarp at a depth of 100-150. m The Makerwal seam is exposed along the scarp overlooking the plain of Isakhel west of the Indus. It is 1 to 3.5 m thick and generally of poor quality.

Laki coal is worked in the Sor Range, at Khost and Mach in the Quetta-Kalat area and in the Lakhra-Jhimpur area in Sind. There are also some seams at Duki, Loralai, Kotli in Mirpur and Johan in Kalat. Some seams occur also in the Kala Chitta Hills. In all these occurrences the coal is lignitic in character and 1 to 3 m thick and of variable quality. The total reserves of all these have been estimated at 175 million tons of which a third is in the Eastern Salt Range. The annual production is around 1.2 million tons.

Limestone.—The Eocene strata contain large resources of excellent limestones in the Nummulitic beds. Such limestones are extensively developed in Western Punjab, in the hills of the Sind-Baluchistan border, as well as in the Sylhet district of East Bengal and in the adjoining parts of the Shillong plateau. Several of these limestone deposits are being used for the manufacture of cement.

Clays.—There are also some clay deposits in the Eocene, though these are not of high quality or of refractory nature. Near Sohrarim and a few other places in the Khasi hills, the sandstones contain a fair amount of white and light coloured clay as matrix which can be recovered by washing and used for the ceramic industry. Some bentonic clays derived from ash beds of the Deccan Traps occur in Kathiawar.

SELECTED BIBLIOGRAPHY

Cossman, M. and Pisarro, G. The Mollusca of the Ranikot Series. *Pal. Ind. N.S.* III (1), 1909 ; X (2), 1927 ; X (4) 1928.
Cotter, G. de. P. The Lamellibrachs of the Eocene of Burma. *Pal Ind. N.S.* VII (2), 1923.
Davies, L.M. The fossil fauna of the Samana Range. *Pal Ind. N.S.* XV, (6-8), 1930.
Davies, L.M. and Pinfold, E.S. The Eocene beds of the Punajb Salt Range. *Pal. Ind. N.S.* XIV (1), 1937.
Furon, R. and Lemoine, P. Sur la presence du Nummulitique a Pondicherry. *C.R. Acad. Sci.* (Paris) CCVII, 1424-26, 1939.
Nuttall, W.L.F. Stratigraphy of the Upper Ranikot Series of Sind. *Rec.* 65, 306-313, 1932.
Nuttall, W.L.F. Stratigraphy of Laki Series. *Q.J.G.S.* 81 (2), 1925.
Pilgrim, G.E. The Perissodactyla of the Eocene of Burma. *Pal. Ind. N.S.* VIII (3), 1925.
Pilgrim, G.E. The Artiodactyla of the Eocene of Burma. *Pal. Ind. N.S.* XIII, 1928.
Pilgrim, G.E. and Cotter, G.de. P. Eocene Mammals from Burma. *Rec.* 47, 27-77, 1916.
Rao, S.R.N. *Current Sci.* 8 (2)78, 1939 ; 8 (4), 167, 1939.
Rao, L.R. and Nagappa, Y. *Curr. Sci.* 9 (8), 1940.

CHAPTER XVIII
OLIGOCENE AND LOWER MIOCENE SYSTEMS

General.—Towards the end of the Eocene there occurred a second great upheaval which contributed to the formation of the Alpine-Himalayan mountain systems. This had the effect of driving out the sea from most of the Himalayan area except along the southern border. The large thickness of sediments which were formed in the shallow seas on the western side of the Baluchistan Arc consisted of calcareous sandstones and greenish shales of singularly uniform appearance. They form the bulk of the ' flysch ' formation similar to the Oligocene flysch of the Alpine region. To the east of the Baluchistan Arc there was a bay, which extended along the foothill region of the Himalayas. This bay was gradually filled up during the rest of the Tertiary.

On the eastern side of India also there was a ridge along the Burma border, on both sides of which were sedimentary basins in which large thickness of sediments were deposited during the Tertiary. As is to be expected, fluviatile sediments were deposited at the head of the bays while brackish water and marine sediments were formed to their south in the direction of the open ocean.

The sedimentation continued more or less uninterrupted until the Middle Miocene when a third mountain building upheaval took place along the Himalayan region and the Baluchistan and Burmese arcs. The Oligocene and Lower to Middle Miocene rocks therefore form one stratigraphic unit. This is represented by the Nari and Gaj beds of Sind, the flysch of Baluchistan, the Murree System of Western Pakistan and the Pegu System of Burma.

The Oligocene was a period of marine regression but sedimentation continued in the deeper parts. Marine transgression took place in the Miocene also along parts of the coast of the peninsula, as for instance, in Orissa and Travancore, and along the eastern coast. The Oligocene-Lower Miocene deposits may therefore be grouped under four types, *viz.*, (*a*) an open-sea calcareous facies, (*b*) a shallow marine flysch facies, (*c*) a lacustrine facies, and (*d*) a coastal facies.

SIND AND BALUCHISTAN (CALCAREOUS FACIES).

The Calcareous facies of the Oligocene-Miocene is developed in the Sind and Baluchistan mountains on the eastern side of the Eocene strata. Two main divisions are recognised, *viz.*, the Nari and Gaj Series both named after rivers on the Sind frontier. Both are characterised by massive limestones but sandstones and shales also occur, especially in the upper portion. When followed northwards, the arenaceous element in the beds increases, showing the approach to land in that direction.

Nari Series.—The Nari Series is well developed on the eastern flanks of the Kirthar range and also to the west of the Laki range throughout Lower

Sind. It is divisible into two sub-series. The Lower Nari is variable in thickness, from 30 to 450 m, and consists mostly of limestones. The lower beds are white and massive but the upper are brown and yellow, inter-bedded with bands of shale and layers of sandstone.

The Upper Nari beds reach a maximum thickness of 1,200 to 1,800 m and consist of thick-bedded grey sandstones and subordinate shales and conglomerate. The rocks are mostly unfossiliferous but certain bands are crowded with *Lepidocyclina* (*L. dilatata* group) of very large size, often 2 inches or more across.

The Nari Series corresponds to the Stampian and Chattian (covering the greater part of the Oligocene) and to a part of Lower Miocene. Amongst its leading fossils, most of which come from the lower division, are :—

Foraminifera	*Nummulites intermedius, N. vascus, Lepidocyclina dilatata.*
Corals	*Montilvaltia vignei.*
Echinoids	*Breynia multituberculata, Eupatagus rostratus, Echinolampas discoideus, Clypeaster simplex.*
Lamellibranchs	*Osrea fraasi, O. orbicularis, O. angulata, Pecten labadyei, P. articulatus, Arca semitorta, Lucina columbella, Crassatella sulcata, Callista splendida, C. exintermedia, Venus puerpera* var. *aglaurae, V. multilamella, Pitar porrectus.*
Gastropods	*Terebra narica, Ancilla indica, Volutsopina sindensis, Lyria anceps, Cypraea subexcisa, Cerithium sindiens ,C. bhagothorense.*

In Kutch the Kirthar Series is followed by the Lower Nari comprising the Lattorfian and Stampian, with *Nummulites fichteli, Spiroclypeus* and *Lepidocyclina dilatata*. The Upper Nari, which follows after a slight break, comprises the Chattian and part of the Aquitanian and contains *Spiroclypeus, Austrotrillina, Miogypsinoides* and *Nephrolepidina*. Another break is probably indicated between the Upper Nari and the Lower Gaj, the latter being also Aquitanian in age.

Gaj Series.—The Nari Series is overlain conformably by the Gaj series which attains a thickness of 450 m and consists of yellow and brown limestones, either massive or rubbly, with intercalations of white arenaceous limestones, clays and gypsum. The lithology indicates that the area of deposition was first marine and later became gradually estuarine. The two divisions of the Gaj have several fossils in common but there are also some species exclusively found in each division.

The following fossils are found :

Foraminifera	*Austrotrillina howchinii, Orbulina universa,* and several species of *Nephrolepidina.*
Echinoids	*Breynia carinata, Eupatagus patellaris, Echinolampas jacquemonti, Clypeaster profundus, Echinodiscus placenta.*
Gastropods	*Vicarya verneuili, Turritella angulata, Telescopium sub-trochleare, Olivancilla nebulosa.*
Lamellibranchs	*Pecten scabrellus, P. senatorius, Dosinia psudoargus, Venus granosa, Clementia papyracea, Discors triformis, Lucina columbella* , etc.

The species found in Lower Gaj are *Ostrea angulata*, *Pecten labadeyi*, *P. articulatus* and *Lepidocyclina marginata*. Those in the Upper Gaj are *Ostrea latimarginata* (characteristic), *O. gajensis*, *O. imbricata*, *O. gingensis*, *O. vestita*, *Pecten placenta*, *P. subcorneus*, *Arca peethensis*, *A. burnesi*, *A. semitorta* and also some remains of Rhinoceros. The age of the Lower Gaj is Aquitanian and that of the Upper Gaj Burdigalian, both being Lower Miocene, and they correspond respectively to the Rembang and Njalindung Series of Java.

BALUCHISTAN (FLYSCH FACIES)

Khojak shales.—Beyond the calcareous zone, in Baluchistan, there occurs a vast series of sandstones, shales, and sandy shales constituting the flysch zone which includes the hills of the Zhob and Pishin valleys the Khwaja Amran range west of Quetta and almost the whole of the Mekran province. This region is occupied by close-set ridges consisting mostly of a monotonous series of sharply folded and sometimes overthrust sandstones and slaty shales of a greenish colour known as the Khojak Shales which resemble the Oligocene flysch of Europe. In the region of northern Mekran the strata are friable clays. The typical Khojak Shales contain fossils only rarely. Amongst them are *Nummulites intermedius*, *N. vascus*, *Lepidocyclina* (*Eulepidina*) *dilatata*, *Rotalia*, *Triloculina*, *Globigerina*, etc. They are apparently the equivalents of the Nari series.

Hinglaj sandstones.—Large masses of sandstones with shale beds rest conformably upon these Oligocene clays and make up the Peninsula of Ormara and Gwadar, the Hinglaj mountains and other hills of the Mekran coast. The shale intercalations sometimes contain fossils, especially in the uppermost and lowermost horizons. The lowest beds contain *Turritella javana*, *Ostrea gingensis*, *Arca burnesi*, *Dosinia pseudo-argus*. The uppermost beds contain *Pecten vasseli* as the commonest fossil and also *Fertagus bonneti*, *Crepidula subcentralis*, *Ostrea frondosa*, *O. cucullata*, *Arca divaricata*, *A. squamosa*, *A. tortuosa*. The beds in the middle contain *Turritella angulata*, *T. bantamensis*, *T. bandoengensis*, *Ostrea virleti*, *O. petrosa*, *O. digitalina*, *Arca inflata*, *Clementia papyracea*, *Circe corrugata*.

A large proportion of the Hinglaj species occurs in the Miocene of Java and Burma, the fauna of the upper portion bearing some resemblance to that of the Karikal beds on the Madras coast. The Hinglaj beds correspond to a part of the Lower Manchhars and are of Burdigalian to Helvetian age.

NORTH-EASTERN BALUCHISTAN

Bugti beds.—In the Bugti hills of Baluchistan the marine element of the Nari series becomes very reduced, being represented by a small thickness of brown arenaceous limestone containing the characteristic *Nummulites*. These are succeeded by a series of fluviatile sandstones with the characteristic Gaj *Ostrea* at the base, the beds above containing a rich vertebrate and fresh-water lamellibranch fauna including the ribbed Unios (*U. cardita*, *U. vicaryi*, *U. cardiformis*).

The vertebrates include :

Anthracotherium bugtiense, A. ingens, Telmatodon bugtiensis, Brachyodus giganteus, B. hyopotamoides, Paraceratherium bugtiensis, Hemimeryx speciosus, Aceratherium bugtiens Teleoceras blanfordi, Cadurcotherium indicum, Baluchitherium sp., Pterodon bugtiensis, Amphicyon shahbazi, Rhinoceros gajensis.

Pilgrim originally considered the Bugti beds as Upper Nari (Up. Oligocene to basal Miocene), on the evidence of vertebrate fossils. He later revised the age to Gaj on the strength of Vredenburg's work on the species of *Ostrea* found in these beds. A re-examination of the evidence by the geologists of the Burmah Oil Co. has now led to the readoption of Upper Nari age.

POTWAR PLATEAU AND JAMMU

Northwestern Punjab and the adjoining regions of Jammu and Kashmir contain one of the most complete Tertiary sequences in India. This region was once a basin of large dimensions in which were laid down very thick deposits of brackish and fresh-water origin during the Oligocene and Miocene times. The earlier deposits (Lower Murrees) are of brackish water origin while the later ones (Upper Murrees) are fresh-water deposits.

Fatehjang zone.—The Chharat Series of Upper Eocene age is overlain, with an unconformity marked by a bed of conglomerate about a foot thick, by the Fatehjang zone which belongs to the basal part of the Murree Series. This zone consists of brown and grey sandstones and pseudo-conglomerates. The numerous large *Nummulites* with which their exposures are covered have been derived from the denudation of the earlier beds. Several mammalian fossils, indicating a lower Burdigalian age, are found in the Fatehjang zone :

Anthracotherium bugtiense, Hemimeryx sp. *Brachyodus* cf. *africanus, Palaeochoerus pascoei, Teleoceras fatehjangense.*

Murree Series.—The Fatehjang zone passes upward into the Lower Murrees which consist of bright purple shales, hard purple and grey sandstones and pseudo-conglomerates. They contain sparse fossils including leaf impressions (e.g., *Sabal major*) and some lamellibranchs.

The upper Murrees are distinct from the lower, especially in the Potwar plateau, being composed of soft, pale coloured sandstones resembling the Chinji (Lower Siwalik) sandstones to some extent. They contain impressions of dicotyledonous leaves and remains of mammalia including primitive rhinoceros.

The Murrees appear to have been derived from a well oxidised terrain along the northern border of the Indian Peninsula, containing Archaean and Pre-Cambrian formations including the Attock-Dogra Slates, the Panjal volcanics and the sedimentary formations of Sub-Himalayan aspect. Some material may have come also from any ridge raised up during the Upper Eocene phase of the Himalayan orogeny, as the early Oligocene marine regression must have left a large land area exposed.

The Murrees are mainly of Lower and Middle Miocene age but may include a part of the Oligocene. The basin of sedimentation was a shallow remnant of the Tethys which had already become brackish by dilution. The strata are typically developed near Murree, in the Potwar Plateau and in Jammu where they attain a thickness of upto 2,500 m. They are succeeded by the Siwalik System, deposited in the fore-deep which was developing in front of the Himalayas.

SIMLA HIMALAYA

When followed eastwards from the Jammu area, the Murrees diminish in thickness and are represented in the Simla hills by the Dagshai and Kasauli beds which are roughly the equivalents of the Lower and Upper Murrees respectively. These beds, together with the underlying Subathus, used to be included formerly under the SIRMUR SYSTEM.

Dagshai Beds.—The Subathus are overlain by the Dagshai beds which comprise a series of very hard, finegrained grey or purplish brown quartzitic sandstones intercalated with seams of red clay. The clays predominate in the lower part but the sandstones gradually increase in proportion and thickness in the upper part. The sandstones are massive, 1.5 to 6 m in thickness, and scarcely show any stratification. The clays are purplish brown, mottled with grey, and are harder than either the Kasauli or Nahan (Lower Siwalik) clays. Though no disconformity is apparent between the Subathus and the Dagshais, the former is Upper Eocene and the latter Lower Miocene in age and there is a stratigraphical gap between. The transition between the two is somewhat abrupt and is marked by pisolitic marl (a red clay containing calcareous concretions) purple shale and white sandstone with ferruginous concretions.

Kasauli Beds.—The Dagshai beds pass conformably up into the Kasauli beds in which there is an absence of bright red clays. The Kasaulis are essentially a sandstone group with minor argillaceous bands, the sandstones being grey to greenish in colour and generally softer, coarser and more micaceous than the Dagshai sandstones. The argillaceous bands are gritty, greenish or brown and weather into angular splintery chips. The Kasaulis are poor in fossils, impressions of palm leaf (*Sabal major*) and *Unio* shells being found. The lithology indicates that the Dagshais are brackish water deposits and the Kasaulis fresh-water ones.

In this region the junction zone between the Kasaulis and the Nahans is generally a thrust plane. It is also possible that some of the rocks considered to be Kasaulis are really Nahans.

ASSAM

The rock groups falling into the Oligocene and Miocene in Assam are the Barail and Surma Series. The Barails extend up to the Chattian (Upper Oligocene) while the Surmas are of Aquitanian and Burdigalian age.

Barail Series.—The formations which represent the Oligocene and Miocene in Assam have also different characters in different areas. The Oligocene is largely included in the Barail Series, the name being derived from the Barail Range which forms the watershed between the Brahmaputra and the Surma valleys. In Upper Assam it consists of three stages :

3. *Tilak Parbat Stage* ... Carbonaceous shales and coal seams (450 m).
2. *Baragoloi Stage* ... Sandstones, Carbonaceous shales and Coal seams (3,000 m).
1. *Naogaon Stage* ... Sandstones (2,400 m).

The NAOGAON STAGE consists of hard, thin-bedded, grey flaggy sandstones forming prominent hills. Recent fossil evidence seems to show that the Naogaon Stage is Eocene and should be put back with the Disang Series as was originally suggested by Mallet. The BARAGOLOI STAGE shows alternating Sandstone and shales with coal seams *e.g.*, in the Baragoloi Colliery. The TIKAK PARBAT STAGE is more or less similar to the Baragoloi, the coal fields of Nazira, Makum, Namdang-Ledo and Tikak being in this series. The boundary between the two upper series is drawn at the base of a thick coal seam. Mallet originally called these the 'Coal Measure Series' of Assam (*Mem.* XII, part 2, 1876). The Baragolai Stage is the thickest of the three and is responsible for the largest number of oil shows. It is in the Barails that the oil horizon of the new Nahorkatiya oil field, some 40 km westsouthwest of Digboi, was struck at a depth of some 3,200 m below the surface. The first bore-hole also showed a good 3 m thick coal seam at about 3,050 m depth.

The Barail Series occupies a large area northeast of the Haflong-Disang thrust but shows a different lithological aspect. It is well developed in the Surma Valley, North Cachar and the Khasi and Jaintia hills, where also it is divided into three series :

Renji Stage ... ... Hard massive sandstones and very subordinate shales (900 m) —Chattian.
Jenam Stage ... ... Sandstones alternating with dominant shales and carbonaceous shales (900-1200 m)— Lattorfian.
Laisong Stage ... ... Hard, thin bedded sandstones and subordinate shales (1,800-2,400 m)—Bartonian to Auversian.

The Laisongs, like the Naogaons, form prominent scarps well exposed in the Barail Range. The Laisongs are roughly equivalent to the Naogaons while the Baragolois may represent Upper Laisong and Lower Jenam. The Tikak Parbat Stage is thought to represent Upper Jenam and Lower Renji. The Barails apparently extend southwards through the almost inaccessible Assam-Arakan mountains into Ramree island.

The Barails in both the areas are mainly arenaceous but the sands increase in coarseness in a northwesterly direction. The argillaceous and carbonaceous contents, especially the latter, increase in a north-easterly direction. There are no coal seams in the Surma valley but they begin to appear east of the main Dhansiri valley, the seams being often 3 m thick northeast of the Dayang valley. The best development is found in the neighbourhood of Ledo. Though thin seams are present at several horizons in the Baragolai and Tikak Parbat Stages, thicker seams are confined to comparatively small parts of these stages.

The Barails are poor in fossils though they are marine to estuarine in large part. Some micro-foraminifera and larger fossils have been found, from which they are regarded as extending from Upper Eocene to Chattian (Upper Oligocene). At the top of the Barails there is a marked unconformity all over Assam, indicating a period of uplift and erosion. The unconformity probably covers a large part of the Oligocene in places and there are great variations in the thickness of the Renji Stage.

Surma Series.—The Surmas, which follow the unconformity over the Barails, are generally comparatively thin in upper Assam, not exceeding 600 m, and sometimes even absent and overlapped by the Tipams. To the southwest they greatly increase in thickness, being 3,000 m thick at the head of the Surma valley and 6,000 m in the Arakan region. The subdivisions recognised are :—

Boka Bil Stage (900 - 1,500 m)		Sandy shales, silts, sandstones, ferruginous sandstones — Burdigalian.
Bhuban Stage (1,200-2,400 m)	Upper :	Conglomerates, sandstones and sandy shales— Aquitanian.
	Middle :	Shales, sandy shales and some conglomerates.— Aquitanian.
	Lower :	Sandstones, sandy shales and conglomerates— Chattian ?

The Bhubans, which take their name from the prominent scarp of the Bhuban Range in North Cacher, are mainly sandstone and shales, with some conglomerates, the relative proportions of the first two varying considerably. The lower series contains roughly equal proportions of the two, while the middle is more shaly and the upper more sandy. In the Mikir hills the unconformity below the Surmas is well seen and they transgress over the Barails and Jaintias on to the metamorphics. The basal conglomerates of the Surmas as well as the Barail rocks are harder than the middle and upper part of the Bhubans and form the more rugged topography which represents roughly the pre-Surma topography of the area.

The Surmas, though also mainly arenaceous, are strikingly different from the overlying massive, coarse, ferruginous, false-bedded sandstones of the Tipam Series. The Surmas are poor in Carbonaceous materials, thus contrasting with the coal-bearing Barails below and the lignite-bearing Tipams above.

The Bhubans are almost devoid of identifiable fossils though shell fragments are found in several exposures. Only at Kanchanpur in the Surma valley was a good collection of molluscan fossils got which are allied to the Upper Pegu forms of Aquitanian age (Lower Gaj). The common genera are *Basilisssa, Cancellaria, Hipponyx* and *Scutus*. Molluscan fossils have also been obtained from localities in the Arakan coast.

The Boka Bil Stage consists of soft sandy shales or alternations of sand and shale layers. In some places they contain also lenticular ferruginous sandstones. In Upper Assam they have not been identified ; they are thin in the Naga Hills and increase to 900 m in Surma valley and 1,500 m in the Arakan region. In some places there is a gradual lateral passage from Boka Bil Stage to the Tipam Sandstones. There are fossiliferous exposures in the Boka Bils in the southwest of the Shillong plateau. Vredenburg (1921) and Mukherjee (1939) have described the fossils from Baghmara and Dalu and the fossils indicate that the Boka Bils are of Burdigalian age. The more important genera in the Garo hills are :—

Lamellibranchs	...	*Arca, Cardium, Chlamys, Dosinia, Barbatia, Drillia, Lucina, Ostrea, Mactra, Pitar, Nucula, Nuculana.*
Gastropods	...	*Architectonica, Natica, Sinum, Mitra, Turritella, Oliva, Terebra, Conus.*

The Upper Boka Bil stage in Tripura has yielded fishes (*Oxyrhina, Prionodon*), reptiles (Crocodiles, *Charialis*) and Mammals (*Trilophodon, Dorcatherium*). The mammalian fossils are assigned a Tortonian age in the Punjab. The Boka Bils may therefore be Helvetian-Tortonian, *i.e.*, distinctly younger than the Burdigalian.

BURMA : PEGU SERIES

The Pegu Series occupies the tract between the Irrawaddy and Sittang rivers including a large part of the Pegu Yomas. It is also found west of the Irrawaddy between the Eocene strata and the later Irrawaddian beds. It is marine in the south but fluviatile to continental in the north, and, because of the lateral variation, the correlation of beds in different areas is not an easy matter. Table 79 shows the classification of the Pegus adopted by the Geological Survey.

TABLE 79—THE PEGU SERIES (G. S. I.).

Akauktaung stage	... Grits, conglomerates, sandstones, and some shales.	Vindobonian (L .Siwalik).
Pyalo Stage (1,500 ft.)	... Sandstones, shales and pebble beds containing *Ostrea latimarginata*	Burdigalian (Up. Gaj.)
Kama Stage (1,500-2,000 ft.)	... Sandstones and shales with rich gastropod fauna.	Aquitanian (Lr. Gaj.)
Singu Stage (1,500 ft.)	... Sandstones and shales with numerous mollusca	Chattian (Up. Nari.)
Sitsayan Stage (1,500-3,000 ft.)	... Shales and sandstones with *Lepidocyclina*	Stampian (Lr. Nari.)
Shwezetaw Stage (3,000 ft.)	... Sandstones with *Ampullina birmanica*.	Lattorfian.

The Shwezetaw Stage consists of shales in Lower Burma but becomes arenaceous when followed northwards. In the Minbu district it is a shallow water sandstone. Thin coal seams of poor quality with numerous sandy partings occur near the Yaw river. Near Shinmadaung, north of Pakokku, the sandstones contain *Ampullina birmanica*. Other fossils found are *Cardita* cf. *mutabilis, Ostrea* spp., *Vicarya* sp., etc.

The Sitsayan Shales are well developed in the Henzada and Prome districts. They are mainly blue clays with poorly developed bedding but contain beds of marl and thin sandstones especially in the upper part. Amongst the fossils are *Tritonidea, Corbula, Pecten*, etc., while *Lepidocyclina theobaldi* occurs in the upper part.

The Padaung Clays of the Minbu district which are blue clays with some grey limestones, are their equivalents and contain *Nucula alcocki, Tellina indifferens, Genota irravadica, Cypraea subexcisa, Clavilithes seminudus, Hindsia pardalis, Athleta theobaldi, Lyria varicosa.*

Singu Stage.—The Pegu rocks exposed in the Singu and Yenangyaung oil-fields are typical of this stage, being sandstones and shales. This stage is represented in the Minbu district by shallow marine deposits and in Pakokku by estuarine deposits. Among the fossils of this stage are :

Corals	... ...	*Dendrophyllia digitalis, D. macroriana.*
Lamellibranchs	...	*Lima protosquamosa, Pteria suessiana, Septifer nicobaricus, Nucula alcocki, Cardita scabrosa, Trachycardium minbuense, Pitar protollacina, Corbula rugosa.*
Gastropods	... ...	*Architectonica maxima, Sigaretus neritoideus, Turritella angulata, Vicarya verneuili, Cassis birmanica, Tritonidea martiniana, Athleta jacobsi, Mitra singuensis, Ancilla birmanica ,Genota irravadica, Conus odengensis.*

Kama Stage.—This stage is named after Kama, 30 km from Prome and consists, in Lower Burma, of blue shales and sandy shales with occasional sandstones. It has been alled the Padaukpin clay in the Thayetmyo district. It consists of soft sandstones and shales with brackish and fresh-water fauna. The fauna of the Padaukpin clays includes — *Leda virgo, Corbula socialis Turritella acuticarinata, T. angulata, Rimella javana, Pyrula promensis, Cassidaria echinophora, Ranella antiqua, Conus odengensis, Drillia protocincta.*

The mammalian fossils found in this stage are :
Anthracotheridae	...	*Telmatodon* sp.
Rhinoceratidae	...	*Cadurcotherium* sp.
Tragulidae	...	*Dorcatherium birmanicum.*

Pyalo Stage.—This is an arenaceous stage found in Lower Burma around Pyalo on the Irrawaddy, and characterised by the occurrence of *Ostrea latimarginata. Terebra promensis, Turritella pinfoldi* and *Terebra myaunguensis* also occur.

Akauktaung Stage.—Named after the Akauktaung hills in the Henzada district, this stage comprises grits, conglomerates and yellow sandstones with bands of blue shales and calcar ous sandstone. The marine fossils in this stage include *Ostrea gingensis, O. virleti, Arca burnesi, Cytherea eryctna, Dione dubiosa, Turritella acuticarinata, T. simplex* and *Conus litera us.* The freshwater representatives of these beds in Upper Burma have yielded *Cyrena petrolei* and *Batissa crawfordi.*

The Geologists of the Burma Oil Company have adopted a different classification of the Pegus in which the two lowermost divisions are the same as those in the Geological Survey classification. An unconformity and palaeontological break occurs in the middle of the series, the portion below it corresponding to the Oligocene and that above to the Miocene. The classification, as published by G.W. Lepper, is shown below :—

TABLE 80.—THE PEGU SERIES (B.O.C.)

MIOCENE	*Obogon alternations* (3,000 ft.). Rapidly alternating thin beds of sands and sandy clay or clay. Often missing in the north.	Vinodobonian
	Kyaukkok sandstones (5,000 ft.). Include the Prome sandstones and the highest Pegu sandstones of the central oilfields. Yellowish brown sandstones and subordinate sandy shales with a rich lamellibranch fauna	Burdigalian
	Pyawbwe clays (3,000 ft.). Concretionary blue sandy clays and thin sandstones with gypsum. Fossils abundant	Aquitanian
———————— Unconformity ————————		
OLIGOCENE	*Okhmintaung sandstones* (3,000 ft.). Massive sandstones, sandy shales with thin grey clays	Chattian
	Padaung clays (2,500 ft.). Blue-grey clays with limestone bands	Stampian
	Shwezetaw sandstone (2,000 - 4,000 ft.). Arenaceous in the north and argillaceous in the south	Lattorfian

The Pegu strata show a great deal of variation in constitution from place to place, the marine facies of the southern areas gradually changing northwards into brackish and fresh-water facies in Upper Burma. The upper beds in the north contain fossil wood, which may be carbonaceous, calcareous, siliceous or ferruginous, but less abundant than in the overlying Irrawaddy system.

COAL

The main coal bearing formations of Upper Assam are the Barails. The coal fields of Makum, Nazira, Namdang and Ledo occur in these formations and are responsible for most of the coal output from Assam. A coal seam was encountered in the Barails in the first borehole in the Nahorkatiya oil field at a depth of about 3,000 m in 1953, some 60 m above the oil horizon.

PETROLEUM

It is now generally agreed that petroleum originates from low marine organisms and that the source rocks are marine shales, silts and limestones. The accumulations of the oil into workable pools depends on the presence of porous sands or limestones folded into anticlines, domes or monoclines and covered with an impermeable shaly cap. Generally, therefore, petroleum and associated natural gas migrate into such suitable structures including stratigraphic traps form which they can ultimately be won.

The Lower Tertiaries of India, Pakistan and Burma constitute the chief source rocks of petroleum. But owing to the fact that the greater part of the likely areas have been violently folded and faulted during the Himalayan orogeny, unbroken petroleum-bearing structures are few and many areas of oil seepages which have been closely examined have been found to be barren of workable petroleum deposits. The Tertiary belt of Burma and a few localities in Upper Assam and in the Potwar plateau are the only areas which were proved to contain useful deposits before the war.

Interest has been roused in recent years in the deltas and alluvial troughs of the Indus, Ganges and Brahmaputra as well as in the coastal areas and the Mesozoic basins, as they are likely to contain suitable hidden structures beneath thick alluvial mantle. The Nahorkatiya field some 40 km S.S.W. of Digboi, in which oil has been struck in the Barail formations at a depth of a little over 3,000 m, is actually in the Brahmaputra valley outside the Tertiary area which has been cut off by the Naga thrust. In the Brahmaputra and Ganges valleys oil pools are likely to be found in stratigraphical traps and unconformities as also in gently folded structures.

Burma.—The Pegus contain the chief petroliferous horizons of Burma. The petroleum has apparently migrated from older beds into the anticlinal crests in the Pegu sandstones. The petroleum and the accompanying natural gas are kept in by beds of impervious argillaceous strata capping the sands.

The most productive oilfields of Burma are those of Upper Chindwin, Yenangyat in the Pakokku district, Singu (Chauk) in Myingyan, Yenangyaung in Magwe and some minor fields in Minbu and Theyetmyo. The main Chindwin-Irrawaddy valley is a syncline with a monocline on the west and series of broad folds on the east. The main oilfields are situated on the first anticline east of the main syncline. Yenangyaung, the most productive field in Burma, maintained its high production (about 130 million gallons per year) for a long period. Singu ranked second with a production of 90 million gallons, the producing sands here being at depths of 420, 550, 900 and 1,400 m depth. Lanywa is on the structural continuation of Singu. Yenangyat and Sabe are on a structure slightly to the east of the Singu one. The Indaw field in upper Chindwan is the most northerly field now worked, while the southernmost group includes the Minbu, Yethaya and Palanyon fields.

The first development in Burma took place in 1887 in Yenangyaung, followed by Yenangyat, Singu and Lanywa. The bulk of the total production of 37.7 million tons upto 1942 has been contributed by Yenangyaung and Singu. The oilfields were put out of action as a war measure in 1942 but they have gradually been rehabilitated since 1950. Most of the oil is of good quality with fairly high content of paraffin wax.

Assam.—Oil and gas shows are fairly numerous in Upper Assam and Surma valley down to the Arakan and a few occur along the southern border of the Shillong plateau. Practically all the major divisions of the Tertiary group below the Boka Bils give such indications, particularly the Barails. The Tipams have oil shows only in the Brahmaputra valley in Upper Assam. The oil sands of the Digboi oil fields are in the Tipams, the producing sands ranging in depth all the way down to 1,500 m. The now abandoned Badarpur field produced from the Bhuban Series. A small production has been obtained from the Barails in the Makum area while the new Nahorkatiya field west of Digboi now produces from the Barails. It is not known whether the Barails are the source rocks in all cases, but it is generally agreed that the oil in the Tipams has migrated from an older formation. Both Badarpur and Digboi are situated on tightly folded asymmetrical anticlines with major thrust faults cutting the steeper flanks. At Badarpur several producing horizons were struck in Lower Bhubans and in the Upper Barails within a depth of 500 m. Some 60 wells were drilled here and the field was producing between 1915 and 1932, yielding about 20,000 tons annually. The Masimpur and Patharia fields, in spite of much attention and drilling expended on them, have given only a very small output.

The Digboi field was discovered in 1889 and until 1921 the production was less than 20,000 tons per annum. Thereafter it rose to over a quarter of a million tons. Most of the producing sands are in the Tipams. The field is about 16 km long and the structures narrow, being cut off on the northwest by the Naga thrust which brings the Tipams against the alluvium. There are several oil sands and several hundred producing wells of which only about 30 are good producers. The oil is of mixed paraffin and asphalt base with an average specific gravity of 0.850 (0.823 to 0.978) and yielding excellent paraffin wax, lubricating oils and some bitumen. The total production to the end of 1965 from Digboi is estimated at about 12 million tons of crude oil. Production in this field is declining.

During the second quarter of the present century a considerable amount of prospecting and drilling have been done in the Naga Hills and neighbouring areas by the Assam Oil Company — Namphuk (27° 24′ : 96° 2′), Namkum-Namdang (27° 16′ : 95° 42′), Barjan (27° 22′ : 95° 31′), Barsilla (26° 56′ : 94° 37′), Tiru Hills (26° 42′ : 94° 29′), Bandarsulia (26° 33′ : 94° 17′), Nichuguard (25° 48′ : 93° 48′) etc., but the results have been disappointing. The Tertiaries in the Naga Hills are steeply folded and the anticlinal structures are almost always broken by faults so that the prospects of finding good oil-pools are poor.

Geophysical prospecting conducted in the Brahmaputra valley to the northwest of the Naga thrust in the early twenties indicated the presence of suitable structures at depth. Before drilling could be undertaken, the War intervened, so that it was possible to test the structures only after 1950. The first drill hole at Nahorkatiya, some 40 km southwest of Digboi, struck oil in the Barails at a little over 3,000 m depth in 1953. Since then a considerable amount of geophysical work has been done in the Brahmaputra valley and several suitable areas have been located. Oil has been proved also in Moran (40 km W.S.W. of Nahorkatiaya) and at Lakwa (20 km S.S.W. of Moran). Rudrasagar near Sibsagar (27° 0′ : 94° 45′) is also promising. Other areas along the Brahmaputra valley are being examined and tested. Nahorkatiya and Moran have been producing for some time from the sands in the Barail Series which is separated from the overlying strata by a well marked unconformity. The producing sands are at a depth of 2,800 to 3,500 m. The full sedimentary column here is over 6,000 m thick ranging from the Cretaceous to the Recent. There are indications of gas and some oil in the Tipam Sandstones but they have not yielded oil so far. The structures are cut by numerous criss-cross faults trending N.N.E.-S.S.W. and E.N.E.—.W.S.W., which are connected with basement faults produced by compression of the area both from the Burmese and Himalayan sides. The reserves in the Nahorkatiya-Moran area are said to be 45 million tonnes of oil and 210 billion cubic metres of gas. The oil produced from the above fields has an average Sp. Gr. of 0.88 (varying from 0.85 to 0.94) and recoverable wax content of 8 per cent.

At Rudrasagar the full Tertiary section was penetrated by a drill hole 3,817 m deep. The crystalline basement here is overlain by 300 m of marine Jaintia series (Mid. to Up. Eocene) succeeded by Barail series (600 m) which contain oil sands. Over an unconformity lie the Surma and Tipam series (1,700 m) the latter containing oil and gas. These are succeeded by the younger Dupi Tila, Dihing and Recent beds. In the Lakwa field, the Tipam sandstones contain thick sands some of which are expected to become good producers.

Gujarat.—Petroleum has been found in the Camby-Kalol area north of the Gulf of Cambay, and in the Ankleswar—Olpad area south of the Narmada river. The Cambay area is a N.N.W.-trending faulted trough in which the Deccan Traps are found at the bottom at about 2,040 m below the Lunej structure. The trough extends N.N.W. to beyond Mehsana, but the sediments therein become thinner, estuarine and carbonaceous. Oil and gas-bearing sands occur in the Oligocene (150 m thick) overlying 500 m of Eocene strata. The oil-sand in the Cambay area is restricted and the oil very viscous, but some gas would be available. The Eocene contains only poor oil-bearing sands, though the strata may have been the mother rocks of the hydrocarbons found in the upper strata. At the northern end of the Lunej structure a gas pool has been located in the Miocene. Other areas where oil and gas are expected are Nawagaon, Sanand, and Kalol, respectively south, W.N.W. and north of Ahmedabad, and Mehsana and other places farther north.

South of the Narmada, an important oil-field has been discovered at Ankleswar near Broach. Exploration is going on at Olpad, Kosamba and other places in the neighbourhood. At Ankleswar, the Tertiary section overlying the Deccan Traps is similar to that at Cambay but the Eocene is thicker (800 - 1000 m). There are three producing zones in the Eocene and the oil is of good quality, with Sp. Gr. of 0.80 and low wax content. It was producing about 3,500 tonnes per day in late 1966.

Kutch and Rajasthan.—Kutch contains a thick Jurassic-Cretaceous succession of a total thickness of over 2,500 m. There are three E.-W trending anticlinal structures in the Jurassic, which are under detailed investigation. In Western Rajasthan, particularly in Jaisalmer and Bikaner, a marine Jurassic-Eocene succession is present, part of the area being covered by desert sands. The sedimentary strata are known to thicken towards the west and this area is under investigation.

Cauvery Basin, Madras.—The coastal region of South Arcot, Tanjore and Ramanathapuram is a marine basin in which the estuarine Upper Gondwana beds are overlain, on the sea-ward side, by marine strata extending from the Albian to the Recent. The strata attain a thickness of 2,500 to 3,000 m near the coast and must be thicker off-shore and in Palk Strait. Boreholes currently being put down near Karikal are reported to have indicated the presence of petroleum.

Ganges Basin.—In the Ganges basin of Punjab — U.P. — Bihar a Tertiary succession overlies, with the interventions of a marked unconformity, Mesozoic rocks of Sub-Himalayan aspect. Though several boreholes have been sunk at various places during 1960 to date there have been no indications of oil or gas except at Jwalamukhi where there is a gas seepage in the Siwaliks. As a large area is involved, it is possible that some oil pools may be discovered at some place or other in this basin.

At Jwalamukhi in the Siwalik hills, a borehole was put down in the Lower Siwaliks. It passed the thrust zone at 1,260 m and reached the Lower Siwaliks again at 1,750 m. It went down to a depth of 3067 m and was still in the Siwaliks or had just gone into the underlying Dharmsala Beds (Dagshai-Kasauli). Gas was met with in this bore hole at 885 m. Another borehole at a locality 1.5 km away down-dip did not find any gas or oil.

The Ganges delta in West Bengal investigated by the Indo-Stanvac organisation proved disappointing after detailed investigation including the drilling of about a dozen holes to depths down to 3,600 m.

The present (1966-67) production of petroleum in India, from Assam and Gujarat, is around 4 million tonnes per year.

West Punjab.—In the Potwar region of West Punjab, which is the only oil producing area in Pakistan, there are five producing oilfields namely Khaur, Dhulian, Joya Mair, Kersal and Balkassar. The first two are elongated domes along the northern side of the Soan basin while the latter two are in the more gently folded southern flank of the basin. The full stratigraphic succession is exposed along the Salt Range which forms the southern edge of this basin. But within the basin itself the Eocene directly overlies beds of Permo-Carboniferous age and is succeeded by conglomerates and sandstones of the Murree and Siwalik systems, ranging from Miocene to Pleistocene. The Murrees and Siwaliks lie unconformably on the Eocene but without marked discordance of dip. The folding of this basin is severe in the north in the Kala Chitta hills, but becomes more and more gentle in the middle and south. The source rocks of petroleum are considered to be of Eocene age, mainly the limestones of the Middle Eocene. Oil has been found in the Eocene (Sakesar Limestone) at Joya Mair and Balkassar. A few oil shows occur on the surface in the Middle Eocene and particularly in the passage beds where they change from Marine to brackish water character.

Khaur is on an anticline 70 km S.W. of Rawalpindi on which the Murrees are exposed. It was discovered in 1915 and has been producing small amounts of oil since 1922 from various horizons in the Eocene and Murree at depths of 50 to 100 m. Better production was obtained from Eocene limestones at a depth of 1,500-17,00 m. The production has been decreasing and is now only 2,000 kilolitres per year. The oil from the upper horizons was heavy (Sp. Gr. 0.88) while that from the lower horizons is 0.84.

Dhalian, 16 km S.S.W. of Khaur is on an open dome. Siwalik rocks are exposed on the crest. Early efforts at drilling were abandoned because of very high water pressure in the Siwaliks. Heavy black oil is obtained from the Murrees and light oil from the Eocene, the latter from 2,300-2,450 m depth. Production began in 1935 from the Murrees but is now from the Eocene. Present output is 156,000 KL of oil and 3,000 million cu. m. of gas per year. The quality of oil has varied according to the horizon, the Sp. Gr. being 0.86 to 0.9 in the Murrees and 0.81 from the Ranikot.

Joya Mair has been producing since 1944 from fractured Eocene limestone, at a depth of 2,100 m. The oil is black and heavy (Sp. Gr. 0.95) containing 75% asphalt. The annual production is 1,400 KL.

Balkassar-Chakwal field is 16 km west of Joya Mair on a gently folded anticline. It has been producing since 1946 from the Eocene Bhadrar beds at a depth of 2,580 m and from Sakesar limestone below. The present production is 180,000 KL per year. The oil is heavy with Sp. Gr. of 0.89 to 0.90.

TABLE 81—GAS FIELDS OF PAKISTAN
(A. H. Khan & J. Azad: ECAFE Min. Res. Dev. Ser. 18 (10), 1963.)

	Depth metres	Pressure Kg/Cm^2	Heating value BTU/CFt.	Reserves Million Cubic Metres
Sui	4,120	137	933	141,580
Zin	906	19	484	2,831
Uch	1,214	142	308	70,792
Khairpur	618	67	130	28,317
Khandkot	1,315	136	842	5,663
Mazaruni	1,897	205	976	849
Mari	689	81	674	141,580
Sylhet	1,124	136	1,052	7,928
Chhatak	472	127	1,007	566
Rasidpur	1,225	...	1,014	20,954

	Hydrocarbon Per Cent	N & CO_2 Per Cent	H_2S gr/100 SCF	Mercaptan Sulphur Per Cent
Sui	90.1	9.8	92.2	3.8
Zin	46.9	53.2	13.3	2.3
Uch	28.8	71.4	33.5	10.2
Khairpur	12.5	87.5	...	2.0
Khandkot	80.9	19.1	30.8	1.2
Mazaruni	90.8	8.3	10.7	2.2
Mari	66.5	33.6	Nil	Nil
Sylhet	99.1	0.8	Nil	0.3
Chhatak	99.3	0.7	Nil	0.8
Rasidpur	99.7	0.4	Nil	Nil

Kersal field was developed in 1957 and the production is small. The present output of all Pakistan oilfields is around 600,000 KL per year. The reserves are estimated at 5 million KL. Several areas in the Tertiary belt are being actively prospected.

Exploration in Pakistan has located over ten important gas fields. Some of them are connected with the subsurface ridge of Khairpur-Jacobabad which is overlain by a full Tertiary succession which is over 3,000 m thick, consisting of Siwaliks, Gaj, Nari, Kirthar-Laki -Ranikot, Parh and Pab formations. The reservoir at Sui gas field is the porous MAIN SUI LIMESTONE of Paleocene to Laki age, or the SUI UPPER LIMESTONE which overlies the former. Sui is a very large gas-field in a dome structure occupying an area of 195 sq. km at a depth of 4,120 m. It has been exploited since 1954. The gas fields of Eastern Pakistan are between the Shillong plateau and Tripura. The gas occurs in the Bokabil Stage (Mid. Miocene) and Bhuban Stage (L. Miocene) of the Surma Series. Though numerous wells were drilled, only a few are productive, viz., Sylhet, Chhatak and Rasidpur. Table 81 gives particulars of all the gas fields in Pakistan. The total gas production in all Pakistan was estimated at 51,000 million c. ft. (about 1,445 million cubic metres) during 1964.

IGNEOUS ACTIVITY IN THE EXTRA-PENINSULA

The extra-Peninsular region bears evidence of considerable igneous activity especially during the latter part of the Cretaceous. Large masses of hornblende and tourmaline granites were intruded in the central Himalayan region mainly during the Miocene and later orogenic phases. Many of them are still designated by the general term Central Himalayan gneiss or granite. The granites of the Trans-Himalaya Mountains are probably of Jurassic age as they have contributed materials to the Eocene conglomerates of Mount Kailas.

THE PENINSULA

Kutch.—The Oligocene is represented here by the LAKHPAT SERIES consisting of 10 to 12 m of yellow and white marls and impure limestones enclosing *Nummulites intermedius N. clypens, Operculina* etc. indicating the Nari age. The Lower Miocene is the WAIOR STAGE, consisting of 10 m of cream coloured marls and oolites with *Spiroclypeus, Miogypsina, Nephrolepidina, etc.* which are typically Aquitanian. The upper part, constituting the KHARI STAGE is 340 m thick and consists of grey and Khaki clays, and variegated siltstones enclosing *Lepidocyclina, Operculina, Miogypsina, Austrotrillina howchini, Taberina malabarica* and also *Sorites, Archais, Ostrea gajensis, Breynia carinata* etc. which are Burdigalian.

Cambay-Broach.—The oligocene shows a regressive marine facies, being only about 160 m thick, composed of alternating shales and sandstones, but the upper part in the south, towards the open sea, shows a few thin intercalations of limestone and calcareous shale. The fossils include some reticulate

Nummulites (*N. fichteli* and *N. intermedius*), *Cassigerinella chipolensis*, *Globigerina officinalis*, *Gumbelina*, *Operculina*, *Rotalia* etc. There is a break above this, and the Miocene is transgressive, filling the channels on the eroded surface of Oligocene strata at the base and attaining a thickness of 800 to 900 m. The Miocene strata are alternating shales (locally carbonaceous) and sandstones. The fossils include *Miogypsina irregularis*, *Virgulina*, *Anomalinella*, *Cassigerinella*, *Elphidium*, and several ostracods. But the characteristic forams like *Austrotrillina howchini*, *Taberina* and *Miogypsina* are absent, though *Ammonia papillosus* is present. In the Broach area, the beds are exposed near Ratanpur and comprise gravel, sandstones and shales. The conglomeratic beds show pebbles derived from the Deccan Trap amygdales with amethyst, carnelian, jasper, agate etc. The Miocene follows on comparatively thin oligocene and shows *Austrotrillina howchini*, *Lepidocyclina canelli*, *Nephrolepidina sumatrensis*, *Miogypsina*, *Baculogypsina*, *Calcarina*, *Actinocyclina* etc. The difference in the fauna indicates that some of the larger forams did not come into the restricted Cambay trough.

Ratnagiri.—Gaj Beds are exposed on the Ratnagiri coast. They are mainly bluish clays with sand and gravel layers and contain streaks of lignite, resin and nodules of pyrite.

Baripada Beds.—At Molia near Baripada in Mayurbhanj (Orissa) are exposed yellow and brownish limestones showing abundant Oyster shells resembling *Ostrea gajensis*. They pass upward into thin-bedded greyish and greenish clays. Similar beds are seen at Baripada under a laterite bed 6 m thick. A boring put down at Baripada to a depth of 45 m revealed several fossiliferous horizons in the Miocene. The fossils include *Ostrea gajensis*, (? *Ammonia papillosus* reported originally as *Rotalia*), *Cibicides*, *Elphidium* etc. The beds are assigned to the Gaj though typical Burdigalian fossils are not present. Perhaps the beds below the depth reached in the borehole may contain them. Beds of about the same age occur near Cuttack and on the Midnapur coast in Bengal.

Bengal Basin.—There is an unconformity above the Eocene. It is followed by the BURDWAN FORMATION, about 160 thick, which consists of fresh water sandstones, lignitic shales and white sandstones. It is mainly of Oligocene age but may include Lower Miocene. It contains brachyhaline foramas and spore-pollen but no normal marine forams. The MEMARI FORMATION, 60 m thick, is estuarine to marine and of Oligocene age. It consists of calcareous shales, shaly limestones and glauconitic sandstones, containing several foraminitera including *Nummulites vascus*, *Lepidocyclina*, *Eulepidina*, *Heterostegina*, etc. Overlying the Burdwan formation is the PANDUA FORMATION (700 m) of sands, silts and shales, somewhat resembling the Durgapur Beds. This formation is of shelf facies and has yielded *Bolivina marginata*, *Miogypsina*, *Uvigerina*, *Globigerina* etc. which indicate Miocene age.

Andaman Islands.—The upper part of the GREYWACKE SERIES is believed to be Oligocene, but is unfossiliferous. The overlying ARCHIPELAGO SERIES consists of a lower arenaceous part containing Aquitanian fossils — *Miogypsina*,

Nephrolepidina, Lepidocyclina (Eulepidina) dilatata, Cycloclypeus etc., and an upper shaly and calcareous part containing *Orbulina universa, Globigerina menardi, Nephrolepidina sumatrensis, Operculinoides*, etc. which indicate Burdigalian age.

Cauvery Basin.—The Oligocene strata found in the Karikal boring between 550 to 710 metres depth are poorly fossiliferous sandstones. They do not contain reticulate Nummulites. The overlying Aquitanian (at 425-550 m depth) consists of sandstones and clays with *Spiroclypeus, Operculina, Bolivina, Textularia, Quinqueloculina* etc. The Burdigalian above these (at 70-425 m depth) are grey shales and silts, which are fossiliferous except in the middle portion, and contain *Austrotrillina, Miogypsina, Operculina, Elphidium, Bolivina, Taberina, Ammonia* etc., very similar to the Gaj fauna.

Quilon Beds.—At Padappakara near Quilon, in Travancore, well sections reveal the existence of limestones, sands and clays a few metres from the surface. The limestones are also exposed in a cliff section near the coast in the vicinity of Quilon. The limestones have yielded abundant *Orbiculina malabarica, Austrotrillina, Operculina* and several other fossils amongst which are :—

Corals	*Stylopora pulcherrima, Leptocyathus* cf. *epithecata.*
Lamellibranchs	*Parallelopipedum proto-tortuosum, Arca theobaldi, Nucula cancellata, Pectunculus sindiensis.*
Gastropods	*Strombus fortisi, Conus catenulatus, C. hanza, Rimella subrimosa, Voluta jugosa.*

A shallow bore hole at Sakthi near Quilon yielded, at a depth between 30 and 120 m, a rich Burdigalian fauna comprising *Bolivina, Miogypsina, Globigerina, Krithe, Sorites, Elphidium* and *Triloculina* in addition to those recorded from the Padappakara limestone. This indicates an extension of the Quilon Beds which are of Burdigalian age. The strata above these are of Mio-Pliocene age.

Ceylon.—Beds of about the same characters as those of Quilon occur along the northwestern coast of Ceylon where they have been called JAFFNA and KUDREMALAI SERIES. They are composed mainly of limestones and mottled sandy shales. The limestones are fairly rich in fossils, which include the foraminifer *Orbiculina malabarica*. Amongst the fossils are :—

Foraminifera	*Orbiculina malabarica, Operculina* sp., *Flosculinella* sp., *Miliolidae* (several species).
Echinoderms	*Clypeaster depressus, Schizaster* sp.
Gastropods	*Trochus cognatus, Phasianella oweni, Natica rostalina, Cerithium* cf. *rude, Ovula ellipsoides, Oliva pupa, Conus brevis.*
Lamellibranchs	*Arca peethensis, Avicula* cf. *suessiana, Spondylus rouaulti, Cardium sharpei, Cardita intermedia, Ostrea virleti.*

The fauna was described by E. J. Wayland and A. M. Davies (*Q.J. G. S.* **79**, p. 577-602, 1923) and assigned an Upper Miocene age. But after a critical examination, Eames (*Geol. Mag.* **87**, p. 53-56, 1950) has come to the conclusion that the fauna of the Quilon and Jaffna beds are not of Upper Miocene but of Burdigalian age and closely allied to that of the Upper Gaj beds.

SELECTED BIBLIOGRAPHY

——————— The Science of Petroleum. Oxford Univ. Press. 1938.

Barber, C.T. Natural Gas resources of Burma. *Mem.* **66**(1), 1935.

Coates. The Progress of oil prospecting in India. *Proc.* 37th *Ind. Sci. Cong.* Part II, 119-133, 1950.
Condit, D. Natural gas and oil in India. *Bull. A.A.P.G.* **18** (3), 283-314, 1934.
Corps, E.V. The Digboi oil-field. *Bull. A.A.P.G.* **33** (1), 1949.
Cotter, G. de P. Notes on the geological structure and distribution of oil bearing rocks of India and Burma. *World Petr. Cong. Proc.* **1**, 7-13, 1933.

Eames, F. E. On the ages of certain Upper Tertiary beds of Peninsular India and Ceylon. *Geol. Mag.* **87**, 233-256, 1950.
Evans, and Sansom, C.A. Geology of the Burmese oil fields. *Geol. Mag.* **78**(5), 321-350, 1941.

Lepper, G.W. Outline of the geology of the oil bearing region of the Chindwin-Irrawaddy valley and of Assam-Arakan. *World Petr. Cong.* I, 15-23, 1933.

Mukherjee, P.N. Fossil fauna of the Tertiary of the Garo Hills. *Pal. Ind. N. S.* **XXVIII** (I), 1939.

Pascoe, E. H. The oil-fields of Burma. *Mem.* **40**, (1) 1912.
Pascoe, E. H. Petroleum occurrences of Assam and Bengal. *Mem.* **40**(2), 1914.
Pascoe, E. H. Petroleum in Punjab and N.W. Frontier Province. *Mem.* **40**(3), 1920.
Pilgrim, G.E. Vertebrate fauna of the Gaj Series in the Bugti Hills and Punjab. *Pal. Ind. N.S.* IV (2), 1912.
Pinfold E.S., Davies T.G.B. and Gill, W.D. Development of the oil-fields of N.W. India. *Trans. Nat. Inst. Sci. India*, II(8), 1947.

Rankine, A.O. and Evans, P. Geophysical prospecting for oil in India. *Trans. Nat. Inst. Sci. India*, II(8), 1947.

Sale H. M. and Evans, P. Geology of the Assam-Arakan oil region *Geol. Mag.*, **77** (5), 337-362, 1940.
Spengler, E. Contributions to the Palaeontology of Assam. *Pal. Ind. N.S.* VII (1), 1923.

Vredenburg, E. Marine fossils collected by Pinfold in the Garo Hills. *Rec.* **51**, 303-337, 1921.

CHAPTER XIX

MIDDLE MIOCENE TO LOWER PLEISTOCENE

North-Western India : The Siwalik System

Introduction.—The termination of the Murree period in the middle Miocene coincided with the third, and perhaps the most violent, episode in mountain building on the northern borders of India. This must have been accompanied by a considerable raising up and folding of the strata laid down in the Tethys into mountain ranges and by large intrusions of igneous rocks into the cores of the folds. A long, narrow depression was formed in front of the rising mountains, *i.e.*, towards the side of the Peninsula. This depression (called the *fore-deep*) was the site of the deposition of the Siwalik strata which commenced in Middle Miocene. Most of the sediments were derived from the denudation of the newly risen mountains. Numerous short streams must have flown from the mountains into the fore-deep in a direction transverse to the latter, and contributed to the water being kept fresh. The rise of high mountain chains to the north of India would have helped to establish the monsoon climate and a high precipitation of rain on the southern flanks of the mountains. The great aggregate thickness (16,000 to 18,000 ft.) of the deposits and their general coarseness give evidence of continuous deposition in a shallow body of water whose depth kept pace with the accumulation of sediments. This sinking was probably aided by the gradual compression to which the crust was subjected, at least intermittently, during the period of sedimentation.

Distribution.—The Siwalik system takes its name from the Siwalik hills of Hardwar region between the Ganges and Jumna rivers. It extends continuously along the foot of the Himalaya from the Brahmaputra valley on the east to the Potwar plateau and the Bannu plains on the west. Its equivalents intervene between the Indus Plains and the early Tertiaries of the Sind-Baluchistan hills. The re-entrant angle near uetta Qexhibits a complete development of these rocks, the Zarghun mountain mass in this region forming a synclinal of these rocks. Similarly, in the Burmese are, we see their equivalents on both sides of the Arakan Yomas, both in Assam and in Burma.

The rock groups have received different names in the different areas. They form the Siwalik System along the outer Himalayas ; the Manchhar System in Sind ; the Makran Series in the Makran region of Baluchistan ; the Dihing Series in Assam and the Irrawaddy System in Burma.

Constitution.—The Siwalik System is made up of sandstones, grits, conglomerates, pseudo-conglomerates, clays and silts having the characters of fluviatile deposits of torrential streams and floods in shallow fresh-water basins. The fossils included in them show that the earlier beds were deposited in a somewhat brackish environment as compared with the later ones. Some of the latest deposits may be continental, *i.e.*, left on dry land by temporary heavy floods. There is a considerable amount of ferruginated matter, especially in some of the older horizons, which indicates that the sediments were derived partly from an old and well oxidised terrain. Coarser and finer sediments alternate. The sandstones show poor stratification and are generally ungraded as to grain size. They are feldspathic, micaceous and current-bedded and some of them have clearly been derived from the breakdown of the central Himalayan granites.

The Siwaliks have been involved in the later phases of Himalayan orogeny, for we find them often folded, faulted, overthrust and lying at steep angles against other formations. Where overthrust, there is often an inversion of the normal order of superposition. The main overthrust in which the Siwaliks are involved used to be called the 'Main Boundary Fault,' but recent work has shown that there are at least three major thrusts in the Himalaya in addition to less important local ones. These thrusts may, in some instances, mark the limits of deposition of the older series involved, but there has been so much movement that even within the Siwaliks the older strata are found thrust over younger ones and lying with very abnormal dispositions.

Conditions of deposition.—The coarse and often ungraded sandstones show that they must have been borne by rapidly flowing and large masses of water and laid down in wide depressions of shallow water or in swampy areas. The alternation of coarse and fine sediments suggests seasonal deposition, the coarse materials during floods of the wet season and the fine sediments during the drier season. The extraordinary similarity of the deposits over long distances along the strike would show that the source rocks were similar and that the basin of deposition was practically continuous. The large thickness of the coarse materials makes us infer that the area of deposition was sinking in pace with the sedimentation. At the same time there was a gradual southward shift of the basin with each fresh pulse of the uplift. It is almost certain that the Siwaliks extend down for several miles underneath the alluvial cover of the Indus and Ganges valleys.

The occurrence of this important strip of fluviatile rocks all along the foot-hill regions of the Himalaya from Assam to Punjab and thence to Sind has led to the view, advocated by Sir E..H. Pascoe and Dr. G.E. Pilgrim that the Siwaliks were laid down in the flood plains of a single large river (the Indobrahm or Siwalik River) which rose in Assam and followed the present line of distribution of these deposits. The present author and N.K.N. Aiyengar have discussed this question (*Rec.* 75, paper No. 6, 1940) and shown that the available evidence points to the basin of deposition being a continuous lagoon or fore-deep formed in front of the Himalayan range.

Climatic Conditions.—The Siwalik deposits give evidence of a warm humid climate through the greater part of the period of sedimentation. The coarse materials, which are often fresh, may have been derived from the north and the finer ferruginous clays from the ancient reninsular area to the south, the one contributing material during the wet flood season and the other during the dry season. Some chemical decay may have also taken place under the swampy conditions in which the sediments were laid down. The earlier Siwalik period — that of the Lower Siwaliks and the lower part 9f the Middle Siwaliks

— was apparently a wet period, or alternatively, the sediments were deposited in shallow water. In Dhok Pathan times there is evidence that the humidity was less and that the sedimentation took place in partly marshy land. The disturbance at the end of the Middle Siwalik times raised the deposits into dry land and shifted the basin southwards. The Upper Siwaliks again show the return of wetter conditions. Towards the end of Tatrot times another uplift took place and the climate became distinctly colder. The animals which lived in the marshes and valleys migrated away or died, as the subsequent deposits were of semi-glacial character.

Organic remains.—The great bulk of the Siwalik formations is unfossiliferous but certain areas are rich in fossils. These include plants, mollusca, fishes, reptiles and mammals. The plant remains consist of leaf impressions in clays and tree trunks in sandstones. The tree trunks are silicified but in most cases the finer woody structures are not preserved. The mammalian remains are the most important fossils as they are of great help in dividing the formations into stages and as they indicate the stages of development through which the animals passed before they disappeared from the scene of life. The present day mammals in India are but the poor remnants of the rich variety that lived formerly in the swamps and forests of the Siwalik basin. The relics consist of hard, bony parts, skulls, jaws and teeth. Their abundance testifies to very favourable conditions of climate and hydrology, abundance of food and suitable environment for entombment of the remains.

The detailed study of the mammalian remains in many countries has thrown much light on the origin, evolution and migration of the animals. Some groups like the pigs, hippopotamus and ancestral elephants are believed to have originated in Africa and later migrated into Asia. The horse is supposed to have come from North America through Alaska over a land bridge across the Behring straits. It is an interesting fact that the horse became extinct in N. America by the Pleistocene and was reintroduced there by man from Europe.

Divisions.—The Siwaliks are divided into three major divisions, ranging in age from Middle Miocene to Lower Pleistocene. The various sub-divisions take their names from localities in the Potwar region. Though the sub-divisions are based on lithology there is much lateral variation in the rocks. There are no marked unconformities within the System but the Upper Siwaliks seem to have been laid down on the Middle Siwaliks after a period of folding, uplift and denudation. The Boulder-conglomerates are also generally transgressive over the previous beds. The characters of the sediments indicate that the basin of deposition was first brackish and that it became increasingly fresh and that there is a variation from lacustrine to fluviatile conditions.

The chief sub-divisions of the Siwaliks are shown in Table 82, together with the nature of the sediments and their age.

Kamlial Stage.—This stage is named after Kamlial (33° 15' : 72° 30') near Khaur oilfield and consists of hard red sandstones with clay nodules (pseudo-conglomerates) and purple shales. The sandstones generally form conspicuous strike ridges and are somewhat finer grained than those of the

TABLE 82—SIWALIK SUCCESSION (N.W. INDIA). (AFTER G.E. PILGRIM)

Upper-Siwalik (6,000-8000 ft.)	Boulder Conglomerate. Coarse conglomerates sands, grits and some clays.	Cromerian
	Pinjor Stage. Coarse grits, sandstones and conglomerates	Villafranchian
	Tatrot Stage. Soft sandstones, drab clays and some conglomerates.	Astian
Middle Siwalik (6,000-8,000 ft.)	Dhok Pathan Stage. Brown sandstones, gravel beds, orange clays and drab shales.	Pontian
	Nagri Stage. Hard grey sandstones and subordinate shales.	Sarmatian
Lower Siwalik (5,000 ft.)	Chinji Stage. Bright red shales and sandstones.	Tortonian
	Kamlial Stage. Hard red sandstones, purple shales and Pseudo-conglomerates.	Helvetian

Murrees. In the Jammu area the Kamlials are not easily separable from the Murree Sandstones as they look alike and contain few fossils. In northern Potwar, the Kamlials contain much tourmaline and only a little epidote, while the Murree and Chinji beds contain abundant epidote, but only a little tourmaline. In the Potwar area the upper part of the Kamlials consists of red shales and some grey or reddish brown sandstones with pseudo-conglomerates containing 'pebbles' of clay and shale. They are about 550 ft. thick and the shales are deeper red in colour than the Chinjis. The chief mammalian remains are —

(Carnivora) *Amphicyon* and *Hyaeneluros* ; (Proboscidea) *Dinotherium* and *Trilophodon* ; (Suidae) *Palaeochoerus* and *Listriodon*. The earlier Proboscidea like *Moeritherium* and *Hemimastodon*, as well as the Rhinocertatids like *Dicratherium, Teleoceras, Baluchitherium* and *Aceratherium*, and Giraffidae, like *Progiraffa* had disappeared.

Chinji Stage.—Taking its name from Chinji (32° 41' : 72° 22'), this stage shows alternating ash-grey sandstones and bright red shales. The sandstones are subordinate in the Potwar area, but dominant in the Jammu area. The thickness varies between 400 to 1,800 m. This stage is apparently of longer duration than the Kamlial and contains a large number of vertebrate fossils and also wood. The mammals include :—

(Primates) *Dryopithecus, Sivapithecus, Brahmapithecus* ; (Carnivora) *Dissopsalis, Amphicyon, Martes, Eomellivora, Sivalictis, Lycyaena, Sansanosmilus, Vishnufelis* ; (Proboscidea) *Dinotherium, Trilophodon, Serridentinus, Synconolophus, Anancus, Stegolophodon* ; (Equidae) *Hipparion* ; (Chalicotheridae) *Macrotherium;* (Rhinoceratidae) *Gaindatherium, Aeratherium* ; (Suidae) *Conohyus, Listriodon, Propotamochoerus, Dicoryphochoerus, Sanitherium* ; (Anthracotheridae) *Hyoboops, Hemimeryx, Telmatodon* ; (Tragulidae) *Dorcabune, Dorcatherium* ; (Giraffidae) *Giraffokeryx, Propalaeomeryx, Giraffa*.

Of these, *Macrotherium, Hipparion* and *Gaindatherium* are supposed to have migrated into India from America. Numerous reptiles like crocodiles, turtles, pythons and lizards are found in these rocks as also shells of *Unio*. In the Hardwar area the Lower Siwaliks have been called *Nahan beds* and they correspond mainly to the Chinji Stage.

Nagri Stage.—This is named after Nagri (32° 40' : 72° 15') in the Attock district and consists of hard grey or buff coloured sandstones with a small proportion of shales and clays. This formation is the starting point of evolution of the major part of the Siwalik fauna. It is poorly fossiliferous, many of the mammals of the previous stage having apparently disappeared. Several Primates — *Brahmapithecus, Sivapithecus,* and *Surgrivapithecus* are present. Among the Carnivores, *Amphicyon* and *Sivanasua* persist and *Crocuta* appears.

No Proboscidea have been found. *Hipparion, Gaindatherium* and *Aceratherium* continue. Several genera of the pig family are found such as *Palaeochoerus, Conohyus, Listriodon, Lophochoerus, Propotamochoerus, Dicoryphochoerus, Hippohyus* and *Sus*. *Hemimeryx* (Anthracotherid), *Dorcabune, Dorcatherium* (Tragulid) and *Giraffokeryx* have also been found. The first primates are found here.

Dhok Pathan Stage.—This is named after Dhok Pathan (33° 8' : 72° 21' on the Soan river and includes brown sandstones, drab shales, orange clays and some beds of gravel. In some areas the sandstones are variegated and brown but there is a variation when they are followed along the strike. The sandstones gradually give place to shales in an easterly direction and the shaly facies extends downwards into the Nagris also. The Middle Siwaliks thin down also in a southerly direction. Thus the source of the sediments was from the northwest, *i.e.*, from the present Indus basin, the drainage being transverse to the basin of deposition. This, incidentally, is evidence against the existence of the Indo-Brahm river.

The Dhok Pathan Stage is the richest fossiliferous stage of the Siwaliks and has yielded a large number of fossils amongst which may be mentioned :—

(Primates) *Macacus, Sivapithecus* ; (Rodentia) *Rhizomys* and *Hystrix* ; (Ursidae) *Agriotherium, Indarctos* ; (Mustelidae) *Promellivora, Enhydriodon, Sivaonyx* ; (Hayenidae) *Ictitherium, Lycyaena, Crocuta* ; (Felidae) *Aleuropsis, Mellivorodon, Paramachoerodus, Felis* ; (Proboscidea) *Dinotherium, Trilophodon, Tetralophodon, Synconolophus, Stegolophoden, Stegodon* ; (Equidae) *Hipparion* ; (Rhinoceratidae) *Aceratherium, Rhinoceros* ; (Suidae) *Listriodon, Tetraconodon, Propotamochoerus, Dicoryphochoerus, Hyosus, Hippohyus, Sus* ; (Anthracotheridae) *Choeromeryx, Merycopotamus* ; (Hipopotamidae) *Hippopotamus* ; (Tragulidae) *Dorcabune, Dorcatherium, Tragulus* ; (Cervidae) *Cervus* ; (Giraffidae) *Hydaspitherium, Vishnutherium, Brahmatherium, Giraffa* ; (Bovidae) *Taurotragus, Perimia, Tragocerus, Boselaphus, Proleptobos*.

Several of the giraffe family and the short-jawed proboscidean *Synconolophus* are confined to this stage while the Bovidae first make their appearance here.

This stage is considered to be of Pontian age by Pilgrim, but Mathew has shown that the fauna, especially horses and giraffes, are more advanced than those of the Pontian beds of Pikermi, Samos and China and therefore of Middle Pliocene age, with which opinion Lewis agrees.

At the end of Dhok Pathan time an uplift occurred and the strata were folded and eroded, before the deposition of the Tatrot beds. This erosion interval is thought to cover the Upper Pliocene period, which is therefore unrepresented by either sediments or fauna. This is attested by the local geological features, for the Upper Siwaliks are found only along the basins and stream channels and not on the Potwar peneplain. This uplift accentuated the Kala Chitta range and extended it eastwards, according to De Terra. A stream called Nandna (which joins the Haro which is a tributary of the Indus) which rises on the Khari-i-Murat and cuts across the Kala Chitta hills is clearly antecedent. The western Potwar was drained by the Indus during the Pliocene.

Tatrot Stage.—This stage named after Tatrot (32° 52' : 73° 21'), consists of conglomerates, soft sandstones and drab and brown clays. The Tatrot beds were laid down in the basins which resulted from the folding movements, the folds having a N.E.-.S.W. axis. They often lie unconformably over the planed surface in which there are prominent upturned ridges of Nummulitic strata. The base is always marked by a coarse conglomerate and the beds show evidences of quick deposition by rivers and delta-like structure. The sediments are coarser than before and contain pebbles derived from Mesozoic as well as the older Siwalik beds and even bones from Dhok Pathan beds are known to have been reimbedded in these sediments. The Tatrot period was marked by heavy rainfall as the beds do not show red colour and enclose remains of elephants, pigs, Hippopotamus, Bovids etc. indicating good moist conditions. The elephants are represented by *Stegodon* and *Pentalophodon*. A true one-toed horse, *Equus sivalensis*, and the pigs *Sus* and *Hippophyus* are also found in these beds.

Pinjor Stage.—Named after Pinjor near Kalka in the Simla foothills this stage shows coarse sediments composed of pebble beds and sandstones. They follow the Tarrot beds without any definite break and therefore belong to one cycle of sedimentation. Variegated sands and pink silts are dominant in this stage, the pink colour of these indicating somewhat drier conditions and also the deposition of much eolian material. The thickness varies from 150 to 450 m. The beds contain a rich fauna particularly in the clays underlying the Boulder-conglomerate and these can be regarded as the immediate ancestors of the present-day mammals :—

(Primates) *Papio, Semnopithecus, Simia* ; (Rodentia) *Rhizomys, Nesokia, Hystrix* ; (Carnivora) *Canis, Agriotherium, Sinictis, Mellivora, Lutra, Enhydriodon, Viverra, Hyaenictis, Crocuta, Megantereon, Panthera, Felis* ; (Proboscidea) *Pentalophodon, Stegodon, Stegolophodon, Archidiskodon, Hypselephas* ; (Equidae) *Equus* ; (Rhinoceratidae) *Rhinoceros, Coelodonta* ; (Suidae) *Tetraconodon, Potamochoerus, Dicoryphochoerus, Hippohyus, Sus* ; (Anthracotheridae) *Mericopotamus, Hippopotamus, Camelus* and *Cervius* ; (Giraffidae) *Sivatherium, Camelopardalis, Giraffa* ; (Bovidae) several including *Cobus, Leptobos, Bubalus, Bos* and *Bison*.

Boulder-Conglomerates (Lei Conglomerate).—At the end of the Pinjor period there was mountain building activity as indicated by the very coarse sediments which constitute this stage. They overlie the older beds disconformably, mainly as fans. The newly risen Pir Panjal supplied much of the sediments and the boulders and pebbles are sometimes facetted, indicating deposition from glaciers. The sediments consist of poorly graded materials and include pebbles of Eocene limestones, with a small porportion of igneous and sedimentary rocks and quartzites. The Conglomerate beds are lenticular and contain a few silt horizons which may indicate eolian material deposited during dry periods. The term Boulder-Conglomerate has been loosely used for all the conglomerates ranging in age from Nagri upwards, and having a total thickness of over 2,200 m. But the stage above the unconformity is only 100 m thick, and is known as the Lei Conglomerate in the Potwar region. This horizon, of Middle Pleistocene age, was deposited during the Second Glaciation in the Pleistocene.

The Himalayan glaciers seem to have descended almost down to the plains making the whole region unsuitable for the existence of highly developed mammalian life. Much of the Siwalik fauna suffered heavily but some species were able to migrate to warmer regions. The giraffes which were abundant in the Upper Siwaliks are now found only in Africa. Though many species of the elephant family lived during the Siwalik times, all that is left of them at the present-day is the Indian Elephant and the Africal elephant. The Boulder Conglomerates contain also relics of pre-historic man whose appearance coincides with the rapid disappearance of mammals.

At the end of the period of the Boulder Bonglomerate there was a fresh diastrophism. Kala Chitta and Khair-i-Murat hills were uplifted along previous faults. The basin suffered compression and the upper Siwaliks were gently folded. The Jhelum changed its course to its present one below Owen by cutting through conglomerate fans. The Indus whose old course was through the Haro channel also shifted its course to the depression above Attock.

Correlation.—Since the Siwaliks are fresh-water (and partly land) deposits, the determination of the age of the stages is a matter of some uncertainty. It is definitely established that the Middle Siwaliks are closely allied to the Pikermi beds of Greece where they are associated with marine Pontian strata.

In Table 83 is give the correlation adopted by G.E. Pilgrim of the Geological Survey of India compared with that of E.H. Colbert and G.E. Lewis. It will be noticed that the ages assigned to the divisions by the American authorities are younger than those adopted by Pilgrim. This is perhaps attributable partly to the hypothesis that the Hipparion (primitive horse) migrated to India from N. America.

SIND : MANCHHAR SERIES

The highest formations of the Tertiary in Sind are called the Manchhar Series, after the Manchhar Lake. They resemble the Siwaliks to a large extent.

Lower Manchhar.—The Manchhar Series attains a thickness of 3,000 m and comprises two divisions. The Lower Manchhars are composed of grey sandstones associated with red sandstones and conglomerates. The conglomerates contain pebbles of sandstone and nodules of clay. The lowest horizon has yielded vertebrate fossils which indicate a Helvetian zone.

Upper Manchhar.—The upper division is well exposed near Larkhana and consists of conglomerates, sandstones and orange and brown clays. It contains pebbles of Gaj and Nummulitic limestones.

The Manchhars of the Kirthar range appear to follow conformably on the older (Gaj) rocks whereas those east of the Laki range lie unconformably on Kirthar beds. They are mainly fluviatile, but as we follow them southwards they become gradually estuarine and marie. Vredenburg divides the Manchhars into three portions which are respectively Vindobonian, Pontian and Pliocene in age.

Makran Series.—The Upper Tertiary rocks developed in the Makran region are called the Makran series and comprise thick pale grey clays with

TABLE—83 CORRELATION OF THE SIWALIK STRATA

	Divisions	Pilgrim 1934-1940	Colbert 1935	Lewis 1937	Europe	North America
PLEISTOCENE	Upper					
	Middle			Tawi	Cromerian	
	Lower		Boulder Cgl.	Break	Villafranchian (=Calabrian)	Rock Creek Sheridan
		Boulder Cgl.	Pinjor	Pinjor Tatrot		
PLIOCENE	Upper	Pinjor	Tatrot	Break	Astian	San Pedro
			Dhok Pathan			
	Middle	Tatrot	Nagri	Dhok Pathan	Plaisancian	Blanco Good Night
	Lower	Dhok Pathan	Chinji	Nagri	Pontian	Republican River Valentine
MIOCENE	Upper	Nagri	Kamlial	Chinji	Sarmatian Tortonian	Barstow
	Middle	Chinji Kamlial		Kamlial	Helvetian Burdigalian	
	Lower				Aquitanian	

thin intercalations of shelly limestone and sandstone. Vredenburg has divided them into a lower TALAR STAGE of Middle Siwalik age and an upper GWADAR STAGE of Upper Siwalik age. The Makran series contains a fairly rich marine fauna which bears a great resemblance to that of Odeng stage (= Talar) and Sonde stage (= Gwadar) of Java. It is inferred that the Indian seas were completely cut off from the European seas during Makran times as there is not much resemblance to the European marine fauna. Many of the Makran species are found in the Karikal beds. The molluscan fauna of the Makran series includes the following :

G—62

| Gastropods | ... | ... | *Surcula tuberculata, Terebra aspera, Pleurotoma haydeni, Drilla mekranica, Lithoconus djarianensis, Clavilithes verbeeki, Melongena ponderosa, Purpura angulata, Cassidia mekranica, Dolium sp., Terebralia mekranica, Natica globosa, Rimella javana.* |
| Lamellibranchs | ... | ... | *Corbula mekranica, Dosinia pseudoargus, Cardium unicolor, Arca newtoni, Pectunculus gwadarensis, Ostrea psudodigitalina, O. virleti, O. frondosa, O. Crenulifera, Pecten vasseli.* |

ASSAM

Tipam Series.—The Siwalik system is represented in Assam by the Tipam and Dihing Series. The Tipams extend from the Arakan coast to Surma Valley and Upper Assam. In Upper Assam they are divisible into three stages, the upper being separated from the middle by an unconformity.

NAMSANG BEDS. Sandstones, grits and conglomerates, containing lignite pebbles (2,500 ft.) } Mio-Pliocene

————————unconformity————————

GIRUJAN CLAYS. Mottled clays and sandy clays and subordinate sandstones, with fossil wood and lignite (3,000 to 6,000 ft.) } Vinodobonian

TIPAM SANDSTONES. Thick coarse gritty ferruginous sandstones, greenish coloured and weathering brown ; some conglomerate and shale partings ; occasional fossil wood and lignite (3,000 to 8,000 ft.) } Burdigalian

Dupi Tila Series.—The NAMSANG STAGE was formerly known as NUM RONG KHU STAGE but as the type section in Mana Bum area (27° 36′ : 96° 10)′, some 35 miles E.N.E. of Digboi, was found to include soft sandstones now included in the Dihings, it has been decided to name it after the Namsang river, a tributary of the Burhi Dihing south of Digobi. The pebbles in the conglomerates in this Series are partly of Barail Sandstones and coal. This is the equivalent of the Dupi Tila.

In the Surma Valley, the Tipam Sandstone-Girujan Clay succession is overlain by the DUPI TILA SERIES, separated by an unconformity. The Dupi Tilas are composed of sandstones and clays about 3,000 m thick. The lower beds are coarse ferruginous sandstones with subordinate mottled clays (600 m). while the upper beds are variegated sandstones, mottled sandstones and clays (2,800 m). The best sections are to be found along the southern margin of Jaintia hills. They are unfossiliferous and are considered to be the equivalents of Namsang Beds and of Mio-Pliocene age.

Dihing Series.—These consist of a great thickness of pebble-beds with subordinate sandstone and clay bands. They overlie the Tipam Series and are named after the type section on the Dihing river near Jaipur (27° 16′ : 95° 24′). The rocks are for the most part poorly consolidated and the pebbles have been derived from the Barails except in the extreme north-east of Assam where the pebbles are of gneisses. They are exposed in Upper Assam, Naga hills and Sylhet. They contain only some carbonised wood and badly preserved leaf impressions and are considered to be of Pliocene to early Pleistocene age, equivalent to the Upper Siwaliks. As they have been involved in the folding and thrusting which have affected the Assam Tertiaries, they are probably mainly of Pliocene age.

BURMA

The Irrawaddy System

Overlying the Pegu series, generally with an unconformity, is a series of fluviatile sandstones of large thickness originally called the *Fossil-wood Group* by Theobold, because of its abundant content of fossil-wood. This name has now been replaced by the term Irrawaddy system, since the sediments lie in the valley of the Irrawaddy river.

Distribution.—The Irrawaddy System is extensively distributed along the central north-to-south tract of Burma. It is found in the north in Katha, Upper Chindwin, and Myitkyina, and extends as far south as Rangoon. Parts of the area are covered by Pleistocene and younger alluvia but it is likely that the Irrawaddy rocks extend under all the alluvium of the delta.

Lithology.—The lower beds often contain ferruginous conglomerates indicative of a period of subaerial weathering. The main formations are sandstones, often current-bedded, with pebbles, boulders and numerous ferruginous concretions. The concretions consist of haematite and limonite with some manganese oxide. Calcareous sandstones and calcareous and siliceous concretions are also met with. Fossil wood occurs profusely in certain areas, often well silicified. The rocks weather into fantastic shapes with formation of intricate gullies and 'bad lands.' The total thickness of the system is of the order of 3,000 m.

Fossils.—The Irrawaddians contain a fairly rich variety of vertebrate fossils, mainly mammalian, but including also crocodiles, tortoises, etc. Both monocotyledonous and dicotyledonous fossil woods are found, the latter including *Dipterocarpoxylon burmense*. The chief vertebrate fossils are :—

Lower Irrawaddy beds :—

(Equidae) *Hipparion antilopinum* ; (Rhinoceratidae) *Aceratherium lydekkeri* ; (Suidae) *Tetraconodon minor* ; (Giraffidae) *Hydaspitherium birmanicum, Vishnutherium iravaticum* ; (Bovidae) *Pachyportax latidens, Proleptobos birmanicus.*

Upper Irrawaddy beds.—

(Proboscidea) *Stegolophondon latidens, Stegedon elephantoides, Stegodon insignis birmanicus, Hypselephas hysudricus* ; *Elephas planifrons.* (Rhinoceratidae) *Rhinoceros sivalensis;* (Anthracotheridae) *Merycopotamus dissimilis* ; (Hippopotamidae) *Hexaprotodon iravaticus, H.* cf. *sivalensis* ; (Bovidae) *Leptobas, Bubalus.*

The Irrawaddy System is the equivalent mainly of the middle and upper Siwaliks. The lower fossiliferous beds are correlated with the Dhok Pathan stage (Pontian) and the upper fossiliferous beds, which are separated from the lower by 1,200 m of strata, are of Pinjor age.

Bengal Basin.—The Plio-Pleistocene strata in the Ganges delta are represented by the Debagram and Pandua formations (Shelf facies), and by Ranaghat and Matla formations (Geosynclinal) farther out. The PANDUA FORMATION has been mentioned in the last chapter as it is mostly Miocene. It is succeeded by the DEBAGRAM FORMATION (900 m) over a slight angular discordance. This is composed of clays, calcareous shales and silts which are deltaic and become more clayey down the dip. These strata pass laterally into the RANAGHAT FORMATION which is marine. The Debagram formation contains *Bolivina hughesi, Hyalina baltica, Globigerina, Gumbelina* etc. The age is mainly Pliocene passing up into Pleistocene. The MATLA FORMATION (1300 m) is alternating marine-deltaic, composed of clays and silts. It passes gradually up into the RANAGHAT FORMATION (2,530 m) which also represents deposits of a marginal marine area. The former may include some Miocene but is mainly Pliocene, while the latter is Plio-Pleistocene.

Along the western border of the Raniganj coalfield and overlapping the upper beds of the Raniganj Series is a strip of lateritised sandstones, feldspathic grits, and bluish grey and mottled clays, called the DURGAPUR BEDS. These are covered by the alluvium some distance east of Durgapur. These sandstones have yielded angiospermous fossils wood near Suri in Birbhum. Similar beds are known to occur to the east and southeast of the Rajmahal Traps, towards the Ganges alluvium. They are probably of Mio-Pliocene age like the Cuddalore Sandstones. The PANDUA FORMATION, named by the geologists of the Stanvac Oil Company after Pandua not far from here, is roughly of the same age.

Kutch.—The Miocene KHAR SERIES is overlain, with a transgressive overlap, by the Pliocene KANKAWATI SERIES. This latter resembles the Manchhar formations of Sind in general and is made up of pink grits enclosing fossil wood, and grey sandstones and clays with *Textularia, Spiroloculina, Rotalia, Nonion* etc. This is followed by the Pleistocene Miliolite Limestone.

Kathiawar.—Fossiliferous variegated shales, marls and impure limestones containing Gaj fossils are exposed near the western end of Kathiawar. Along the eastern coast there are the thinbedded sandstones and grits of Gogha which are also of the same age. In the Piram Island in the Gulf of Cambay there are conglomerates, grits and clays (PIRAM BEDS) which contain the remains of vertebrate fauna resembling those of the Middle Siwaliks. The presence of such fauna here in the Cambay trough implies that there must have been a low-lying island or a small peninsula in the Gulf during the Mio-Pliocene, which has been uplifted later.

The DWARKA BEDS occur at the western end of Kathiawar overlying the Gaj strata. They consist of gypsiferous clays and sandy formainiferal limestones which are also believed to be Mio-Pliocene.

Warkalli (Varkala) Beds.—The Warkalli Beds exposed in and around Varkala and other places along the coast of Kerala (Travancore-Cochin-Malabar) consist of rather ferruginous sandstones with some clay intercalations the thickness being of the order of 90 to 120 m. They grade down into grey to dark clays and carbonaceous sandstones containing carbonised plant fragments and tree stems, thin seams of lignite, pieces of resin and marcasite.

These are underlain by the Quilon Limestone which is a grey, compact, hard limestone containing fossils. Intercalated with the limestones are thin beds of stiff blue-black clays. The limestone is exposed only in a very few places near Quilon but has been found in several boreholes near Varkala. It is high in calcium and low in magnesia (about 2 per cent) with varying silica (6-15 per cent). Its age as determined from the fossils is Gaj (Burdigalian or upper part of Lower Miocene). The age of the Warkalli Beds may be slightly younger (Upper Miocene).

The Geological Department of the former State of Travancore assumed that there was a continuous bed of lignite in this formation over an area of 650 sq. km. and estimated the reserves as 270 million tons. As out 15 drill holes put down near Varkala by the Geological Survey of India in 1957 in an area of 130 sq. km. indicated the presence of only very thin lenticular beds of lignite or lignitised fragments of wood, but no good seam. The rest of the area remains to be investigated.

The Warkalli Beds are lateritised at the top to a depth of 10 to 13 m. Where the sandstones directly overlie the feldspathic gneisses, the latter are kaolinised to 10 m or more. Deposits of kaolin are worked at Kundara near Quilon and at several places in Malabar. Some of the clays in the Warkalli Beds are refractory clays and ball clays. Lignite seams have been noted in these formations at the bottom of the sea-cliff at Cannanore, in a stream section near Beypore and at Kasaragod in Malabar. There are therefore good prospects of finding workable seams of lignite in these formations.

Cuddalore and Rajaamhendri Sandstones. —A wide stretch of sandstones generally somewhat lateritised and ferruginous, extends intermittently from near Rameshwaram through Pudukottah, Tanjore, Cuddalore, Pondicherry, Madras, Nellore etc. to Rajamahendri (Rajahmundry). The Rajamahendri Sandstones exposed in the Godavari districts are 300 m thick and consist of fedspathic and ferruginous sandstones with thin pebble-beds. The Cuddalore Sandstones occur in the southern districts of Madras. They are overlain in the coastal tract and in the various river valleys by deltaic alluvium and coastal sands but may be expected to continue down to the coast. They consist of soft red, yellow and mottled ferruginous sandstones, sandy clays, sands, clays and pebble beds. They have a gentle easterly or E.S.E. dip. The beds have the characters of shallow estuarine deposits, showing marked current-bedding and lenticularity. They generally lie unconformably over the Cretaceous rocks, Upper Gondwanas or gneisses. It is thought that in the Pudukottah, Tanjore, Cuddalore, and Pondicherry areas they overlie a fairly complete Lower Tertiary succession, though there is a stratigraphical gap of varying magnitude beneath them, as they undoubtedly represent a marine transgression in the Mio-Pliocene. In some places they show lumps and veins of chert. Near the surface they are lateritic in character. Artesian aquifers occur in several places as near Karaikudi, Neyveli, Cuddalore, Pondicherry and also in some places in West Godavari district. They contain algal, foraminiferal and Molluscan remains. (*e.g.*, *Ostrea, Fusus, Terebra, Oliva, Conus*, etc). They are generally considered to be of Upper Miocene or Pliocene age though the question has to be decided by detailed work. At Tiruvakkarai near Pondicherry, they show numerous large silicified trunks of angiospermous trees (*Peuce schmidiana*) some of which are 18-31 m long and 1-1.6 m in diameter. Recent work by C.G.K. Ramanujam (*Jour. Sci. Ind. Res.* **B13**, 146 ; *Current Sci.* **22**, 336, 1953) shows that these woods include also the families Leguminosae, Guttiferae, Anacardiaceae, Euphorbiaceae, Sonnaratiaceae and that some of them are referable to the genera *Palmoxylon, Albizzia, Mangifera, Shorea, Terminalia*, etc. The plant remains appear to indicate an Upper Miocene age. Eames has however come to the conclusion that the molluscan fossils contained in these formations indicate a Pliocene age.

The Cuddalore Sandstones in South Arcot and Pondicherry contain thick lignite beds. At and around Neyveli, about 35 km west of Cuddalore, a seam has been proved which extends over an area about 250 sq. km and has thickness varying up to 25 m. The reserves of lignite in this field are estimated at about 2,000 million tons. Two seams of lignite have been resorcded at Bahur and some other places in the Pondicherry area, with a thickness varying up to 9 m. It is likely that if extensive drilling is undertaken, other lignite deposits may come to light in these strata at other places. The Lower Tertiary rocks underlying them may also contain petroleum.

Cauvery Basin.—Towards the end of the last century a borehole was sunk at Karikal on the Tanjore coast by the French administration to a depth of 105 metres. The rich fossil collection made from the borehole material, which consisted entirely of gastropods, has been described by Cossmann (*Jour. Conchyliologie*, XLVIII, p. 14, 1900 : LI, p. 106, 1903 ; LXVIII, p. 34, 1910). Vredenburg thought that the fauna of these KARIKAL BEDS indicated their equivalence to the (Upper) Gwadar Stage of the Makran Series. More or less similar fauna has been obtained from the Tipam Sandstones of the Arakan which overlie beds containing Burdigalian fossils. The Karikal Beds are therefore of Mio-Pliocene or Pliocene age, and may be the continuation of the Cuddalore Sandstones.

A borehole put down near Karikal by the Oil and Natural Gas Commission in 1965 went through the whole Tertiary succession to the basement crystallines at a depth of 1760 m. In this borehole the Miocene was cut through between 550 and 70 m, the section between 550 and 425 m being assigned to the Aquitanian and the rest to the Burdigalian on the evidence of foraminifera. The top 70 m is therefore Mio-Pliocene and Pleistocene and part of it must correspond to the Karikal beds, as some molluscan shell horizons occur in this section. Similar sections are expected to be present along most of the coastal tract of this basin which extends from Pondicherry in the north to Ramanathapuram in the south.

Conjeevaram Gravels.—A large area north of Conjeevaram (Kanchipuram) and east of the exposures of the Upper Gondwanas, is occupied by gravels, shingles and grits called Conjeevaram Gravels. They probably represent fluviatile deposits. Their age may be anything between the Upper Miocene and Pleistocene.

ECONOMIC MINERALS

The Warkalli beds and the Cuddalore sandstones contain some useful deposits of lignite intercalated with the sandstones and clays. Two seams have been met with in the Warkalli beds with thickness upto 7 m. The seams are not continuous.

A good seam of lignite, varying in thickness up to 24 m has recenlty been found at and around Neyveli some 35 km west of Cuddalore. The seam occupies an area of about 250 sq. km and has been estimated to contain reserves of about 2,000 million tons. Two or three seams have been recorded in boreholes in the Pondicherry territory, north of Cuddalore. The extent of these requires investigation. Borings made in 1965 near Karaikudi near the southern end of the Cauvery Basin in the Cuddalore formation showed varicoloured clays down to 70 m., and black carbonaceous shales and calcareous sandstones containing molluscan shells from 70 to 180 m. The section between 50 and 120 m depth showed streaks and thin beds of lignite. The extent of the lignite beds and their workability is yet to be proved. These deposits are thought to be Pliocene in age.

The Neyveli lignite is being exploited by open cast mining on a large scale. As mined, it contains about 50 per cent moisture. Air-dried material shows (all figures per cent) 12-15 moisture, 37-44 volatiles, 32-40 fixed carbon and 3-7 ash. Airdried Karaikudi lignite shows 10-15 mostiure, 32-35 V.M., 36-42 fixed C and 7-10 ash. The Calorific value is between 8,200 and 9,000 B.T.U. per lb. The ultimate composition of Neyveli lignite (pure coal basis) is 65-70 carbon, 20-25 oxygen, 5 hydrogen, and' 1.5 nitrogen plus sulphur.

The same strata, both in the west coast and in some east coast districts of Madras, contain ochres and clays of various descriptions — fire-clay, ball-clay and terra-cotta clay. In several places near the western coast, the feldspathic gneisses below the Tertiary beds have been kaolinised to a depth of 9-12 m and yield excellent kaolin on washing. Some deposits are being worked for the ceramic industry near Quilon and for the textile industry in Malabar and South Kanara. These deposits may be expected to be more extensive than is suggested by the few outcrops which are exposed now.

SELECTED BIBLIOGRAPHY

Colbert, E.H. Siwalik mammals in the American Museum of Natural History. *Trans. Amer. Phil. Soc.* **26**, 1935.

De Terra H. and Paterson, T. Studies on the Ice Age in India and associated human cultures. *Carnagie Inst. Washngton.* Publ. No. 493, 1939.

Eames, F.E. On the ages of Ccertain Upper Tertiary beds of Peninsular India. *Geol. Mag.* **87**, 233-252, 1950.

Falconer and Cautley. Fauna antiqua sivalensis. London, 1846.

Gill, W.D. The stratigraphy of the Siwalik Series in Northern Potwar, Punjab, *Q.J.G.S.* **107** (4), 375-394, 1953.

Lewis, G.E. A new Siwalik correlation. *Amer. Jour. Sci.* **33**, 191-204, 1937.

Lydekker, R. Indian Tertiary and post-Tertiary vertebrata. *Pal. Ind. Ser.* X, Vols. 1 to IV, 1874, 1887.

Matthew, W.D.   Critical observations on Siwalik mammals.   *Bull. Amer. Mus. Nat. Hist.* **56,** 437-460, 1929.

Pilgrim G.E.   Tertiary and post-Tertiary fresh water deposits of Baluchistan and Sind.   *Rec.* **37,** 139-155, 1908.

Pilgrim. G.E.   The fossil giraffidae of India.   *Pal. Ind. N.S.* IV, (1), 1911.

Pilgrim, G.E.   Revised classification of the Tertiary fresh water deposits of India,   *Rec.* **40,** 185-205, 1910.

Pilgrim, G.E.   Correlation of the Siwaliks with the mammal horizons of Europe,   *Rec.* **43,** 262-326, 1913.

Pilgrim, G.E.   New Siwalik Primates, *Rec.* **45,** 1-74, 1915.

Pilgrim, G.E.   The fossil Suidae of India.   *Pal. Ind. N.S.* VIII (4), 1926.

Pilgrim, G.E.   Thd fossil Carnivora of India.   *Pal. Ind. N.S.* XVIII, 1932.

Pilgrim, G.E.   The fossil Bovidae of India.   *Pal. Ind. N.S.* XXVI, 1939.

Vredenburg, E.W.   Considerations regarding the age of the Cuddal9re Series.   *Rec.* **36,** 321-323, 1908.

Wadia, D.N. and Aiyengar, N.K.N.   Fossil Anthropoids of India.   *Rec.* **72,** 467-494, 1938.

CHAPTER XX

THE PLEISTOCENE AND RECENT

General.—The Quaternary era was heralded by a general lowering of temperature in the northern hemisphere, connected with the formation of great ice-sheets of continental dimensions over the North Pole and large parts of the neighbouring areas. Many areas now in the temperate zone were covered by ice-sheets — *e.g.*, North America down to the northern parts of the United States and the northern parts of Europe. The Alpine — Himalayan mountain regions were also covered by extensive and thick ice masses which descended to lower altitudes compared to the present day. The glaciers in the Himalayas are known to have descended as low down as 5,000 ft. altitude, as evidenced by the presence of moraines and other features.

The history of glaciation in Kashmir has been studied by De Terra and Paterson (as described briefly on a later page) who have made out five periods of ice advance and four inter-glacial periods. Man had already appeared in the Potwar area, Kashmir valley and Narmada valley by the Second Interglacial period, and perhaps even somewhat earlier.

The prolific mammalian fauna of the Upper Tertiary suffered grievously with the on-coming of the Ice Age, for these highly specialised orgainsms were not fitted to withstand the extreme cold. Many genera and species died off while some managed to escape to warmer tropical regions and to exist there.

As evidences of the mild temperate climate of the Peninsula of India during the Pleistocene, W. T. Blanford cites the rpesence, in the plains, of certain plants and animals which are now confined entirely to the Himalayas or the higher mountains in the lower latitudes. Some species of rhododendron and other plants are now found isolated in the higher hills of the Peninsula though they should have been formerly present over large areas of lower elevations, having migrated to the Peninsula from the Himalayan region during the Ice Age.

Divisions.—The beginning of the Pliocene is taken as coinciding with commencement of the *Pontian*, as a fairly important change in fauna occurs at about that time. *Hipparion* is the most easily recognised guide fossil of the Pontian and it does not occur before this age. The Pontian was also a period of Marine regression marking the end of a period of sedimentation. The Pontian is followed by the *Plaisancian* and *Astian* (both of which are Pliocene) which are respectively off-shore and coastal deposits. But these do not indicate any marked difference in fauna. These are followed by the *Calabrian* which is marine. The fresh-water and lacustrine equivalent of the Calabrian is called the *Villafranchian*. The Calabrian is followed by the *Sicilian* which is also marine and occurs in the form of well marked terraces in Southern Europe.

The Plaisancian and Astian strata contain a large proportion of the fauna stilll iving, which lived in somewhat warmer conditions than at present. During the Calabrian the percentage of living marine forms increased and it is noticed that the fauna included certain species which had migrated from colder regions (*e.g.*, *Cyprina islandica*) indicating the beginning of a colder climate. The Sicilian can scarcely be differentiated in faunal content from the Calabrian, but contains a few more cold-water forms. It is therefore clear that the first indication of glaciation in Western and Southern Europe appears at the beginning of the Calabrian. The Villafranchian and Val d'Arno deposits also contain plant remains belonging to a climate colder than that of today. There are also evidences of an uplift of the Appennines just before the beginning of the Calabrian.

Many geologists therefore agree that the beginning of the Pleistocene should coincide with the beginning of the Calabrian-Villafranchian as this is ushered in by an orogenic uplift, a distinct cooling down of the climate and a change in the fauna.

TABLE 84—PLEISTOCENE CORRELATION

Age	Culture	Climate	W. Europe	India
Recent	Present Chalcolithic	Warming up		Newer alluvium
Holocene	Neolithic / Mesolithic	Fluctuating	Monasterian (Flandrian)	Older alluvium
Upper Pleistocene	Up. Palaeolithic / Mousterian / Levalloisian	IV Glacial / III Interglacial	Wurm / Tyrrhenian	Potwar loess and silts
Middle Pleistocene	Chellean / Acheulian / Clactonian / Abbevillean	III Glacial / II Interglacial / II Glacial / I Interglacial	Riss / Milazzan / Mindel / Sicilian	Narmada alluvium / Up. Boulder-conglomerate / Lr. Boulder-conglomerate
Lower Pleistocene	Kafuen	I Glacial	Gunz Villafranchian	Bain Boulderbed, Pinjor

The Villafranchian deposits are characterised by the appearance of *Equus* and *Elephas*, some of the important species being *Elephas planifrons, E. meridionalis, E. antiquus, Mastodon borsoni, Equus robustus, E. caballus, Rhinoceros etruscus, Sus arvernensis, Leptobos elatus, Bison priscus*, etc. In the Upper Val d'Arno they consist of a thickness of 150 m of fluviatile and lacustrine sediments which indicate deposition over a fairly long period of time.

A rough correlation of the different glacial advances and inter-glacial deposits together with their stratigraphic equivalents in India and the prehistoric human cultures is given in Table No. 84. The heights of the marine terraces of the different inter-glacial in Southern Europe are shown below :

	Metres
Late Monastirian	7.5
Main Monastirian	18
Tyrrhenian	32
Milazzan	56
Sicilian	103
Calabrian	200

The Calabrian forms a fairly broad and well-marked terrace about 200 m above the present sea-level. The period immediately following this may represent the Gunz or first glaciation, after which the Sicilian was laid down in the First Inter-Glacial period. Then followed the Mindel (or second) glaciation and the Milazzan deposits which represent the Second Inter-Glacial. Important changes occurred in the Mammalian fauna during the Mindel glaciation. This was succeeded by the Riss (or third) glaciation and the Tyrrhenian deposits representing the Third Inter-Glacial. The fourth glaciation, which is of Upper Pleistocene age, is called the Wurm and this is succeeded by the Monastirian deposits which enclose Mesolithic and Neolithic implements.

The East Anglian crags (Red crags etc.) of England contain a fauna similar to that in the Val d'Arno. They are considered to be the equivalents of the Sicilian. The Cromer Forest beds are considered to represent the First Interglacial.

There has been some difference of opinion as to where the beginning of the Pleistocene should be shown in the Siwalik succession. The typical elephant genus *Archidiskodon* first appears in the Tatrot beds. But the Tatrots also contain some nine genera which continue on from Dhok Pathan and six genera which are new. The Tatrots are therefore transitional form Pliocene to Pleistocene. The Pinjor Beds are the true equivalents of the Villafranchian (now agreed to be included in Pleistocene). *Equus* and *Leptobos* appear in them and not in Tatrot Beds.

HUMAN EVOLUTION AND CULTURE

Though man-like apes are found in the Miocene and Pliocene, their anatomy differs markedly from that of man. Fossil man appears only in the Pleistocene. Such remains have been found in several countries of Western Europe, Southern Russia, Palestine, Iraq, India, Burma, Java, China, East and South Africa and North America. They are chance finds in quarries and caves but systematic search in late Tertiary strata in these and other parts of the world are sure to yield further rich material. More common, however, are the stone implements used by early man, these being made of hard materials like flint, chert, quartzite and hard slaty shale, and in later stages bone and ivory. Table 85 shows the relationship of the Pleistocene glacial periods to the human culture levels.

TABLE 85—HUMAN CULTURE LEVELS

Age	Glaciation.	Human Culture.		
RECENT	Post-Glacial	2000 B.C.	Iron Age	
		3500 B.C.	Bronze Age—Use of copper and bronze; early civilizations.	
		8000 B.C.	Nelithic—Animals domesticated; agriculture and settled communities.	
		20000 B.C.	Mesolithic—Azilian, Maglemosian, etc.	
UPPER PLEISTOCENE	IV Glacial	Upper Palaeolithic	Auringnacian, Solutrian, Magdalenian etc.	
	III Inter-Glacial	Middle Palaeolithic	Mousterian / Levalloisian	
MIDDLE PLEISTOCENE	III Glacial	Lower Palaeolithic	Early Levalloisian.	
	II Inter-Glacial		Acheulian—Chellean	
	II Glacial		Abbevillean—Clactonian	
LOWER PLEISTOCENE	I Inter-Glacial	Pre-Palaeolithic	Earliest stone age with very crude implements.	
	I Glacial			

The four major periods of glaciation are the Gunz, Mindel, Riss and Wurm, named after places in the Alpine region of Europe. Some authorities opine that there were really only two major periods of ice-advance, the II Interglacial being the major ice retreat, while there were also minor retreats within the two major advances.

There are evidences of human implements in the I Interglacial, consisting of pre-Palaeolithic or Eolithic hand-axes, very crude in design. These may be considered Lower Pleistocene or ear y Middle Pleistocene.

The *Lower Palaeolithic* culture (Middle Pleistocene) which succeeded it in the II Interglacial, is characterised by rather heavy hand-axes, with sharp edges. These are found in the terraces near Abbeville and St. Acheul in Northern France and therefore named after those places. The Abbevillean is slightly earlier than the Acheulian (also known as Chellean). Similar implements have been found in India, South Africa, Burma and China. In the Soan valley (Potwar plateau in West Punjab) crudely shaped choppers of this age have been found. Of about the same age are the Clactonian and early Levalloisian implements which are flakes chipped off from the heavier ones. This stage of culture is apparently associated with *Pithecanthropus* and *Sinanthropus*. *Pithecanthropus* (Java man) was discovered by Eugene Dubois in 1892 in the terrace bordering the Solo River in Java which has also yielded several other specimens. The Peking man, *Sinanthropus*, was found in a cave near Chou Kou Tien, some 45 km N.W. of Peking and the original very meagre remains were named by Davidson Black; this identification has since been fully authenticated by numerous full skeletons and skulls unearthed in the same place. From the associated materials it is surmised that the Peking man used stone implements and fire, and was probably a cannibal. The Heidelberg man found in Germany was probably a contemporary. These were all hunters who used natural rock shelters and caves.

The *Middle Palaeolithic*, which may be assigned to the lower part of the Upper Pleistocene and to the III Interglacial, is characterised by a variety of flaked stone implements of Mousterian culture (after Le Moustiere in S. France), though some of the Lower Palaeozoic implements also persist. This culture is also found in other countries such as Palestine, Turkestan, and East and South Africa. Some sites have yielded the Neanderthal man who may be a descendant of Heidelberg man. The Piltdown man found in a gravel quarry in Sussex, England, may probably belong to this age though he has some of the characteristics of the Peking man.

The *Upper Palaeolithic* is contemporaneous with the last phase of Wurm glaciation, *i.e.*, the end of Pleistocene. It is characterised by long thin sharp flakes ('willow leaf' and 'laurel leaf' flakes), stone knives, bone blades and statuettes. Cave paintings appear in the Pyrenees and Southern France. The culture levels are called Aurignacian, closely followed by Solutrian and Magdalenian. The Cro-Magnon man is mainly Upper Palaeolithic and approximates to *Homo sapiens*. It is thought that the Aurignacian and Solutrian cultures were those of a cave-dweller while the Magdalenian was that of a nomad.

The *Mesolithic* is not very well defined and has not been indetified in some areas. It centered around 20,000 B.C. In Europe, the ice cap had definitely retreated far to the north, leaving damp forests. The present day desert-belt was still forested. Man was still a hunter and nomad but had domesticated the dog. The stone implements of this age are small and sharp and carefully made (microliths). This culture has been called Azilian and by other names.

The next stage is the *Neolithic* by which time great cultural advances had been made. It may have been around 8,000 B.C. The pig, cow, sheep and perhaps even the ass and the horse had been domesticated. Man had begun to live in groups in river valleys such as the Nile, Euphrates, Indus, Oxus. Agriculture began during this period by the cultivation of certain edible grasses. Copper, and somewhat later, bronze, began to be used. By about 4,000 B.C. the bronze age had been ushered in and pottery and bricks had already been made and used for 1,000 years or more. Sometime later, perhaps by about 2,000 B.C. man had discovered the art of making iron in India and this knowledge spread to Persia, Mesopotamia and Asia Minor by 1,500 B.C.

Glaciation in Kashmir.

In the Sind and Lidar valleys in Kashmir, De Terra and Paterson have distinguished four or five periods of glaciation with three inter-glacial periods. Each glaciation consisted of a series of pulses — *e.g.*, two advances of ice are noted in the Second Glaciation and four advances in the Third Glaciation.

These pulsations are attributable to climatic changes. The largest advance was during the Second Glaciation when man had already appeared on the scene, for implements of the nature of primitive flakes are found in the deposits of this period. The Third Glaciation is correlated with the last phases of the Acheulian stage of civilization. The inter-glacial periods were generally longer than the glacial periods, allowing a certain amount of weathering and erosion of the deposits. The first and third Glacial periods are known to have been marked by the uplift of the Pir Panjal, the latter being the more impressive one.

TABLE 86—CHRONOLOGY OF PLEISTOCENE GLACIATION

	Years ago
End of last (V) glacial	15,000
Beginning of :—	
V Glacial	30,000
IV Interglacial	80,000
IV Glacial—Terrace 4 — Loam	150,000
III Interglacial — Terrace 3 (Soan industry) ...	250,000
III Glacial— Terrace 2 — Potwar Silt (Soan Industry)...	300,000
II Interglacial — Terrace 1.— Upper Terrace gravel. (Chelleo-Acheulian and early Soan Industry)	400,000
II Glacial— Boulder Conglomerate (Oldest flake industry)	500,000
I Interglacial— Pinjor. (Early Pleistocene fauna of Up. Siwalik age) ...	800,000
I Glacial— ? Tatrot (Up. Siwalik fauna)	1,000,000

According to De Terra and Paterson, the Lower Pleistocene in Kashmir embraces the First Glacial and Inter-glacial periods. These are considered to be the equivalents in age of the Tatrot and Pinjor Stages of the Siwaliks and of the Villafranchian. The Middle Pleistocene included the Second Glacial and the Second Inter-glacial periods of erosion. They contain Cromerian (pre-Soan) type of flakes in the Punjab Foothills. This period was marked by heavy sedimentation in the Kashmir valley, consisting of boulder fans and thick fluvo-glacial deposits. The Upper Karewas are thought to represent the Second Interglacial deposits. Of the same age are the Lower Narmada beds containing similar flakes and the remains of a straight-tusked elephant. The Upper Pleistocene includes the Third and Fourth Glaciations and the intervening Third Interglacial. The end of the Fourth Glaciation marks the final phase of the terrace deposits of Kashmir.

There was an uplift of the Pir Panjal towards the end of the First Glaciation. The tilted gravel fans underlying the Lower Karewas are of this age. This was the period of maximum inundation of the Kashmir valley. The land supported great pine and oak forests and more than 100 species of mammals have been found in the Pinjor deposits of the foothill region. A sharp uplift

THE PLEISTOCENE AND RECENT

took place just before or during the Second Glaciation, which was responsible for the elevation and tilting up of the lake deposits. The Boulder-conglomerates are of this age, according to De Terra.

During the Second Interglacial deposition continued, helped by periodical uplifts. In the adjoining Jammu and Potwar areas Palaeolithic man was living. During the Third Glaciation, the disturbance was less than before. The third interglacial was fairly long and continued to give. The third terrace is of this age. By or before the Fourth Interglacial, Pir Panjal had risen to its present height.

Two layers of post-Karewa age have been found in Rampur and other places in Kashmir. These include one swamp deposit and the latter one alluvial and eolian. The latter yielded charcoal and clay figurines, the former some flakes of Levalloisian Middle Palaeolithic). It has not been possible to date these in Kashmir, but Neolithic culture in Mesopotamia, associated with evidences of agriculture, has been assigned an age of 5000 to 6000 B.C.

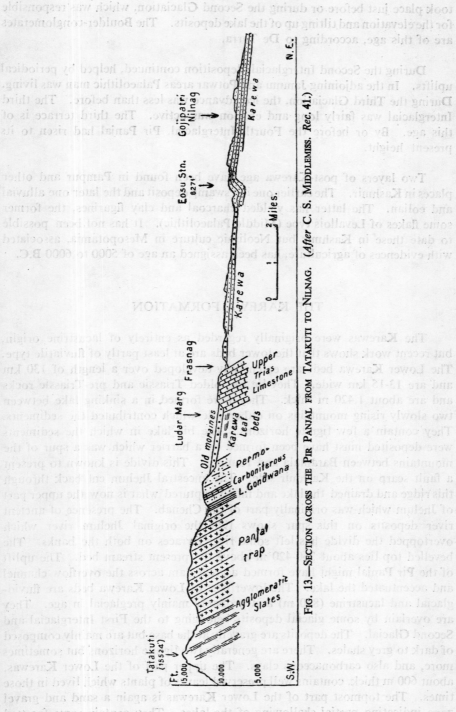

FIG. 13.—SECTION ACROSS THE PIR PANJAL FROM TATAKUTI TO NILNAG. (*After* C. S. MIDDLEMISS, *Rec.* 41.)

took place just before or during the Second Glaciation, which was responsible for the elevation and tilting up of the lake deposits. The Boulder-conglomerates are of this age, according to De Terra.

During the Second Interglacial, deposition continued, helped by periodical uplifts. In the adjoining Jammu and Potwar areas Palaeolithic man was living. During the Third Glaciation, the ice advance was less than before. The third Interglacial was fairly long and erosion was active. The third terrace is of this age. By or before the Fourth Interglacial, Pir Panjal had risen to its present height.

Two layers of post-Karewa age have been found in Pampur and other places in Kashmir. The earlier one is a swamp deposit and the later one alluvial and eolian. The latter has yielded charcoal and clay figurines, the former some flakes of Levallois type (Middle Palaeolithic). It has not been possible to date these in Kashmir but Neolithic culture in Mesopotamia, associated with evidences of agriculture, has been assigned an age of 5000 to 6000 B.C.

THE KAREWA FORMATION

The Karewas were originally regarded as entirely of lacustrine origin, but recent work shows that the lower beds are at least partly of fluviatile type. The Lower Karewa beds are extensively developed over a length of 130 km and are 13-15 km wide. They lie on folded Triassic and pre-Triassic rocks and are about 1,420 m thick. They were formed in a sinking lake between two slowly rising mountains on either side which contributed the sediments. They contain a few lignite horizons. The big lake in which the sediments were deposited must have been formed by a barrier which was a spur of the mountains between Baramula and Rampur. This divide is known to present a fault scarp on the Kashmir side. The ancestral Jhelum cut back through this ridge and drained the lake and finally captured what is now the upper part of Jhelum which was originally part of the Chenab. The presence of ancient river deposits on this spur shows that the original Jhelum river which overtopped the divide has left some river terraces on both the banks. The bevelled top lies about 360-420 m above the present stream bed. The uplift of the Pir Panjal might have formed a great dam across the overflow channel and accentuated the lake. The lower part of Lower Karewa beds are fluvio-glacial and lacustrine (820 m) and probably mainly preglacial in age. They are overlain by some glacial deposits belonging to the First Interglacial and Second Glacial. The deposits are gravelly at the base but are mainly composed of dark to grey shales. There are generally two lignite horizons but sometimes more, and also carbonaceous clays. The upper part of the Lower Karewas, about 600 m thick, contains well preserved leaves of plants which lived in those times. The topmost part of the Lower Karewas is again a sand and gravel zone indicating partial shallowing of the lakes. They contain some facetted boulders.

TABLE 87—THE KAREWA FORMATION

Upper	{	Erratics, Sands, Clays and varved clays containing remains of molluscs and some plants.
		Second Glacial
Lower	{	Sands, gravels, buff and blue clays and varved clays with leaf impressions of mainly sub-tropical plants and also shells.
		First Glacial
		Conglomerates and carbonaceous shales with lignite (*Elephas, Rhinoceros* etc.)
		Pre-glacial beds (Birds, fish, plants etc.)

A disconformity separates the Lower Karewas from the shell and plant bearing clays of Upper Karewas which are 300 m thick. The following beds indicate a fall of the water level and the deposition of alternating laminated yellow marls and silts and sands. In the last stages, the deposits consist mainly of lacustrine and eolian silty materials. These beds are often found lying over moraines of the second glaciation. Though the Upper Karewas contain plant remains, they do not include well preserved leaves such as are found in the Lower Karewas.

The Karewa lake has left 3 sets of terraces, the highest being at about 1,800 m. An older terrace appears at a height of about 1,630 m. During the period between the Second and Third Glaciation there are indications that the lake became deeper again and that an uplift of the Pir Panjal took place. After this, the Jhelum seems to have gradually cut back through the barrier and largely emptied the lake.

The Karewas are found generally to dip towards the Kashmir valley and the dip increases up to about 40° in the lower parts of the flanks of the Pir Panjal. The last stage of the uplift of the Pir Panjal was post-Karewa in age and is believed to have taken place after the appearance of early man. This uplift folded and tilted the Karewas and also the rocks of the Potwar plateau on its southern side.

According to G.S. Puri (*Q.J.G.M.M.S.* XX, 61-66, 1948) the plant fossils of the Karewas have been investigated in three areas *viz.*, Liddermarg, Laredura and Ningal nala. By far the largest group of the flora consists of flowering plants of which nearly 70 genera have been identified. They are predominantly dicotyledons with only a few monocotyledons. Gymnosperms are restricted to barely half a dozen species. Several of the plants are of aquatic habit. The flora shows tropical, subtropical and temperate species. Many of the plants which are now found living only below an altitude of 1,500 m have been found as fossils at 3,200 m, indicating the uplift of the deposits to the present position : The plants include species of *Acer, Berberis, Indigofera, Rhamnus, Prunus, Rosa, Pyrus, Viburnum, Betula, Quercus, Juglans, Salix, Populus, Pinus, Picea, Juniperus, Nelumbium*, etc. The lignite beds show several genera of diatoms. Some of the species are still living in the Himalayas, while others are extinct. The Liddermarg flora is mostly tropical while the Ningal nala flora is exclusively temperate containing willow, poplar, cherry, walnut, maple, alder, sruce, fir, pine and cedar. The modern representatives of these are still found in the Kashmir Valley at an altitude of 2,100-3,000 m. Amongst the animal remains found in the Lower Karewas are fresh-water shells, fishes, and mammalia, the last including *Equus, Elephas namadicus, Bos, Sivatherium, Rhinoceros, Cervus, Sus, Felis*, etc. These indicate correlation with the Pinjor Stage of the Upper Siwaliks.

POTWAR SILTS AND LOESS

The Potwar plateau is covered by a mantle of yellow to pinkish silt, which is thick in the valley depressions attaining a thickness of 100 m. In places it shows typical loess landscape, for example vertical sided canyons near Chakwal and Rawalpindi. The silts are generally laminated though occasionally without structure. They are very fine grained and uniform and are composed of angular grains, due to the material being wind-borne. Locally there are thin layers of marl or small concretions, some fragments of teeth of vertebrates and a few land and fresh water molluscs. (*Planorbis, Limnaea, Unio, Vivipara, Melania* etc.) are the only fossils found in them. At their base there are gravels containing Early Soan implements.

Though these silts resemble loess, they were apparently laid down in shallow fresh water lakes or rearranged by streams. The bottom 6 m of the loess contain early man's relics of Levallois type which is more advanced than the Soan Implements in the gravels below. The tools in the loess are not of fluviatile origin. The wind borne nature and the high content of calcium carbonate in the deposits account for the absence of fossils bones. Alluvial silt also is a very poor medium for preservation of fossilised bones.

In the Soan valley in Potwar, there is a system of five terraces of which the third corresponds to and is composed of Potwar Silts. This terrance is cut into the moraine of the Third Glaciation in Jammu and Poonch, so that the Potwar Silts may be regarded as contemporaneous with the Third Interglacial period. The Third Glaciation is younger than the Upper Karewas, according to De Terra.

Bain Boulder-bed.—Associated with the Siwaliks of the Marwat Kundi and Shekh Budin hills in the Trans-Indus continuation of the Salt Range there is a boulder-bed about 21 m thick (the Bain boulder-bed), intercalated with the Marwat formation. The boulder-bed is of glacial origin and the associated formation contains a fauna which is of Villafranchian to Lower Pleistocene age according to T.O. Morris (*Q.J.G.S.* XVIV, p. 385, 1938.)

Erratics of the Potwar Plateau.—There are several localities in the Attock district where large blocks of rock are found amidst boulders, gravel and finer sediments. Several of these occur near Nurpur Kamalia, a few miles from Campbellpur, one being a granite block of 15 m girth and another a basalt block with a girth of 14 m. Some of the blocks show grooved surfaces. They

THE PLEISTOCENE AND RECENT

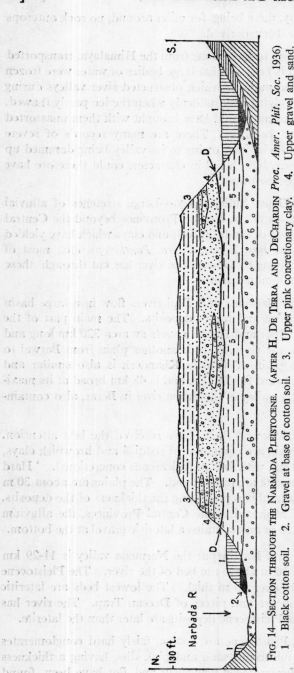

FIG. 14.—SECTION THROUGH THE NARMADA PLEISTOCENE. (AFTER H. DE TERRA AND DE CHARDIN *Proc. Amer. Phil. Soc.* 1936)
1. Black cotton soil.   2. Gravel at base of cotton soil.   3. Upper pink concretionary clay.   4. Upper gravel and sand.
5. Lower red concretionary clay.   6. Basal conglomerate.   7. Laterite.   D. Disconformable break.

now lie amidst alluvial country, there being, for miles around, no rock outcrops from which they could have been derived.

These ' erratic ' blocks seem to have come from the Himalaya, transported through the agency of ice. It is believed that large bodies of water were frozen up behind barriers of rock and moraine which obstructed river valleys during the glacial times, the barriers giving way suddenly when the ice partly thawed. The glacial floods which spread out would have brought with them unassorted material including large sized boulders. There are many records of severe floods of the Indus during historic times owing to its valley being dammed up by rock debris. Similar floods, but glacial in character, could therefore have taken place during the Pleistocene times.

Alluvial deposits of the Upper Sutlej valley.—Large stretches of alluvial terraces exist in the Sutlej valley in the Hundes Pronvince beyond the Central Himalaya. These are composed of sands, gravels and clays which have yielded remains of mammalia such as *Bos. Equus, Capra, Pantholops*, etc., most of which belong to genera now living in Tibet. The river has cut through these deposits to a depth of a few hundred metres.

Narmada Alluvium.—The Narmada and Tapi rivers flow in a large basin covered by extensive Pleistocene and recent deposits. The main part of the Narmada plain, between Jabalpur and Harda, covers an area 320 km long and 56 km wide. Further down the river, there is another plain from Barwai to Harin Pal near Bagh. The plain of Tapi in Khandesh is also similar and extends from Burhanpur westward for 240 km and is 48 km broad at its maximum. To its south-east is the valley of the Purna river in Berar, also containing Pleistocene deposits.

The Narmada deposits, amongst these, have received the best attention. The deposits in the Harda plain are composed of reddish and brownish clays, with intercalations of gravel and with *kankar* (calcareous concretions). ' Hard pans ' of calcareous conglomerate are often found. The plains are about 30 m above the bed of the Narmada, this representing the thickness of the deposits. Near Gadarwada, north of Mohpani in the Central Provinces, the alluvium attains a thickness of over 150 m and contains a lateritic gravel at the bottom.

Between Hoshangabad and Narsingpur the Narmada valley is 11-29 km wide and a terrace is found 36 m above the bed of the river. The Pleistocene and recent deposits here are over 45 m thick. The lowest beds are lateritic with lumps of hematitic iron ore and pieces of Deccan Trap. The river has cut through the laterite ; the Pleistocene deposits are later than the laterite.

The basal beds of the Pleistocene are coarse, fairly hard conglomerates with intercalations of grey micaceous sand and pink silts, having a thickness of up to 3 m. *Elephas, Hexaprotodon namadicus* and *Bos* have been found in these, as also Pre-Soan flakes and Abbevillean hand-axes.

The conglomerate is succeeded by a red silty clay with lime concretions, having a thickness of 10 m. This appears to be a river deposit, containing unrolled flakes and Palaeolithic tools.

These are separated from the upper group of beds by a disconformity. Then follow red concretionary clays containing remains of *Bos namadicus, Bubalus palaeindicus,* and *Elephas namadicus.* A rich bone-bed in these yielded : *Trionyx gangeticus, Ursus namadicus, Sus* sp., *Leptobos fraasi, Cervus, Rhinoceras, Hexaprotodon palindicus, Elephas namadicus, E. insignis, Stegodon ganesa, S. insignis,* and some reptiles. This fauna is younger than Pinjor in age. Above these beds are found 10-20 m of pink to brown clays with concretions, which have yielded early Palaeolithic flakes and cores of late Soan Type. Above these clay beds are brown coloured regur or 'black soil.' These contain small blades and scrapers of flint and jasper, of proto-Neolithic type.

TABLE 88—CORRELATION OF THE NARMADA AND N.W. INDIA PLEISTOCENE
(*After* DE TERRA AND P. TEILHARD DE CHARDIN.)

Age		N.W. Punjab	Kashmir Valley	Narmada Valley
PLEISTOCENE	UPPER	Redeposited Potwar and Second loess	IV Glacial	Black Cotton Soil
		Erosion interval	III Inter-glacial	Eorsion interval
		Potwar silt	III Glacial	Upper zone
	MIDDLE	Long Erosion interval	II Inter-glacial	Erosion interval
		Boulder Conglomerate	Upper Karewa II Glacial Karewa gravels	Lower Zone
	LOWER	Pinjor stage	I Inter-glacial	
		Totrot stage	I Glacial	Narmada laterite
Pliocene		Dhok Pathan stage	—	

The laterite may be of early Pleistocene age. The lower group is correlated by De Terra with the Boulder Conglomerate ; the upper group with Potwar silts ; and the black soil with the redeposited Potwar Silts and the Second Loess. Table 88 shows the correlation suggested by De Terra and De Chardin.

Godavari alluvium.—There are thick deposits of alluvium along the upper Godavari in the Maharashtra and Andhra States. They are composed mainly of brown clay and sandy silts with nodules of *kankar* and beds of gravel. In some places west of Chanda they are saline. The gravels are composed of pebbles of Deccan trap, agate and chalcedony. Fossil wood trunks have been found in them between Wardha and Enchapalli and between the latter and Albaka. Animal remains including *Elephas namadicus* and *Bos* sp., have been found near Mungi, Paitan, Hingoli and a few other places.

Kistna (Krishna) alluvium.—Gravels, sandy silts and calcareous conglomerates of Pleistocene age occur in the upper Krishna valley 18-24 m above the present river bed. Parts of the cranium of *Rhinoceros deccanensis* have been found near Gokak and also remains of *Bos*. The alluvia of the tributaries of this river in South Maharashta have yielded stone implements.

Madras.—Overlying the conglomerates of the Satyavedu (Upper Gondwana) beds, in the valley of the Kortalayar river, there is a series of four terraces respectively 30 m, 18 m, 6 m, and 2.5 m above the river bed. These contain implements which are referable to Abbevillian-Acheulian (first terrace), Acheulian (second terrace) Late Acheulian and Levalloisian (third terrace) and Upper Palaeolithic (fourth terrace). These were discovered by Bruce Foote.

In the Vadamadurai area south-west of Madras, Pleistocene boulder-conglomerates containing Abbevillian tools overlie the Upper Gondwanas. Above this is detrital laterite and alluvium, the former containing Acheulian type of tools.

Irrawaddy alluvium.—Four terraces have been recognised in the Irrawaddy valley overlying the Irrawaddy System. The oldest terrace is seen in a group of hills 90 m above the present river-bed near the Chauk oil-field. This contains gravel and boulder-beds and sandstones which are tilted. The fossils found in them are *Elephas namadicus*, *Bos* cf. *namadicus* and *Hippopotamus*. The second and third terraces lie 54 and 30 m respectively above the river and contain Middle Pleistocene fauna and Palaeolithic tools near Mandalay. The fourth terrace of sandy gravel and sand contains a late type of Palaeolithic implements. The fifth terrace grades into recent river deposits. The whole of the terrace system is assigned a Middle to Upper Pleistocene age by De Terra, the uppermost Irrawaddys being regarded as Lower Pleistocene by the same authority.

THE INDO-GANGETIC ALLUVIUM

General.—The great alluvial tract of the Ganges, Brahmaputra and the Indus forms one of three main physiographic divisions of Indo-Pakistan, separating the Peninsular from the Extra-Peninsular region, and covering an area estimated at over 250,000 sq. miles (about 850,000 sq. km). The deposits of this tract belong, so to say, to the last chapter of earth's history and conceal beneath them the northern fringes of the Peninsular formations and the southern fringes of the Extra-Peninsular formations. The area is geologically uninteresting but, being a rich agricultural tract, is of great interest and importance in human history.

The alluvial tract is of the nature of a synclinal basin formed concomitantly with the elevation of the Himalaya to its north. One view, due to Eduard Suess, the great Austrian geologist, is that it is a 'fore-deep' formed in front of the resistant mass of the Peninsula when the Tethyan sediments were thrust southward and compressed against them. The Peninsula is regarded as a stable mass and Central Asia as the moving segment of the crust. According to a second view, due to Sir Sydney Burrard (formerly Surveyor-General of India) the plains represent a rift-valley bounded by parallel faults on either side. A third and more recent view regards this region as a sag in the crust formed between the northward drifting Indian continent and the comparatively soft sediments accumulated in the Tethyan basin when the latter were crumpled up and lifted up into a mountain system.

The generally accepted view at present is that it has been formed by the buckling down of the northern border of the peninsular shield beneath the sediments thrust over it from the north. These sediments originally accumulated in a broad coastal area and ocean basin which comprised a southern zone of terrestrial Gondwana strata (tillite, coal-bearing beds and volcanics of the Punjal volcanic activity) with occasional Permo-Carboniferous and Permian marine intercalations. A more northerly zone contained unfossiliferous (except for spores and pollen) Mesozoics of the Blaini-Krol-Tal type including some shallow-water current-bedded sandstones, coloured shales and gypsiferous deposits. A still more northern zone comprises richly fossiliferous geosynclinal Tethyan sediments. The whole of this great sedimentary belt was compressed by the coming together of the Indian and Central Asian continental masses and its contents had to be accommodated in the narrow region now occupied by the Himalayan ranges. The southern zones have over-ridden the northern border of the Indian shield and the more northern ones have not only been uplifted to great heights but have also been thrust in a series of *nappes* over the rocks of the southern zones. In this process, parts of the marine deposits have been buried deep, as is the case with the strata now represented only by the few 'exotic blocks' which have been brought up to the surface by igneous rocks coming up from depth.

The Indo-gangetic depression must have been formed in the later stages of Himalayan orogeny when the Indian shield underthrust the Asian continent. The well-consolidated crust of the shield became engulfed under the light, soft, moist sediments. As the northward drift of India continued, the more consolidated and metamorphosed older strata and the granitic magmas intrusive into them slid southwards, impelled partly by gravity and partly by compressive force. The northern border of the Peninsular shield in the foot-hills region is now overlain by 500-1000 m. of Pleistocene to Recent strata, and by a varying thickness of Siwaliks (upto 5000 m thick) or Dharmsalas (upto 4000 m), the deepest section showing perhaps 5000-6000 m of sedimentary strata. This was first estimated by aero-magnetic surveys and later confirmed by borings at Janauri, Jwalamukhi and Raxaul. More details about the shape of the Basin will be available as deep borings for petroleum are put down at various places. The basement surface under this basin is now known to slope down from south to north at an average angle of 1° to 3°, but that surface is rather irregular and must locally contain hills and valleys. Both longitudinal and transverse faults are present in the basement and some of them are loci of the earthquakes occasionally felt in the region.

The Pleistocene and Recent deposits covering the Indo-Gangetic Basin are up to 1,000 m thick. They are composed of gravels, sands and clays with remains of animals and plants. The sands and gravels constitute aquifers. The older alluvium (called *Bhangar* in the Ganges valley) is rather dark coloured and generally rich in concretions and nodules of impure calcium carbonate known as *kankar* in Northern India. The *kankar* concretions are of all shapes and sizes from small grains to lumps as large as a man's head. The older alluvium forms slightly elevated terraces, generally above the flood level, the river having cut through it to a lower level. It is of Middle to Upper Pleistocene age.

The Newer alluvium (called *Khadar* in the Punjab) is light coloured and poor in calcareous matter. It contains lenticular beds of sand and gravel and peat beds. It merges by insensible gradations into the recent or deltaic alluvia and should be assigned an Upper Pleistocene to Recent age.

The older alluvium contains the remains of extinct species of animals including *Rhinoceros*, *Hippopotamus*, *Palaeoloxoden*, *Elephas* and *Equus*. The fossils in the Newer Alluvium are mostly those of animals still living.

Depth.—The shape of the depression is known in a general way, though not accurately. It is deepest within a few miles of the Himalayan foot-hills and gradually shelves up towards the Peninsula. Borings put down for petroleum have penetrated it in many places. Aero-magnetic surveys of the Ganges basin indicate that the basement rocks lie at a depth of about 7,000 m.

In the Nahorkatiya area near Digboi in Assam the prospecting work of the Assam Oil Co. has revealed that the alluvia and Dihings are about 2,700 m thick. Below them come the Tipams down to about 1,500 m underlain by the Barails which extend to a depth of more 3,700 m. Geophysical indications of the basement are at depths of 6,000-7,5000 m below the surface.

The Bihar earthquake of 1934 gave indications of the presence of zones of disturbance, evidently faults, in the basement below the alluvium, parallel to the trend of the Himalayas.

COASTAL DEPOSITS

Eastern Coast.—Raised beaches, some of them elevated as much as 15-30 m above the present sea level, occur at several places along the coast. They contain molluscan shells of recent or present day species. Such deposits are common along the Orissa, Nellore, Madras, Madura and Tinnevelly coasts.

The Chilka Lake.—The Chilka lake on the Ganjam coast dates back to the Pleistocene. It has been rendered shallow by the deposits from the mouths of the Mahanadi, while a sand spit has been thrown across the mouth by the monsoon winds. Near the south-western end of the spit there is a deposit of estuarine shells 6 to 9 m above the high tide level. The shells include *Cytherea costa* and *Arca granosa*, neither of which live in Chilka Lake at present, but the former is known in the estuary connecting the lake with the sea. The Chilka lake is be gradually diminishing in size.

South-western Coast.—Along the Travancore and Malabar coasts, mud banks have been formed which separate the lake-like expanses of water (backwaters or *Kayals*) from the sea. The backwaters are used for coastal communication by small boats. The mud banks are Pleistocene to Recent in age.

Gujarat.—The lowlying tract connecting the Kathiawar Peninsula with the mainland near Ahmedabad shows recent deposits containing *Cerithium*, *Potamidees*, etc., indicating that this was an estuarine area in Pleistocene times.

Kathiawar.—On the coast of Kathiawar is found a marine limestone composed of the remains of the foraminifer *Miliolite*, around which calcite grains have been formed. It is usually sandy in the vicinity of the coast, and attains a thickness of 30 m in western Kathiawar, but is thinner and less extensive in eastern Kathiawar. This Miliolite Limestone, also called Porbandar Stone, is found on the top of Chotila hills (357·5 m altitude) which provides clear proof of the elevation of the coast in recent times. This is locally used as a building stone in Kathiawar.

There are also raised beaches, oyster beds and coral reefs on the Kathiawar coast which have been elevated a few meters since the Pleistocene.

Rann of Kutch.—During the Pleistocene the Rann was a shallow arm of the sea. Even in historic times it was so, as the Indus and the Sarasvati of Vedic times flowed into it. It is now silted up and forms an extensive and desolate salt marsh during the dry part of the year and a tidal flat covered with a little sea water during the monsoon.

Makran Coast.—Raised beaches containing shells of living species are found about 30 m above the sea level in the Makran coast of Baluchistan. The commonest shells are *Ostrea hyotis*, *Anomya archaeus*, *Pecten crassicostatus*, *Arca antiquata*, *A. nivea*, etc.

Detailed studies have not been made in India on the succession of marine terraces, their heights above sea-level, their fossil content and other characters. Until such studies are undertaken it will not be possible to attempt the correlation of the Pleistocene strata.

AEOLIAN AND OTHER DEPOSITS

Loess.—Large stretches of wind-blown dust of sandy to clayey constitution are found in Punjab, Kashmir, Sind and Baluchistan. This material, called LOESS is fine grained buff or grey coloured and with little signs of stratification. It covers the land surface irrespective of barriers and altitude, and deposits are particularly well seen in the Potwar plateau in the Salt Range, and in Thal Chotiali in Baluchistan. It is essentially a deposit of arid regions liable to strong winds carrying much dust. The irregular distribution of loess has in some measure been responsible for the formation of shallow lakes at the top of the Salt Range. Recent work in America shows that loess may be a flood plain deposit in major valleys, consisting of two-thirds silt and one-third clay. The silt is rich in quartz and contains calcium carbonate. Bedding or lamination may be seen sometimes. The composition is similar to that of siliceous shale. The clay mineral is dominantly montmorillonite, whereas glacial clay is dominantly illite.

Desert sands.—A large tract of western and south-western Rajasthan and Sind, 640 km long and 160 km wide, constitutes the Thar desert. It is, in general, covered by a depth of several metres of sands which are constantly being shifted by winds blowing from the south-west. The sands cover an irregular rocky floor, but occasionally local prominences and ridges rise above the level of the sand. Over the greater part of the area the sands are piled up into dunes. The longitudinal ridge-like dunes are common in the more southern parts where the winds ar strong, while the crescentic type (Barchane) is more common in the interior.

The desert grades in the more eastern and north-eastern parts into a semi-arid region covered by shrubs and by stunted xerophytic vegetation. It is also studded with the remains of abondoned villages and towns. In the area where the sand cover is thin, it is still possible to do some cultivation, but the main difficulty is non-availability of sufficient water for agricultural purposes.

The sub-soil water varies in quality from place to place and in the more easterly areas it can be used for human consumption and for agriculture. In some parts of the semi-desert areas well waters are distinctly brackish to saline, and during the monsoon, the new sweet water percolating down from the surface may float on top of the brackish water till it is exhausted or becomes mixed with the latter. The salinity is largely due to inland drainage characterising the area. Some have attributed it to fine particles of salt brought by the south-west monsoon from the Gulf of Cambay and Rann of Kutch. This view, however, does not seem to be fully supported by facts, for recent investigations seem to prove that at least in the more interior parts of Rajasthan the wind carries very little saline matter from the sea.

The desert conditions seem to have grown gradually only during the last 3,000 to 4,000 years. It is known that Sind and Baluchistan and the adjoining parts of Rajasthan were wooded and had a much more favourable climate in pre-historic and early historic times. The use of burnt bricks for building and the presence of well designed drainage systems in the settlements of the Mohenjo daro period apparently indicate that the climate was fairly humid. It is also known that during historic times wild elephants lived in Rajasthan which also indicates more moist conditions than those obtaining at present. But, between that period and the time of invasion of the Punjab by Alexander the Great (4th century B.C.), conditions had become definitely worse and Baluchistan was already becoming arid.

The growth of the desert conditions is due to the fact that this region lies along the well-known northern desert belt. Though moisture-bearing winds of the south-west monsoon blow over Rajasthan for four months in the year, there are no hills across the direction of the winds to impede their progress and precipitate some rain. The Aravallis are aligned parallel to the direction of the winds and do not form a serious obstacle to their course across Rajasthan.

The monsoon winds therefore bring very scanty, if any, rainfall to the area west of the Aravallis ; this area receiving only 10 to 20 cm of rain per year and most of it precipitated in a few sudden and violent cloud-bursts. The increasing aridity and the large diurnal variations of temperature are instrumental in distintegrating the rocks and helping in the accumulation of sand which is distributed by the action of winds.

In a Symposium held under the auspices of the National Institute of Sciences of India in 1953, various aspects of the physical features, environments and resources of Rajasthan have been discussed and the reader is referred to the Special Bulletin on Rajasthan Desert published by that Institute for more information.

Daman Slopes.—In the hilly region of Baluchistan and Sind there are important gravel and talus deposits of Pleistocene and recent ages. These materials, being loose and highly porous, are generally good aquifers. The water contained in them is drawn by driving tunnels and wells through them. In favourable conditions the water rises up in wells under semiartesian pressure The long, nearly horizontal tunnels which are driven into these deposits in order to obtain the water contained in them are called *Karezes*.

Bhabar and Terai.—In the foot-hill region of the Himalaya, the hills are fringed on the side of the plains by talus fans. The upper portion of the talus fans is composed of rock fragments, gravel and soil and supports good forests. This zone, known as the Bhabar, has a vertical extent of less than 300 m between its upper and lower limits.

The Terai tract lies immediately below the Bhabar. It is composed of gravel and soil often forming a marshy tract overgrown with grass and thick jungle. It is an unhealthy zone, uninhabitable unless well drained.

Cave Deposits.—Though a number of caves exist in various parts of the country, especially in limestones, few of them have been investigated in detail. Some are in unfrequented areas and even the more easily accessible ones have not received the attention they deserve from the scientists.

A group of caves (Billa Surgam) near Betamcherla in the Kurnool district, Andhra, was examined by Bruce Foote and his son who found a rich Pleistocene fauna well preserved in the red marl of the floor of the caves. The majority of the fossils belong to species still living, but a few which are extinct in India are still found in Africa, such as *Papio* sp., *Equus asinus*, *Hyaena crocuta*, *Manis* cf. *gigantea*. Amongst the others are :

Mammalia	*Presbytis, Macacus, Cherioptera, Sorex, Felis tigris, Viverra karnulensis, Canis, Herpestes, Hystrix crassidens, Mus, Nesokia, Sus karnulensis, Lepus, Rhinoceros karnulensis, Rusa, Memimma, Gazella, Portax, Copra, Ovis, Bos.*
Reptiles	*Crocodilus, Varanus, Lacerta, Ophidia, Emys, Phelona.*
Amphibia	*Rana, Bufo.*

RECENT DEPOSITS

Coastal Dunes.—Several parts of the eastern coast of India are covered by sand which is massed into small sized dunes by winds, rarely atatining a height of 9 m. Such are seen on the coasts of Ganjam, Vizagapatam, Nellore, Ramnad and Tinnevelly. The sand dunes slowly march inward and are a menace to cultivation. Their progress can be controlled by plantations of trees.

River Alluvia.—All the important rivers have an erosive action in their upper courses and deposit their sediments in the delta region. The deposits consist of lenticular beds of sand and clay.

Along the upper courses of North Indian rivers there are deposits of blown sand and fine dust. These are laid down where there are obstacles to wind movement such as clumps of trees or shrubs. In course of time these deposits near river banks become consolidated into mounds and form good sites for villages above the flood level. They are called *bhur* lands.

In the drier portions of the Indus and Ganges valleys, where there is no good outward drainage, the soil becomes water-logged and the accumulated salts in the ground water are drawn up to the surface to appear as efflorescences. Such efflorescences are called *reh* in the Madhya Pradesh and *kallar* in the Punjab. The term *usar* is applied to alkali-laden land unfit for cultivation. The salts in these are mainly sodium chloride, sodium sulphate and sodium carbonate. Water-logging and concentration of salts is also a feature in parts of the canal-irrigated areas wherefrom there is no adequate outward drainage.

The salts become concentrated comparatively near the ground level especially within a depth of some 10m from the surface. The water from this part is unpalatable and often injurious to the health of plants and animals. Further down, however, sweet potable water is usually met with.

USEFUL MINERAL DEPOSITS

The Pleistocene and Recent deposits contain peat beds in the deltas of most of the large rivers — the Indus, Ganges-Brahmaputra, Mahanadi, Godavari and Cauvery. Peaty material is encountered frequently in these in pits and boreholes. The occasional reports of natural gas when borings are made in these formations, are to be attributed to methane-rich gas escaping from them.

Workable beds of peaty lignite have been reported from Faridapur, Sylhet and Moulvi Bazar in East Pakistan. Most of the total estimated reserves of 130 million tons in East Bengal is in Faridapur. Comparatively small occurrences of peat have been reported to occur near Vedaranyam in the southeast of the Cauvery delta and in parts of the Nilgiri and Palni hills.

Brick-clay and other types of clay, gravel and sand are obtained in large quantities from the Pleistocene and Recent deposits. In some places, the alluvial soil is underlain by an one-metre layer of KANKAR which is used for making building lime for local use. Kankar deposits, though of small dimensions, are extensively found all over the country.

LATERITE

No account of the stratigraphy of India is complete without a reference to the peculiar ferruginous material called laterite which is a product of tropical alteration suffered by some rocks. It is typically developed in tropical lands, such as India, Malaya, East Indies, West Indies, and Tropical America, though similar formations of warm climates in some past ages are known in the temperate regions of the present day also.

Characters.—Laterite is a porous, pitted, clay-like rock with red, yellow brown, grey and mottled colours, depending in some measure on the composition. It has a hard protective limonitic crust on the exposed surface which is generally irregular and rough. When dug up, the fresh material is comparatively soft and easily cut by a spade or a saw. In this state it has often variegated colours and shows vermicular cavities which are irregular and tortuous. Laterite is often pisolitic, the pisolites having a concentric structure and being cemented together by ferruginous or aluminous material. When the fresh soft rock is exposed to air, it is quickly dehydrated and becomes quite hard.

The term was first used for material from Malabar in South India by Francis Buchaman in 1800. The following extrats from his diary (P. 436) are of interest in this connection.

"The ore is found forming beds, veins, or detached masses in the stratum of indurated clay that is to be afterwards described, and of which the greate part of the hills of Malabar consist. This ore is composed of clay, quartz in the form of sand, and of the common black iron sand. This mixture forms amall, angular nodules closely compacted together and very friable. It is dug out with a pick-axe."

Continuing, he wrote on a later page (p. 440) :—

"It is diffused in immense masses without any appearance of stratification and is placed over the granite that forms the basis of Malayala. It is full of cavities and pores and contains a very large quantity of iron in the form of red and yellow ochres. In the mass, while excluded from the air, it is so soft that any iron instrument readily cuts it and is dug up in square masses with a pick-axe and immediately cut into the shape wanted with a trowel or a large knife. It very soon after becomes as hard as a brick and resists the air and water much better than any bricks that I have seen in India.

As it is usually cut into the form of bricks for building, in several of the native dialects it is called the brick-stone (*itica-callu*). Where however by the washing away of the soil, part of it has been exposed to air and has hardened into rock, its colour becomes black, and its pores and inequalities give it a kind of resemblance to the skin of a person affected with cutaneous disorder ; hence in the Tamil language it is called *shuri-cull* or itch-stone. The most proper English name would be laterite, from *lateritis*, the appellation that may be given to it in Science."

Composition.—Laterite is composed mainly of hydrated oxides of iron and alumina together with those of certain elements which form the group of hydrolysates such as manganese, titanium, vanadium, zirconium etc. The silica along with magnesia, alumina etc. contained in the original rock are removed in solution leaving behind hydroxides of iron and alumina, manganese etc. Some analyses of gneisses and khondalites in different stages of weathering will be found in the papers by C. S. Fox and M. S. Krishnan cited in the bibliography.

In accordance with the relative amounts of the elements present, laterite may be called ferruginous, aluminous and manganiferous. In general, ferruginous laterite is red to red-brown in colour, the aluminous one grey or cream, and the manganiferous one dark brown to black.

Relation to Parent Rock.—Laterite may be derived from a variety of rocks. These include alkali rocks like nepheline-syenite, trachyte; intermediate and bzsic igneous rocks like diorites and basalts; gneissic rocks rich in feldspars; and sedimentary rocks including shales and impure limestones. The bauxite deposits of Arkansas in the United States and of Pocos de Caldas in Brazil represent the type derived from alkali rocks. Most of the deposits of India are derived from dolerite and basalt of the Deccan Trap formation. The laterites of French West Africa and Portuguese Guinea are derived mainly from dolerite and diabase. Many deposits in the world belong to the type derived from feldspathic rocks containing only a moderate amount of alumina and some silica; these include gneisses and granitic rocks which have given rise to deposits in South India, in the Guianas in South America and other places. Aluminous clays have given rise to bauxite and laterite in Georgia and South Carolina in the U.S.A. Impure limestones and dolomites have given rise to laterite in the Mediterranean region and also in the Caribbean region.

In India large deposits of laterite are found as cappings over the Deccan Trap, the thickness of the capping being sometimes as much as 100 feet. There is usually a layer of highly ferruginous material at the surface, below which there is a bed of aluminous laterite or bauxite. These grade further below into lithomargic clay which gradually merges into the unaltered rock. The upper portion consists of laterite and bauxite with little or no combined silica.

Laterisation is seen also in the Eastern Ghats where the prevalent rock is khondalite (garnet-sililmanite-feldspar gneiss). The material is mostly ferruginous laterite but occasionally rich enough in alumina to be called bauxite. Similar is the case with the laterite found on gneissic rocks in Malabar. This latter material still contains some combined silica. In typical sections in Malabar, the surface material is gravelly or pellety in structure underlain by mottled vermicular laterite containing clay minerals, iron oxides and some free silica. Below this comes the soft, pale coloured, lithomargic clay which preserves to some extent the gneissic structure of the original rock which underlies it. This is really kaolinised gneiss gradually changing to laterite.

In this connection Sir John Harrison's observations are interesting :—

"Under tropical conditions, acid rocks, such as aplites, pegmatites or granites or granitic gneiss, do not undergo laterisation but gradually change through katamorphism into pipe or pot-clays, or more or less qurtziferous impure kaolins."

"Under tropical conditions, the katamorphism of basic and intermediate rocks, at or close to the water-table, under conditions of more or less perfect drainage, is accompanied by the almost complete removal of silica and of calcium, magnesium, potassium and sodium oxides, leaving an earthy residuum of aluminium trihydrate (in its crystalline form of gibbsite), limonite, a few unaltered fragments of feldspars, and in some cases, secondary quartz, and the various resistant minerals originally present in the rock. The residuum is termed primary laterite."

Distribution.—Laterite is extensively distributed in Peninsular India. It is common over the Deccan traps in the greater part of Bombay, Madhya Pradesh and Bihar. There are several occurrences on the khondalites of Eastern Ghats and gneisses of the Western Ghats in Malabar and Travancore. There are also numerous occurrences of thin crusts of laterite in many other parts of India on rocks of almost every description. Thin beds of laterite have been found over the Eocene rocks of Western Pakistan, and also at the top of the Upper Gondwanas at their junction with the Cretaceous beds in the Trichinopoly district, Madras.

High-level and Low-level Laterite.—All the more important occurrences of laterite form massive beds which generally are found capping hills in the Deccan trap country. The laterite cap varies considerably in thickness and may be up to 50 or 60 m. Laterite also occurs in the plains and at the base of the hills, these being in most cases of secondary origin, derived from the high-level laterite and recemented after deposition in the valleys or plains. Low-level laterite is therefore mainly of detrital origin while high-level oaterite is primary material. On the whole, the primary material is compact and fairly uniform in composition while the detrital laterite is heterogeneous.

Age.—There are no definite criteria for determining the age of the laterite. The existing deposits in most parts of India may have been formed during the Upper Tertiary, probably mainly during the Pleistocene. The process is active even at the present day. As mentioned above, there are also laterites belonging to other ages not only in this country, but also elsewhere even in countries which are now in the cool temperate zone.

Origin.—Laterite consists mainly of the oxides of iron and of alumina, those rich in alumina grading towards bauxite. It is now generally agreed that the conditions favouring the formation of laterite are a warm, humid climate with plentiful and well-distributed rainfall and good drainage. The weathering process may produce either laterite and bauxite or clay minerals from aluminous and iron-bearing rocks. Under condition of poor drainage, clay minerals are formed such as kaolinite, montmorillonite and illite. If there are well-defined wet and dry seasons and fairly good drainage, the clay minerals are decomposed to form laterite. But with evenly distributed rainfall all through the year, iron oxide tends to be removed in solution because of aeration, leaving behind mainly aluminium hydroxides. An alkaline condition of the waters (a fairly high pH value) is favourable for the formation of bauxite and laterite, but an acid environment produces only silicates in the form of clay minerals. Eevn though the percolating waters may be alakaline, an acid condition may often be brought about by the presence of humus acids in the beds undergoing weathering.

It also happens that when weathering has produced a zone of clay minerals, subsequent uplift of the area resulting in good drainage may bring about the formation of bauxite and the downward migration of the clay zone. The clay minerals are formed in the zone where there is poor drainage.

V. M. Goldschmidt has shown that the elements concentrated in the laterite type of weathering are those with intermediate ionic potential (the ratio of ionic charge to the ionic radius) ranging between 3 and 10. Those with low potential, *e.g.*, the alkalies and alkaline earths, form the soluble cations while those with high potential like phosphorus, nitrogen, sulphur, silica etc. form the soluble anions. The elements of intermediate ionic potential are hydrolyxed and form the hydroxides of the laterite zone. The minerals found in bauxite are diaspore, boehmite, bayerite and gibbsite. The former two are monohydrates while the other two are trihydrates. Aluminium hydroxide gels or one of the trihydrates can be converted into diaspore by heat or low grade metamorphism. The gels may also form one of the hydroxides, depending on the prevalent conditions, exposure to carbon dioxide and ageing being favourable for the formation of bayerite and gibbiste.

There is a fairly voluminous literature on laterite and bauxite and the reader is referred to the references at the end of the chapter.

Uses.—Ferruginous laterite is used extensively in some parts of India for building houses, culverts, bridges and other structures. It is easily dressed when freshly quarried but hardens on exposure and is a fairly good and durable building stone.

Aluminous laterite (or bauxite) is used as an ore of aluminium, *i.e.*, for the preparation of high grade alumina which is fused and electrolysed for the preparation of the metal. It is also used in oil refining as its colloidal constituents have the property of decolourising oils. It can be employed for the manufacture of salts of aluminium and for making high-alumina cement. There is, in the ferruginous laterites of India and other countries, a vast store of iron which it should be possible to smelt cheaply at some future date when the high grade haematite deposits become scarce or costly to mine or when technological advances enable laterite to be used economically as an ore of iron.

SOILS

The two major types of soil are *residual* and *transporteed*. The first one is derived *in situ* from the rocks present in the area and the second is brought in by flowing water or wind from elsewhere. The soils in river valleys, deltas and mountain valleys belong to the second type, while those of the other areas are mainly of the first type.

The soils of India have been classified into lateritic, red, black, forest, alluvial, marshy, saline-alkaline and desert soils. This classification, depending on climate and rainfall as well as the drainage characteristic of the area, has been adopted in a Publication entitled " Report of the All-India Soil Survey Scheme " under the Indian Council of Agricultural Research and published as its Bulletin No. 73 in 1953. Other methods of classifying soils according to their physical characteristics and chemical composition may also be adopted.

Lateritic Soil.—This is rich in iron and aluminium with some titanium and manganese. It is generally red and clayey and is fairly common in the areas occupied by the Deccan Traps and some Archaean gneisses, particularly in the Western Ghats of Mysore, Malabar and Travancore.

Red soils form a large group and occupy large areas in India. They are light and porous and contain no soluble salts, *kankar* or free carbonate. These soils are extensively developed over Archaean gneisses and are generaoly deficient in phosphorus, lime and nitrogen. The red soils are not always necessarily red in colour though frequently light red to brown. The colour is due to the oxidation and wide diffusion of the iron content. They are moderately fertile for agricultural purposes.

Black soil or **Regur** is a clayey to loamy soil composed largely of clay material. It is the well-known ' black cotton soil ' and is the same as the Russian *chernozem*. It is generally black and contains high alumina, lime and magnesia with a variable amount of potash, low nitrogen and phosphorus. It is generally porous and swells considerably on addition of water and dries up with ocnspicuous cracks on losing the moisture. The swelling property is due to the high content of montmorillonite and beidellite groups of clay minerals. Though sticky when wet and practically impassable in the rainy season, the black soil is not a compact or heavy clay. It is fairly widely spread on the Deccan Traps and on some areas of gneissic and calcareous rocks, as for example in Hyderabad and in the Central and Southern districts of Madras. It appears to be prevalent in areas with rather low rainfall (50-80 cm).

Alluvial soils do not really form a definite group. They represent both transported and residual soils which may have been re-worked to some extent by water. Most of the alluvial soils are found in valleys and deltas and some may be present in forest and semi-desert areas also.

Forest soils may be divided into two groups — one composed of acid soils and humus with low base status favourable for the formation of *podsol*, while the other consists of neutral soils with high base status. Forest soils are generally quite rich in humus.

Peaty and marshy soils are found in areas which are water-logged due to impeded drainage. They are generally rich in organic matter and may be associated with peaty material.

Saline and alkaline soils are found in areas of poor drainage with high evaporation or in areas of excessive irrigation without proper flushing out of the salts by excess water. They often show efforescences of sodium, calcium and magnesium salts as these salts are drawn up to the surface by capillary action and dry up at the surface. Some of these soils contain fair amounts of exchangeable sodium.

Desert soils are those which are found in arid regions under ocnditions of poor water supply. They often contain some soluble salts wihch are concentrated by inland drainage.

Soils can also be classified according to the parent rocks from which they have been derived, but in such cases the transported soils will have to be studied in relation to the areas from which they were originally derived. The soils of India, on the basis of their geological and mineralogical origin, have been described by Wadia, Krishnan and Mukherji. In this case emphasis was given to the nature of the original rocks and their chemical and mineralogical and weathering characteristics.

The soils of the Indian Peninsula have, in a large measure, attained a high degree of maturity as they have been under cultivation for many centuries and, therefore, represent the products of weathering over long ages. In contrast with this the alluvial soils are generally not so mature. They are, however, rich and support a large agricultural population snd livestock.

There is naturally a considerable variation in the nature, origin and characteristics of soils depending on their origin and the changes to which they have been subjected during their evolution. Their study now belongs to the realm of agriculture, although their mineralogical charaeters can best be elucidated by geologists.

RECENT CHANGES OF LEVEL ALONG THE COAST

Numerous changes have occurred along the coasts during Recent and historic times. These comprise both submergence and emergence.

As evidences of emergence, we may cite the elevation of parts of the coast of Kathiawar, of the Rann of Kutch and of the eastern coast of Southern India. In the districts of Tinnevelly and Ramnad, Pleistocene and recent grits and clays are noticed to form raised beaches. Several places which were on the sea some centuries ago are now a few miles inland. For instance, Coringa near the mouth of Godavari, Kaveripatnam in the Cauvery delta and Korkai on the coast of Tinnevelly were all flourishing sea ports about 1,000 to 2,000 years ago. Their present position some distance inland may be attributed to the gradual growth of deltas of the rivers at whose mouths they lie.

In the Princes Dock at Bombay tree stumps were found standing *in situ* at a depth of 9 m below high water level. On the Tinnevelly coast also, in the Valinokkam Bay, a similar submerged forest has been noticed, with numerous tree stumps of about 0.6 m diameter at the base showing up at low tide over a bed of black clay containing oyxter and other marine shells. A large part of the Gulf of Mannar and Palk Strait is very shallow and has apparently been submerged only in recent geological times. Similarly also a part of the former town of Mahabalipuram near Madras is known to have been submerged several feet under the sea. In the city of Madras itself, the sea is vigorously invading the area to the north of the harbour, but this is evidently largely due to the construction of the harbour, for the coastal current sweeps past the pier and whirls in towards the land on the northern side, the prevailing direction of current being from south to north.

The Arakan coast as well as the islands of the Andaman and Nicobar group have undergone submergence in the Pleistocene and Recent times. Indications of submergence of the Arakan and Tenasserim coasts are furnished by the deep creeks and inlets which are noticeable in the topographical map. The submergence is also very clearly seen from the air when flying along these coasts. On the other hand, the eastern cost of the Malay Peninsula has been slightly uplifted during the same period.

Changes of level have also occurred as a result of earthquakes. A large area bordering the Rann of Kutch was suddenly submerged to the esxtent of 5 m while the adjoining area to the north was elevated, after the earthquake of 1819. Sevel cases of change of level have been recorded after great earthquakes like those of Assam in 1897 and 1950 and of Bihar in 1934. The Madhupur jungle in eastern Bengal is known to have been elevated by as much as 15 to 30 m in historic times, inducing thereby a westward shift of the course of the Brahmaputra in Bengal.

Changes in the courses of rivers

Some of the important changes in the courses of the rivers of the Brahmaputra, Ganges and Indus systems have already been referred to. Some of them are attributable to the building up of alluvial layers and of subsidiary deltas along the courses of the rivers. All the rivers emerging from the Himalayas have built large talus fans at their debouchures in the plains, through which they repeatedly cut new channels. It is also noticed that there is a tendency, particularly amongst the rivers of the Indus system, gradually to shift their courses towards the west. It has been suggested that this is connected with the rotation of the earth, but in such a case similar phenomena should be observable in the case of all rivers in the northern hemisphere.

Amongst the most notable changes is the drying up of the rivers which once flowed through Rajasthan. This is mainly due to the gradual desiccation of the region, aided to a large extent by deforestation. Deforestation always contributes to a decrease in percolation into the strata underground and to an increase in the run-off, the latter leading to frequent floods. It also leads to the shallowing up of the river beds and encourages the tendency of the rivers to overflow their banks almost at every opportunity when the run-off is fairly large.

SELECTED BIBLIOGRAPHY

(Pleistocene and Recent)

Auden, J.B. Report on the *Reh* soils of the U.P. *Rec.* **76**, paper 1, 1942.

De Terra, H. Evidence of recent climatic changes shown by Tibetan highlands. *Georgr. Jour.* **84**(4), 1934.

De Terra H. and De Chardin T. Observations on the Upper Siwalik formation and later Pleistocene deposits in India. *Proc. Amer. Phil. Sec.* **76**, 791-822, 1936.

De Terra H. and Paterson T. Studies on Ice Age in India, and associated human cultures. *Garnegie Inst. of Washington.* Publ. No. 493, 1939.

Foote, R.B. Billa Surgam and other caves in Kurnool. *Rec.* **17**, 27-34, 1884 ; *Rec.* **18**, 227-235, 1885.

Hora, S.L. On fossil fish remains in the Karewas. *Rec.* **72**(2), 1938.

Iyengar, M.O.P. and Subramanyan, R. Fossil diatoms from the Karewas. *Proc. Nat. Ac. Sci. India* **13**, 225-237, 1943.

La Touche, T.D. Submerged forests of Bombay. *Rec.* **49**, 214-219, 1918.

Middlemiss, C.S. Lignite coalfields in Karewa formations. *Rec.* **55**, 241-253, 1924.

Morris, T.O. The Bain boulder-bed. *Q.J.G.S.* **94**, 385-421, 1938.

Prashad, B. On a collection of land and fresh-water molluscs from the Karewas. *Rec*, **55**, 356-361, 1924.

Puri, G.S. Preliminary note on the Pleistocene flora of the Karewas. *Q.J.G. M.M.S.* **XX**, 61-66, 1948.

Sahni, B. The Karewas of Kashmir. *Curr. Sci.* (Bangalore) **5**, 10-16, 1936.

Wadia, D.N. Pleistocene Ice age deposits in Kashmir. *Proc. Nat. Inst. Sco.* **7**, 45-51, 1941.

Wadia, D.N., Krishnan, M.S. and Mukerjee, P.N. Geological foundations of the soils of India. *Rec.* **68**, 363-391, 1935.

Wright, W.B. The Quternary Ice Age. (London), 1937.

Zeuner, F.E. The Pleistocene Period. London, 1945.

——— Symposium on the Pliocene — Pleistocene boundary. *Proc. 18th Int. Geol. Cpng.* (London) Part IX, 1948.

——— Report of the all-India asoil Survey Scheme. *Indian Council Agr. Rec. Bull.* **73**, 1953.

LATERITE

——— Laterite and Laterite Soils. *Imp. Bur. Soil. Sci.* (London) *Techn. Comm.* No. 24, 1932.

——— Problems of Clay and Laterite geneiss (Symposium). *Amer. Inst. Min. Met. Eng.* (New York), 1952.

Buchanan, Francis. A journey from Madras through the countries of Mysore, Canara, Malabar etc. (London 1807, 3 Vols.) Vol. 2 pp. 436-436.

De Lapparent J. Les Bauxites de la France meridionale. *Mem. Expl. Carte Geol. de France*, 1930.

Fermor, L.L. What is Laterite ? *Geol. Mag.* **48**, 454-462 ; 507-516 & 559-566, 1911.

Fox, C.S. Bauxite and aluminous laterite. London, 1932.

Fox, C.S. Buchanan's laterite of Malabar and Kanara. *Rec.* **69**, 389-422, 1936.

Goldman, M.A. and Tracey, J.I. Relations of bauxite and kaolin in the Arkansas bauxite deposits. *Econ. Geol.* **41**, 567-575, 1946.

Goldschmidt, V.M. Principles of distribution of the Chemical elements in minerals and rocks. *Jour. Chem. Soc.* 1937, pp. 655-673.

Harder, E.C. Stratigraphy and origin of bauxite deposits. *Bull. Geol. Soc. Amer.* **60**, 887-908, 1949.

Harrison, J. The katamorphism of igneous rocks under humid tropical conditions. *Imp. Bur. Soil. Sci.*, London, 1934.

Holland, T.H. On the constitution, origin and dehydration of laterite. *Geol. Mag.* **40**, 59-66, 1903.

Krishnan, M.S. Lateritisation of khondalite. *Rec.* **68**, 392-399, 1934.

Lacroix, A.F.A. Les phenomenes d'alteration superficielle des roches silicatees alumineuses des payes tropicaux. *Bur. Etudes Geol. Min. Col.* (Paris). Publ. No. 25, 1934.

Lake, P. Geology of Malabar. *Mem.* **24**, 217-233, 1890.

Middlemiss, C.S. Bauxite deposits of Jammu. *Kashmir Mineral Survey Report*, 1928

Mohr, E.C.J. Soils of equatorial regions with special reference to Netherland East Indies, (Trans. by Pendleton). Edwards & Co. Ann Arbor, 1944.

Rumbold, W.G. Bauxite and aluminium. *Imp. Inst.* (London) *Monograph series*, 1925.

INDEX

A

Abor Hills, 12, 179, 245, 247
Abur beds, 362
Aden, Gulf of, 67
Ages, radiometric, 51, 88, 112, 419, 425
Agglomeratic Slate, 301
Airy, 72
Aiyengar, N. K. N., 483
Ajabgarh Series, 172
Aka Hills, 12, 140, 179, 247
Akanktaung Stage, 471
Albaka Stage, 168
Algonkian, 82
Alkali Rocks, 101, 104, 108, 122, 141
Alluvial deposits, 2, 24, 56, 508
Almod beds, 248
Alum Shales, 358, 446
Alveolina Limestone, 447
Alwar Series, 172
Amb Beds, 308
Amgaon Series, 112
Amjori Sill, 130
Anaimalai, 4
Anceps Beds, 367
Andaman Islands, 13, 63, 401, 402, 432, 456, 479
Angara Flora, 280
Angaraland, 55, 292
Anorthite, 103
Anorthosite, 103, 133
Antarctica, 94, 234 272
Antimony, 147
Arabian Sea, 66
Arakan Coast, 45, 399, 429
Arakan Yomas, 44, 346, 399, 429, 456
Aravalli Mountains, 6, 7, 48
Aravalli Strike, 48
Aravalli System, 120
Archaean Group, 87
Archipelago Series, 432, 479
Argentina, 270
Ariyalur Stage, 392
Arkasani Granophyre, 126, 129
Arkell, W.J., 241, 257, 336, 351, 368, 376
Arsenic, 147
Artesian Water, 493, 512
Artini and Melzi, 107
Asbestos, 156, 165, 180

Assam : Archaean, 133 ; Permian 300 ; Cretaceous, 395 ; Tertiary, 428 ; Eocene 453 ; Oligo-Miocene 467; Mio-Pliocene, 490
Assam Plateau. 43, 58, 133
Assam Wedge, 51
Astian, 497
Athgarh Sandstone, 255
Athleta Beds, 367
Attock Slates, 177
Auden, J. B., 54, 139, 407
Auk Shales, 188, 195
Australia 260
Axial Group, 346, 401

B

Badhaura formation, 244
Bagh Beds, 385
Bagra Stage, 250
Bailadila Iron-Ore Series, 109
Bain Boulder-bed, 506
Bairenkonda Quartzite, 165
Baisakhi Beds, 362
Balasubramanian, M. N., 104
Ball, V., 193
Ballore, M.de, 43, 63
Baluchistan — Trias, 345 ; Juras, 360 ; Cretaceous, 381 ; Tertiary, 425 ; Eocene, 439 ; Oligocene, 463 ; Miocene, 488
Banded ferruginous quartzite (B.H.Q.), 102, 109, 125
Banded Gneissic Complex, 118, 120
Banerjee, A.K., 126
Banganapalle Series, 188
Bap beds, 241
Baragoloi Stage, 430, 468
Barail Series, 430, 468
Barakar Series, 245
Baripada beds, 432, 479
Barite, 165, 181, 386
Barker, P. F. ,66
Barmer Sandstone, 254
Barren Island, 43, 63
Barren Measures, 247
Bauxite, 420, 520
Bawdwin ore deposits, 216

INDEX

Bawdwin Volcanics, 216
Baxa Series, 178
Bay of Bengal, 3, 63
Beacon Group, 273
Beas River, 21
Beardmore group, 273
Beaufort Series, 251, 265
Bedesir beds, 362
Belemnite Shales, 381, 384
Bellary Gneiss, 101
Bengal Gneiss, 132
Bengpal Series, 109
Berinag Quartzite, 191
Berwali Series, 436, 460
Beryl, 132, 156
Betwa River, 27
Bhabar, 515
Bhadrar beds, 427, 441
Bhander Series, 186, 194
Bhangar, 512
Bhattacharjee, D. S., 111
Bhima Series, 190
Bhola, K. L., 120
Bhuban Stage, 430, 469
Bhuj Stage, 254, 368
Bhutan, 140,1 78
Bichua Stage, 113
Bijawar Series, 116, 169
Bijawar Traps, 170
Bijaigarh Shale, 185, 193
Bijli rhyolite, 112
Bijori Stage, 248, 249
Bilara Formation, 186, 244
Binota Shales, 121
Bintenne Gneiss, 105
Birmania Formation, 244
Biswas, B., 431
Biralve beds, 344
Blaini Series, 313, 314
Blaini boulder-beds, 243, 313
Blanford, W.T., 117, 497
Block faulting, 55, 56, 58, 64, 67, 276
Boileau, V. H., 192
Bokabil Stage, 430, 469
Bolan Limestone, 441
Bolpur formation, 432
Bose, P.N., 112, 170
Bournon, Count de, 103
Boulder-Conglomerate, 428, 487
Brahmani River, 16
Brahmaputra River, 19, 29
Brown, J.C., 140, 218
Buchanan, F., 517
Budavada Sandstone, 256
Budhi Schists, 176
Bugti Beds, 465
Building stones, 156, 181, 194, 282, 362, 419, 462

Bundelkhand Gneiss, 118, 119, 124
Burdwan formation, 431, 479
Burma — Archaean, 141 ; Cambrian, 216 ; Ordovician-silurian, 223; Devonian, 228 ; Carboniferous, 235 ; Permian, 315 ; Triassic, 345 ; Jurassic, 369 ; Cretaceous, 399 ; Tertiary, 429, 458, 470, 491, 510
Burmese Arc, 12, 43, 63
Burrard, S.G., 19, 72, 511
Burton, R.C., 110
Byans, 228, 298, 331, 354

C

Calabrian, 497
Calcareous Zone, 62, 360, 425
Cambay, trough, 64, 73, 460
Cambrian System, 197
Cape Comorin, 2, 5, 386
Carboniferous System, 230
Cardita beaumonti beds, 417
Cardita subcomplanata beds, 382
Carlsberg Ridge, 66
Carnatic Gneiss, 101
Carnic Stage, 324, 328, 331, 342
Caster, K. K., 274
Cathysian Flora, 280
Cauvery River, 17
Cauvery Basin, 384, 433, 461, 480, 494
Cave deposits, 515
Celestite, 386
Cenomanian transgression, 63, 370, 386
Central Gneiss, 137
Ceratite Beds, 342
Ceratite Marls, 343
Ceratite Sandstone, 343
Ceylon, 104, 258, 480
Chagai Zone, 62, 427
Chail Series, 138, 175
Chalk Hills, 102
Chaman Fault, 67
Chambal River, 27
Champaner Series, 116
Champion Gneiss, 93
Chandarpur Sandstone, 189
Chandpur Series, 176, 236
Chandragiri Limestone, 192
Chari Series, 363, 367
Charnockite, 64, 93, 105, 107, 386
Charmuria Stage, 189
Chasilaha Gneiss, 178
Chatterjee, S. C., 133, 408
Chatterjee, S. K., 113
Chaugan Stage, 253
Chaung Magyi Series, 143, 180, 216
Chenab River, 20

Cherrapunji, 3
Cherra Sandstone, 455
Cheyair Series, 165
Chharat Series, 428, 449
Chidamu Beds, 351
Chidru Stage, 310
Chikiala Stage, 253
Chikkim Series, 373, 376
Chilka Lake, 37, 512
Chilpi Ghat Series, 110
Chindwin River, 33
Chinji Stage, 428, 485
Chin Shales, 346
Chintalpudi Sandstone, 248, 255
Chiplonkar, G. W., 386
Chitichun facies, 298, 335, 375, 376
Chitral, 228, 234, 381
Chitral Slates, 234
Chocolate Series, 334
Chomolhari Granite, 178
Chorbaoli Stage, 113
Chor Granite, 138
Chota Nagpur Gneiss, 129
Christie, W. A. K., 37, 206, 208
Chromite, 68, 102, 147, 402
Cimmerian Orogeny, 348
Clays, 158, 181, 282, 462, 493, 516
Clegg, E. L. G., 142
Closepet Granite, 100
Coal, 193, 282, 358, 429, 455, 461, 465, 472
Coast, eastern, 63, 372
Coast, Western, 64
Coastal deposits, 512
Coates, J. S., 105
Cobalt, 172, 154, 180
Colair Lake, 37
Colbert, E.H., 488
Columbite-tantalite, 148
Conjeevaram gravels, 494
Conularia bed, 243, 305
Cooray, 105
Copper ore, 118, 127, 148, 169, 172, 180
Copper belt (Singhbhum), 68, 126
Coral Limestone, 325
Cordierite, 104, 109, 141
Corundum, 100, 103, 108, 133, 142, 158
Coulson, A.L., 117
Cretaceous System, 370
Crinoidal Limestone, 221
Crustal shortening, 73
Cuddalore Sandstone, 433, 493
Cuddapah Basin, 93, 163
Cuddapah Slabs, 194
Cuddapah System, 163
Cuddapah Traps, 165,167
Cumbum Slates, 166
Cutch (see Kutch).

D

Dachsteinkalk, 325, 331, 335, 355
Dagshai Beds, 57, 429, 467
Daling Series, 139, 140, 178
Dalma Volcanics, 127, 128, 131
Damodar River (Valley), 16, 35, 284
Damuda System, 245, 278
Daonella Shales, 324
Daonella Limestone, 324
Darjeeling Gneiss, 139, 140
Das Gupta, H. C., 395, 415
Davies, L. M., 441
Dauki Fault, 52, 58
Debagram formation, 431, 491
Deccan Traps, 405-414
Delhi System, 54, 170
Delhi (Alwar) Quartzite, 172, 173
Denwa Stage, 250, 251
Deoban Limestone, 139, 191
Deola Marl, 385
Deoli Beds, 250
Desert Deposits, 514
Devonian System, 226
Dhak Pass beds, 427, 446
Dhands, 39
Dhandraul Quartzite, 185
Dhanjori orogeny, 128
Dhanjori Stage, 127
Dhansiri River, 30
Dharwar Strike, 48, 89
Dharwar System, 88
Dhok Pathan Stage, 428, 486
Dhosa oolite, 367
Dhrangadhra Sandstone, 384
Diadematum Zone, 367
Diamonds, 184, 188, 189, 192
Diener, C., 138, 299
Dihing Series, 430, 490
Dinosaurs, 251, 265, 393, 415, 451
Disang Series, 430, 455
Disang Thrust, 52
Dogra Slate, 138, 177
Dome Gneiss, 132
Dongargarh Granite, 112
Dravidian Group, 82
Dras Volcanics, 378, 401
Dubey, V. S., 193, 408
Dubrajpur Sandstone, 250
Dunes, 514, 516
Dunghan Limestone, 440
Dunghan Shale, 383
Dunn, J. A., 127, 130, 218
Dupitila Series, 430, 490
Durgapur Beds, 492
Dutt, N. V. B. S., 188
Dutta, P.N., 112
Dwarka Beds, 345, 492

INDEX

Dwyka Tillite, 263
Dzong Buk Shales, 453

E

Eames, F. E., 480
Early man, 499, 500
Earthquakes, 41, 66
Eastern Ghats, 45, 49, 107
Eastern Ghats Strike, 49
Ecca Series, 264
Enderbite, 99
Eocene System, 439
Erinpura granite, 174
Erosion surfaces, 74
Erratics, 15, 506
Eskola, P., 108
Euramerican flora, 280
Eurydesma Bed, 243, 305
Evans, P., 73
Everest, Mount, 12, 26, 299, 355
Everest Limestone, 299
Everest Pelitic Series, 300
Everest's spheroid, 70
Exotic Blocks, 73, 298, 376, 401

F

Falkland Isles, 271
Farrar group, 274
Fatehjang Zone, 466
Faults—Cambay, 64 ; Chaman, 67 ; Dauki, 52 ; Deccan Trap, 407 ; Gondwana, 276 ; Great Boundary, 117 119, 187 ; Kathiawar, 66; Kishanganj, 58 ; Machkund River, 108 ; Main Boundary, 58, 483; Makran, 63, 66 ; Morabad 67 ; Narmada, 18, 57 ; Owen, 66 ; Purnea, 58 ; West Coastn,, 18, 67
Fawn Limestone, 184
Fenestella Shales, 234, 244
Fermor, L.L., 5, 7, 89, 107, 108, 112, 113, 118, 137, 143, 407, 408, 410
Fireclay, 195
Floods, 35
Fluorite, 101, 385
Flysch, 218, 373, 374, 402, 465
Foote, R.B., 88, 190, 257, 258, 510
Fore-deep, 72, 424, 482, 510
Fossil wood, 248, 252, 392, 491, 494
Fossil-wood Group, 491
Fox, C.S., 191, 193, 207, 240, 307, 409, 455
Fuller's earth, 436, 449
Furon and Lemoine, 395

G

Gaj Series, 426, 435, 464
Gajansar Beds, 367
Gandak River, 26
Gangamopteris Beds, 243, 345, 302
Ganges Basin, 72, 253, 431
Ganges River, 24, 35
Gangpur Series, 131
Gansser, A., 68, 134, 178, 378
Ganurgarh Shales, 184
Garbyang Series, 216, 237
Gastropod Limestone, 453
Gee, E.R., 201, 207, 208
Gemstones, 141, 142, 159, 419
Geodetic data, 70
Geologic Time-Scale, 78
Geologic divisions, standard, 79, 80
Ghatal formation, 432
Ghazaband Limestone, 443
Ghazij Shales, 441
Ghosh, P.K., 107, 117
Ghosh, A. M. N., 140, 246
Gigantopteris flora, 280, 281
Gill, W.D., 199
Giri Limestone, 379
Girnar Hill, 408, 409
Girujan Clay, 490
Giumal Sandstone, 373, 402
Glaciers, 8, 14, 501
Glaciation — Carboniferous, 241, 277, 304; Pleistocene, 424, 497, 498, 501
Glennie, E.A., 12
Glossopteris flora, 240, 276, 280
Gneiss — Bezwada, 107 ; Bellary, 101 ; Bengal, 112 ; Bundelkhand, 118, 124 ; Carnatic ,101 ; Central, 137; Champion, 93 ; Chota Nagpur, 127, 132 ; Dajreeling, 139 ; Dome, 132 ; Kailassa, 107 ; Nilgiri, 93 ; Peninsular, 93, 101; Wanni, 105
Godavari River (Valley), 16, 255, 509
Gogha Beds, 435
Gogra River, 26
Golapilli Sandstone, 255
Gold, 101, 148
Goldschmidt, V. M., 520
Golden oolite, 358, 367
Gondite, 113, 115, 131
Gondwanaland, 239, 277, 280, 293
Gondwana Group, 54, 239
Gopal, V., 258
Granite—Berach, 118 ; Central Himalayan, 378 ; Chor, 138 ; Chakradharpur, 128; Closepet, 100; Erinpura, 123, 124, 174 ; Idar, 174 ; Jalor-Siwana, 174 ; Kabaing, 142 ; Kailas, 452 ;

Mylliem, 132 ; Romapahari, 128 ;
   Tawng Peng, 218 ; Tonigala, 105
Graphite, 158
Graptolite Beds, 225
Gravity anomalies, 67, 70, 72, 73
Great Limestone, 313
Grey Shales, 324
Greywacke Series, 432, 479
Griesbach, C.L., 138, 214, 221, 335
Groves, A. W., 99
Gulcheru Quartzites, 165
Gunderdehi Stage, 189
Gunz glaciation, 499
Gwadar Stage, 489
Gwalior System, 121, 122, 169
Gwalior Tfaps, 121, 169
Gypsum, 204, 208, 362, 388, 439, 492

H

Hackett, C. A., 116, 117, 169
Haflong-Disang Thrust, 52
Haimanta System, 138, 214
Hallstat, marble, 299, 332, 334, 355
Halobia Beds, 324, 328
Halorites Beds, 328
Hangu Shales, 447
Hangu Breccia, 449
Harrison, J., 519
Hayden, H. H., 19, 138, 214
Hayford, 71
Hazara, 7, 222, 313, 356, 380, 450
Hazara Slates, 177
Hedenstroemia Beds, 322
Heim, A., and Gansser, A., 73, 192, 237, 299,
   334, 351, 354, 370, 375, 483
Heiskanen, 71
Hemiaster Beds, 379
Hemipneustes Besd, 381
Hercynian revolution, 55, 239
Heron, A. M., 117, 119, 124, 174, 188, 408
Highland Series, 105
Hill Limestone, 428, 449
Himalayan Arc, 7, 12, 70
Himalaya Mountains, 9, 58
Himalayan Orogeny, 73, 402, 423
Himgir Beds, 248, 249
Himmatnagar Sandstone, 254, 384
Hinglaj Sandstone, 465
Holland, T. H., 37, 93, 94, 99, 104, 207
Hooghly River, 28, 35
Hora, S. L., 418
Horizon problematica, 335
Hornstone Breccia, 172
Howie, R. A., 94, 99
Huene, F. Von, 239, 415
Hundes (Malla Johar), 298, 335, 355, 374

I

Idar Granite, 174
Indo-Gangetic basin, 42, 56, 69, 510
Indo-Brahm, 483
Indocephalites Beds, 361
Indravati Series, 188
Indus Flysch, 378
Indus Suture, 378
Indus River, 19, 36
Infrakrol Series, 313
Infra-Trappean Beds, 415, 416, 418, 461
Infra-Trias, 313, 336, 344
Inter-Trappean Beds, 415, 416, 418, 461
Irlakonda Quartzite, 117
Iron ores, 122, 151, 282
Iron-ore Series, 127, 128
Iron-ore Series (Bailadila), 109
Iron-ore Orogeny, 127
Ironstone Shales, 247
Irrawaddy River, 32, 510
Irrawaddy System, 431, 491
Isostasy, 71
Iyer, L. A. N., 130, 142, 256
Iyengar, S. V. P., and Alwar, M. A., 128

J

Jabalpur Series, 253
Jabbi Beds, 310
Jacob, K., 4, 258
Jadeite, 402
Jaffna Beds, 63, 480
Jagdalpur Stage, 188
Jain, J. L., 251
Jainti Quartzite, 178
Jaintia Series, 453
Jaisalmer Limestone, 362
Jalangi formation, 432, 456
Jalor granite, 174
Jammalamadugu Series, 188
Jaunsar Series, 236
Jenam Stage, 430, 468
Jharia Coalfield, 245, 289
Jhelum River, 20
Jhiri Shales, 185
Jiran Sandstone, 121, 173
Jones, H. C., 125
Jumna River, 25
Junewani Stage, 113
Jurassic System, 347
Jutogh Series, 138
Juvavites Beds, 325

INDEX

K

Kabaing Granite, 142
Kailas Range (Mt.), 8, 19, 29, 59
Kailas conglomerate, 59, 452
Kailas Granite, 452
Kaimur Series, 184
Kajrahat Limestone, 184
Kakdi Series, 436, 460
Kalabagh Beds, 308
Kaladgi Series, 167
Kalapani Limestone, 334
Kali River, 25
Kallar, 204
Kama Stage, 471
Kamawkala Limestone, 345
Kamlial Stage, 428, 484
Kampa Shale, 379
Kampa System, 377, 379, 452
Kamthi Beds, 248
Kanawar System, 233
Kandra Volcanics, 102
Kankawati Series, 436, 492
Kanger Stage, 188
Kankar, 386, 512, 516
Kantkote Sandstone, 367
Kaolin, 493
Karakorum Range, 7, 8, 53
Karewa formation, 504
Karharbari Series, 245
Karikal Beds, 433, 494
Karnali River, 25
Karroo System, 263
Kasauli Beds, 57, 428, 467
Kashmir Wedge, 52, 198
Kashmir, 137, 212, 222, 228, 234, 245, 301, 302, 313, 336, 356, 378, 450, 504, 497, 501
Katrol Series, 363, 367
Katta Beds, 308
Khairagarh Series, 112, 189
Khader, 512
Khairabad Limestone, 427, 446
Khardeola Grits, 122
Khari Series, 436, 478
Khasi Greenstone, 134
Kheinjua Stage, 183
Khewra Trap, 204
Khojak Shales, 62
Khojak Zone, 62, 402, 425, 465
Khondalite, 100, 107, 108
Kimberlite, 193
King, W., 101, 110, 164, 168, 190, 256
Kiogar facies, 335, 375
Kiogar Limestone, 355
Kioto Limestone, 324, 329, 331, 335, 341, 342, 344
Kirthar Series, 426, 443

Kiseri formation, 57
Kistna Series, 167
Kodurite, 5, 107, 153
Kohat, 199, 203, 206, 210
Kohat Shale, 447
Koilkuntla Limestone, 188
Kolamnala Shales, 167
Kolar Goldfields, 91, 149
Kolhan Series, 127, 130, 168
Kopili Stage, 432, 453, 456
Kortalaiyar River, 510
Kosi River, 26, 32, 35
Kossmat, F., 394
Kota Stage, 251
Krafft, A. Von, 138, 298, 374
Krishnan, M.S., 118, 130, 131, 483
Krol Belt, 59, 313
Krol Series, 57, 313
Kuar Bet Beds, 364
Kudremalai Series, 480
Kuldana Beds, 449
Kuling System, 296
Kumaon, 138, 216, 221, 227, 298, 327, 334, 373
Kundair Series, 188
Kundghat Beds, 310
Kurnool System, 187
Kushalgarh Limestone, 172
Kuti Shale, 335
Kutch — 254, 363, 436, 459, 464, 459, 492, 513
Kyanite, 127, 159
Kyaukkok Sandstone, 472
Kyi-chu Granite, 452

L

Laccadives, 7, 48, 67
Lachi Series, 300
Ladakh Range, 8
Ladinic Stage, 324, 327, 342
Laisong Stage, 430, 465
Lakadong Sandstone, 455
Lake, P., 69, 93
Lakes, 36-39
Laki Series, 426, 440
Lakhpat Series, 436, 478
Lameta Beds, 414
Langpar Stage, 397
Laptal Series, 351
Laskar, B., 247
Laterite, 412, 420, 517
Lathi formation, 244, 362
Latouche, T. D., 117
Laungshe Shales, 458
Lavender Clays, 307
Lead-Zinc ores, 151, 164, 166, 216
Leptynite, 95

Lei Conglomerate, 487
Lewis, G.E., 486, 488
Lignite, 436, 459, 492, 495, 516
Lilang System, 319
Limestone—Ajabgarh, 172 ; Alveolina, 447 ; Bhander, 186 ; Bhima, 190 ; Bolan, 441 ; Ceratite, 343 ; Chikkim, 373 ; Coral, 325 ; Daonella, 324 ; Deoban, 191 ; Dunghan, 440 ; Everest, 299 ; Ghazaband, 443 ; Giri, 379; Great, 313 ; Halobia, 324, 328 ; Hemipneustes, 381 ; Hill, 449 ; Infra-Trias, 313 ; Jaisalmer, 362 ; Kioto, 325, 329, 335 ; Kaladgi, 167 ; Kajrahat, 184 ; Kalapani, 334 ; Kamawkala, 346 ; Kiogar, 325, 335, 355, 376 ; Khairabad, 446 ; Kheinjua, 183 ; Koilkuntla, 188; Krol, 314; Kushalgarh, 172 ; Lockhart, 449 ; Meting, 440 ; Miliolite, 435 ; Mogok, 141 ; Moulmein, 143 ; Lungma, 355 ; Narji, 188 ; Nimbahera, 184 ; Niti, 327 ; Nodular, 385 ; Nummulitic, 462 ; Nyaungbaw, 225 ; Padaukpin, 230 ; Pithoragarh, 191 ; Palnad, 188 ; Parh, 381 ; Phalodi, 186, 244 ; Plateau, 228, 235 ; Porbander, 435 ; Prang, 455 ; Productus, 308 ; Quilon, 480 ; Raialo, 123, 173 ; Raipur, 189 ; Rohtas, 12; Sakesar, 427; Scarp, 379 Shali, 391 ; Shekhan, 447 ; Siju, 455; Spintangi, 443 ; Sylhet, 453 ; Syringothyris, 234 ; Tirohan, 183 ; Traumatocrinus, 328 ; Tropites, 331 ; Tuna, 379 ; Umlatdoh, 455 ; Vempalle, 165
Lipak Series, 233
Lochambal Beds, 351
Loess, 506, 513
Lohangi Stage, 113
Loi-An Series, 369
Lonar Lake, 38
Lumachelle, 352
Luni River, 17
Lungma Limestone, 355

M

MacLaren, J. M., 247, 300
Macrocephalus Beds, 367
Madh Series, 436, 459
Madhan Slates, 314
Madhupur Jungle, 35, 523
Magnesiun Sandstones, 211
Magnesite, 68, 102, 160
Mahadek (Mahadeo) Stage, 397

Mahadeva Hills, 6, 250
Mahadevan, C., 54, 168, 190
Mahadeva Series, 250
Mahanadi River, 16, 255
Mahanadi Strike, 49
Main Sandstone Series, 380
Makran Coast, 8, 44, 66
Makran Series, 488
Majhgawan diamond pipe, 193
Makrana Marble, 123
Malani Igneous Suite, 172, 174
Maleri Stage, 251
Mallet, F. R., 112
Manasarowar Lake, 21, 29
Manchhar Lake, 39
Manchhar Series, 426, 488
Mandhali Beds, 236, 243
Manganese ore, 113, 114, 131, 152
Mallet, F. R., 112, 170
Mangli Beds, 249
Mansar Stage, 113, 114
Martoli Series, 176, 237
Mathur, K. K., 408
Mathur, S. M., 193, 194
Matla formation, 431, 492
Mawson Series, 225
Mayakor formation, 244
Medlicott, H. B., 169, 239, 249
Memari formation, 431, 479
Mergui Series, 193
Meting Limestone, 440
Meting Shale, 440, 441
Mica, 102, 160
Mica belt (Bihar), 129
Mica-peridotite, 289
Middlemiss, C. S., 117, 206, 208, 384
Miju thrust, 52, 180
Miliolite Limestone, 66, 513
Mindel glaciation, 499
Mishmi Hills, 140, 179
Mogok Series, 141
Molybdenite, 153
Monazite, 102, 107, 132, 153
Mong Long Schists, 142
Monotis Shale, 325
Moonstone, 107, 142
Morar Series, 169
Morris, T.O., 506
Motur Stage, 247
Moulmein Limestone, 143, 235
Mud Volcanoes, 44
Murray Ridge, 66, 67
Murree Series, 427, 466
Muschelkalk, 322, 327, 331, 342
Muschelkalk, 322, 327, 331, 342
Musikohla formation 57
Muth Quartzite, 227, 237

N

Naga Thrust, 52
Nagari Quartzite, 165
Nagmarg Beds, 307
Nagri Stage, 428, 485
Nagthat Beds, 176, 191, 236
Nal Lake, 38
Nallamalai Series, 165
Namcha Barwa, 12, 13, 29, 61, 180
Nammal Stage, 427, 447
Namsang Beds, 430, 490
Namshim Beds, 226
Namyau Series, 369
Nandgaon Series, 112
Nandyal Shales, 188
Nanga Parbat, 12, 19, 36, 62, 68, 378
Naogaon Stage, 430, 468
Napeng Beds, 345
Nappes 2, 114, 191
Narayanaswami, S, 4, 115
Narmada trough, 67
Narmada River (Valley), 17, 18, 507
Nari Series, 426, 463
Narji Limestone, 188, 194
Naunkangyi Stage, 223
Negrais Series, 399
Neobolus Beds, 211
Neolithic Culture, 500
Newer Dolerite, 129, 130
Nickel-Cobalt ores, 154
Nilgiri Mountains, 4, 5, 45, 93
Nilgiri Gneiss, 93
Nimar Sandstone, 384
Nimbahera Limestone, 184
Ninetyeast Ridge, 63
Niniyur Stage, 393
Nishpa formation, 383
Nodular Limestone, 385
Noetling, F. 207, 360
Noric Stage, 328, 331, 342
Nuttall, W. L. F., 441, 444
Nyaungbaw Limestone, 225

O

Ochre, 195
Obogon Beds, 472
Older Metamorphic Series, 125
Oldham, R. D., 66, 182, 191, 417
Oldham, T., 43, 69, 134
Olive Series, 305
Oman, 13, 62, 66, 402
Ophiceras zone, 320
Ordovician System, 219
Operculina Limestone, 453
Orthoceras Beds, 225
Otoceras Zone, 320, 322
Owen Fracture, 66, 67
Orbitolina Limestone, 453
Orogenic belts, 68

P

Pab Sandstone, 382
Pachmarhi Stage, 250
Padaukpin Limestone, 230
Padaung Clays, 471
Pakhal Series, 168
Palaeolithic culture, 499
Palghat Gap, 4
Pali Beds, 248, 249
Palnad Limestone 188
Pamir plateau, 42, 53, 69
Panchet Series, 249
Pandua formation, 431, 479
Paniam Series, 188
Panjal Volcanics, 301, 336
Panna Shales, 185
Panvel flexure, 64
Papaghni Series, 165
Par Series, 169
Para Stage, 351
Parahio Series, 215
Parh Limestone, 381
Parihar Sandstone, 362
Parsora Stage, 250
Pascoe, E. H., 206, 207, 483
Patala Shales, 427, 446
Patcham Series, 365
Paunggyi Conglomerate, 458
Pavagad Hill, 117, 408
Pavalur Sandstone, 256
Pegu Series, 431, 470
Pellatispira beds, 460
Penganga Beds, 168
Peninsular Gneiss, 93
Penner River, 16
Permian System, 292
Petroleum, 429, 460, 468, 472
Phalodi Limestone, 186, 244
Phosphate, 156
Phosphatic nodules, 388
Pichamuthu, C. S., 94
Pilgrim, G. E., 466, 483, 486, 488
Pindaya Beds, 225
Pinfold, E.S., 208, 210
Pinjor Stage, 428, 487, 506
Pinnacled Quartzite, 187
Piram Beds, 345, 492
Piram Beds, 345, 492
Pir Panjal, 9, 12, 336, 502
Pitchblende, 132
Plaisancian, 497

Plateau Basalt, 408, 411
Plateau Limestone, 228, 235
Plateau Quartzite, 187
Pleistocene 49y
Po Series, 233
Pokharan Boulder-bed, 241
Pondaung Sandstone, 459
Pontian, 497
Porbander Stone, 435
Porcellanite Stage, 183
Port Blair Series, 432
Potash Salts, 206
Potwar Plateau, 39, 52, 56, 198, 427
Potwar Silts, 506
Pratt, 71
Productus Limestone Series, 308
Productus Shales, 293
Ptilophyllum flora, 240, 279
Ptychites Beds, 327, 347
Pulicat Lake, 37
Pulivendla Stage, 165
Pullampet Slates, 165
Purana Group, 53, 82
Puri, G.S., 505
Purple Sandstones, 210
Pyabwe Clays, 472
Pyalo Stage, 471
Pyrite, 185, 193

Q

Quartzite Series, 325
Quartz-magnetite rocks, 101
Quensel, 94
Quilon Limestone, 64, 434, 480

R

Radhakrishna, B...P., 74, 100
Raghavapuram Shales, 258
Raghunathpali Conglomerate, 131
Raialo Series, 122
Rainfall, 3
Raipur Limestone, 189
Rajagopalan, N., 394
Rajasthan, 117, 112, 170, 183, 361, 459
Rajamahendri Sandstone, 493
Rajmahal Hills, 6, 252
Rajmahal Series, 252
Rajmahal Traps, 252, 259, 432
Rajnath, 254
Rakshas Lake, 21, 22, 40
Ralam Series, 192, 237
Ramberg, H., 100
Ramganga River, 25
Ramri Island, 44, 45, 399

Ranaghat formation, 431, 492
Rangit Valley, 246, 300
Raniganj Coalfield, 245, 285
Raniganj Series, 248
Ranikot Series, 426, 439
Rann, 34, 37, 39, 513
Ranthambhor Quartzite, 121
Rao, B. Rama, 89, 91, 93, 99
Rao, L. Rama, 394, 461
Rao, K. Srıpada, 395, 416
Rao, S. R. N., 384, 388, 395, 416, 460
Reed, F. R. C., 212, 243, 311
Regur, 412, 420
Rehmanni Beds, 367
Renji Stage, 430, 468
Rewa Series, 185
Rewak Stage, 455
Rhyolite, 174, 336, 408
Rikba Beds, 307
Riss glaciation, 499
Rivers, 15-36
Robinson, P.L., 251
Rock Salt ,204, 206
Rohtas Stage, 183
Roy Chowdhury, M.K., 170
Roy Chowdhury, T.K., 251
Roy, Supriya, 116

S

Safed Koh Mts., 7, 12, 13, 62
Saha, A.K., 129
Sahni, B., 207, 253
Sahni, M.R., 182, 244
Sakarsanite, 93
Sakesar Limestone, 427, 441
Sakoli Series, 110   112
Saline Series (Salt Marl), 203-208
Salt Pseudomorph Shales, 186, 212
Salt Range, 56, 198, 203, 304, 342, 358, 384, 427, 444
Salween River, 33
Samana Range, 360, 380, 447
Sambhar Lake, 7, 37
Santa Caterina System, 268, 270
Sapphirine, 104, 108
Sarikhol Shales, 237
Sarasvati River, 22, 34
Sarkar, S. N., 111, 112, 116
Sarkar and Saha, 128
Satpura Mts. 6
Satpura Strike, 49
Satyavedu Beds, 257
Sausar Series, 111, 112, 114
Sawa grits, 173
Seismicity, 69
Semri Series, 183

INDEX

Serpentine, 67, 124, 180, 299
Seward, A.C., 250, 258
Sewell, R.B.S., 63, 66
Shali Limestone, 191
Sharma, N.L., 120, 175
Shell Limestone, 352, 362, 392
Shekhan Limestone, 447
Shiala Series, 237
Shillong Series, 134
Shimoga belt, 92
Shwezetaw Stage, 470
Sibirites spiniger Zone, 331
Sicilian, 497
Siju Limestines Stage, 455
Sillimanite, 107, 109, 113, 133, 141, 159
Simla Slates, 176
Sind, 381, 425, 439, 463, 488
Singhbhum granite, 129
Singhbhum orogeny, 127, 128
Singu Stage, 471
Singu Stage, 471
Sinor, K. P., 243
Sinchula Stage, 178
Sirmur System, 467
Sitsayan Shales, 471
Sittampundi Complex, 103
Sittang River, 33
Sivaganga Beds, 258
Sivamalai Series, 104
Siwalik System, 427, 492
Siwana granite, 174
Slates, 166, 181
Smith, F.H., 110
Smith-Woodward, 418
Smeeth, W. F., 89
Soils, 520
Sonakhan Beds, 110
Sonawani Series, 111
Sone River (Valley), 127, 132, 183
Spath, L. F., 256
Speckled Sandstone, 307
Spengler, E., 397
Spintangi Limestone, 443
Spiti, 138, 214, 220, 227, 233, 290, 319, 350, 355, 373
Spiti Shales, 350, 351, 355
Spondylus Shales, 453
Sriperumbudur Beds, 257
Srisailam Quartzites 167
Stachella Beds, 343
Steatite, 161, 181
Stillwell, F. L., 99
Stone implements, 499
Stormberg Series, 265
Straczek, J. A., 113
Stratigraphy, India, review, 81
Structural trends, Precambrian, 48-50
Stuart, M., 203, 206, 208

Subansiri River, 30, 247, 300
Subarnarekha River, 16
Subathu Beds, 429, 452
Subramaniam, A.P., 94, 99, 103
Suess, E., 55, 293, 299, 510
Suket Shales, 182
Sukma Series, 109
Sulaiman Range, 7, 13, 62, 383
Sulcacutus Beds, 351, 354
Sullavai Series, 190
Sulphur, 43
Surma River, 31
Surma Series, 469
Susnai Breccia, 185
Sutlej River, 21, 22, 34
Sylhet Limestone, 432, 453, 456
Sylhet Trap, 134
Syringothyris Limestone, 234

T

Tabbowa Series, 258
Tabyin Clays, 458
Tadpatri Shales, 165
Tagling Stage, 351
Tal Series, 355
Talar Stage, 489
Talchir Tillite, 208, 241
Talchir Series, 241
Tanakki boulder-bed, 313
Tanawal Series, 237
Tapti River, 17, 18
Tasmania, 262
Tatrot Stage, 428, 486
Taungnyo Series, 143
Tawng Peng granite, 218
Tawng Peng System, 142
Terai, 515
Terra, H. de, 486, 497, 502, 506
Tethys, 55, 73, 281, 293, 372
Thabo Stage, 233
Thar desert, 3, 514
Therria Stage, 453
Thorium,, 107
Thrust— Chail, 68, 175 ; Copper Belt, 125 ; Disang, 52 ; Giri, 58, 314 ; Krol, 58, 314 ; Miju, 52 ; Jutogh, 58 ; Murree, 58 ; Naga, 52 ; Salt Range, 207, 210
Tibet, 8, 40, 290, 355, 377, 452
Tikak Parbat Stage, 430, 468
Tilin Sandstone, 458
Tilley, C.E., 99, 412
Tin ore, 154
Tipam Series, 430, 490
Tiratgarh Stage, 188
Tirodi Gneiss, 114